Physical Inorganic Chemistry

Physical Inorganic Chemistry

A Coordination Chemistry Approach

S. F. A. KETTLE

*Professorial Fellow, University of East Anglia, and
Adjunct Professor, Royal Military College, Kingston, Ontario*

OXFORD · NEW YORK · TOKYO

OXFORD UNIVERSITY PRESS

Oxford University Press, Great Clarendon Street, Oxford OX2 6DP

Oxford New York

Athens Auckland Bangkok Bogota Buenos Aires Calcutta
Cape Town Chennai Dar es Salaam Delhi Florence Hong Kong Istanbul
Karachi Kuala Lumpur Madrid Melbourne Mexico City Mumbai
Nairobi Paris São Paolo Singapore Taipei Tokyo Toronto Warsaw
and associated companies in
Berlin Ibadan

Oxford is a registered trade mark of Oxford University Press

Published in the United States by
Oxford University Press Inc., New York

© *S. F. A. Kettle, 1996, 1998*

First published by Spektrum Academic Publishers, 1996
First published by Oxford University Press, 1998

A catalogue record for this book is available from the British Library

Library of Congress Cataloging in Publication Data
(Data available)

ISBN 0 19 850405 5 (Hbk)
ISBN 0 19 850404 7 (Pbk)

Typeset by KEYWORD Publishing Services, London
Printed in Great Britain by Biddles Ltd,
Guildford and King's Lynn

In memory of Doreen, 1929–1994

Contents

Foreword

GEORGE CHRISTOU

Indiana University, Bloomington

I am no doubt representative of a large number of current inorganic chemists in having obtained my undergraduate and postgraduate degrees in the 1970s. It was during this period that I began my continuing love affair with this subject, and the fact that it happened while I was a student in an organic laboratory is beside the point. I was always enchanted by the more physical aspects of inorganic chemistry; while being captivated from an early stage by the synthetic side, and the measure of creation with a small c that it entails, I nevertheless found the application of various theoretical, spectroscopic and physicochemical techniques to inorganic compounds to be fascinating, stimulating, educational and downright exciting. The various bonding theories, for example, and their use to explain or interpret spectroscopic observations were more or less universally accepted as belonging within the realm of inorganic chemistry, and textbooks of the day had whole sections on bonding theories, magnetism, kinetics, electron-transfer mechanisms and so on. However, things changed, and subsequent inorganic chemistry teaching texts tended to emphasize the more synthetic and descriptive side of the field. There are a number of reasons for this, and they no doubt include the rise of diamagnetic organometallic chemistry as the dominant subdiscipline within inorganic chemistry and its relative narrowness *vis-à-vis* physical methods required for its prosecution.

These days inorganic chemistry is again changing dramatically with the resurgence of coordination chemistry, fuelled by the increasing importance of metals in biology and medicine and the new and explosive thrusts into inorganic materials encompassing a wide variety of types and areas of application, of which high-temperature super-conductors, molecular ferromagnets and metallomesogens are merely the tip-of-the-iceberg. Modern-day, neo-coordination chemistry is thus a much broader discipline and one that now demands greater knowledge and expertise in a much larger range of theoretical or spectroscopic techniques and physicochemical methods, and to a higher level of sophistication.

At Indiana University, as at most universities I am sure, we have assigned a high priority to modifying our inorganic chemistry curriculum to accurately reflect the changing nature of the field and to better prepare our students for the demands on them of the new century. The general paucity of suitable texts *directed towards the inorganic chemistry student* is a problem. There are, of course, many advanced texts available for consultation but, on the theoretical/physical side at least, these are frequently directed at the more quantum mechanically and mathematically competent reader. In my experience as an instructor, the average student of inorganic chemistry picking up an advanced text on magnetochemistry, for example, will probably not survive the initial jump into the deep waters of quantum mechanics.

This present work by Sid Kettle represents a wonderful bridge for the student. It is designed as an intermediate-level text that can serve both as a user-friendly introduction to a large number of topics and techniques of importance to the student of coordination and physical inorganic chemistry, and also as a springboard to more advanced texts and studies. It is written in a style that is appropriate for a teaching text, anticipating and answering the questions that students will typically have on encountering the topic for the first time, and introduces a large number of theoretical, spectroscopic and physicochemical techniques without sacrificing the more classical content of a coordination chemistry text. In this regard, it is a wonderful hybrid of the classic and modern aspects of coordination and physical inorganic chemistry and is consequently an admirable text for the student of this area.

Preface

Some twenty years ago, theoretical aspects of inorganic chemistry formed a major component of any inorganic textbook. Today, this component is much less evident. No doubt, this shift in emphasis is a proper response both to the undue weight then given to theoretical aspects and to the developments that have taken place elsewhere in the subject. However, in the interval there have been theoretical developments that deserve a place; further, it has probably become more difficult for the interested student to access the older work. There seemed to me to be a real need for an easy-to-read, and so largely non-mathematical, text that would bridge the gap between the relatively low-level treatments currently available and the research level paper, review, monograph or text. The present book was written with the object of providing a bridge for this gap. Although the motivation for writing it is seen in its theoretical content, it was recognized that there are advantages in placing this in a broader context. So, what has resulted is a book which contains an overview of the relatively traditional and elementary along with contemporary research areas, wherever possible viewed from an integrated theoretical perspective. Because a text on physical inorganic chemistry can easily become a series of apparently disconnected topics, I have given the subject a focus, that of coordination chemistry, and have included chapters which should enable the book to double as a text in that area. To keep the size of the book manageable, to recognize that it is aimed at the intermediate stage reader, and because the topic is covered so extensively elsewhere, I have assumed a knowledge of the most elementary aspects of bonding theory.

In a book such as this it is impossible to avoid cross-references between chapters. However, it is equally difficult to ensure that such cross-references supply the answers expected of them. I have therefore attempted to make each chapter as free-standing as possible and have used the resulting duplications as a mechanism for deepening the discussion. This strategy can produce its own problems as well as benefits; I hope that the index will provide direction to sufficient additional material to deal with the problems!

I am indebted to many institutions which provided the hospitality that enabled most of the book to be written—Chalmers University and the University of Gothenburg, Sweden; the Royal Military College, Kingston, Canada; the University of Turin, Italy; the University of Nairobi, Kenya; Tokyo Institute of Technology, Japan; the University of Szeged, Hungary and Northwestern University, USA. Of the numerous individuals who have provided helpful comments on sections of the book, and often offered material for inclusion, I am grateful to Professor R. Archer and his students at the University of Massachusetts, Amherst, USA, who made many detailed comments on an early edition of the text, to Professor K. Burger of the University of Szeged, Hungary—his contributions were very helpful—and to Dr. S. Cotton, who was a constant fund of comment and information. I am particularly indebted to the Rev. Dr. Iain Paul who, in his own inimitable manner, worked through every sentence and made a multitude of suggestions for improvement and clarification. Defects, errors and omissions, of course, are my own responsibility.

S.F.A.K.

1

Introduction

Textbooks on physical inorganic chemistry can, during their preparation, easily evolve into compilations of apparently unrelated topics. In writing the present book, therefore, it was decided to circumvent this problem by adopting a single, unifying theme: coordination chemistry. The benefit of this approach is that the theme spans almost all aspects of physical inorganic chemistry; furthermore, the resulting book also doubles up as a text on coordination chemistry itself. In achieving this duality, some of the material present might appear out of place in a book devoted to physical inorganic chemistry alone. However, it is probably no bad thing that, for example, in addition to a discussion about the chemical bonding within a particular exotic species, reference can also be found to its preparation. Since, then, the theme of this book is that of coordination compounds (or, as they are often called, coordination complexes), our first task is to define the term *coordination compound*. This is not straightforward, for the use of the term is determined as much by history and tradition as by chemistry. In practice, however, confusion seldom arises. Let us consider an example.

When boron trifluoride, a gas, is passed into trimethylamine, a liquid, a highly exothermic reaction occurs and a creamy-white solid separates. This solid has been shown to be a 1:1 adduct of the two reactants, of which the molecules have the structure shown in Fig. 1.1, the boron atom of the boron trifluoride being bonded to the nitrogen atom of the trimethylamine. The adduct, resulting from the combination of two independently stable molecules, is an example of a coordination compound. An electron count shows that the boron atom in boron trifluoride possesses an empty valence shell orbital, whilst the nitrogen of the trimethylamine has two valence shell electrons in an orbital not involved in bonding. It is believed that the bond between the boron and nitrogen atoms in the complex results from the donation of these nitrogen lone pair electrons into the empty boron orbital, so that they are shared by both atoms. Coordination compounds in which such electron-transfer appears to be largely responsible for the bonding are sometimes also called *donor–acceptor complexes*,

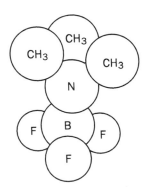

Fig. 1.1 The structure of the coordination compound formed between boron trifluoride and trimethylamine.

although it is to be emphasized that, once formed, there is no difference in kind between these and ordinary covalent bonds; the difference is in our approach to them. In the boron trifluoride–trimethylamine adduct, the nitrogen atom of the trimethylamine molecule is said to be *coordinated* to the boron atom. That is, the electron donor is said to be coordinated to the electron acceptor. A coordinating group (usually called a *ligand*—it is ligated, 'tied to', the electron acceptor) need not be a molecule and need not be uncharged. For example, boron trifluoride reacts with ammonium fluoride to give the salt $NH_4[BF_4]$ which contains the complex anion $[BF_4]^-$. Here we adopt the convention of placing the complex species of interest within square brackets, a convention that will almost invariably be adopted in this book. In the $[BF_4]^-$ anion the boron atom is tetrahedrally surrounded by ligands, just as it is to a first approximation in $[BF_3 \cdot NMe_3]$ (Fig. 1.1). Notice that, for non-transition metals and metalloids, complex formation is associated with a change (usually an increase) in the number of groups to which the central atom is attached. Boron trifluoride, BF_3, is not normally thought of as a complex, but its adduct with trimethylamine certainly is.

Most workers regard both trimethylamine and the fluorides as ligands in the adduct (a pattern that has just been followed). It would be a logical deduction from the picture just presented to conclude that the maximum number of ligands which can be added to form a complex is determined by the number of empty valence shell orbitals on the acceptor atom. Whilst this is generally true, an indication of the difficulty of rigorously defining 'a complex' is given by the fact that, in practice, the criterion of *change* in number of bonded atoms outweighs all others for these elements. Thus, phosphorus pentachloride exists in the gas phase as discrete PCl_5 molecules. The solid, however, is an ionic lattice, containing $[PCl_4]^+$ and $[PCl_6]^-$ ions. These two species are usually classed as complex ions, although the molecule in the gas phase is not.

The detailed geometry of a complex molecule is not simply a combination of the geometries of its components. In the trimethylamine–boron trifluoride adduct, for instance, the B–F bond length is 1.39 Å and the F–B–F bond angle 170° compared with 1.30 Å and 120° in the isolated BF_3 molecule. Similarly, the geometry of the bound trimethylamine fragment differs from that of the free amine. Information about the bonding within a complex may, in favourable cases, be obtained by a detailed consideration of these bond length and angle changes.[1] It is not surprising, then, that a recurrent theme throughout this book will be the relationship between molecular

---

[1] But there are traps for the unwary. In the simpler compound $H_3B–NH_3$ it was found that a discrepancy exists between the B–N bond length determined by X-ray crystallography (1.564 Å) and by microwave spectroscopy (1.672 Å). Some detailed theoretical calculations have been carried out on the problem and have shown that the energy difference between these two bond lengths is rather small for the isolated molecule. Simulation of the molecular environment showed that the longer bond length in the crystal almost certainly arises from environmental effects and therefore carries no great bonding significance—except that over a short range the total bonding energy is rather insensitive to the precise internuclear distance. A second trap arises from the observation that the (stabilization) energies of complex formation increase in the order $BBr_3 > BCl_3 > BF_3$, an observation that has been related to π bonding between boron and the halogens (being greatest for the bromide). In fact, accurate calculations have shown that the difference in stabilities results from variations in the simple donor–acceptor bonding described in the text.

geometry and electronic structure, the link between the two commonly being provided by group theory.

Complexes are formed by both transition metal and non-transition elements. Indeed, at the present time all compounds of transition metal ions, with very few exceptions, are regarded as complexes. However, despite the argument given above, the simple donor–acceptor bond approach does not seem immediately applicable to coordination complexes of the transition metals, since the molecular geometry does not depend greatly on the number of valence shell electrons—and, so, on the number of empty orbitals. As will be seen in Chapter 7, in the simplest model of the bonding in transition metal complexes electron donation is not even considered to be involved, a molecule being regarded as held together by electrostatic attraction between a central transition metal cation and the surrounding anions or dipolar ligands. However, in more sophisticated discussions of the bonding (Chapters 6 and 10) the donor–acceptor concept is largely reinstated for these compounds. So we may conveniently (but not always correctly) regard a coordination compound as composed of (a) an electron donor (*ligand* or *Lewis base*), an individual atom or molecule which possesses non-bonding lone-pair electrons but no low-lying empty orbitals; and (b) an electron acceptor (*metal atom, cation* or *Lewis acid*) which possesses a low-lying empty orbital. As in many other areas of chemistry, we shall often be particularly concerned with the pair of electrons that occupy the highest occupied molecular orbital (the HOMO) of the electron donor. This is matched by an interest in the lowest unoccupied molecular orbital of the electron acceptor (the LUMO).[2] The donor atom of a ligand is usually of relatively high electronegativity and the acceptor atom is either a metal or metalloid element.

Chapters 2–4 are full of examples of ligands and coordination compounds and the reader can gain an impression of the field by quickly thumbing through them. The field is not as complicated as it may appear, although it will rapidly become evident that at the present time some rather unusual organic molecules are increasingly being used as ligands and that neither the methods of preparation nor the molecular geometries formed need be quite as simple as for the examples given above. Indeed, part of the current fascination of the subject lies in the elegance of many of the complexes which are currently being studied. Complexes in which the metal atom is totally encapsulated, as in the *sepulchrates*; those in which it is at the centre of a crown (*crown ether* complexes, for instance); those in which it is surrounded by two ligands which interleave each other (complexes of *catenands*); those in which it is at the centre of a stockade-like ligand (*picket-fence* complexes) and so on. By such means it is proving possible to design highly metal-specific ligands, which offer the prospect of selective ion extraction from, for example, low-grade ores or recycled materials. The future importance of such possibilities in the face of ever-declining natural resources can scarcely be overestimated. Similarly, the use of such complexes in small-molecule activation will surely be of vital importance— for instance, in the fixation of gaseous nitrogen and the synthetic use of hydrocarbon species which would otherwise be used as fuels.

[2] Because the basics of the subject were developed before use of the HOMO and LUMO terminology became widespread these labels are scarcely to be found in the relevant literature.

Inevitably, current research tends to focus on the unusual and the exotic and so, since a book such as this attempts to reflect something of current work, tends to make the subject appear less straightforward than it really is. Perhaps it is helpful to recognize that even at a simple level, problems of definition can occur. Thus, an uncharged compound containing a main group metal or metalloid element bonded to a methyl group is not usually viewed as a complex in which CH_3 functions as a ligand, although the CH_3 group is isoelectronic with ammonia, a molecule which is frequently a ligand. So, compounds such as $Zn(C_2H_5)_2$ and $Si(CH_3)_4$ would not usually be considered complexes. However, successes in the synthesis of transition metal–methyl compounds means that there has been a change in attitude and that these, too, are now regarded as complexes containing the CH_3 group as ligand. The question of whether they should be considered as complexes of CH_3^- is not usually regarded as of great importance. A similar ambiguity is that although the manganate ion MnO_4^{2-} would be considered a coordination complex (of Mn^{6+} and O^{2-}) the sulfate anion SO_4^{2-} would not. Evidently, we have reached the point at which history and tradition, as well as utility, colour the definition of a coordination compound.

The father of modern coordination chemistry was Alfred Werner, who was born in 1866 and lived most of his life in Zürich. At the time it was known that the oxidation of cobalt(II) (cobaltous) salts made alkaline with aqueous ammonia led to the formation of cobalt(III) (cobaltic) salts containing up to six ammonia molecules per cobalt atom. These ammonia molecules were evidently strongly bonded because very extreme conditions— boiling sulfuric acid, for example—were needed to separate them from the cobalt. There had been considerable speculation about the cobalt–ammonia bonding and structures such as

$$Co \begin{array}{l} \diagup NH_3\!-\!Cl \\ -\!NH_3\!\cdot\!NH_3\!\cdot\!NH_3\!\cdot\!NH_2\!-\!Cl \\ \diagdown NH_3\!-\!Cl \end{array}$$

which today look quite ridiculous (although based on the not unreasonable hypothesis that, like carbon, nitrogen can form linear chains) had been proposed for the cobalt(III) salt $CoN_6H_{18}Cl_3$ (which we would now write as $[Co(NH_3)_6]Cl_3$). Werner's greatest contribution to coordination chemistry came in a flash of inspiration (in 1893, at two o'clock in the morning) when he recognized that the number of groups attached to an atom (something that he referred to as its *secondary valency*) need not equal its oxidation number (he called it *primary valency*). Further, he speculated that for any element, primary and secondary valencies could vary independently of each other. The chemistry of the cobalt(III)–ammonia adducts could be rationalized if in them cobalt had a primary valency of three, as in $CoCl_3$, but a secondary valency of six, as in $[Co(NH_3)_6]Cl_3$. The term secondary valence has now been replaced by *coordination number* and primary valency by *oxidation state* but Werner's ideas otherwise stand largely unchanged.

Subsequently, Werner and his students obtained a vast body of experimental evidence, all supporting his basic ideas. They further showed that in the complexes they were studying the six coordinated ligands were

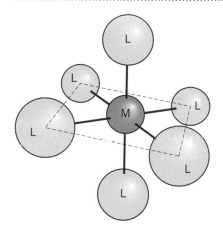

Fig. 1.2 An octahedral complex ML_6 where M is represented by the central white atom and the ligands L each by a shaded atom. A regular octahedron—one is shown in Fig. 1.3—has eight faces (each an equilateral triangle) and six equivalent vertices. In an octahedral complex the ligands are placed at the vertices. In Fig. 1.2, and in similar diagrams throughout this book, the perspective is exaggerated (the central four ligands lie at the corners of a square) and all ligand atoms are the same size. In this example, all six ligands are identical. Even if they are not, provided the geometrical arrangement shown in Fig. 1.2 is more-or-less maintained, the complex is still referred to as octahedral. As molecular symmetry is important for the arguments to be presented in many of the following chapters, it will often prove convenient to emphasize this by including in structural diagrams, lines which remind the reader of the molecular symmetry. Commonly, such lines will link ligands together and, clearly, should not be interpreted as bonds between ligands.

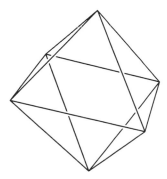

Fig. 1.3 An octahedron, a regular figure in which all vertices are equivalent, as are all faces and all edges.

arranged octahedrally about the central atom (Figs. 1.2 and 1.3). Werner was awarded the Nobel prize for chemistry for this work in 1913. Some measure of his stature and work is provided by the fact that in one field (that of polynuclear cobaltammine complexes) there has, to this day, been scarcely any addition to the list of compounds he prepared.

Most textbooks discuss transition metal complexes separately from those of the main group elements. There is, in fact, much in common between the two classes and, whenever possible, we shall treat them as one. However, complexes of the transition metal ions may possess an incomplete shell of d electrons which necessitate separate discussion. This characteristic makes it particularly useful to determine the magnetic and spectral properties of members of this class of complexes and the exploration of these properties will require separate chapters devoted to them. In a similar way, complexes of the lanthanide and actinide elements, with, typically, an incomplete shell of f electrons tucked rather well inside the atom and away from the ligands—and so behaving rather as if they are in an isolated atom—require their own discussion.

The water-soluble ionic species of transition elements such as chromium, manganese, iron and copper seem to exist in aqueous solution as, for example, $[Cr(H_2O)_6]^{3+}$, $[Mn(H_2O)_6]^{2+}$ and $[Fe(H_2O)_6]^{2+}$. That is, it is more accurate to talk of 'the aqueous chemistry of the $[Cr(H_2O)_6]^{3+}$ ion' than of 'the aqueous chemistry of the Cr^{3+} ion'. Similarly, in solid $FeCl_3$, the iron atoms are not attached to three chlorines but, octahedrally, to six (each chlorine is bonded to two iron atoms). We have already encountered the fact that solid PCl_5 is really $[PCl_4]^+[PCl_6]^-$. The lesson to be learnt from all this is that coordination compounds are much more common than one might at first think. The colour of many gemstones and minerals, the chemistry carried out within an oil refinery, element deficiency diseases in animals, the reprocessing of nuclear fuel rods, the manufacture of integrated circuits, the chlorophyll in plants, the colours of a television screen—all involve complexes, though we shall not be able to cover all of these diverse topics in the present book. Although the first example we gave in this chapter portrayed complexes as being formed between independently stable species, and this is often the case, there are also many fascinating examples of molecules which are only stable when they exist as part of a complex; even independently stable species have their chemical as well as their physical properties drastically changed as a result of coordination.

In summary, there is no precise and time-constant definition of a coordination compound—at one extreme methane could be regarded as one—and the usage of the term is extended to all compounds to which some of the concepts developed in the following chapters can usefully be applied. Indeed, one could argue that the value of the concept lies in its flexibility and adaptability so that the absence of a fixed and agreed definition is no handicap. We shall find that the study of coordination compounds excludes few elements—the sodium ion forms complexes—and overlaps with biochemistry and organic chemistry. Further, it will involve some fairly detailed theoretical interpretations, although in this book the powerful but surprisingly simple concepts of symmetry are used to reduce theoretical complexities to a minimum.

Further reading

Most inorganic texts published before the mid-1950s give historical overviews of the development of coordination chemistry. The language used sometimes seems strange in terms of modern usage and should not be allowed to distract the reader unduly. A browse through the older books in a good library should be adequate; as an indication of the variety available the following are worthy of mention:

- *Modern Aspects of Inorganic Chemistry* H. J. Eméleus and J. S. Anderson, Routledge and Kegan Paul, London, 1938/1952.
- *The Chemistry of the Coordination Compounds* J. C. Bailar (ed.), Reinhold, New York, 1956.
- *An Introduction to the Chemistry of Complex Compounds* A. A. Grinberg, 1951, English translation by J. R. Leach, Pergamon, Oxford, 1962.

For the person interested in exploring the historical aspect in depth, one book is essential reading. This is *Werner Centennial*, published in 1967 by the American Chemical Society as No. 62 in their Advances in Chemistry series (R. F. Gould, editor). This book contains over 40 chapters on historical and current (in 1967) chemistry, including chapters devoted to the Werner–Jørgensen controversy (the nitrogen-chain structure for cobaltammines was originally proposed by Blomstrand and Jørgensen, who was his student at the time), polynuclear complexes of Co^{III} ammines and so on.

A useful source is Volume 1 of *Comprehensive Coordination Chemistry* G. Wilkinson, R. D. Gillard and J. A. McCleverty (eds.), Pergamon Press, Oxford, 1987. Chapter 1.1 is 'General Historical Survey to 1930' by G. B. Kauffman and Chapter 1.2 is 'Development of Coordination Chemistry since 1930' by J. C. Bailar Jr. Although they are frequently too specialized to warrant inclusion as further reading in the following chapters, the contents of the volumes of *Comprehensive Coordination Chemistry* provide a wealth of information on the details of coordination chemistry. The volumes assume, however, knowledge of the basic language and concepts of the subject, such as can be gained from the following chapters.

Question

1.1 In the book *Principles of Chemistry* published in English in 1881 Mendeleeff wrote:

'The admixture of ammonium chloride prevents the precipitation of cobalt salts by ammonia, and then, if ammonia be added, a brown solution is obtained from which potassium hydroxide does not separate cobaltous oxide. Peculiar compounds are produced in this solution.'

At about the same time, of course, Werner's work was providing an understanding of these peculiar compounds. Write a one-page letter to Mendeleeff on behalf of Werner outlining the key points in Werner's understanding.

2

Typical ligands, typical complexes

2.1 Classical ligands, classical complexes

As was seen in the previous chapter, ligands are atoms or molecules which, at least formally, may be regarded as containing electrons which can be donated to an atom which functions as an electron acceptor. The presence of such a pair of electrons is a necessary, but not a sufficient, condition. Thus, the halide anions are typical simple ligands but halogen compounds such as CH_3Cl or C_6H_5F are very seldom ligands, although the halogen atoms in them still possesses electron pairs which could be donated. The reason lies, at least in part, in the high electronegativity of the halogens. When the halogen possesses a negative charge there is little energetic cost in reducing this charge, a cost that can be paid for by the exothermicity of the bond formed. When the halogen is uncharged, the energetic cost of it becoming positively charged (as it would if it donated electrons) is too great. However, electronegativity cannot be the sole reason because, as we shell see, organic oxides and sulfides form many complexes—and yet the electronegativity of oxygen is greater than that of chlorine. An illustration of the complexing ability of ethers, for example, which is more spectacular than dangerous—although appropriate precautions should be taken—is to add a drop of diethylether, Et_2O, to tin(IV) chloride (stannic chloride, a liquid). The solid complex $[SnCl_4(Et_2O)_2]$ is instantly formed. The heat of reaction is sufficiently great to boil off some of the diethylether, sending clouds of the white complex into the air. As evident from the previous chapter, ammonia and related compounds such as trimethylamine form many complexes. Organophosphorus ligands are also widely used in synthetic chemistry and we shall meet them in the contexts of organometallic chemistry (that of complexes in which a ligand which would be regarded as part of organic chemistry is bonded through carbon to a metal) and catalysis, in particular.

Table 2.1 Some classical ligands which are common in the complexes of either transition metal and/or main group elements. The names given follow the rules to be detailed in Chapter 3. Note that some species are shown twice, when they can coordinate in more than one way. Note, too, that some ions shown once can, in fact, coordinate in more than one way; examples are provided by CN^-, $S_2O_3^{2-}$ and OCN^-

	Ligand	Name
Donor atoms from group 17(7) of the periodic table	F^-	fluoro
	Cl^-	chloro
	Br^-	bromo
	I^-	iodo
Donor atoms from group 16(6) of the periodic table	O^{2-}	oxo
	OH^-	hydroxo
	$\cdot O_2^{2-}$	peroxo
	CO_3^{2-}	carbonato
	$CH_3CO_2^-$	acetato
	ONO^-	nitrito
	SO_4^{2-}	sulfato
	SO_3^{2-}	sulfito
	$S_2O_3^{2-}$	thiosulfito
	S^{2-}	thio
	CH_3S^-	methylthio
	H_2O	aqua
	CH_3OH	methanol
	$(NH_2)_2CS$	thiourea
	$(C_2H_5)_2O$	diethylether
Donor atoms from group 15(5) of the periodic table	CN^-	cyano
	OCN^-	cyanato
	SCN^-	thiocyanato (note, bonded through N)
	NO_2^-	nitro
	N_3^-	azido
	NH_3	ammine[a]
	CH_3NH_2	methylamine
	$N(CH_3)_3$	trimethylamine
	C_6H_5N	pyridine (usually abbreviated as py)

[a] Note the spelling, 'mm', not 'm'.

A list of some simple and common ligands is given in Table 2.1. The entries in this table are confined to *classical* ligands, such as could well have been studied by Werner. There are other ligands, many also simple and common but *non-classical*—such as the organophosphines, which will be covered shortly. Inevitably, the distinction we are making is an arbitrary one. In Table 2.2 are listed representative examples of complexes formed by some of the ligands in Table 2.1. The detailed molecular geometries of the complexes in Table 2.2 will not be discussed because for many of them there are ambiguities. These problems will be dealt with in Chapter 3, where many of the examples given in Table 2.2 will reappear.

Complexes of most of the ligands that have so far been mentioned have been studied for almost a century. Although one might expect the field to be exhausted, each year there are a few new surprises: the discovery of a method for the easy preparation of complexes of a metal in a valence state

Table 2.2 Some of the complexes formed by some of the ligands in Table 2.1. This table endeavours to demonstrate that any one metal may well form complexes with different numbers of ligands and that many complexes are not monomeric (there are sulfur and iodine bridging atoms in the two cases not explicitly detailed). Note that when two different complexes contain the same number of ligands it does not necessarily mean that the geometrical arrangements of the ligands is the same in the two cases. Note, too, that the attempt to show variety means that this table does not properly reflect the fact that the majority of complexes contain metal ions bonded to six ligands

$[Co(NCS)_4]^{2-}$
$[Co(CN)_5]^{3-}$
$[Co(NO_2)_6]^{3-}$
$[Co(NH_3)_5N_3]^{2+}$

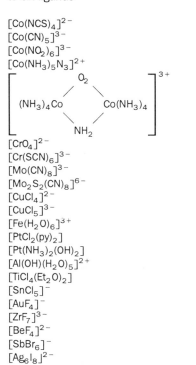

$[CrO_4]^{2-}$
$[Cr(SCN)_6]^{3-}$
$[Mo(CN)_8]^{3-}$
$[Mo_2S_2(CN)_8]^{6-}$
$[CuCl_4]^{2-}$
$[CuCl_5]^{3-}$
$[Fe(H_2O)_6]^{3+}$
$[PtCl_2(py)_2]$
$[Pt(NH_3)_2(OH)_2]$
$[Al(OH)(H_2O)_5]^{2+}$
$[TiCl_4(Et_2O)_2]$
$[SnCl_5]^-$
$[AuF_4]^-$
$[ZrF_7]^{3-}$
$[BeF_4]^{2-}$
$[SbBr_6]^-$
$[Ag_6I_8]^{2-}$

that was previously regarded as difficult, or the preparation and characterization of a complex which was previously believed not to exist. In this field, however, the main work that is still to be done is not that of preparation, but rather work at a deeper level, deeper perhaps than the fields covered in this book. For instance, in the solid state, how does a complex ion interact with its environment? How are these interactions changed with different counterions; to what extent are the properties that we observe those of complex plus environment and different from those of the isolated complex itself? Why, for instance, does the $[Ni(CN)_5]^{3-}$ anion have two different geometries in the crystals of one of its salts? There has been some considerable interest generated by the recent discovery that some anions which have always been regarded as having very little tendency to coordinate can actually do so—perhaps the best example is provided by the anion $[B(C_6H_5)_4]^-$. The current thrust of preparative work exploits the fact that several donor atoms, oxygen and nitrogen in particular, can be strung together with a web of carbon (and sometimes boron or phosphorus) atoms. An almost infinite variety of exotic ligands becomes possible. This field has many attractions. By choosing the ligand to be one with a rather rigid backbone it is possible to impose an unusual coordination geometry on a metal atom. A very popular strategy at the present time is to chose a ligand which has very bulky, and so sterically demanding, substituent groups. In the complexes it forms there simply is not enough space to fit very many ligands around the central atom and so a low coordination number or unusual geometry results. It is found that metal ions in unusual coordination geometries often have unusual reactions and/or properties and this makes them of particular interest. Thus, with suitable choice of ligand it is possible to make volatile compounds of sodium! Alternatively, by careful tuning of the ligand geometry it may be possible to make it highly specific for a particular metal. This produces visions of metal recovery from low-grade ore and even gold from sea water (such schemes tend to fail because the cost of the ligand and its recovery for reuse exceeds the value of the metal obtained).

Next, it may be possible to produce a ligand which closely mimics, in its geometry and composition, that of one found in nature in a complex of biological importance. The biological compound is almost certainly only available in small quantities, difficult to purify and unstable under most laboratory conditions. Working with a model compound is much easier than working with the real thing! This topic will be covered in much more detail in Chapter 16.

There is one further advantage to working with ligands containing more than one donor atom. This is that the (thermodynamic) stability of a complex in which two or more donor atoms are part of the same ligand molecule often appears much greater than if the same atoms were in separate ligand molecules. There has been much debate on the origin of this co-called *chelate* effect. Chemists like to use their imaginations and to compare a metal ion held between two donor atoms on a ligand with a crab holding its prey in its claws—hence chelate (Greek *chelos*—a claw). It is common to talk of chelating ligands and of chelate complexes. Complexes are often conveniently divided into two classes, *labile* and *inert*. In the former, consisting of most complexes of main group metals and many of the more familiar transition metals, ligands are readily replaced. In the latter, for example complexes of Cr^{III} and Co^{III}, ligand replacement is very slow except under forcing conditions. The chelate effect is a phenomenon which increases the inertness of complexes (which may mean making a complex less labile). We shall return to the chelate and related effects in Section 5.5. At this point all that will be added is a word of caution: although the occurrence of a chelate effect is common it is not invariably present. So, organic isocyanides, RNC, form many complexes (bonding to the metal through the terminal C). The R–N–C sequence remains essentially linear in complexes and the metal atom also bonds colinearly. The result is that if a bidentate organic ligand containing two RNC groups is synthesized then there has to be a sizeable number of carbon atoms (seven CH_2 units, for instance) between the two –NC groups if they are both to coordinate to the same metal atom and thus form a chelate. This means a 12-membered ring system and, as will become evident in Chapter 5, 12-membered rings show no hint of a chelate effect.

In Table 2.3 are listed some of the more common, classical, polydentate ligands. Again, imagination. The ligand is now pictured as biting, and thus holding onto, the metal with several teeth (bidentate[1] = two donor atoms, tridentate = three donor atoms). The (minimum) distance between two donor atoms in a bidentate ligand is sometimes referred to as the *bite* of the ligand and the angle subtended at the metal atom the *bite angle*. In Table 2.4 are detailed a selection of some of the more exotic ligand species under current study. The systematic names of these molecules are usually so horrendous that trivial, often physically descriptive, names are preferred. Typical examples are *picket fence*, *crown* and *tripod*, some of which are given in Table 2.4. Table 2.5 shows a selection of the complexes formed by the ligands contained in the previous two tables.

2.2 Novel ligands, novel complexes

Since the 1950s it has been clear that the simple 'lone electron pair donor' picture of a ligand and 'lone electron pair acceptor' picture of a metal in a complex is inadequate. This is nowhere clearer than in the field of organo-metallic chemistry, where a host of organic molecules, in which all of the valence electrons are involved in bonding within the organic molecule, form complexes with metal atoms. Notice the use of the word atom—the metal is commonly, formally, zero-valent in these compounds and so any simple electrostatic model for ligand–metal bonding which might be applied to their classical counterparts seems rather implausible. The same conclusion is also forced upon us by the nature of the ligands commonly involved—molecules such as hydrocarbons, carbon monoxide and all sorts of unexpected species, even, on rare occasions, the H_2 molecule. However, it is also clear that there are links with the more classical complexes—with ligands such as the halides, sulfides and organic phosphines, being common to both sets. As is so often the case, there is a continuous gradation. We have looked at the classical case

[1] The current recommendation is that the word bidentate be replaced by didentate but this recommendation has yet to gain general acceptance.

Table 2.3 Some common polydentate ligands[a] (charges on anions are omitted)

Name	Common abbreviation	Structure
Bidentate[a] ligands		
Acetylacetonato or 2,4-pentanedionato	acac	
2,2'-Bipyridine	2,2'-bpy often written as bpy	
Oxalato	ox	
Ethylenediamine or 1,2-ethanediamine	en	
o-Phenylenenebis(dimethylarsine) or 1,2-phenylenebis(dimethyarsine)	diars	
Glycinato	gly	
8-Hydroxyquinolinato	oxinate	
1,10-Phenanthroline	phen	
Dimethylglyoximato or 2,3-butanedione dioximato (see Table 2.5)	dmg	

(continued)

Table 2.3 (continued)

Name	Common abbreviation	Structure
Tridentate[a] ligands		
2,2′,6′,2″-Terpyridyl	terpy	
Diethylenetriamine	dien	
1,2,3-Triaminopropane	tap	
Tetradentate[a] ligands		
Triethylenetetraamine	trien	
Tris(2-aminoethyl)amine	tren	
Nitrilotriacetato	NTA	
Tris(2-diphenylarsineophenyl)arsine	QAS	

Table 2.3 *(continued)*

Name	Common abbreviation	Structure
Porphyrino	p; porphyrins are usually labelled according to their substituents – thus, tpp = tetraphenylporphyrin, see Table 2.4.	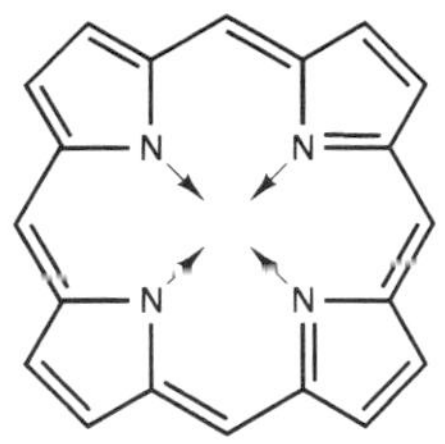
Phthalocyanino	pc	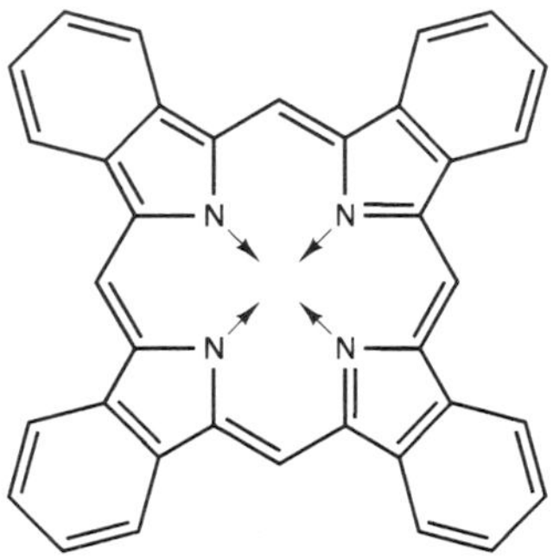

Hexadentate[a] ligand

Ethylenediaminetetraacetato	EDTA	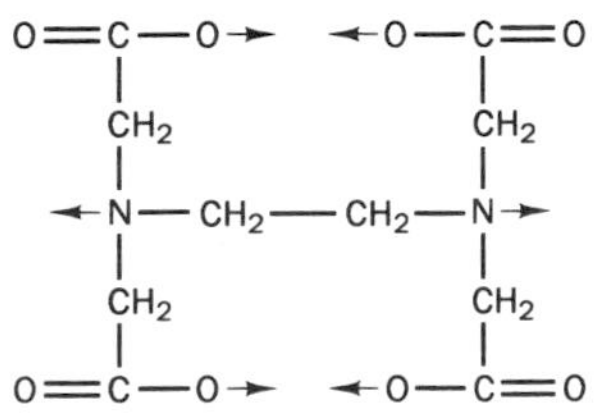

[a] This notation is probably self-explanatory because of the examples given. If not, it is described towards the end of Section 3.1.

in some detail if not depth (that will come in the next few chapters). What of the complexes formed by non-classical ligands and, in particular, what is the ligand–metal bonding involved? The answer to this question has become so important in inorganic chemistry that an outline answer will be given here, although it will have to be refined later. The pattern is illustrated by a discussion of the carbon monoxide molecule and the way that it bonds to a transition metal; with not much modification a general picture emerges. At the heart of it is the simultaneous—and interdependent— coexistence of two distinct bonding mechanisms.

At first sight one might well expect CO to bond to a metal through its oxygen atom—this oxygen has a lone pair of σ electrons and oxygen donors form many complexes; we would expect the oxygen to be more negatively charged than the carbon because of the difference in their electronegativities. Although oxygen-bonded CO complexes are known, the almost invariant mode of bonding of CO to a transition metal is through the carbon atom.

Table 2.4 Some more exotic polydentate ligands

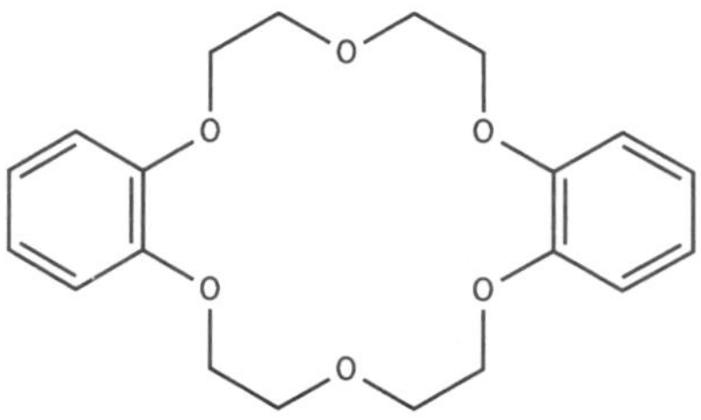

Dibenzo-18-crown-6
A crown ether with 18 atoms in the ring of which 6 are oxygens

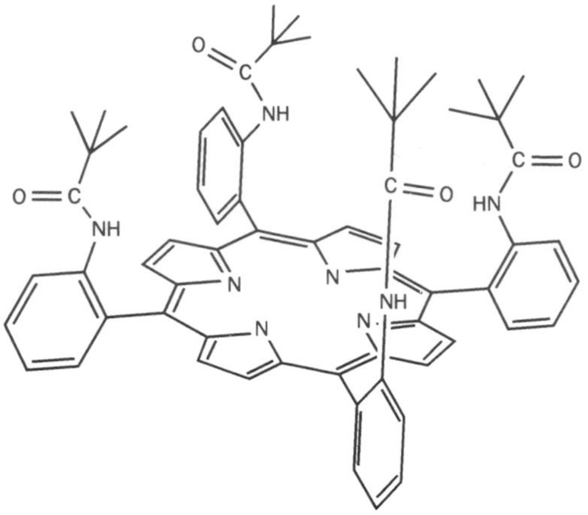

A *picket-fence* porphyrin

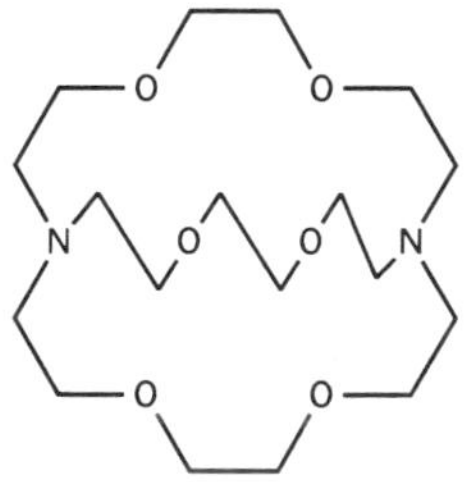

2,2,2-crypt
A cryptand (= Greek *hidden*); the 2s indicate the number of
oxygens in each N···N chain

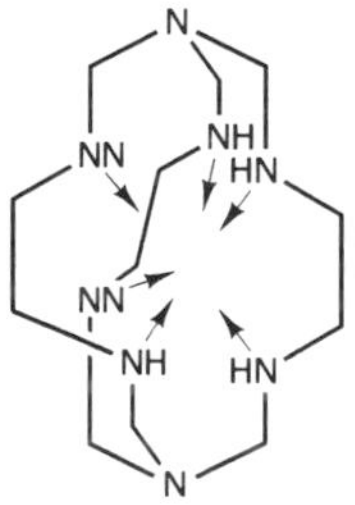

The sepulchrate ligand, which is hexadentate (the top
and bottom nitrogens do not normally coordinate)

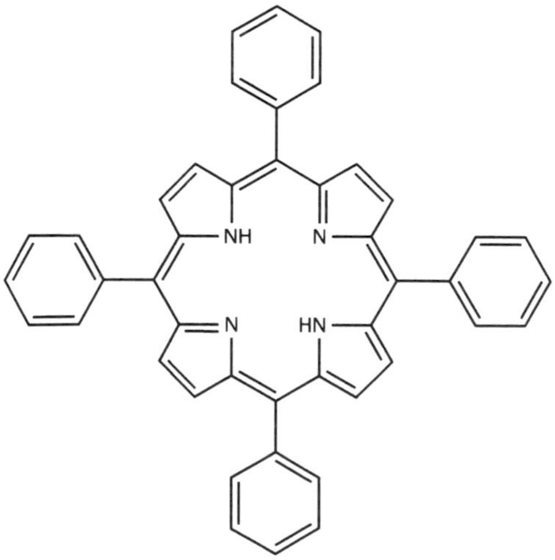

tpp
Tetraphenylporphyrin

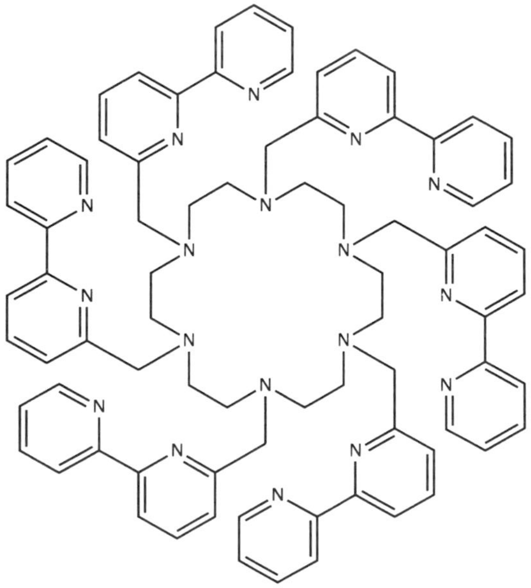

Bipyridyl groups grafted onto a cyclic hexamine

Table 2.4 (continued)

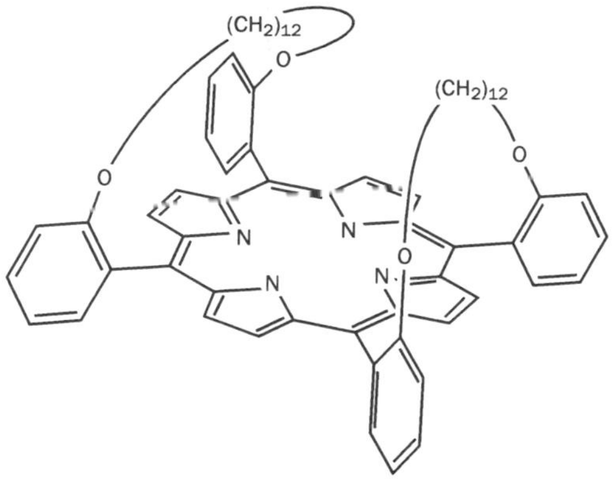

A *basket-handle* porphyrin

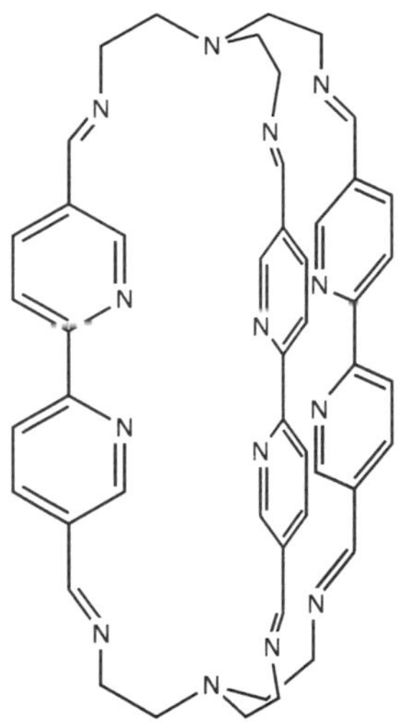

When three bipyridine ligands are capped (twice) by a triamine the cage ligand that results can complex three metal atoms simultaneously (in this example the 'caps' are, essentially, the ligand tren, tris(2-aminoethyl)amine)

Table 2.5 Selection of complexes formed by some polydentate ligands

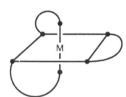

The general structure of [M(bidentate)$_3$]$^{n+}$ complexes where (bidentate) is one of the bidentate ligands of Table 2.3

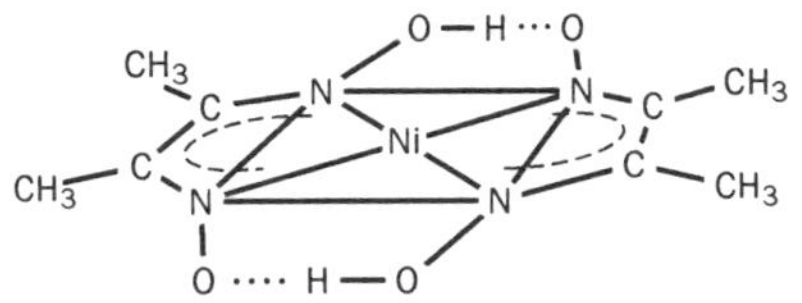

The square planar complex [Ni(dmg)$_2$]: note the interligand hydrogen bonds (also indicated in Table 2.3)

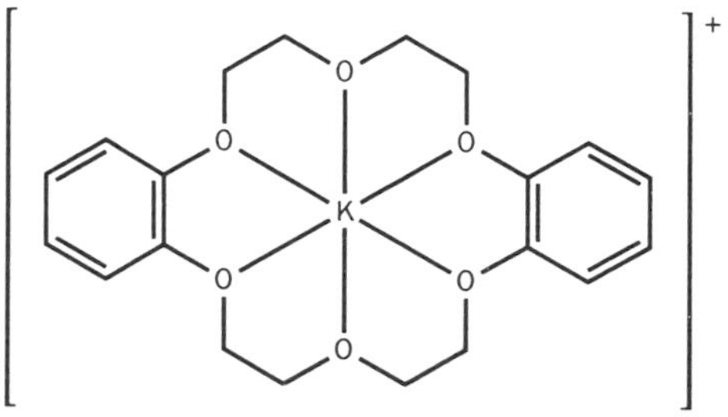

The K$^+$ complex of dibenzo-18-crown-6

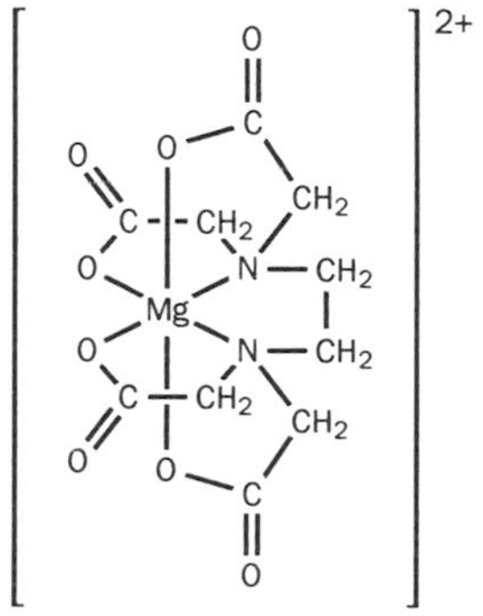

The Mg^{2+} EDTA complex

(continued)

Table 2.5 (continued)

The Co^{II} complex of tpp: one of two equivalent structures (cannonical forms) of the ligand is shown

The crystal structure of the M = Ag⁺ species with this ligand has been determined

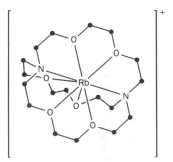

Rb⁺ 2,2,2-cryptate: oxygen and nitrogen donor atoms are shown, black dots indicate CH$_2$ groups

[Co(sepulchrate)]$^{3+}$

Cu^{2+} phthalocyanin: a complex which is used as a blue pigment

There is also a lone pair of σ electrons on the carbon which is larger, and so more available for bonding, than that on the oxygen. This, then, is the first member of the bonding partnership, a perfectly normal σ donation from carbon to the metal. But, given that a carbon atom is involved and that carbon atoms seldom act as ligands in classical complexes unless there is a negative charge around (as in CN$^-$), it would be surprising if this σ donation above were to lead to a strong bond. It needs reinforcing. The CO molecule is well able to provide this reinforcement because not only is the lone pair σ orbital larger on carbon than on oxygen, so too are the lobes of the orbitals

Fig. 2.1 (a) The σ donation OC → metal: here, the transition metal d_{z^2} orbital is shown as the acceptor but it could be some mixture of s, p_z and d_{z^2}. Note that there is lone-pair σ electron density on both O and C. That on C is the larger and so has the greater overlap with the empty metal orbital. In this, and all other figures in this chapter, filled orbitals are shaded; the phases of orbitals are given explicitly. (b) The π back-donation metal → CO: the metal orbital is almost pure d, the CO orbital is an empty antibonding π orbital. Note that for a linear triatomic OCM system there is second, equivalent, interaction to that shown above (it is like that shown but rotated 90° about the OCM axis and so is located above and below the plane of the paper).

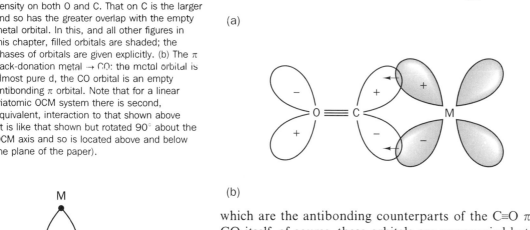

(a)

(b)

which are the antibonding counterparts of the C≡O π bonding orbitals. In CO itself, of course, these orbitals are unoccupied but that is no reason for ignoring them in complexes. They are LUMOs and a lesson which has been well learnt in recent years is that LUMOs are seldom disinterested spectators—they commonly play a key role in determining the outcome of chemical reactions (and that means that they get involved in the bonding somewhere along the way). It is therefore not surprising to learn that there is both experimental and theoretical evidence that electron donation from transition metal d orbitals to the CO π antibonding orbitals is of vital importance in the M–CO bonding. That is, the metal to carbon bonding consists of two parts; σ electron donation from the carbon to the transition metal and π electron back-donation from the transition metal to the carbon atom. This bonding is pictured in Fig. 2.1. Either bonding mechanism, on its own, would lead to charge buildup on the metal (σ) or the carbon (π); by acting together, the resultant charge buildup is small and each charge transfer can proceed further than would be possible in the absence of the other. One talks of *synergic* bonding (*synergismus* is Greek for 'working together').

Whenever one learns that a ligand bonds strongly to a transition metal ion but much more weakly to a main group element, the involvement of some π-mediated back-bonding synergic mechanism is likely. Indeed, such ligands are usually referred to as π *bonding ligands*, not only because of the existence of this mechanism but also because of its constancy, the bonding always involving ligand empty π orbitals of one sort or another. In contrast, there can be more variability in the σ-bonding mechanism. Thus, for the latter the electrons involved can be those that are associated with a chemical bond rather than with a particular atom. An excellent example is provided by the ethene (ethylene) molecule, C_2H_4. Its mechanism of bonding (the usual pattern of σ donation and π acceptance) is shown in Fig. 2.2. The orientation of the bonded molecule is such that it is clear that its orbital involved in σ bonding is the C=C π bonding orbital (which, of course, is occupied). The π antibonding orbital is also that associated with the C=C π bond (and which, of course, is empty in the isolated ethene

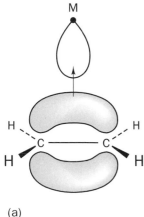

(a)

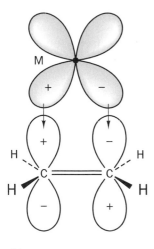

(b)

Fig. 2.2 (a) The σ donation C_2H_4 π → metal empty orbital. (b) The π back-donation metal → C_2H_4 π antibonding orbital.

Table 2.6 Some of the relatively simple ligands which occur in the organometallic chemistry of transition metals. Ligands such as CO frequently occur along with such ligands

Alkenes   Alkynes

But-1,3-diene (*cis*-butadiene)

1,5-Cyclooctadiene (cod)

The allyl group, a 3-electron donor; alternatively, it may be helpful to think of it as (allyl)⁻, a 4-electron donor

Benzene

The cyclopentadienyl group C_5H_5, usually written Cp. It is a 5-electron donor (some prefer to regard it as $C_5H_5^-$ and as a 6-electron donor)

molecule). If ethene seems strange as a ligand it is salutary to learn that the first compound containing ethene π-bonded to a metal was prepared (largely accidentally) by Zeise, a Danish chemist, in 1827. One of the two compounds he prepared, $K[Pt(C_2H_4)Cl_3]$, bears his name to this day—it is almost invariably referred to as Zeise's salt.

In Table 2.6 are listed a few of the simpler π bonding ligands and in Table 2.7 some of the species in which they are found. The bonding in these compounds will be covered in more detail in Chapter 10, but, for the moment, one important point suffices. This arises from the fact that the σ donor–π acceptor model of ligand–metal bonding involves different metal orbitals. It must, because, seen from the ligand, the metal σ and π orbitals have different nodality. This will be true for each π bonding ligand in the compound. The maximum number of such ligands which can be attached will be limited by the availability of suitable metal orbitals—either empty orbitals into which to donate or filled orbitals from which to accept. Eventually, we will run out of metal orbitals. Now, the valence shell orbitals of the (transition) metal are nd, $(n+1)s$ and $(n+1)p$ (for first row elements 3d, 4s and 4p). This is a total of $5+1+3=9$ orbitals so it seems likely that 18 electrons will be the maximum that will be involved in bonding, one way or another. It is not clear whether or not this hand-waving argument is generally valid; what is clear is that the vast majority of complexes of π bonding ligands obey the so-called '18-electron rule'. A book-keeping exercise of the number of electrons associated with the metal in the final complex (be they metal-originating or ligand-originating) usually leads to the number 18. In Table 2.8 this is demonstrated for some of the complexes of Table 2.7, but, to maintain a balance, two exceptions are included.

The choice of ligands in Table 2.6 has been made rather selectively. Although at the present time it would be difficult to find examples similar to the exotic ligands of Table 2.4, there are many unusual ligands that could have been included in Table 2.6. Thus, not all of the ligands in low-valence state complexes are independently stable molecules. For example, complexes in which cyclobutadiene (C_4H_4) is coordinated to a transition metal are commercially available, yet organic chemists have sought for generations to prepare this molecule. Further, in a sense, it would have been useful to have included many of the ligands in Table 2.6 several times over. This is because many of the ligands listed can bond in several different ways, each way giving rise to different complexes. In that a classic ligand such as Cl^- can bond not only to one but also to two metal ions (a bridging chloride ligand) or to three which lie at the corners of a triangle (a face-bonding chloride ligand) it is not surprising that a ligand such as CO should do the same. How CO differs is that, much more than Cl^-, it seems to form an almost continuous range of intermediate bonding patterns between these three main types. A ligand such as C_5H_5 may have one, two, three, four or five of its carbons bonded to a transition metal; when there are five they may not all be equally bonded, and so on. Again, the range is enormous. The existence of such a range strongly suggests that the energy differences between the various arrangements may be small. It seems that this, indeed, is so. There are several possible consequences. First, the thermal vibrations of the molecule may contain sufficient energy to enable such a ligand to hop

Table 2.7 Examples of complexes formed by some of the ligands in Table 2.6 and related ligands

The most famous complex of cyclopentadiene, ferrocene (**1**)

Dibenzenechromium; the corresponding (**2**) cation $[Cr(C_6H_6)_2]^+$ was made in 1919 but not recognized as such

A complex of a cyclopentadiene derivative. (**3**) $P(C_6H_5)_3$ is a ligand that plays a similar role to CO

(**4**)

Two of the compounds formed by reaction of $Fe(CO)_5$ with alkynes (**5**)

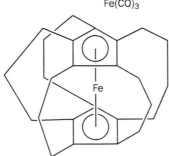

Some more exotic complexes which, despite appearances, are not significantly different in kind from those above.

A fulvalene complex (**6**)

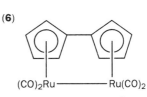

A diphenylethyne (diphenylacetylene) complex (**7**)

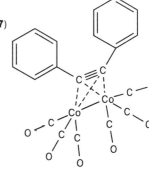

A *flyover* complex in which a single hydrocarbon ligand straddles two cobalt atoms which are bonded to each other. Each cobalt is also bonded to an allyl group (these bonds are shown large dotted) and to a carbon atom. (**8**)

A bisallyl palladium complex (**9**)

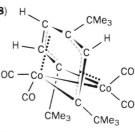

1,5-Cyclooctadiene (cod) complex of platinum(II) (**10**)

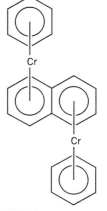

Table 2.8 Electron counts of complexes (**1–10**) in Table 2.7. Although it is some-times convenient to assign a formal charge to the metal when making these counts, experience is that fewer mistakes are made if metal atoms are treated as uncharged, if at all possible. This in turn means that ligands must also normally be regarded as uncharged. So, C_5H_5 is a five-electron donor. Note that, although in the text there is talk of two metal–CO bonding mechanisms, it is only the σ donation from the carbon of the CO that increases the number of electrons formally associated with the metal. So, CO and ligands such as phosphines which are believed to behave similarly, are regarded as two-electron donors

Complex	Metal	Formal charge	Number of metal valence electrons (m)	Ligands	Number of ligand electrons (l)	Total electrons $m + \sum l$
(**1**)	Fe	0	8	2 C_5H_5	2 × 5	18
(**2**)	Cr	0	6	2 C_6H_6	2 × 6	18
(**3**)	Fe	0	8	C_5H_5R 2 CO $P(C_6H_5)_3$	4 2 × 2 2	18
(**4**)	Fe	0	8	C_5H_4O 3 CO	4 3 × 2	18
(**5**)	Fe	0	8	C_6Me_4O 3 CO	4 3 × 3	18

(**6**) Consider each Ru separately: the Ru–Ru bond means that each Ru gains one electron from the other for electron counting purposes:

	Ru	0	8	C_5H_4R Ru–Ru 2 CO	5 1 2 × 2	18

(**7**) Either the diphenylethyne (diphenylacetylene) ligand may be regarded as donating two π electrons to each Co, or as forming two C–Co bonds (and thus contributing two electrons also). Again, the metal–metal bond contributes one electron to the electron count of each metal. Consider each Co separately:

	Co	0	9	$C_2(C_6H_5)_2$ Co–Co 3 CO	2 1 3 × 2	18

(**8**) Here, counting Co as Co^0 means that the allyl group is a three-electron donor. The cobalt–carbon and cobalt–cobalt bonds each contribute one additional electron to each cobalt:

	Co	0	9	$C_3H_2RR'R''$ Co–Co Co–C 3 CO	3 1 1 3 × 2	18

(**9**) Again, is it simplest if the pattern adopted in (**8**) is followed:

	Pd	0	10	2 C_3H_4R	2 × 3	16

This is an example of a complex which does not follow the 18-electron rule; square planar complexes of d^8 ions usually deviate, giving 16 instead.

(**10**) In this final example it is difficult to avoid a formal charge on the metal because to work with Pt^0 would mean working with Br rather than Br^-:

	Pt	2+	10	cod 2 Br^-	2 × 2 2 × 2	16

Another d^8 square planar complex that does not follow the 18-electron rule.

from one arrangement to another, perhaps equivalent to the first but perhaps not, at quite a high frequency. Such molecular gymnastics of so-called *fluxional* species have been the subject of extensive study, most notably by NMR. This is a topic which will be covered in Section 14.8.

Second, in a chemical reaction it may happen that the most stable orientation of a ligand in the starting material is not the most stable over some parts of the reaction pathway, nor, perhaps, in the reaction product. The third point stems from the second. When attached to a transition metal, a ligand may be stable in a range of convoluted geometric orientations that are not readily available to it as a free ligand. This may well mean that some new, otherwise impossible, reaction pathways become available. Further, their availability and nature are likely to be influenced by the other ligands present, both for bonding and steric reasons. The consequence is that the use of transition metal complexes in catalytic reactions has been an enormous field of study, one in which it is of interest to study large numbers of closely related compounds. With just the right, carefully tuned, catalytic molecule one may hit the jackpot and become able to turn an otherwise useless byproduct into a highly valuable compound! At a more realistic level, organometallic compounds are the catalysts in several important industrial processes. They do have a problem, though. If they involve an expensive metal, platinum for instance, should catalyst recovery be difficult or incomplete, the cost of replacement catalyst may make uneconomic an otherwise viable process. This is just one reason that heterogeneous catalysts—involving surface complexes—are more important than homogeneous—involving dissolved complexes—unless the metal involved in the homogeneous catalysis is cheap, the products volatile—and so easily separated—or the products obtained of particular value. There will be more on this topic in Chapter 15.

2.3 Some final comments

The object of the present chapter has been that of introducing the reader to something of the breadth of complexes which are there to be studied, without going into great detail at any point. In such a survey it is difficult to avoid neglect of the majority, but commonplace, and instead to emphasize the novel. Some final comments will help to complete the picture without unduly distorting it.

First, so far in this book we have given the impression that coordination compounds are obtained by reaction between a ligand or ligands and a metal ion or compound. We shall see in Chapter 4 that, whilst this is commonly the case, alternatives are possible. In particular, a ligand coordinated to a metal may undergo an organic-chemistry-type reaction with another (organic) molecule, leading to an extension of the ligand with, often, an increase in the number of atoms of the—enlarged—ligand coordinated. Alternatively, two ligands independently bonded to the metal may be linked by a similar reaction. Such reactions, in which ligands which are held anchored to a metal ion undergo ligand-extending reactions at a site remote from the metal ion, are called *template reactions*. Although this particular discussion is relevant to classical complexes, it probably is even more important for organometallics. In the early years of organometallic chemistry it was commonplace to take a transition metal carbonyl such as $Fe(CO)_5$[2] with some likely-looking organic compound—an alkene or alkyne derivative, for instance—and to heat them together in an autoclave for hours. Unholy mixtures of fascinating compounds resulted, some of which have not been properly studied to this day. It is likely that the novel ligands found in most, if not all, of the products resulted from template-type building-up reactions involving organic molecules (and often CO, too) coordinated to the metal. However, their formation is seldom regarded as providing examples of template reactions. Template syntheses are planned and usually of high yield; in contrast, the stabilization of an organic molecule in a speculatively prepared complex is often adventitious and each compound represents a minor component in a complicated reaction product mixture.

Secondly, we have tried to keep our examples simple, although hinting at the existence of some points of a greater complexity. Thus, in recent years there has been enormous study of compounds containing many metal atoms, linked directly to each other, linked indirectly through ligands, or both. So wide has this field become that there is real discussion about whether or not compounds with perhaps 40 or 50 metal atoms or more, directly bonded one to another, should be treated as fragments of a metal as much as chemical molecules. Metal clusters will be the subject of Chapter 15 and metals in Chapter 17.

Finally, whilst the ideas about bonding in complexes so far presented have been rather simple, it has already become clear that there are aspects of real molecules that they do not even begin to cover. Chemists tend to start with the usual and common and wonder what happens when new constraints are added. A popular, and fruitful, line of current enquiry has already been mentioned. What happens if groups attached to the ligating atom are made very bulky? Will the increase in steric interactions lead to a weaker metal–ligand bonding (which could mean interesting properties) or to a low coordination number (which could also mean interesting properties)? Generally speaking, the answer to both these questions is 'yes' and so a need has developed to quantify in some way the steric demands imposed by individual ligands—and none of the simple bonding theories so far mentioned addresses this question. This point will be returned to in Chapter 5 where an angle, the *cone angle*, will be described, which provides a measure of steric effects[3] in some complexes, at least. The molecular modelling to be described in Chapter 13 is one technique which aims to give an accurate description of steric effects.

---

[2] Note that, although this compound is a complex, the rule that complexes are written in square brackets is often applied rather sloppily in organometallic chemistry.

[3] See C. A. Tolman, *Chem. Rev.* (1977) 77, 313.

Further reading

The contents of this chapter have been rather broad-brush, stretching from the traditional through to the contemporary. Further reading in this vein can be obtained by thumbing through any contemporary text in inorganic chemistry or, to come really up-to-date, any current issue of any journal devoted to inorganic chemistry.

Much of the further reading suggested for Chapter 1 is relevant for Chapter 2 also, although to include transition metal organometallic chemistry, the browse through the introductory coordination chemistry section in general inorganic texts will need to cover texts published from the 1960s through to the early 1980s. More recent texts tend to include material relevant to the later chapters of the present book but which has only briefly been alluded to in Chapter 2.

Behind the content of this chapter, but not discussed within it, is the way that the information was obtained. Sometimes, there is a helpful story to be told. Two that may be mentioned are the discovery of the first complex containing coordinated N_2: 'The Discovery of $[Ru(NH_3)_5N_2]^{2+}$' by C. V. Senoff, *J. Chem. Educ.* (1990) *67*, 368 and, rather older, the nature of the species giving rise to the intense blood-red colour obtained when thiocyanate, SCN^-, ions are added to iron(III) solutions, 'The nature of iron(III) thiocyanate in solution' by S. Z. Lewin and R. S. Wagner, *J. Chem. Educ.* (1953) *30*, 445. Although both make reference to ideas developed later in this book, this problem should not unduly inhibit understanding.

There are many sources of information on the exotic ligands discussed in the text. Samples which give easy-to-read insights are:

- 'Coordination Chemistry of Alkali and Alkaline-earth cations with Macrocyclic ligands' by B. Dietrich, *J. Chem. Educ.* (1985) *62*, 854.

- 'Powerful new metal chelating agents developed' in *Chem. Eng. News* (August 1st 1988) 21.
- 'Molecules with Large Cavities in Supramolecular Chemistry' by C. Seel and F. Vögtle, *Angew. Chem., Int. Ed.* (1992) *31*, 528.
- 'Calixranes—supramolecular pursuits' by A. McKervey and V. Böhmer, *Chemistry in Britain* (1992) 724.
- 'The Specification of Bonding Cavities in Macrocyclic Ligands' by K. Hendrick and P. A. Tasker, *Prog. Inorg. Chem.* (1985) *33*, 1.
- *Supramolecular Chemistry, an Introduction* by F. Vögtle, J. Wiley, Chichester, 1991.

A Nobel lecture by the major authority in the area, which spans a wide range of topics from the history of the subject through to its applications in catalysis, photochemistry and biochemistry is 'Supramolecular Chemistry—Scope and Perspectives, Molecules, Supramolecules and Molecular Devices' by J-M. Lehn, *Angew. Chem., Int. Ed.* (1988) *27*, 89.

Questions

2.1 Use Table 2.2 to compile a list of atoms (which may be part of a polyatomic ligand) which coordinate to metal ions and relate the list to the region of the periodic table where the atoms fall. If in doubt about whether an atom can be a ligand, the answer is almost certainly 'yes', although even so the final list will be far from comprehensive.

2.2 Unlike Table 2.2, Table 2.3 contains no examples of ligands in which a halogen atom is the donor atom. Suggest a reason why such species are rare.

2.3 Although Table 2.2 contains examples of ligands in which nitrogen or arsenic are donor atoms there are no examples which contain phosphorus. Suggest a few possible phosphorus-containing ligands (if you wish, some of the ligands in Table 2.2 might be appropriately modified).

2.4 Using the ligands in Table 2.4 as a guide, suggest probable structures for:

- 15-crown-5

- 18-crown-6
- 21-crown-7
- dibenzo-12-crown-4.

2.5 Cryptand ligands have been prepared in which up to half of the oxygens in 2,2,2-crypt (Table 2.4) have been replaced by either S or NCH_3. Suggest some ligands of this type which might well have been studied.

2.6 Select any three of the complexes shown in Table 2.7 and carry out a valence electron count on them. Compare your results with those given in Table 2.8 (if any of the last three complexes in Table 2.7 were chosen some thought will be needed in using Table 2.8).

2.7 As Table 2.6 hints, complexes are known in which 1,5-cyclooctadiene (cod) bonds through both of its double bonds to a single metal atom. It may be that Fig. 2.2 can be applied to each double bond separately. Equally, it could be argued that although this approach may be valid for the π back-bonding, it is inadequate for the σ donation from the cod. Suggest a reason for this reservation.

3

Nomenclature, geometrical structure and isomerism of coordination compounds

3.1 Nomenclature

In order to facilitate communication between chemists it is desirable that a convention for naming coordination compounds be followed. This section contains an outline of the system suggested by a Nomenclature Committee of the International Union of Pure and Applied Chemistry (IUPAC).[1] Although this convention is commonly adopted—the Russian literature contains some variants and each language uses its own words, or spelling (as in 'sulphate' and 'sulfate' for UK and USA, respectively, although the UK has recently agreed to adopt the American version)—it is often simpler to give a structural formula, e.g. $[Co(NH_3)_4Cl(NO_2)]^+$, than to write the name in full and this will frequently be done in the following chapters. Apart from this device for side-stepping the problem, in this book we generally follow the most recent IUPAC recommendations except for those which have yet to gain general acceptance.

The IUPAC system has both advantages and disadvantages. One major advantage is that the most recent recommendations are just that, recom-

---

[1] Perhaps the most generally available complete system is the American version because it is contained in the *Handbook of Chemistry and Physics* published annually by the CRC Press and which can be found in most libraries. Inevitably, there is a time lag before the most recent recommendations appear so it is as well to check on this. The greatest detail and the most current recommendations have been published in *Nomenclature of Inorganic Chemistry, Recommendations 1990* (IUPAC) ed. G. J. Leigh, Blackwell, Oxford and it is these that have been followed in the present text. At the same time as the IUPAC book appeared so, too, did another, *Inorganic Chemical Nomenclature* by W. H. Powell and W. C. Fernelius, published by the American Chemical Society (the publication is the outcome of the deliberations of this Society's nomenclature committee), 1990. Whilst there are minor points of disagreement with the IUPAC book, with its greater emphasis on the notations that will be found in the older literature, the ACS publication may be seen as an acceptable complement to it.

mendations. Previously, they had been 'rules', a situation that led one group of writers to comment 'such rules often represent a compromise of conflicting views and may not be completely acceptable (or convenient) to all'. In what follows, in the spirit in which they are presented, the current recommendations are not regarded as above criticism.

Having mentioned one advantage, let us get the disadvantages out of the way. Names are often quite long, and involve numbers, brackets, subscripts, superscripts and Greek letters. When reading a name all of these details have to be remembered because it is not until the name is completed that one can really start building up the complex. The name of the metal is given at the end, whereas the thinking of most chemists starts with the metal. It is not surprising that 'nickel tetracarbonyl' is in much more common usage than the IUPAC name 'tetracarbonylnickel(0)'. So, whilst the IUPAC notation is that generally encountered in the scientific literature, when chemists talk to each other they either greatly simplify it or use one that has grown up in their specialist field. It is therefore only to be expected that trivial names persist, so, although one should talk of the hexacyanoferrate(II) and hexacyanoferrate(III) anions, most workers bow to common usage and continue to speak of them as ferrocyanide and ferricyanide, respectively. The practice of naming what has at some time or other proved an important coordination compound after the person who first prepared it is widely followed, thus: $NH_4[Cr(NH_3)_2(NCS)_4]$, Reinecke's salt; $[Pt(NH_3)_4][PtCl_4]$, Magnus's green salt; $[IrCOCl(P(C_6H_5)_3)_2]$, Vaska's compound; $[RhCl(P(C_6H_5)_3)_3]$, Wilkinson's compound and, one met in Chapter 2 $K[Pt(C_2H_4)Cl_3]$, Zeise's salt. A system which has mercifully disappeared, but which will be found in the very old literature, is that of naming a compound according to the colour of the corresponding cobalt(III) complex (no matter the colour of the complex itself!). Thus, 'purpureo' salts meant ions with the general formula $[M(NH_3)_5Cl]^{n+}$.

The following rules summarize the more important recommendations of the IUPAC committee and contained in their 1990 publication. It should be recognized that when these differ from the previous rules the earlier version is the one likely to be met by the reader—the new recommendations are only just being accepted and used.

Although in writing the formula of a complex the central atom is given first, in the corresponding name it is given last, thus $[Fe(CN)_6]^{3-}$, hexacyanoferrate(III). For anionic complexes the characteristic ending is '-ate' (as hexacyanoferrate(III)), but for neutral or cationic complexes the name of the central element (normally a metal) is not modified: $[Fe(H_2O)_6]^{2+}$, hexaaquairon(II). A distinction is made for anionic complexes so that the corresponding acids can be systematically named, the characteristic ending for the acid being '-ic', as for $H_4[Fe(CN)_6]$, hexacyanoferric(II) acid. As indicated in these examples, the formal oxidation state of the central atom (Werner's primary valency) is indicated by a Roman numeral in parentheses after the name of the complex, but with no space between them. A formal oxidation state of zero is indicated by (0) and a negative state by a minus sign, e.g. ($-$I).

The *name* of the complex species is written as one word. Ligands which may be regarded as carrying a negative charge (Cl^-, SO_4^{2-} etc.) all end in '-o' (chloro, sulfato etc.). Whereas, previously, the negatively charged ligands preceded uncharged ligands, now ligands are listed in alphabetical order, without regard to charge. Prefixes such as di-, tri-, bis-, tris- (see later) are disregarded in determining the alphabetical order. Previously, ligands in each class (anionic, neutral) were written in order of increasing complexity. A problem with an alphabetical order arises when different languages are used—the name can become language dependent. In practice, since English has become the international language of science, the problem is not severe.

A different pattern is recommended for writing the *formulae* of compounds. The central atom is written first (in contrast to its name, where it comes last), followed by the anionic ligands in alphabetical order and then the neutral ligands, again in alphabetical order. Polyatomic ligands are enclosed in parenthesis (otherwise there could sometimes be confusion about which atom belongs to which ligand). This latter rule also applies when abbreviated symbols are used for ligands (e.g. (py) for pyridine). Some ligands have names dictated by common usage: H_2O aqua, CO carbonyl, NH_3 ammine, NO nitrosyl. Some names which have been used in the past are no longer recommended. One is aquo (which has been replaced by aqua), another is mercapto for $(SH)^-$. In future sulfanido should be used instead, but it remains to be seen whether this gains common acceptance. Other names can beg questions. Thus, when an H atom coordinates it is regarded as H^-, and so named hydrido, even if there is experimental evidence that the ligand behaves chemically more as H^+ than as H^-. When there are several different possible attachment modes of a ligand, for simple cases the attached ligand can be indicated by a bond:

$(-ONO)^-$, nitrito-O; $(-NO_2)^-$, nitrito-N

(although nitro is also acceptable for this ligand, N-coordinated).

For more complicated ligands a Greek kappa (κ) is placed before the coordinated atoms(s). Thus, although one would normally write thiocyanato-N for $(-NCS)^-$, it could be called thiocyanato-κN. Similarly, $(-SCN)^-$ could be called either thiocyanato-S or thiocyanato-κS. Ligands which commonly coordinate to one coordination position but can do so in more than one way, and we have just met two simple examples, are called *ambidentate* ligands—more examples of the use of the word dentate will be given shortly; it was first mentioned in Chapter 2. Ligands which bond to more than one coordination site and can do so in more than one way are termed *flexidentate*. When several identical ligands are coordinated to the same central atom, two cases arise. If the ligand is simple and with a simple name, the number of ligands is indicated by the appropriate prefix: di-, tri, tetra-, penta-, or hexa-. Several examples of this usage have already been given. When the ligand is so complicated that it has a polysyllabic name—perhaps already including one of the above prefixes—or ambiguity might arise for some other reason, the name is enclosed in parentheses and the number of ligands present indicated by the prefix bis-, tris-, tetrakis-, pentakis-, or hexakis-. Examples are ethylenediamine,[2] $NH_2-CH_2-CH_2-NH_2$, which gives rise to the complex tris(ethylenediamine)nickel(II).

---

[2] The correct name for this ligand is 1,2-diaminoethane, a name which is being increasingly used. However, in the literature and in texts ethylenediamine is the name which is usually encountered and for this reason is used in the present book also.

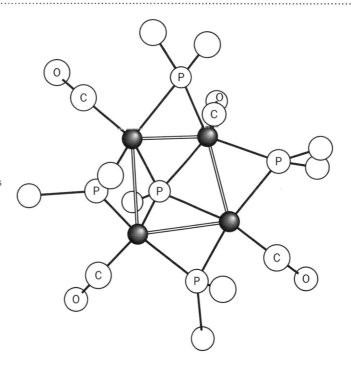

Fig. 3.1 The complex [Rh$_4$(CO)$_4$(μ_4-P(C$_6$H$_{11}$) (μ-P(C$_6$H$_{11}$)$_2$)$_4$]. The rhodium atoms are indicated in black. The carbon atom of each cyclohexyl group, C$_6$H$_{11}$, that is attached to P is the only one shown and is drawn as an open circle.

[Ni(en)$_3$]$^{2+}$ (en is the usual abbreviation for ethylenediamine), and triphenylphosphine, P(C$_6$H$_5$)$_3$, in the complex [NiCl$_2$(P(C$_6$H$_5$)$_3$)$_2$], dichlorobis(triphenylphosphine)nickel(II). A name which is increasingly being used to describe complexes in which all the ligands are identical is to say that they are *isoleptic*. So, [Ni(en)$_3$]$^{2+}$ is an isoleptic complex but [Ni(P(C$_6$H$_5$)$_3$)$_2$Cl$_2$] is not.

Bridging groups attached to two distinct coordination centres are indicated by the prefix μ (mu):

$$\left[(NH_3)_4Co \underset{NO_2}{\overset{NH_2}{\diagup \diagdown}} Co(NH_3)_4 \right]^{4+}$$

μ-amido-μ-nitrobis(tetraamminecobalt(III))

When more than two metal atoms are spanned by a single ligand then the number of centres spanned is indicated by a subscript: μ_3, μ_4, μ_5 (although the reader should be warned that some authors prefer to use superscripts: μ^3, μ^4, etc.). In such cases, since each metal usually has its own set of ligands, the formal name can stretch over several lines and, not surprisingly, its use is avoided; a formula is given instead. So, the molecule shown in Fig. 3.1, in which P(C$_6$H$_{11}$)$_2$ groups bridge four Rh(CO) units which themselves lie at the corners of a square, the whole being capped by a P(C$_6$H$_{11}$) group, is written [Rh$_4$(CO)$_4$(μ_4-P(C$_6$H$_{11}$)(μ-P(C$_6$H$_{11}$)$_2$)$_4$], although, written like this, the 1:1 relationship between the Rh and CO is not explicitly stated. Note that μs appear in sequence of decreasing suffix values, although when a ligand is both bridging and non-bridging the non-bridging is listed first (e.g. Cl$^-$ as both bridging and terminal ligand).

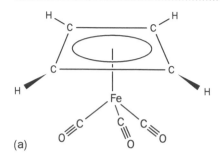

(a)

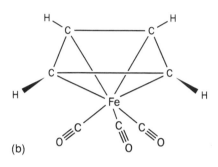

(b)

Fig. 3.2 Alternative symbolic representations of the bonding between C_4H_4 and $Fe(CO)_3$ groups. In (a) the delocalized nature of the π electron system of the C_4H_4 group is emphasized. In (b) the η^4 aspect of the bonding is made evident.

Not only can a ligand be attached to more than one metal atom, a given ligand may be bonded through several atoms to a single metal. This is described by the prefix η (eta), with a superscript to indicate the number of atoms bonded. One talks of the *hapticity* (Greek: *hapto* to fasten) of the ligand, thus: η^1 monohapto, η^2 dihapto, η^5 pentahapto etc. This nomenclature is particularly important in organometallic chemistry, where, as mentioned in Chapter 2, a given ligand may often coordinate in a variety of ways and jump from one mode to another. So, the compound in Fig. 3.2 is tricarbonyl(η^4-cyclobutadienyl)iron(0). Although its name would lead one to adopt the representation shown in Fig. 3.2(b), for simplicity that given in Fig. 3.2(a) is usually preferred, unless a specific point has to be made (for instance, if, in a reaction, one of the four Fe–C bonds breaks). A detailed discussion of the bonding in such compounds will be given in Chapter 10. The above paragraphs have been written with the intention of providing a comprehensive overview, not detail. The latter is contained in Tables 3.1–3.3 which give specifics and examples.

Names such as those we have given above indicate the ligands attached to a central metal atom but do not detail the positions of the ligands relative to one another. If this information is important and has to be included, an extension of the nomenclature is necessary. It often happens that the prefixes *cis* or *trans* are adequate, as in the complex *cis*-dichloro-di(pyridine)platinum(II) in which the platinum and the four atoms co-ordinated to it are coplanar:

$$
\begin{array}{ccc}
\text{Cl} & & \text{py} \\
& \diagdown \diagup & \\
& \text{Pt} & \qquad \text{(py = pyridine)} \\
& \diagup \diagdown & \\
\text{Cl} & & \text{py}
\end{array}
$$

The most recent IUPAC recommendations on this point represent a considerable departure from those previously made. We first outline the new and then review the old, for as already mentioned, it is the latter which will be encountered in all but the most recent literature and texts.

Table 3.1 Anions and their names when acting as ligands

Free anion	Coordinated anion
Amide (NH_2^-)	amido (or azanido)
Azide (N_3^-)	nitrido (azido will also be met)
Bromide (Br^-)	bromo
Carbonate (CO_3^{2-})	carbonato
Cyanate (CNO^-)	cyanato
Fluoride (F^-)	fluoro (*not* fluo)
Hydroxide (OH^-)	hydroxo (or hydroxido or hydroxy)
Nitrite (NO_2^-)	nitro or nitrito-N (see text)
Oxide (O^{2-})	oxo (or oxido)
Thiocyanate $(SCN)^-$	thiocyanato-N (N-bonded), thiocyanato-S (S-bonded)

Table 3.2 Examples of the nomenclature of simple coordination compounds. Some of these examples contain, and adequately define, points not explicitly covered in the text

Compound	Nomenclature
$K_2[ReF_8]$	potassium octafluororhenate (note: only 'potassium')
$[Cu(NH_3)_4]SO_4$	tetraamminecobalt(II) sulfate (note: 'aa' and 'mm')
$[CuCl_2(py)_2]$	dichlorobispyridinecopper(II) (note: bipyridine is the present name for the 2,2'-bipyridine ligand—see Table 2.3. More strictly, and as in the text, di(pyridine) should be used to give dichlorodi(pyridine)copper(II). However, in the spoken language an ambiguity can arise)
$[Hg(C_2H_5)_2]$	diethylmercury(II)
$[Ni(PPh_3)_4]$	tetra(triphenylphosphine)nickel(0)
$[Ru(NH_3)_5(N_2)]^{2+}$	pentaamminedinitrogenruthenium(II) (note: similarly, O_2 is dioxygen, but beware confusion with O_2^-, superoxo and O_2^{2-}, peroxo)
$K_2[FeCl_4]$	potassium tetrachloroferrate(II)
$(NH_4)_2[SnCl_6]$	ammonium hexachlorostannate(IV)

Table 3.3 The nomenclature of compounds containing bridging groups

$[(NH_3)_2Pt\underset{Cl}{\overset{Cl}{\diamond}}Pt(NH_3)_2]Cl_2$

di-μ-chloro-bis[diammineplatinum(II)] chloride

$\left[(NH_3)_4Co\underset{NH_2}{\overset{O-O}{\diamond}}Co(NH_3)_4\right]^{3-}(ClO_4)_3$

μ-amino-μ-peroxo-bis[tetraamminecobalt(III)] perchlorate

bis[carbonyl(μ-carbonyl)η^5-cyclopentadienyliron(0)]
(one could add *trans* at the front as the *cis* form also exists)

η^3-allylpalladium(II)di-μ-chlorodichloroaluminium(III)

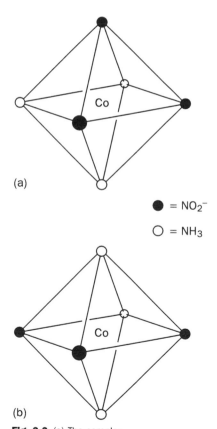

(a)

● = NO_2^-

○ = NH_3

(b)

Fig. 3.3 (a) The complex
(OC-6-22)-triamminotrinitrocobalt(III)—this is
commonly referred to as the *fac* (facial) isomer,
i.e. *fac*-triamminotrinitrocobalt(III). (b) The
complex (OC-6-21)-triamminotrinitrocobalt(III)—
this isomer is commonly referred to as the *mer*
(meridional) isomer.

Each symmetry-distinguishable geometry about the central metal is given a so-called *polyhedral symbol*. The symbol is based on the name for the geometry and these are discussed later in this chapter: OC for octahedral, T for tetrahedron, SP for square planar, SPY for square pyramid and TBPY for trigonal bipyramid are the most important. Some of the others are contained in Question 3.9. Because there are cases where ambiguity would otherwise arise, the coordination number is given immediately after the polyhedral symbol: OC-6, T-4, SP-4, SPY-5, TPBY-5. Then, ligating atoms are assigned a priority, the highest atomic number having highest priority, the lowest atomic number the lowest. When the ligating atoms are identical but with different substituents, the same rule is applied to the substituents. So, $-NO_2^-$ has a higher priority than $-NH_3$, but $-OH^-$ has a higher priority than either. Consider complexes with the two geometries shown in Fig. 3.3. They are distinguished as follows (the general case is similar). Consider the ligand of highest priority (add the words 'which lies on the axis of highest rotational symmetry' if ambiguity persists). Write down the priority number of the ligand *trans* to it. Here, $-NO_2^-$ is of highest priority and has priority number 1 and $-NH_3$ has priority number 2 (had the complex contained $-OH^-$ these two numbers would have been 2 and 3). So, both the isomers, so far, have the symbol (OC-6-2). This is because for that in Fig. 3.3(b), of the two axes containing $-NO_2^-$, the rules require that we choose that axis for which the priority number difference is a maximum. We now move to the plane perpendicular to the axis already chosen and again select the ligand of highest priority number. We again write down the priority number of the ligand *trans* to it. So, the two isomers above become (OC-6-22) and (OC-6-21), respectively, and they are distinguished. This completes the process so that the isomer in Fig. 3.3(a) is called:

(OC-6-22)-triamminetrinitrocobalt(III)

Optical isomers are not distinguished by the notation so far presented but it can be extended to cover this requirement.

In the past, the system adopted has been to number the coordination positions. The numbering system adopted for square planar complexes was

$$
\begin{array}{c}
1 \\
| \\
4-M-2 \\
| \\
3
\end{array}
$$

so that an alternative to *cis*-dichlorodi(pyridine)platinum(II) is 1,2-dichlorodi(pyridine)platinum(II). Notice the brackets around (pyridine). This practice is recommended and in the present case serves to remove any confusion with the ligand 2,2'-bipyridine (previously dipyridine). For octahedral complexes the *cis* and *trans* nomenclature is often simplest, but for complicated cases the numbering systems shown in Fig. 3.4 has been adopted. The ligands which are at the corners of the front face of the octahedron are cyclically numbered 1 → 3 and those on the back face numbered 4 → 6. There are just two ways of arranging two sets of three identical ligands, $[ML_3L_3']$, in an octahedral complex. When identical ligands lie at the corners of a face of the octahedron, one has what traditionally has been called the facial—denoted *fac*—arrangement (Fig. 3.3(a)). When the three identical ligands lie on a plane that bisects the octahedron, the meridional—denoted *mer*—arrangement results (Fig. 3.3(b)).

As has been recognized, two atoms of the same ligand may coordinate to the same metal, leading to the formation of a ring structure. As mentioned in Chapter 2, the formation of such rings by coordination is termed chelation and the ligand is called a chelating ligand. Historically, the language used to describe chelates depended on which side of the Atlantic you lived. Traditionally, Americans talked of bidentate, tridentate, tetradentate, pentadentate and hexadentate, the whole class being called *polydentate ligands* (ligands attached by a single atom being monodentate). British textbooks preferred bidentate, terdentate, quadridentate, quiquedentate and sexadentate, calling the whole set *multidentate ligands*; single atom attachment being unidentate. However, the latest IUPAC recommendations are for didentate, tridentate, tetradentate and so on. Whether didentate will replace the one name common to both American and British notations—bidentate—remains to be seen. In this book the American usage will be followed, this being the more commonly met. (This notation was used in Table 2.3.) A ligand which coordinates four atoms to each of two metals is called *bisquadridentate*, thus combining both notations, and so on. Note that a polydentate ligand is not *necessarily* a chelating ligand

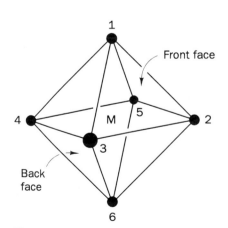

Fig. 3.4 Ligand numbering system for an octahedral complex, indicating how the numbering is related to the selection of two opposite faces of the octahedron.

(although commonly it is), because the coordinating atoms of the ligand may be so arranged that they cannot be coordinated to the same metal atom. So, a polydentate ligand it is not necessarily a chelating one. The example of multidentate isocyanide ligands, which have to be rather special if they are to coordinate to a single metal atom, was mentioned in Section 2.1.

Ligands, and polydentate ligands in particular, can be rather complicated. To simplify the description of complexes involving such ligands, abbreviated forms of the ligands' names are used—we have already used cn as a shorthand for ethylenediamine and py for pyridine. Although there are IUPAC recommendations, these are not always followed and, in practice, there is only an approximate system of standard abbreviations. Fortunately, it is normal for authors of papers to define the abbreviations they use, so little confusion arises. A representative selection of such ligands was given in Table 2.3 together with commonly used abbreviations.

There are two other areas covered in this book for which notations are needed. These are in the fields of cluster compounds and of bioinorganic chemistry. Both commonly involve very complicated molecules and attempts to produce systematic names such as those met in this chapter can defeat the whole purpose—to make communication easy. In both cases it is difficult to separate the notations currently used from developments in our understanding of the topic. Discussions of nomenclature will therefore be given in Chapters 15 and 16 rather than introduced here.

So far our concern has been entirely with a ligand and its attachment to a metal atom. It has been assumed that we could ignore atoms of the ligand that are not bonded to the metal. In some cases this is not justified. The reason is that, for chelates in particular, there is more than one geometric arrangement possible for the ligand atoms between those bonded to the metal. The subject of the conformation of chelate rings and the consequences of the various conformations has been much studied. It is an interesting and elegant subject area; a brief introduction to it is given in Appendix 1.

3.2 Coordination numbers

Werner was the first to recognize that one characteristic of a coordination compound is the number of ligands directly bonded to the central atom. He called this number the secondary valency of the central atom, but this usage has not persisted and it is now called the *coordination number*. The coordination number need not have a unique value for a particular metal ion. For example, in pink cobalt(II) chloride it is six and in the blue form it is four (see Fig. 3.5). In particular, when a coordination compound is participating in a reaction in which one ligand is being replaced by another, there is overwhelming evidence that the coordination number in the reaction intermediate is different from that in either the initial or product compound.

Coordination number is more than just a convenient method of classifying coordination compounds—complexes of a given transition metal ion with the same coordination number often also have closely related magnetic properties and electronic spectra. In later chapters of this book, these

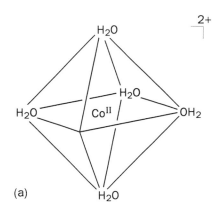

(a)

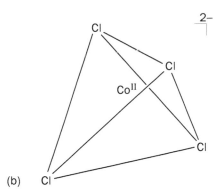

(b)

Fig. 3.5 When cobalt(II) salts dissolve in water they commonly give pale pink solutions, normally associated with the octahedral hexaaqua species (a). Addition of concentrated hydrochloric acid gives a deeper, beautiful blue colour, 'cobalt blue', normally associated with the tetrahedral tetrachloro species (b). In fact, the colours are perhaps more associated with the two different geometries than ligands—replacing an H_2O by Cl^- in (a), for example, does not greatly change the observed colour of the solution.

properties are discussed in some detail. In the present section are described some of the geometrical arrangements of ligands which have been found for various coordination numbers and examples of each are given.

First, however, we set the scene by noting that the frequency of occurrence of coordination numbers for some ions of the first transition metal series is roughly as follows:

chromium(III)   6(oct) ≫ 5, others very rare
iron(III)   6(oct) > 4(tet) > 5 ≈ 7
cobalt(II)   6(oct) > 4(tet) > 5 > 4(planar)
cobalt(III)   6(oct) ⋙ 5 > 4
nickel(II)   6(oct) > 4(planar) > 4(tet) ≈ 5
copper(II)   6(oct)† > 4(planar) > 5† ≈ 4(tet)†

† usually distorted

Although not exhaustive, this series illustrates the fact that there is no fixed coordination number for any ion. It cannot be emphasized too strongly that the empirical formula of a compound often has little connection with either the coordination number or geometry of any complex species it describes—this recognition was Werner's breakthrough.

3.2.1 Complexes with coordination numbers one, two or three

Coordination numbers of one, two and three are rare. Although it is possible to conceive of complexes with coordination number one— presumably a very bulky ligand with only one coordinating atom tucked well into its centre might form such complexes—it is only recently that two have been reported. The compounds are actually organometallic complexes of copper and silver with a ligand which has three phenyl groups symmetrically attached to a central phenyl, which is bonded to the metal through another carbon of the phenyl ring: 2,4,6-triphenylphenylcopper and -silver.

The best-known example of coordination number two is the complex ion formed when silver salts dissolve in aqueous ammonia, $[Ag(NH_3)_2]^+$. This, like all other known cases of this coordination number, is linear, $[H_3N–Ag–NH_3]^+$ although it should be possible to obtain bent examples, perhaps when two different but sterically-demanding ligands are involved. Other complexes of this coordination number, which is almost entirely confined to copper(I), silver(I), gold(I), and mercury(II), are $[CuCl_2]^-$ and $[Hg(CN)_2]$. One way of reducing the coordination below the normal for a particular metal ion is to choose ligands that are so bulky that they block the entry of further ligands. So, the ligand $P(C_6H_5)_3$ causes zerovalent Pt—usually either three- or four-coordinate—to have coordination number two in $Pt(P(C_5H_5)_3)_2$, where the P–Pt–P sequence is linear.

Examples of coordination number three are few, the $[HgI_3]^-$ anion perhaps being one of the best characterized. In this anion the iodide ions are arranged at the corners of a slightly distorted equilateral triangle which has the mercury atom at its centre. In the anion $[Sn_2F_5]^-$ two SnF_2 units are bridged by the fifth F, leading to a (distorted) three-coordinated structure around each tin atom. Examples of three-coordination in transition metal chemistry are the iron(III) complex [Fe(N(SiMe$_3$)$_2$)$_3$]—the chromium compound is similar—and the complex formed when copper(I) halides dissolve in aqueous solutions of thiourea, [Cu(SC(NH$_2$)$_2$)$_3$]. Although these mostly involve a planar coordination around the metal ion, in some examples the metal ion is slightly out of the plane.

The [AgCl$_3$]$^{2-}$ anion has recently been found to have a perfectly planar, equilateral D_{3h} structure. It provides an example of how large ligands can lead to interesting structures. The anion occurs in the complex dibenzo-18-crown-6-KCl·AgCl (the ligand is that shown in Table 2.4). In the crystal the ligand complexes the potassium, the resulting complex resembling a slightly buckled wheel with the potassium at its centre. Perpendicular to the wheel is a Cl$^-$ of the [AgCl$_3$]$^{2-}$ anion, coordinated to the potassium. As a result, three of the wheels surround the anion (Fig. 3.6); presumably steric interactions between the wheels impose the threefold symmetry in which the bulk of the wheels prevents any other ligand having access to the silver ion and thus increasing its coordination number.

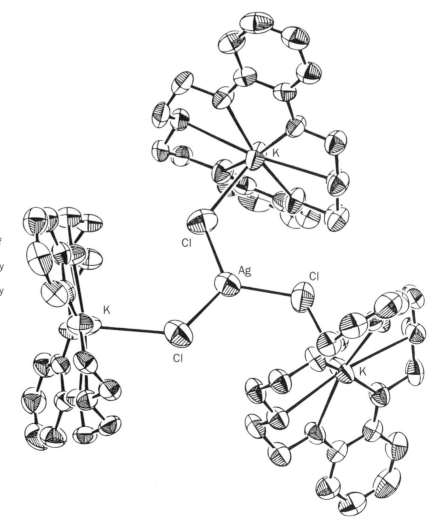

Fig. 3.6 The complex formed between dibenzo-18-crown-6-KCl, a complex of potassium, and AgCl which contains an [AgCl$_3$]$^{2-}$ anion. The potassium atoms are shown bonded to the oxygens of the crown ether, the other atoms are carbons. The ellipsoids give an indication of the vibrations of individual atoms. If the ellipsoids associated with the chlorine atoms, for example, were very large (which they are not) one might have doubts about whether the equilibrium geometry of [AgCl$_3$]$^{2-}$ is equilateral triangular (reproduced courtesy of Prof. S. Jagner).

3.2.2 Complexes with coordination number four

A tetrahedral arrangement of ligands is commonly exhibited by complexes with coordination number four. It is found for both transition metal and non-transition elements; for the latter it is rather common. In Chapter 1 the two species $[BF_3 \cdot (NMe_3)]$ and $[BF_4]^-$ were mentioned, in both of which the boron is tetrahedrally coordinated; other examples amongst main group elements are the $[BeF_4]^{2-}$, $[ZnCl_4]^{2-}$, and $[Cd(CN)_4]^{2-}$ anions. Complexes of transition metals in their higher oxidation states are often tetrahedral and often also anionic—$TiCl_4$, $[CrO_4]^{2-}$ and $[MnO_4]^{2-}$ are examples—but the same geometry is found for other valence states also. Transition metal chlorides, for instance, quite often give tetrahedral anionic species when dissolved in concentrated hydrochloric acid: iron(III) chloride gives the yellow ion $[FeCl_4]^-$ and cobalt(II) chloride gives the well-known blue ion $[CoCl_4]^{2-}$.

The four-coordinate arrangement in which the ligands lie at the vertices of a square (square planar complexes) is almost entirely confined to transition metal complexes (but XeF_4 also has this structure), where it is common and dominant for ions of the second and third transition series having d^8 configurations—rhodium(I), iridium(I), palladium(II), platinum(II) and gold(III). Examples are the $[PtCl_4]^{2-}$, $[PdCl_4]^{2-}$ and $[AuF_4]^-$ anions. Nickel(II), also a d^8 ion, is interesting in that it forms both tetrahedral and square planar complexes. (The red precipitate obtained in the gravimetric analysis of nickel salts and made by adding dimethylglyoxime to nickel(II) solution is a planar complex of Ni^{II}. Square planar complexes of this ion are often yellow, orange, brown or red.)

As Fig. 3.7 shows, it is possible in principle, if not in practice, to distort a tetrahedral arrangement of ligands so that they eventually assume the square planar structure, and vice versa. This suggests that complexes may exist with structures which are neither tetrahedral nor square planar, but intermediate between the two. Indeed, such is the case with the $[CuCl_4]^{2-}$ anion, made by dissolving $CuCl_2$ in concentrated hydrochloric acid. However, caution is needed because symmetry arguments can be invoked

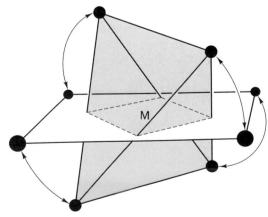

Fig. 3.7 At least in principle, by a smooth, coupled motion of all ligands (that motion indicated by the arrows) it is possible to change a square planar complex into a tetrahedral one. It is therefore not surprising to find that complexes with intermediate geometries do indeed exist.

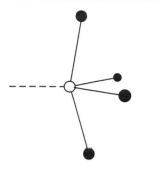

Fig. 3.8 The mode of fourfold coordination (C_{2v}) which seems to be related to the trigonal bipyramidal mode of five-coordination (Fig. 3.9(a)). The distortion of the ligands away from colinearity can be understood in terms of VSEPR theory (Appendix 2).

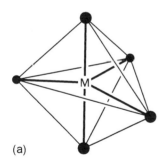

(a)

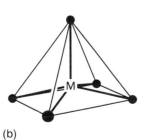

(b)

Fig. 3.9 (a) Trigonal bipyramidal (D_{3h}) and (b) square pyramidal (C_{4v}) modes of five-coordination.

which suggest that tetrahedral $\rightleftharpoons$ square planar interconversions are unlikely.

A related, but rather different, structure (Fig. 3.8[3]) is found for some main group compounds, for example $[SbCl_4]^-$ and $[AsCl_4]^-$. The co-ordination geometry here may be regarded as derived from trigonal bipyramidal five-coordination (Fig. 3.9 and Section 3.2.3) by omitting one of the three equatorial ligands. It has been suggested that the empty coordination position of the trigonal bipyramid is not really vacant but is occupied by a non-bonding lone pair of electrons of the central atom (Sb, As).

This does not exhaust the list of possible four-coordinate geometries. A trigonal monopyramidal, C_{3v}, arrangement is known when it is imposed by the ligand. So, complexes of M^{III} (M = Ti, V, Cr, Mn and Fe) are known for the ligand R_3N, where R– is $(t\text{-BuMe}_2Si)NCH_2CH_2$–, all four nitrogens being coordinated. The three bulky (t-BuMe$_2$Si) groups block the second axial position; the formal negative charge carried by the N in R– means that the complexes, $[R_3NM]$, are electrically neutral and so offer little attraction for most ligands. Ligand-induced geometries exist for the other coordination geometries that now follow but, in general, will not be included in the discussion.

3.2.3 Complexes with coordination number five

Many examples of five-coordination have been found and it is now clear that this coordination number is much more common than was once supposed. Although in practice they are usually found to be distorted, there are two idealized five-coordinate structures, the trigonal bipyramidal and the square pyramidal arrangements (Fig. 3.9). These structures are energetically similar and there seems to be no general way of anticipating which is adopted by a particular complex.[4] Indeed, it is possible that the structure is determined by intermolecular forces within the crystal (almost all structures have been determined in the solid state). In some five-coordinate compounds it has been shown that there is a facile interchange of ligands between the non-equivalent sites in either structure. The most probable mechanism for this is shown in Fig. 3.10(a, b). Only relatively small angular displacements are needed to interconvert the square pyramid and the trigonal bipyramid, and alternation between the two would lead to the observed interchange of ligand positions. Moreover, geometries between the two extremes are possible and are those commonly observed, particularly for complexes containing chelating ligands.

The interconversion shown in Fig. 3.10 is called a Berry pseudorotation—Berry because this is the name of the person who first suggested the mechanism and pseudorotation for the reason shown in Fig. 3.10(c, d). Whereas the top half of the figure shows the motions required to turn a

[3] The caption to this figure makes use of group theoretical notation. The reader who is unfamiliar with this notation should at least read Appendix 3, which includes an outline of it, at this point..

[4] Two partial exceptions should be noted. First, transition metal complexes containing strongly π-bonding ligands tend to adopt the trigonal bipyramidal configuration. Secondly, it is possible to make approximate predictions for main group complexes, although a delicate interplay of factors is involved.

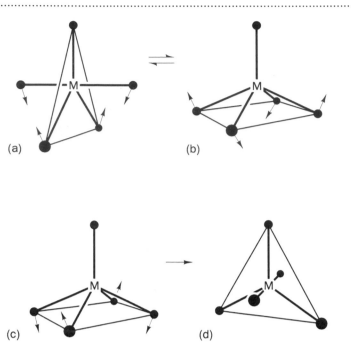

Fig. 3.10 The synchronous ligand motion which serves to turn (a) a trigonal bipyramidal into a square pyramidal complex and (b) a square pyramidal complex into a trigonal bipyramidal one. If a complex undergoes the displacement shown in (a) but does not become locked into the square pyramidal geometry (c) then, as the ligands continue their original motion a trigonal bipyramidal geometry is reattained (d). Comparison of the geometries in (a) and (d) might lead an observer to think that (a) has been rotated to give (d), although of course it has not. For this reason the sequence is called *pseudorotation*. Pseudorotation serves to interchange axial and equatorial ligands of a trigonal bipyramidal complex.

pentagonal bipyramid into a square pyramid and vice versa, the bottom half shows what happens if the amplitude of the vibration shown in the top left hand corner is so great that the atoms carry on beyond the square pyramid arrangement to give another trigonal bipyramid. A casual observer, knowing nothing of the square pyramidal intermediate and seeing only the before (a) and after (d) arrangements, might well conclude that the original molecule had simply been rotated to give the final one. Of course, no rotation has occurred, it just looks as if one has. Hence the use of the word pseudorotation. Other mechanisms, some rather ingenious, have been proposed as involved in equational–axial interconversions in trigonal bipyramidal complexes but the Berry mechanism is believed to be the most important. Indeed, it has become quite common in crystallographic work in which the structure of a five-coordinate complex is reported, for the authors to comment on the position on the Berry reaction pathway that their particular complex occupies.

The small energy difference between the two modes of five-coordination is demonstrated in the crystal structure of the compound $[Cr(en)_3][Ni(CN)_5]1.5H_2O$, where there are two distinct types of $[Ni(CN)_5]^{3-}$ anions, one square pyramidal and the other approximately trigonal bipyramidal. Were one form to be appreciably more stable than the other, then that would be the only one present in the crystal. Although it was not evident from the crystal structure work, it seems that the $1.5H_2O$ play a key role presumably by hydrogen bonding to the anions; on dehydration of the compound there is spectroscopic evidence that all the anions become square pyramidal.

Examples of trigonal bipyramidal structures are the $[Co(NCCH_3)_5]^+$ and $[Cu(bpy)_2I]^+$ cations. In the latter, one nitrogen of each bipyridyl is in an axial position. Anionic examples are $[CuCl_5]^{3-}$, $[SnCl_5]^-$ and

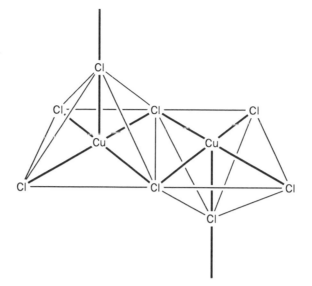

Fig 3.11 The C_{2h} geometry of the complex anion $[Cu_2Cl_6]^{2-}$.

$[Pt(SnCl_3)_5]^{3-}$. The latter, with Pt–Sn bonds, is formed when acidic tin(II) chloride solution is added to many platinum salts.

Some main group halides have trigonal bipyramidal structures but care is needed—structures may well differ from gas to solid (with solutions being different again). PF_5 and $SbCl_5$ retain the structure in gas and solution but PCl_5 in the solid is better described as $[PCl_4]^+[PCl_6]^-$; $SbCl_5$ remains a trigonal bipyramid in the solid; $NbCl_5$, $TaCl_5$ and $MoCl_5$ exist as dimers in the solid state, the two chlorine bridges making each metal atom six-coordinate.

Perhaps the best-known example of square pyramidal coordination is the compound bisacetylacetonatovanadyl, $[VO(acac)_2]$, where acac is acetylacetone[5] (CH_3–CO–CH_2–CO–CH_3) less one of the central (acidic) hydrogens, and in which the oxygen atom directly bound to the vanadium occupies the unique position.[6] In one salt of the $[Cu_2Cl_6]^{2-}$ anion, bridges between adjacent anions lead to a square pyramidal configuration about each copper atom (Fig. 3.11); compare this example with $[CuCl_5]^{3-}$ mentioned above. Among the main group elements, the $[SbCl_5]^{2-}$ anion provides an example of square pyramidal coordination.

A feature of square pyramidal structures is that there is the possibility of an additional ligand occupying the vacant axial site to produce a six-coordinate complex. Some of the small variations that have been observed in the electronic spectrum of $[VO(acac)_2]$ in different solvents are believed to be caused by a solvent molecule being weakly bound at the sixth coordination position. There is evidence that good donor solvents sometimes also introduce a ligating atom *cis* to the vanadyl oxygen.

[5] The correct name for this ligand is pentane-2,4-dionate, a name that is being increasingly used. However, in the literature and most texts the name acetylacetone is the one which will be encountered and for this reason is the one used in this book.

[6] Note the use of the ending yl in vanadyl. The vanadyl cation is VO^{2+}; the vanadium and oxygen are strongly bonded. The fact that this bond remains intact through most reactions and that we are dealing with a cation containing oxygen bonded to a metal is indicated by the yl. Another example is uranyl, UO_2^{2+}. These two ions are sometimes quoted as examples of one- and two-coordination, respectively (see Section 3.2).

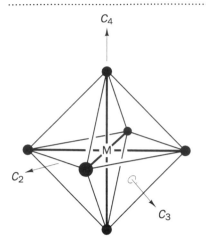

Fig. 3.12 An octahedral complex. Distorted octahedral complexes are usually described by an axis along which a compression, elongation or other change may be regarded as occurring. Possibilities are a tetragonal distortion (along a C_4 axis), a trigonal distortion (along a C_3) and, more rare, a digonal distortion (along a C_2).

3.2.4 Complexes with coordination number six

The majority of coordination compounds that one encounters are six-coordinate, the structure adopted being that of a regular or slightly distorted octahedron (Fig. 3.12). It is important to recognize that the octahedral geometry is found for complexes of both transition metal and main group elements (although much less common for the later lanthanides and actinides). The emphasis which will be placed on transition metal complexes in some of the following chapters may tend to obscure this fact. Examples of octahedral complexes of main group elements are $[Al(acac)_3]$, $[InCl_6]^{3-}$ and $[PCl_6]^-$.

An alternative but rare form of six-coordination is the trigonal pyramidal arrangement which is found in some sulfur ligand complexes such as $[Re(S_2C_2Ph_2)_3]$ (Fig. 3.13). The most noteworthy example of this form of six-coordination is the compound $[W(CH_3)_6]$; in that all complexes $[M(NH_3)_6]^{n+}$ are octahedral it seems that hexamethyltungsten must have a trigonal pyramidal, D_{3h}, arrangement for electronic reasons. These are not at present understood, although it may be that the discussion below on the bonding in molecules with cubic eight-coordination hints at the explanation. This form of six-coordination is also the configuration about the metal atoms in MoS_2 and WS_2 (the crystals of these compounds contain layer lattices and not discrete molecules).

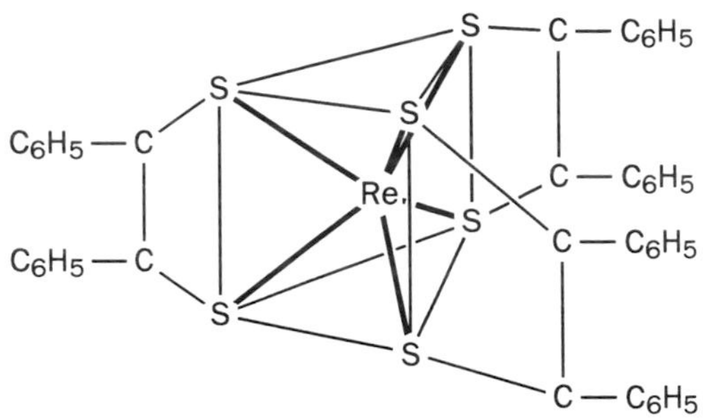

Fig. 3.13 An example of a molecule showing the trigonal prismatic mode of six-coordination.

Another possible six-coordinate arrangement is that of six ligands at the corners of a regular hexagon with the metal atom at the centre. It has been found, but only when the geometry is imposed by the structure of the ligand as for the K^+ at the centre of the 2-dibenzo-18-crown-6 ligand in Fig. 3.6.

3.2.5 Complexes with coordination number seven

There are three main structures adopted by complexes with coordination number seven; as is commonly the case with the higher coordination numbers there appears to be no great energy different between them. In the salt $Na_3[ZrF_7]$, the anion has the structure of a pentagonal bipyramid (Fig. 3.14), but in $(NH_4)_3[ZrF_7]$, it has the structure shown in Fig. 3.15, in which a seventh ligand caps one rectangular face of what would otherwise be an approximately trigonal pyramidal six-coordinate complex. No doubt the hydrogen bonding in the ammonium salt is a factor contributing to the difference in geometry,[7] just as in the case of the $[Ni(CN)_5]^{3-}$ anion discussed above. The anion $[NbOF_6]^{3-}$, which is isoelectronic with $[ZrF_7]^{3+}$, adopts the third mode of seven-coordination, shown in Fig. 3.16. This is derived from an approximately octahedral six-coordinate arrangement by an additional ligand capping one face.

A nomenclature which is sometimes used to distinguish these three forms of seven-coordination is to proceed down the axes indicated by arrows in Figs. 3.14–3.16 (these are the axes of highest symmetry) and to list the number of ligands lying in planes perpendicular to those axes. In this nomenclature the geometries are called the 1:5:1 (Fig. 3.14), 1:4:2 (Fig. 3.15) and 1:3:3 (Fig. 3.16) modes of seven-coordination.

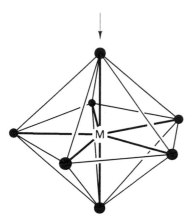

Fig. 3.14 The pentagonal bipyramidal (1:5:1, D_{5h}) mode of seven-coordination.

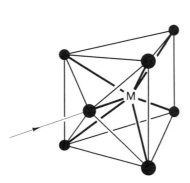

Fig. 3.15 The one-face centred trigonal prismatic (1:4:2, C_{2v}) mode of seven-coordination.

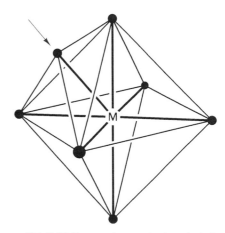

Fig. 3.16 The one-face centred octahedral (1:3:3, C_{3v}) mode of seven-coordination.

3.2.6 Complexes with coordination number eight

There are two common arrangements of eight ligands about a central atom, the square antiprismatic and the dodecahedral arrangements. Consider an array of eight ligands at the corners of a square-based box (not necessarily a cube). If the top set is rotated by 45° about the four-fold rotation axis a square antiprism results (Fig. 3.17). Dodecahedral coordination is more difficult to describe. Consider two pieces of cardboard cut and marked as shown in Fig. 3.18(a). If these are interleaved as shown in Fig. 3.18(b), the eight points lie at the corners of a dodecahedron (Fig. 3.18(c)). (A dodecahedron has twelve faces and eight vertices—if you are in doubt, make a model and count them.) There appears to be little energetic difference between the two structures so, whilst $[Zr(acac)_4]$ is square antiprismatic, $[Zr(ox)_4]^{4-}$, involving the oxalate anion $C_2O_4^{2-}$, is dodecahedral. Similarly, the $[Mo(CN)_8]^{4-}$ anion may have either arrangement in the crystal, the shape adopted varying with the cation. Other examples

[7] It is interesting that the salt $(NH_4)_3(HfF_7)$ is actually $(NH_4)_2[HfF_6] + NH_4F$; this is one of the few clear-cut distinctions between the chemistries of zirconium and hafnium—the zirconium compound contains the $[ZrF_7]^-$ anion.

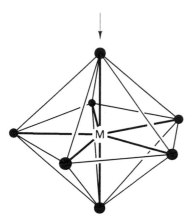

Fig. 3.17 The square antiprismatic (D_{4d}) mode of eight-coordination.

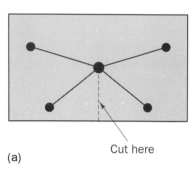

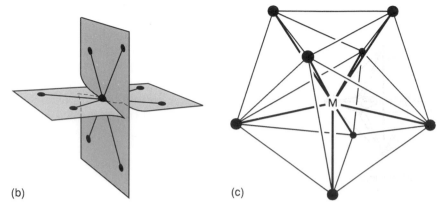

(a) Cut here

(b)

(c)

Fig. 3.18 The dodecahedral (D_{2d}) mode of eight-coordination (a dodecahedron is perhaps best thought of as a solid figure with 12 faces, which can be, but are not required to be, equilateral triangular). The arrangement is shown in (c)—note the approximate pentagon subtended by this projection of the five 'outer' ligands. The arrangement is best understood by its construction from two pieces of card (a), interleaved as in (b). The diagram (c) is drawn such that one of the planes in (b) is approximately in the plane of the paper.

of square antiprismatic coordination are the $[TaF_8]^{3-}$ and $[ReF_8]^{2-}$ anions. Typically, bidentate ligands with relatively short separations between the two coordinating atoms, i.e. with a short bite, form dodecahedral complexes. Examples are $[Co(NO_3)_4]^{2-}$, in which two oxygens from each nitrate coordinate to give four-membered rings, and $[Cr(O_2)_4]^{3-}$ in which both atoms of the peroxy O_2^{2-} anions coordinate to give three-membered rings.

Both the dodecahedron and square antiprism may be regarded as distortions of a cubic arrangement of ligands (Fig. 3.19). They are favoured because a cubic configuration would involve greater interligand steric

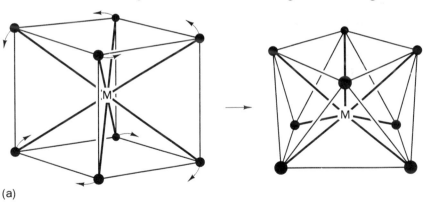

(a)

Fig. 3.19 By an appropriate concerted motion of the ligands, a cubic arrangement can be converted into (a) a square antiprism or (b) a dodecahedron. A cubic intermediate therefore offers one explanation of the interconversion between square antiprismatic and dodecahedral arrangements (this interconversion seems to occur rather readily).

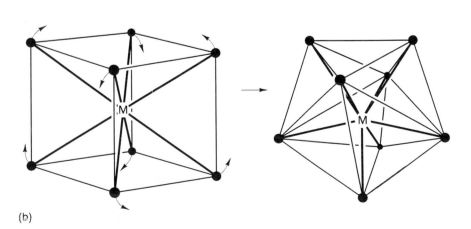

(b)

Fig. 3.20 The ideal hexagonal bipyramidal (D_{6h}) mode of eight-coordination.

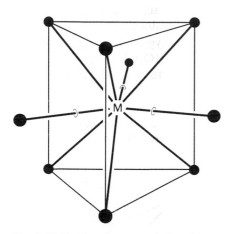

Fig. 3.21 The three-face centred trigonal prismatic (D_{3h}) mode of nine-coordination.

interactions. However, a similar argument favours the octahedron as opposed to a trigonal prism (for six-coordination) and yet the latter arrangement has been found. This suggests that the cubic eight-coordination should exist. In fact, it has been found for the $[PaF_8]^{3-}$ anion in Na_3PaF_8, as well as in $[UF_8]^{3-}$ and $[NpF_8]^{3-}$. Comparison of the symmetries of the antiprismatic and cubic geometries suggests that the involvement of f orbitals in the bonding is required for the cubic arrangement to be stabilized.[8]

Another form of eight-coordination, largely confined to the actinide series, is the hexagonal bipyramidal arrangement of ligands (Fig. 3.20). The ideal geometry has not yet been observed. When the distance between adjacent pairs of the six equivalent ligands are equal then the hexagon is found to be puckered. If the hexagon is planar then these distances are alternately long and short. The two axial ligands are usually oxygen atoms which are strongly bonded to the central metal, as in the trisacetatouranyl, commonly called uranylacetate, anion, $[UO_2(acetate)_3]^-$. A pattern closer to the ideal is found in some complexes containing crown ethers. For instance, in its 18-crown-6 complex the K^+ ion is surrounded by a near-regular hexagon of oxygens; additional ligands can be accommodated above and below this hexagon to give eight-coordination.

3.2.7 Complexes with coordination number nine

A spectacular example of nine-coordination is that of the $[ReH_9]^{2-}$ anion. This has the structure commonly found for nine-coordination, a trigonal prismatic arrangement of six ligands, each of the three rectangular faces of the prism being capped by an additional ligand (Fig. 3.21). Many hydrated salts of the lanthanide elements (for example $[Nd(H_2O)_9]^{3+}$) adopt this coordination. It is also found for salts such as $PbCl_2$ and UCl_4 in their extended lattices.

3.2.8 Complexes of higher coordination number

Although examples exist, coordination numbers of 10 and above are relatively rare. Further, it seems that the concept of coordination geometry becomes less applicable. The reason is that, whilst idealized geometries can be identified, most real structures show distortions and there may be some arbitrariness about which of the ideal structures the distorted structure is derived from. Examples of idealised coordination geometries are given in Figs. 3.22 (coordination number 10), 3.23 (coordination number 11) and 3.24 (coordination number 12). The captions to these figures describe the construction of the polyhedra.

---

[8] Because it is the fluoride ligand which is involved we confine our discussion to σ bonding. In the square antiprism (D_{4d}), the set of eight ligand σ orbitals spans the irreducible representations $A_1 + B_2 + E_1 + E_2 + E_3$—a set which matches the s, p and d orbitals on the central metal: $s(A_1)$; $p_z(B_2)$; $p_x, p_y(E_1)$; $d_{xy}, d_{x^2-y^2}(E_2)$; $d_{zx}, d_{yz}(E_3)$. In the cube (O_h) the eight ligand σ orbitals span $A_{1g} + T_{2g} + A_{2u} + T_{1u}$. Only if an f orbital is included is this set spanned by the orbitals of the metal atom: $s(A_{1g})$; $d_{xy}, d_{yz}, d_{zx}(T_{2g})$; $f_{xyz}(A_{2u})$; $p_x, p_y, p_z(T_{1u})$. In neither set is the metal d_{z^2} involved in the σ bonding; in the cube, $d_{x^2-y^2}$ is not involved either.

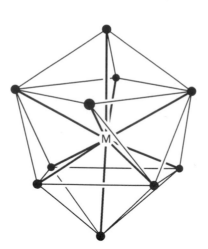

Fig. 3.22 One mode of 10-coordination—the bicapped square antiprism (D_{4d}). The two capping ligands are those at the top and bottom of the diagram. At the present time no complex is known which contains ten monodentate ligands.

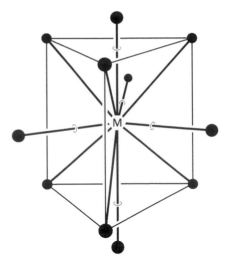

Fig. 3.23 Eleven-coordination is very rare. One possible mode of coordination is the all-face capped trigonal prism (D_{3h}). This differs from Fig. 3.21 by the addition of ligands at the top and bottom of the figure.

3.3 What determines coordination number and geometry?

So far in this chapter the question posed by the title of this section has largely been ignored. Rather, an attempt has been made to assemble the available data in a reasonably compact and accurate form. However, what has been said so far gives us little confidence in our ability to answer it. The question implicitly separates out the central atom and the ligands bonded to it from more remote atoms. Yet in our discussion of two- and three-coordination we met cases where remote steric effects clearly seem to be of dominant importance. Similarly, in our discussion of five-coordination we suggested that in one case hydrogen-bonding involving non-coordinated water molecules plays a determining role. Clearly, what is involved is a delicate balancing of interactions, some of which may not be immediately obvious. For instance, in forming an aqua complex in aqueous solution, there is a cost in removing each coordinated water molecule from the bulk solvent which has to be included somewhere in the balance sheet.

Part of the difficulty with the question is an assumption about the form of the answer. We are conditioned to seek an answer in a simple language involving orbitals, their overlap, their bonding and repulsion. We look for an answer in terms of individual electrons, or at least individual one-electron orbitals. Unfortunately, these are approximations. In reality the behaviour of all of the electrons within a molecule is indivisible. We shall, at several points in this book, arrive at the conclusion that an orbital model may well not be capable of providing an answer to a question. One has to carry out detailed and accurate calculations for a variety of geometries and compare the results. Whilst such calculations are available for lighter

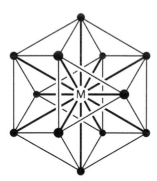

Fig. 3.24 There are several modes of 12-coordination, all of high symmetry (although no complex containing the twelve monodentate ligands needed to give these symmetries is known). That shown is the cuboctahedron (O_h. It is best known as the cubic close packing arrangement in metals. The solid figure may be regarded as derived from the octahedron (Fig. 3.12). The equilateral triangular face evident in the cuboctahedron is derived from a face of the parent octahedron, but is smaller. This is because a square face has been obtained by cutting the solid octahedron perpendicular to each fourfold axis to give the cuboctahedron. This has the effect of removing the corners of the octahedron face, so that the corners of the triangular faces of the cuboctahedron are at the mid-points of faces of the parent octahedron.

(with not too many electrons) main group elements, calculations on transition metal compounds are less accurate.

In Appendix 2 is outlined the most popular and successful simple model for predicting molecular geometry of main group compounds, the valence shell electron pair repulsion (VSEPR) model. However, alongside it are presented the results of some detailed calculations which prompt the comment 'the VSEPR model usually makes correct predictions, but there is no simple reason why'. The problem of the bonding in transition metal complexes will be the subject of models presented in Chapters 6, 7 and 10; this last chapter reviews the current situation. At this point it is sufficient to comment that the most useful applications of current simple theory are those that start with the observed structure and work from there. In the opinion of the author, the general answer to the question posed at the head of this section is that we really do not know.

3.4 Isomerism in coordination compounds

The evidence used by Werner to conclude that six-coordinate complexes are almost invariably octahedral was obtained from a study of the isomerism of these compounds. Although it is a topic that sometimes excites considerable attention, as when recently a new form of isomerism was suggested (see below), there are few studies in inorganic chemistry nowadays of isomerism per se (except optical isomerism). Nonetheless, it remains an important aspect of the chemistry of coordination compounds because ligand interchange often occurs readily in solution. That is, when a pure complex is dissolved, the solution may contain a variety of compounds, including isomers of the original solid-state structure. To work intelligently, one must be aware of what may have happened so that a scheme can be devised to check out the possibilities, should it be necessary. Hence, in the following pages the various forms of isomerism which have been recognized for octahedral complexes are outlined. It should be remembered that the categories are not mutually exclusive and that two or more of the classes we define may have to be invoked to describe fully the isomerism between two given compounds. The differences between isomers are evident crystallographically and, usually, spectroscopically. In some cases analytical differences may also exist.

3.4.1 Conformation isomerism

This is a simple form of isomerism in which the isomers have different stereochemistries but are otherwise identical, e.g. *trans* planar and tetrahedral $NiCl_2(Ph_2PCH_2PPh_2)_2$ (cf. Fig. 3.7). Its occurrence is confined to relatively few metal ions, usually those with a d^8 configuration.

The flurry of recent interest in isomerism arose because of the suggested existence of an isomerism related to conformation isomerism, a isomerism which is variously called *distortional isomerism* or *bond-stretch isomerism*. The suggestion was made that a bond length in a given complex could have either of two very different values. This could arise, for instance, if there were two different bonding interactions, each leading to stability at a different internuclear distance. What was the ground state at one distance would correspond to a low lying excited state at the other, and vice versa. Although it was at first found that the suggested example, in the compound $[Mo(O)Cl_2(PMe_2Ph)_3]$, was erroneous (one crystal structure determination was on an impure crystal and gave misleading results), it has stimulated great interest in the possible existence of this form of isomerism. A subsequent reinvestigation has revealed two (pure!) crystal forms of the compound in which the rather asymmetrical phosphine ligands adopt rather different conformations. The Mo–O bond length is 1.663 Å in one isomer and 1.682 Å in the other. There *are* different bond lengths but the name *distortional isomerism*, the original one, perhaps is the more appropriate if the phenomenon is regarded as a form of conformational isomerism. However, there are some clearly established examples of bond length differences in some dimeric ruthenium complexes such as $[\eta^5\text{-}Cp^*RuCl(\mu\text{-}Cl)]_2$ in which Cp* is the sterically demanding ligand $C_5(CH_3)_5$; apart from its steric effects, which are currently giving rise to considerable study,[9] it behaves like C_5H_5. In these, one isomer is diamagnetic, has no unpaired electrons, and has a Ru–Ru separation of 2.9 Å. The other isomer is paramagnetic, it has unpaired electrons, and a Ru–Ru distance of 3.8 Å. However, because of the magnetic differences between the two isomers, they are perhaps better regarded as *spin isomers*, a type which will be described later.

3.4.2 Geometrical isomerism

This form of isomer has already been met when discussing nomenclature; *cis* and *trans* isomers are examples of geometrical isomers. Interconversion between two geometric isomers is often an important step in mechanisms postulated as those by which coordination compounds catalyse reactions, particularly those involving unsaturated organic molecules.

3.4.3 Coordination position isomerism

In this form of isomerism the distribution of ligands between two coordination centres differs; an example is shown below.

Note that each of these two cations exists in a number of isomeric forms. The reader may find it a useful exercise to draw pictures of all of the forms and to enquire into the isomeric relationship between pairs.

3.4.4 Coordination isomerism

This may occur only when the cation and anion of a salt are both complex,

---

[9] Similarly, transition metal complexes of the pentaphenylcyclopentadienyl ligand have been studied. Unfortunately, they tend to be rather insoluble, a clear disadvantage. It is likely that the addition of alkyl groups to the phenyl rings would increase the solubility of the complexes formed. However, if such substitution is not symmetrical then further complications ensue. The recent synthesis of complexes of the penta-*p*-tolylcyclopentadienyl ligand suggests that this may become a much studied ligand in the future.

the two isomers differing in the distribution of ligands between the cation and anion:

$$[Co(NH_3)_6][Cr(ox)_3] \quad \text{and} \quad [Cr(NH_3)_6][Co(ox)_3]$$

The same metal may be the coordination centre in both cation and anion:

$$[Cr(NH_3)_6][Cr(SCN)_6] \quad \text{and} \quad [Cr(NH_3)_4(SCN)_2][Cr(NH_3)_2(SCN)_4]$$

3.4.5 Ionization isomerism

Two coordination compounds which differ in the distribution of ions between those directly coordinated and counterions present in the crystal structure are called ionization isomers:

$$[Co(NH_3)_5Br]SO_4 \quad \text{and} \quad [Co(NH_3)_5(SO_4)]Br$$

The difference between these isomers is analytically apparent—an aqueous solution of the first gives an immediate precipitate with barium chloride solution and the second with silver nitrate.

3.4.6 Hydrate isomerism

Hydrate isomerism is similar to ionization isomerism except that it really only applies to crystals. An uncharged ligand changes from being co-ordinated to being in the crystal but uncoordinated whilst another ligand moves in the opposite sense. Although the uncharged ligand need not be a water molecule, in practice it almost always is (and hence the term hydrate isomerism), for example

$$[Cr(H_2O)_6]Cl_3, \quad [Cr(H_2O)_5Cl]Cl_2 \cdot H_2O \quad \text{and} \quad [Cr(H_2O)_4Cl_2]Cl \cdot 2H_2O$$

3.4.7 Linkage isomerism

In our discussion on nomenclature the problem that some ligands may coordinate in two or more ways was encountered. As has been mentioned, such ligands are sometimes called *ambidentate* ligands. Corresponding to this is the phenomenon of linkage isomerism, for example

$$[Cr(H_2O)_5(SCN)]^{2+} \quad \text{and} \quad [Cr(H_2O)_5(NCS)]^{2+}$$
$$[Co(NH_3)_5(NO_2)]^{2+} \quad \text{and} \quad [Co(NH_3)_5(ONO)]^{2+}$$
$$[Co(NH_3)_5(SSO_3)]^+ \quad \text{and} \quad [Co(NH_3)_5(OSO_2S)]^+$$

3.4.8 Polymerization isomerism

Strictly speaking, polymerization isomerism, in which n varies in the complex $[ML_m]_n$ (the Ls need not all be identical), is not isomerism. It is included in this list because it represents an additional way in which an empirical formula may give incomplete information about the nature of complex. For example, all members of the following series are polymerization isomers of $[Co(NH_3)_3(NO_2)_3]_n$.

$$[Co(NH_3)_3(NO_2)_3] \qquad n = 1$$

$$[Co(NH_3)_6][Co(NO_2)_6] \qquad n = 2$$

$$[Co(NH_3)_4(NO_2)_2][Co(NH_3)_2(NO_2)_4] \qquad n = 2$$

$$[Co(NH_3)_5(NO_2)][Co(NH_3)_2(NO_2)_4]_2 \qquad n = 3$$

$$[Co(NH_3)_6][Co(NH_3)_2(NO_2)_4]_3 \qquad n = 4$$

$$[Co(NH_3)_4(NO_2)_2]_3[Co(NO_2)_6] \qquad n = 4$$

$$[Co(NH_3)_5(NO_2)_6]_3[Co(NO_2)_6]_2 \qquad n = 5$$

3.4.9 Ligand isomerism

If two ligands are isomers, the corresponding complexes are isomers also; for example

CH$_2$—CH—CH$_3$	and	CH$_2$—CH$_2$—CH$_2$
NH$_2$ NH$_2$		NH$_2$ NH$_2$
propylenediamine (pn)		trimethylenediamine (tn)

are isomers, both of which form complexes of the type shown in Fig. 3.25 (where a convenient representation has been adopted for the two isomeric ligands which shows only the coordinated atoms). In this situation, the two isomers, indistinguishable by elemental analysis, are termed *ligand isomers*.

A special form of ligand isomerism arises when two different sites in a ligand can be protonated. If only one proton is added then two different species result, sometimes called protonation isomers. They are important because both in the case that the proton is replaced by a metal and in the case that the unprotonated site coordinates, different complexes result. Such differences are important in some biochemical systems.

A special case of ligand isomerism also arises when the ligands are optical isomers—*enantiomorphs*—of each other. One interesting problem is the extent to which electron absorption bands, which, as a first approximation, are supposed to be localized on a transition metal ion, acquire optical activity because of the activity of a coordinated ligand. An example of this is provided by the ligand mentioned above, propylenediamine (pn) which exists in optically isomeric forms.

3.4.10 Optical isomerism

A molecule is optically active when it cannot be superimposed on its mirror image. Although this condition is met by an octahedral complex such as $ML_aL_bL_cL_dL_eL_f$ it is rare indeed to be able to resolve such a complex. In practice, optical activity is largely confined to octahedral complexes of chelating ligands. Optical activity has also been observed for chelated tetrahedral and square planar complexes but only rarely. It is necessary for the chelated complex to be stable kinetically; to permit resolution, it must retain its configuration for at least a matter of minutes. This confines attention to complexes of a few ions, of which cobalt(III), chromium(III)

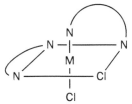

Fig. 3.25 The bidentate ligands are shown here very schematically. If in one complex they represent a particular ligand but in another complex an isomeric ligand, then the two complexes provide an example of ligand isomerism.

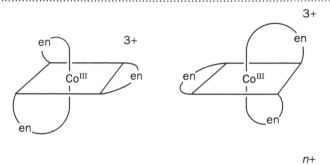

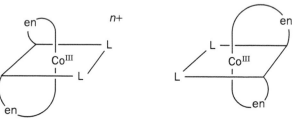

Fig. 3.26 The complex $[Co(en)_3]^{3+}$ and cis-$[Co(en)_2L_2]^+$ (L = anion⁻). The two molecules represented in this figure for each species are mirror images which cannot be superimposed

and rhodium(III) are examples. Although the optical activity of complexes of many polydentate chelating ligands has been studied, for simplicity our discussion will be confined to the bidentate case.

Two classes of optically active complexes formed by bidentate chelating ligands which have been the subject of much work are $[M(L_2)_3]$ and cis-$[M(L_2)_2L'_2]$, where (L_2) is a bidentate ligand and L' a monodentate. Figure 3.26 shows the pairs of isomers for M = Co and L_2 = ethylene-diamine (en). At this point the discussion contained in Appendix 1 becomes relevant, because amongst the molecules considered there are the $[M(L_2)_3]$.

If a compound is optically active it cannot crystallize in a centrosymmetric space group (the action of a centre of symmetry serves to convert one optical isomer into the other). A consequence is that in X-ray diffraction crystallographic studies, pairs of related diffracted beams (h, k, l and $-h, -k, -l$), which in a centrosymmetric crystal would have identical intensities, no longer do so. An analysis of the difference in intensities in such pairs provides the absolute configuration of the optically active species.

3.4.11 Structural and fluxional isomerism

For simplicity, almost all of the forms of isomerism discussed above concerned classical octahedral complexes. It was implicitly assumed that each complex has a single structure and that this structure does not change with time. Other forms of isomerism are recognized if we remove one or both of these restrictions. For instance, in Section 3.2.3, we met the fact that the anion $[Ni(CN)_5]^{3-}$ can exist in two different geometries, trigonal bipyramidal and square pyramidal (a phenomenon which is usually classified under the heading of structural isomerism). A rather more extreme example is provided by $Co_2(CO)_8$ of which at least two forms coexist in solution:

$$(CO)_3Co\underset{CO}{\overset{CO}{\diamond}}Co(CO)_3 \quad \text{and} \quad (CO)_4Co-Co(CO)_4$$

Fig. 3.27 Potential energy profiles in η^5-$C_5H_5Mn(CO)_3$. Consider a microscopic probe, shown at the right hand side of the dotted circle surrounding the manganese atom. If the molecule is held rigid and the probe rotated around the dotted circle, the top end of the probe will experience five bumps per circuit and the bottom end will experience three. That is, the η^5-C_5H_5 and $Mn(CO)_3$ units generate fivefold and threefold potentials, respectively. Now forget the probe and consider just one carbon atom of the C_5H_5 ring as the ring is rotated against a rigid $Mn(CO)_3$ unit. The carbon atom will experience three bumps per circuit. So too will each of the other four carbon atoms in the C_5H_5 ring. But the geometry is such that, although all of these bumps will be equal, none will coincide. That is, in a complete circuit there will be a total of $5 \times 3 = 15$ bumps, a 15-fold potential.

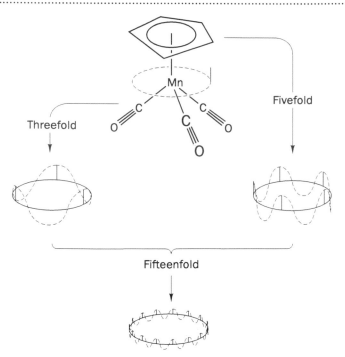

Indeed, the ability to undergo such molecular gymnastics is a characteristic of many organometallic compounds, particularly of transition metals. As another example consider the compound η^5-$C_5H_5Mn(CO)_3$, shown in Fig. 3.27. Here, the η^5-C_5H_5Mn unit has a local fivefold rotational axis and the $Mn(CO)_3$ unit a local threefold. Bringing the two together, one has a $5 \times 3 = 15$-fold rotational barrier. The logic behind this arithmetic is contained in the answer to Question 3.1. However, a 15-fold rotational barrier means $360°/15 = 24°$ between equivalent potentials—and so just $12°$ between maxima and minima in the potential surface. This is also illustrated in Fig. 3.27. Apart from the fact that for such a small angle the difference between maxima and minima must be small, the atoms involved are big relative to the corrugations of the potential energy surface. Not surprisingly, if the molecule is labelled in some way (perhaps by inclusion of a ^{13}C in the C_5H_5 and another in the $(CO)_3$) it is found that rotation of the two halves of the molecule relative to each other is rather free. It is often convenient to think of the rotation as confined to the C_5H_5 system, particularly when the ring is less symmetrically bonded than in our example, because the rotation then makes the time average of all ring positions identical. One talks of *ring whizzers*, evocative of a firework display—chemists have a sense of humour too. The collective name given to such interchange phenomena is to talk of fluxionality and of fluxional molecules. We shall meet them again in Chapter 14.

3.4.12 Spin isomerism

As will be seen in Chapter 7, octahedral complexes of Fe^{III} can exist in one of two spin states, high spin and low spin. As will be explained in that chapter, the difference can be attributed to different magnitudes of splitting, Δ, between two sets of d orbitals:

high spin   low spin

```
 – ↑ – –      – – – –
 – ↑ – –      – – – –
    Δ            Δ
 – ↑ – –      – ↑ – –
 – ↑ – –      – ↑ ↓ –
 – ↑ – –      – ↑ ↓ –
```

In some complexes, usually of Fe^{III} but also for other ions, most notably Fe^{II} and Co^{II}, it seems that the magnitude of the splitting Δ is such that both forms occur; spin isomers coexist in the same sample An example is provided by the octahedral complex [Fe(S$_2$CNMe$_2$)$_3$], where the *N,N*-dimethyldithiocarbonate ligand is

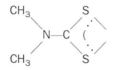

Here, one has to be careful because spin isomerism behaviour in solution may well be different to that in the solid state. In the solid state the individual magnetic ions couple weakly together but often sufficiently strongly for the phenomenon to have a cooperative aspect and to show hysteresis; the stronger the cooperativity the more abrupt the transition. The spin-crossover may be induced not only thermally but also by application of pressure, for small structural changes accompany the spin change.

For both of the last two classes of isomerism we have detailed—fluxional and spin isomerism—the lifetime of individual isomers may be rather short. Spin isomers, for instance, typically live for about 10^{-7} s (but see Section 3.4.1 and the Further Reading at the end of this chapter). Some would argue that classical isomerism refers only to species capable of physical separation, and so of long lifetime. However, with the increasing use of methods which explore short lifetimes—NMR and EPR, particularly, in the present context—it seems sensible to ignore this limitation.

Further reading

Progr. Inorg. Chem. has featured a number of articles devoted to specific coordination numbers. They are:

- 4- and 5-coordination   M. C. Favas and D. L. Kepert *27*, 325
- 5-coordination   R. R. Holmes *32*, 119
- 5-coordination   J. S. Wood *16*, 227
- 6-coordination   D. L. Kepert *23*, 1
- 7-coordination   M. G. B. Drew *23*, 67
- 7-coordination   D. L. Kepert *25*, 41
- 7- and 8-coordination   S. J. Lippard *21*, 91
- 8-coordination   D. L. Kepert *24*, 179
- 8-coordination   S. J. Lippard *8*, 109

Other discussions exist, for instance, 11-coordination in W. O. Milligan, D. F. Mullica, H. O. Perkins D. A. Grossie and C. K. C. Lok, *Inorg. Chim. Acta* (1984) *86*, 33.

Isomerism in general has been reviewed in 'The isomerism of complex compounds' R. G. Wilkins and M. J. G. Williams in *Modern Coordination Chemistry*, R. G. Wilkins and J. Lewis (eds.), Interscience, New York, 1960. Although old, this remains a useful review.

The fact that spin equilibria (spin isomerism) has been the subject of much recent work is indicated by the almost simultaneous appearance of three reviews on the subject:

'Dynamics of spin equilibria in metal complexes' by J. K. Beattie in *Adv. Inorg. Chem.* (1988) *32*, 1; 'Static and dynamic effects in spin equilibrium systems' by M. Bacci, *Coord. Chem. Rev.* (1988) *86*, 245; and 'Spin equilibria in iron(II) complexes' by H. Toftlund, *Coord. Chem. Rev.* (1989) *94*, 67. See also

Section 9.12. Sometimes it is possible to photoexcite a molecule to the less stable spin state, whereupon it may be stable for weeks! An example, which shows how the methods introduced in later chapters of this book may be used to investigate such systems, is in a paper by Gutlich and Poganiuch in *Angew. Chem., Int. Ed.* (1991) *30*, 975.

A modern general review on isomerism is to be found in an article by J. C. Bailar in '*Coord. Chem. Rev.*' (1990) *100*, 1.

For those interested in learning more of the bond-stretch isomer story there is detailed review by V. C. Gibson and M. McPartlin in *J. Chem. Soc., Dalton Trans.* (1992) 947 and a shorter one by J. M. Mayer, *Angew. Chem., Int. Ed.* (1992) *31*, 286. Both became slightly out of date because of another article: A. P. Bashall, S. W. A. Bligh, A. J. Edwards, V. C. Gibson, M. McPartlin and O. B. Robinson *Angew. Chem., Int. Ed.* (1992) *31*, 1607. Even more recent is 'Bond Stretch Isomers: Fact not Fiction' by P. Gütlich, H.A. Goodwin and D. N.

Hendrickson *Angew. Chem., Int. Ed.* (1994) *33*, 425 (discusses the isomerization in the context of spin-crossovers and 'show that it is a reality'). A paper that looks back at the origins of the controversy, and comes up with the suprising answer contained in its title, is 'Studies of Distortional Isomers. 2. Evidence That Green [LWOCl₂]PF₆ is a Ternary Mixture' by P. J. Desrochers, K. W. Nebesny, M. J. LaBarre, M. A. Bruck, G. F. Neilson, R. P. Sperline, J. H. Enemark, G. Backes and K. Wieghardt *Inorg. Chem.* (1994) *33*, 15; see also the footnote on page 2.

A useful source is Volume 1 of *Comprehensive Coordination Chemistry* G. Wilkinson, R. D. Gillard and J. A. McCleverty (eds.), Pergamon Press, Oxford, 1987, and in particular Chapters 2 ('Coordination Numbers and Geometries' by D. L. Kepert), 3 ('Nomenclature of Coordination Compounds' by T. E. Sloan) and 5 ('Isomerism in Coordination Chemistry' by J. MacB. Harrowfield and S. B. Wild).

Questions

3.1 Give systematic names for the complexes listed in Table 2.2 (in some cases, because full structural details are not given in the table, there will be some ambiguity about the correct name).

3.2 Use a piece of paper to cover up the formulae in Table 3.2 leaving just the names visible. Write down the corresponding formulae and check your answers by removing the paper. Repeat the exercise, this time covering up the names and attempting to write them out.

3.3 Repeat the exercise of Question 3.2, this time using Table 3.3.

3.4 When VCl_3 is dissolved in acetonitrile (CH_3CN), there is evidence that above 50 °C two non-ionic monomeric octahedral complexes co-exist. Suggest reasonable structures for these two complexes.

3.5 Figure 3.26 shows *cis*-$[Co(en)_2L_2]^+$ (L is anionic) as an example of a complex which can be optically active. Could either (a) *trans*-$[Co(en)_2L_2]^+$ or (b) $[Co(NH_3)_4L_2]^+$ be optically active?

3.6 If you were given the task of attempting to prepare a complex with an unusual coordination number, which coordination numbers should you seek to avoid, and why?

3.7 You are asked to attempt to prepare as many isomers as possible of $[Co(NH_3)BrCl(en)]_2SO_4$. As a first step, prepare as complete a list as possible of these isomers.

3.8 The discussion in the text concerned the combination of a fivefold and threefold rotation. As a simpler example show that in the molecule $[(py)Ni(CO)_3]$, in which the pyridine is bonded to the nickel through its nitrogen atom, the twofold rotational symmetry of the py–Ni unit combines with the threefold rotational symmetry of the $Ni(CO)_3$ to give a sixfold rotational barrier. Hint: draw each of the six equivalent arrangements; invent a system of labelling which enables equivalent arrangements to be distinguished.

3.9. The author of this book believes that a better (including a language-independent) polyhedral system than that recommended by IUPAC would be to give a point group (symmetry) symbol followed by the coordination number. There follow two lists, one in the author's and, in a different sequence, one of the corresponding IUPAC symbols. None of the latter have been used in the present chapter, although all of the polyhedra have been mentioned. Pair off corresponding symbols. This problem should give more familiarity with both the polyhedra and the use of point group notation.

$D_{\infty h}$-2	C_{2v}-2	TP-3	PBPY-7
D_{3h}-3	C_{3v}-3	TPRS-7	HBPY-8
D_{5h}-7	C_{3v}-7	TPY-3	A-2
C_{2v}-7	O_h-8	OCF-7	SAPR-8
D_{4d}-8	D_{2d}-8	DD-8	CU-8
D_{6h}-8		L-2	

Preparation of coordination compounds

4.1 Introduction

This chapter reviews the most common methods by which coordination compounds are prepared. However, current research is almost invariably aimed at producing the unusual and exotic, not the common. So, a contemporary research journal would describe methods rather less simple than most of those covered here. A flavour of the current has therefore been included, although the reader is unlikely to meet some of the compounds outside the research laboratory. In the reactions described in this chapter, there are two important variables—coordination number and oxidation number (the latter is often called the valence state). In principle, either may increase, decrease or remain unchanged in a reaction, and the reader may find it helpful to classify the preparative methods described according to changes in these two numbers. In practice it is not always possible to be certain of either without more information than that contained within a chemical equation or chemical formula. A ligand which is potentially tridentate may, for example, act as a bidentate ligand and so the coordination number differs from that expected. Similarly, is the complex ion $[Co(NH_3)_5NO]^{2+}$ a complex of cobalt(II) or one of cobalt(III)? It depends on whether you believe that the NO is better represented as $NO^{\cdot}$ (where, in the complex, the odd electron is paired with a cobalt electron) or as NO^-. This problem of formal valence states will reappear later in this chapter and again in Chapter 6.

Complications apart, reactions in which the coordination number of an electron acceptor is increased are called *addition reactions*, and when it is unchanged they are called *substitution reactions*. The coordination number decreases for *dissociation reactions*. Reactions involving valence state changes are called *oxidation* or *reduction reactions*, as appropriate.

An important classification of complexes depends on the speed with which they undergo substitution reactions. When excess of aqueous ammonia is added to a solution of copper(II) sulfate in water the change in colour from pale to deep blue is almost instantaneous, because an ammine complex is formed very rapidly (in this reaction ammonia replaces some of the water molecules coordinated to the copper(II) ion). This is an example of the generalization that copper(II) forms *kinetically labile* complexes. On the other hand, it takes hours (or even days at room temperature) to replace water molecules coordinated to a chromium(III) ion by other ligands. Again we can generalize: chromium(III) forms *kinetically inert* complexes. It is important to recognize the distinction between kinetic and thermodynamic stability at this point. The thermodynamic stability of a complex (which will be discussed at length in the next chapter) refers to the concentrations of complex species and ligands at equilibrium. Kinetic stability refers to the speed at which equilibrium conditions are reached. As one would expect, the preparations of kinetically inert and labile complexes present quite different problems. In general, the ions of the second- and third-row transition elements usually form kinetically inert complexes. With the exception of chromium(III) and cobalt(III), the common ions of first-row transition elements usually form kinetically labile complexes. Metallic main group elements usually form labile complexes.

Complexes involving low valence states, organometallic complexes for instance, are usually inert. However, inertness relates to kinetics and kinetics depend on mechanism. An organometallic compound which normally reacts slowly may spontaneously catch fire, or, less dramatically, rapidly oxidize, if exposed to air. Not surprisingly, special inert atmosphere techniques have to be used in preparing such compounds. Gaseous oxygen, of course, is a diradical, with two unpaired electrons, and so it is not unexpected that it should react rather differently to many other potential reactants.

4.2   Preparative methods

It is difficult to present reaction techniques in an order which is obviously logical and sequential. In the following pages the pattern usually adopted is to move from the simple to the complicated, although simplicity has its own complications. So, the first reaction considered is a simple gas-phase reaction between molecules, but one for which a quite complicated glass vacuum line would be needed. The reaction between aqueous Cu^{II} and aqueous ammonia, considered later, can be carried out using a couple of test tubes, but is chemically quite complex.

4.2.1   Simple addition reactions

The most direct method of preparing $[BF_3(NH_3)]$ is by gas-phase addition, in which a carefully controlled flow of each of the gaseous reactants is led into a large evacuated flask, where the product deposits as a white powder:

$$BF_3 + NH_3 \rightarrow [BF_3(NH_3)]$$

When one reactant is a liquid and the other a gas at room temperature, a different technique is usually followed. In the preparation of $[BF_3(OEt_2)]$, for instance, the diethylether and boron trifluoride, stored in separate bulbs on a vacuum line, are condensed separately into an evacuated flask cooled in liquid nitrogen. When the flask is warmed slowly, a controlled reaction takes place:

$$BF_3 + Et_2O \rightarrow [BF_3(OEt_2)]$$

Reactions between liquids or solids are best carried out by mixing solutions of them in a readily removable inert solvent, e.g.

$$SnCl_4 + 2NMe_3 \xrightarrow[\text{petroleum ether}]{40-60\ °C\ bp} trans[SnCl_4(NMe_3)_2]$$

If at all possible, the presence of a solid reactant should be avoided unless it is one of those reactions in which an otherwise insoluble compound dissolves in the presence of a complexing agent. Many of these reactions occur with no change in valence state, as when silver chloride dissolves in aqueous ammonia:

$$AgCl(s) + 2NH_3(aq) \rightarrow [Ag(NH_3)_2]^+(aq) + Cl^-(aq)$$

or when the gelatinous precipitates formed on adding alkali metal cyanides to aqueous solutions of many metal ions, dissolve in excess cyanide, for instance

$$Zn(CN)_2(s) + 2CN^-(aq) \rightarrow [Zn(CN)_4]^{2-}(aq)$$

Solids may dissolve in complexing agents with a change in valence state (so it is debatable whether such reactions should be classified as simple addition). In the case of the dissolution of metallic silver or gold in water in the presence of cyanide ion, the oxygen of the air acts as the oxidizing agent:

$$2M(s) + 4CN^-(aq) + \tfrac{1}{2}O_2 + H_2O \rightarrow [M(CN_2)]^- + 2OH^- \qquad (M = Ag\ or\ Au)$$

More commonly, however, the oxidizing agent is carefully chosen, as when the sparingly soluble $PbCl_2$ dissolves in aqueous hydrochloric acid through which chlorine is bubbled:

$$PbCl_2(s) + 2HCl(aq) + Cl_2(g) \rightarrow H_2[PbCl_6]$$

The product of this reaction is relatively unstable, decomposing by the reverse of the formation reaction. A trick which is widely used in such cases is to add a large, poorly polarizing counterion (a cation such as pyridinium, $C_5H_5NH^+$, added as the chloride, in the present case). One then obtains either a precipitate or crystals in which the unstable cation or anion is less prone to decompose. Other counterions which are commonly used in this way are $[As(C_6H_5)_4]^+$, $(t-C_4H_9)_3NH^+$, $[Co(NH_3)_6]^{3+}$, $[B(C_6H_5)_4]^-$ and $[Cr(SCN)_6]^{3-}$. There have even been attempts to place this on a quantitative footing by defining size and shape parameters for such species.

An example of a reaction, involving a solid, which would be avoided if at all possible is the apparently simple reaction

$$NH_4F(s) + BF_3(g) \rightarrow NH_4[BF_4](s)$$

because it is difficult to ensure that reaction is complete. Further, purification of the product may be difficult; however, a large number of anionic complexes of formula $[MX_n]^{m-}$, where X is a halogen (usually F or Cl), have been made in this way, for example

$$2KCl + TiCl_4 \rightarrow K_2[TiCl_6]$$

The important factor is whether the product forms an impenetrable layer around the crystals of the solid reactant.

As one would expect, good examples of simple addition reactions of transition metal complexes are confined to those ions which readily change their coordination number. Copper(II) provides many examples provided that excess of incoming ligand is used so that a mixture of products is avoided. An example is provided by the addition of pyridine to $[Cu(acac)_2]$, a four-coordinate complex becoming five-coordinate:

$$[Cu(acac)_2] + py \rightarrow [Cu(acac)_2(py)]$$

As kinetic studies show (Chapter 14), many reactions in solution proceed through a reaction intermediate in which the solvent is coordinated. What may, on paper, appear to be an addition reaction may in fact be a ligand substitution reaction.

4.2.2 Substitution reactions

The majority of complexes, both of transition and non-transition elements, may be prepared by substitution reactions. The mechanisms of some of these reactions have been extensively investigated and will be discussed in Chapter 14. Although the coordination number of the atom at the coordination centre in both reactant and product species is the same in these reactions, it must be emphasized that only limited information can be inferred about the reaction mechanism from a study of the products of a reaction. In particular, phrases such as 'the ligand A displaces ligand B' should be avoided in detailed discussions unless the reaction has been properly investigated. For the non-transition elements in particular, where substitution reactions usually proceed if thermodynamically favourable, a study of these reactions enables both qualitative and quantitative assessments to be made of the relative strength of donor–acceptor bonds. Thus, because ammonia displaces diethylether from $BF_3 \cdot OEt_2$, even in ether solution, to give crystals of $BF_3 \cdot NH_3$ it has been concluded that the B–N bond is stronger than the B–O bond, although this argument is open to the objection that if $BF_3 \cdot NH_3$ has a high lattice energy, perhaps because of hydrogen bonding, then it could be this fact that leads to the formation of $BF_3 \cdot NH_3$ rather than a higher B–N bond strength.

As noted earlier, there is an experimental distinction between the substitution reactions of labile and inert complexes. The formation of labile complexes is virtually instantaneous upon mixing of the reactants, so that there are few practical difficulties in their preparation, but three points must be remembered. First, for classical, Werner-type, complexes it is found in practice that it is difficult to prepare such complexes with several different non-ionic ligands bonded to the same metal atom, although it is much easier to prepare complexes in which an anionic species is coordinated together with a neutral ligand. Secondly, although it may be possible to isolate and characterize a solid complex, quite a different complex may be the predominant species in solution. So, the blue complex $Cs_2[CoCl_4]$ crystallizes from pink aqueous solutions containing octahedral Co^{II} and CsCl. The third point, that some complex ions display *incongruent solubility*, as will be seen, is related to the second.

If an aqueous solution containing iron(II) sulfate and ammonium sulfate in a 1:1 molar ratio is allowed to crystallize, then a compound which historically is variously known as Mohr's salt and as ferrous ammonium sulfate, $[Fe(H_2O)_6]SO_4(NH_4)_2SO_4$, is obtained. Mohr's salt is said to show congruent solubility. On the other hand, if an aqueous solution containing a 2:1 molar ratio of potassium chloride and copper(II) chloride crystallizes, crystals of potassium chloride are obtained first. Only later does the complex $K_2[Cu(H_2O)_2Cl_4]$ crystallize. Similarly, attempts to recrystallize the salt will lead to the initial deposition of potassium chloride. The complex is said to display incongruent solubility; it can only be obtained from aqueous solutions containing excess of copper(II) chloride. A system which displays incongruent solubility at one temperature may display congruent solubility at another.

Examples of the formation of complex ions by substitution reactions of labile complexes are the following.

1. The action of excess of ammonia on aqueous solutions of copper(II) salts:

$$[Cu(H_2O)_4]^{2+} + 4NH_3(aq) \rightarrow [Cu(NH_3)_4]^{2+} + 4H_2O$$

Although this equation[1] shows the complete substitution of coordinated water by ammonia all such reactions occur in steps and the species $[Cu(H_2O)_4]^{2+}$, $[Cu(H_2O)_3NH_3]^{2+}$, $[Cu(H_2O)_2(NH_3)_2]^{3+}$, $[Cu(H_2O)(NH_3)_3]^{2+}$, and $[Cu(NH_3)_4]^{2+}$ are all present in the solution, although the concentrations of some are low. By a suitable choice of concentration (using stability-constant data of the sort discussed in Chapter 5) it is possible to ensure that the concentration of one particular component, $[Cu(H_2O)_2(NH_3)_2]^{2+}$ say, is a maximum in the solution. However, it does not follow that if crystallization is induced (for example, by adding ethanol to the solution and so decreasing the solubility of the complex species) the complex which crystallizes will contain the $[Cu(H_2O)_2(NH_3)_2]^{2+}$ cation. There are many labile complexes which may be studied readily in solution but which are very difficult to obtain in the solid state. The converse is also true. Copper(I) bromide reacts in ethanol with Br^- to give solutions in which only the $[CuBr_2]^-$ anion has been identified. From such solutions, crystals of salts containing anions such as $[Cu_2Br_5]^{3+}$ and $[Cu_4Br_6]^{2-}$, as well as $[CuBr_2]^-$, have been obtained. When a salt such as $[N(CH_3)_4]_3[Cu_2Br_5]$ is dissolved in nitromethane, the dominant species in solution is again $[CuBr_2]^-$.

2. The reaction between aqueous solutions of thiourea and lead nitrate:

$$[Pb(H_2O)_6]^{2+} + 6SC(NH_2)_2 \rightarrow [Pb(SC(NH_2)_2)_6]^{2+} + 6H_2O$$

The lead(II) ion in aqueous solutions exchanges water between its coordination sphere and the bulk very rapidly but is probably best regarded as six-coordinate (although some evidence indicates that the coordination number

---

[1] A common coordination geometry for the copper(II) ion is to be surrounded by four ligands in a plane which, together with two ligands one above and one below this plane but further from the copper atom, form a tetragonally distorted octahedron. In this discussion these two, more weakly bonded, ligands have been neglected.

may be as high as eight). However, an aqueous solution of lead nitrate, say, may also contain polymeric species and the reaction given above is, therefore, oversimplified both for this reason and because it makes no mention of species intermediate between $[Pb(H_2O)_6]^{2+}$ and $[Pb(SC(NH_2)_2)_6]^{2+}$.

3. If an uncharged complex is prepared in aqueous solution from ionic species it is often precipitated from aqueous solution and, unless highly polymeric, may usually be recrystallized from organic solvents, e.g.

$$[Fe(H_2O)_6]^{3+}; + 3acac^- \rightarrow [Fe(acac)_3] + 6H_2O$$

$$\text{insoluble}$$
$$\text{in water}$$

Some examples of preparations involving substitution reactions of inert complexes are given below.

1. The oxidation of (labile) cobalt(II) salts in aqueous solution containing both ammonia and ammonium carbonate by air bubbled through the mixture leads to the formation of the (inert) $[Co(NH_3)_5(CO_3)]^+$ cation. It is only on heating with aqueous ammonium hydrogen fluoride solution at 90 °C for 1 h that this is converted into the $[Co(NH_3)_5F]^{2+}$ cation:

$$[Co(NH_3)_5CO_3]^+ + 2HF \rightarrow [Co(NH_3)_5F]^{2+} + F^- + CO_2 + H_2O$$

The species $[Co(NH_3)_5H_2O]^{3+}$ is almost certainly an intermediate in the reaction. Surprisingly, cobalt(III) complexes containing coordinated carbonate ion lose CO_2 rather easily—they fizz when dilute acid is poured onto them—but this reaction does not involve breaking the Co–O bond; isotopic studies show that the oxygen atom in the final Co–OH$_2$ bond is the same as that in the original Co–OCO$_2$.

Reactions of Co^{III} complexes played a key role in the development of coordination chemistry and so mention of a few more of these reactions is appropriate. In the preparation of Co^{III} salts from Co^{II}, the composition of the reaction mixture, the choice of oxidant (H_2O_2, PbO_2, perhaps charcoal being added as a catalyst) and temperature are the key variables. A molar ratio Co^{II}:NH$_4$Cl:NH$_3$:NaNO$_2$ of 1:1:2:3 with air as the oxidant at room temperature gives *mer*-$[Co(NH_3)_3(NO_2)_3]$ as the major product. Notwithstanding what has been said above, this is the product most soluble in water (presumably the ligands strongly hydrogen bond with the water) and is thus separated. $[Co(NH_3)_3(NO_2)_3]$ reacts at room temperature with concentrated hydrochloric acid over a day with evolution of brown fumes of nitrogen oxides to give $[Co(NH_3)_3(H_2O)Cl_2]^+$. This, in turn, reacts with ice-cold aqueous ammonia over about 2 h to give a dimeric compound containing three hydroxyl bridges:

$$\left[(NH_3)_3Co \overset{\overset{\displaystyle OH}{\diagup\diagdown}}{\underset{\underset{\displaystyle OH}{\diagdown\diagup}}{-OH-}} Co(NH_3)_3 \right]^{3+}$$

The contrast between these reactions and those of labile complexes, where reaction is complete almost as soon as the reactants are mixed, is very evident.

2. Potassium hexanitritocobalt(III), potassium cobaltinitrite, reacts with an aqueous solution of ethylenediamine fairly rapidly at ca 70 °C to give *cis*-dinitrobis(ethylenediamine)cobalt(III):

$$[Co(NO_2)_6]^{3-} + 2en \rightarrow cis\text{-}[Co(en)_2(NO_2)_2]^+ + 4NO_2^-$$

In this reaction the complex $[Co(en)(NO_2)_4]^-$ is presumably an inter- mediate. However, the solid obtained by removing the solvent from a solution in which the major component is *cis* $[Co(en)_2(NO_2)_2]^+$ will also consist largely of this complex ion, because it is kinetically inert.

In the preparation of some complexes, particularly organometallic com- plexes, the presence of water must be avoided. An important example from classical coordination chemistry is that the action of ammonia (either as a gas or in solution) on hydrated chromium(III) salts—those commercially available—leads to the precipitation of insoluble hydroxy complexes and not to the formation of $[Cr(NH_3)_6]^{3+}$. This complex is prepared by reaction between liquid ammonia and anhydrous chromium(III) chloride.

A variety of methods have been developed for the preparation of such anhydrous chlorides. Reaction of the heated metal with chlorine is one obvious procedure, but is difficult to control. Better is removal of water from the hydrated salt by chemical reaction on heating with thionyl chloride (unpleasant), dimethyoxypropane or triethylorthoformate:

$$H_2O + SOCl_2 \rightarrow SO_2 + 2HCl$$
$$H_2O + (CH_3O)_2C(CH_3)_2 \rightarrow 2CH_3OH + (CH_3)_2CO$$
$$H_2O + (C_2H_5O)_3CH \rightarrow 2C_2H_5OH + HC(O)OC_2H_5$$

Probably most useful is reaction of a metal oxide with a chlorinated hydrocarbon; the high boiling hexachloropropene, $C(Cl)_2C(Cl)CCl_3$, is favoured, the terminal $-CCl_3$ becoming $-COCl$ in the reaction.

Important though anhydrous halides are in synthetic coordination chem- istry, they suffer from one disadvantage. They tend to have low solubilities and to react slowly. In such cases, an alternative is to form a complex which, whilst stable, is one from which the ligands are readily displaced. For instance, if triethylorthoformate is used as a dehydrating agent with ethanol as a solvent, complexes such as $[Mg(C_2H_5OH)_6]^{2+}$, $[Co(C_2H_5OH)_6]^{2+}$ and $[Ni(C_2H_5OH)_6]^{2+}$, from which the ligands are readily displaced, are obtained. A ligand which has become increasingly popular for this purpose since it became commercially available is the trifluoromethansulphonate anion, $CF_3-SO_3^-$, more usually called triflate (hence triflic acid and, as a ligand, triflato). The pure acid itself is very corrosive and care has to be taken in its use. However, it is probably much less dangerous than the perchlorate anion which was previously similarly used, for the latter has a well-known tendency to destroy apparatus and to remove parts of the anatomy—perchlorates are prone to explode. Although substitution of $CF_3-SO_3^-$ into Co^{III} is slow; the preparation of *cis*-$[Co(en)_2(OSO_2CF_3)_2]^+$ from from $[Co(en)_2Cl_2]^+$ requires use of triflic acid at 100 °C for 3 h the triflato ligand is readily displaced by reaction with a replacement ligand in a relatively inert solvent such as acetone. Anhydrous metal triflates may, usually with advantage, replace anhydrous metal chlorides in organometallic chemistry. These compounds may be made, for example, by refluxing the anhydrous chloride with triflic acid.

We have now moved to the area of coordination chemistry in which water is to be avoided as a solvent. This is a general area with many facets. Examples of the preparation of complexes by substitution reactions in non-aqueous media are the following.

1. Potassium thiocyanate melts at 173 °C and may be used as a solvent at temperatures above this. For example, in this medium, water is readily displaced from the $[Cr(H_2O)_6]^{3+}$ ion:

$$[Cr(H_2O)_6]^{3+} + 6NCS^- \xrightarrow[\text{molten KNCS}]{180\,°C} [Cr(NCS)_6]^{3-} + 6H_2O$$

2. As has been seen, refluxing thionyl chloride reacts with water and may be used to prepare anhydrous metal chlorides from the hydrates. Additionally, it is a suitable solvent for the preparation of the chloro anions of metals:

$$2NEt_4Cl + NiCl_2 \xrightarrow[\text{reflux}]{SOCl_2} (NEt_4)_2[NiCl_4]$$

At high temperatures thionyl chloride slowly decomposes to give chlorine and this reduces its usefulness as a solvent because the chlorine may become a reactant.

3. Most salts are converted by bromine trifluoride into the highest fluoride of the element, or, if an alkali metal salt is present, into a fluoro anion. It is so powerful a fluorinating agent that it will even react with metals and alloys. For example, with a 1:1 alloy of silver and gold,

$$AgAu(alloy) \xrightarrow{BrF_3} Ag[AuF_4]$$

4. As will be discussed in more detail in Chapter 15, there are three distinct bonding mechanisms which contribute to the metal–metal bonding in $[Cl_4Re–ReCl_4]^{2-}$, σ, π and δ (the chlorines are all terminal and eclipsed), and so it is an anion of particular interest. It is readily prepared from the commercially available $ReCl_3$, which consists of molecules containing a triangle of rhenium atoms, by fusion in molten (220 °C) diethylammonium chloride:

$$2Re_3Cl_9 + 6(C_2H_5)_2NH_2Cl \rightarrow 3[(C_2H_5)_2NH_2]_2[Re_2Cl_8]$$

4.2.3 Oxidation–reduction reactions

As has been seen, inert complexes of the transition metals may be interconverted by substitution reactions, but such methods cannot generally be relied upon and it is preferable to prepare inert complexes by a different method. The chosen method is to take a compound containing the metal in a different oxidation state and oxidize or reduce it, as appropriate, in the presence of the selected coordinating ligand. This technique is used extensively in the preparation of oxalato complexes of chromium(III). Other chromium-(III) complexes are prepared by the oxidation of chromium(II) salts.

The success of this general preparative method rests on two factors. First, although the product is an inert complex, the starting material is one which is relatively labile. Other things being equal, concentrations used in the preparation approximate to those which maximize the concentration of a complex species identical in composition with the desired product but differing from it in charge. Electron addition or removal (i.e. reduction or oxidation) then gives the product. Secondly, as has been mentioned earlier, there will be several labile complexes in equilibria, each of which can undergo oxidation (or reduction) to give an inert product. In general, the product actually obtained will be derived from that labile complex which is the most readily oxidized (or reduced).

Examples of complexes prepared by oxidation–reduction reactions

1. Some examples of complexes of cobalt(III) prepared by oxidation–reduction reactions have been given earlier, as part of the description of the substitution reactions of the inert compounds formed. However, the preparation of what may be regarded as the parent complex, hexaamminecobalt(III) chloride, was not described. As Werner found, this is made by hydrogen peroxide oxidation of an aqueous solution of cobalt(II) chloride made alkaline with ammonia in the presence of ammonium chloride:

$$2[Co(H_2O)_6]Cl_2 + 2NH_4Cl + 10NH_3 + H_2O_2 \xrightarrow{\text{charcoal}} 2[Co(NH_3)_6]Cl_3 + 14H_2O$$

This reaction is catalysed by the presence of charcoal; in its absence the product consists largely of pentaamminecobalt(III) complexes, the sixth coordination site being occupied by either H_2O or Cl^-. The function of the charcoal is not known with certainty but it is believed that it may act by donating an electron to a pentaamminecobalt(III) ion, converting it momentarily into a labile pentaamminecobalt(II) species into which a further ammonia molecule substitutes.

2. An aqueous solution of oxalic acid and potassium oxalate reduces potassium dichromate to the trisoxalatochromium(III) anion:

$$K_2Cr_2O_7 + 7H_2C_2O_4 + 2K_2C_2O_4 \rightarrow 2K_3[Cr(C_2O_4)_3] + 6CO_2 + 7H_2O$$

3. Complexes containing manganese in less-common formal valence states may be made either by reduction of the permanganate anion, $[MnO_4]^-$, or by oxidation of the hexaquamanganese(II) cation. Sometimes, as in the preparation of the $[MnF_5(H_2O)]^{2-}$ anion, containing manganese(III), the two are combined:

$$8[Mn(H_2O)_6]^{2+} + 2[MnO_4]^- + 25HF_2^- \rightarrow 10[MnF_5(H_2O)]^{2-} + 9H^+ + 46H_2O$$

In this example is seen another reason for using an oxidation–reduction reaction for the preparation of a complex, the non-availability of suitable precursors in which a metal is in the desired valence state.

Systems in which a series of complex ions of identical stoichiometry are interrelated by a series of one-electron oxidations or reductions have been extensively studied. The existence of such a related series is conveniently

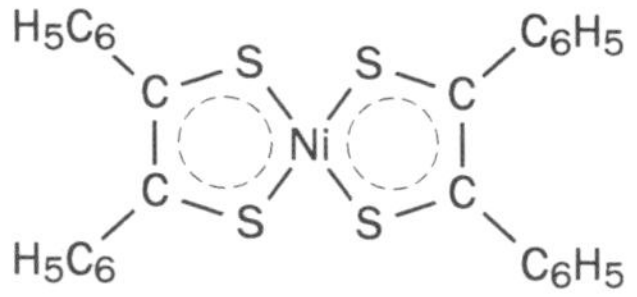

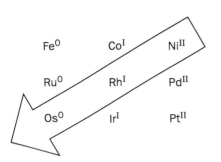

Fig. 4.1 The square planar complex nickel(II) bis(stilbenedithiolate).

investigated by electrochemical methods, of which one is polarography,[2] the half-wave potentials obtained suggesting suitable chemical oxidizing or reducing agents for the bulk preparation of the species. An alternative technique is to carry out the oxidations or reduction electrolytically, preferably at a potential which is held constant despite fluctuations in current. This technique is called *controlled potential electrolysis*; some related techniques will be considered in more detail in Chapter 12. The method can throw up some interesting problems. For example, it has been shown polarographically that the square planar complex of nickel given in Fig. 4.1, undergoes reduction to give the species $[Ni(S_2C_2(C_6H_5)_2)_2]^-$ and $[Ni(S_2C_2(C_6H_5)_2)_2]^{2-}$. There has been considerable discussion of the valence state of the metal atom in these and similar complexes. Is it the metal or is it the ligand which is being reduced, or is it both? It seems that the electrons added in the reduction are delocalized over the whole complex (the stilbenedithiolate ligand is planar and has a delocalized π system) and so it becomes difficult to define the valency state of either the metal or ligand. Such problems will be considered in more depth in Chapter 6.

Although in the above discussion it has been clear that there can be a change of coordination number associated with a change of valence state—tetrahedral chromium and manganese in $[Cr_2O_7]^{2-}$ and $[MnO_4]^-$ becoming octahedral in complexes of the trivalent metals—there was no clear link between the two changes. However, there exist many reactions where there is such a link, one that is easy to see. As has been mentioned, square planar complexes tend to be formed by transition metal ions with the d^8 configuration. There exists a whole series of reactions of such complexes in which the metal is, formally, oxidized by two units, and so becomes d^6, and, simultaneously, increases its coordination number by two, becoming octahedral. The linking of these two changes is signalled by the name given to this type of reaction—*oxidative addition reactions*. A typical example involves Vaska's compound *trans*-$[Ir(PPh_3)_2(CO)Cl]$ (Fig. 4.2).

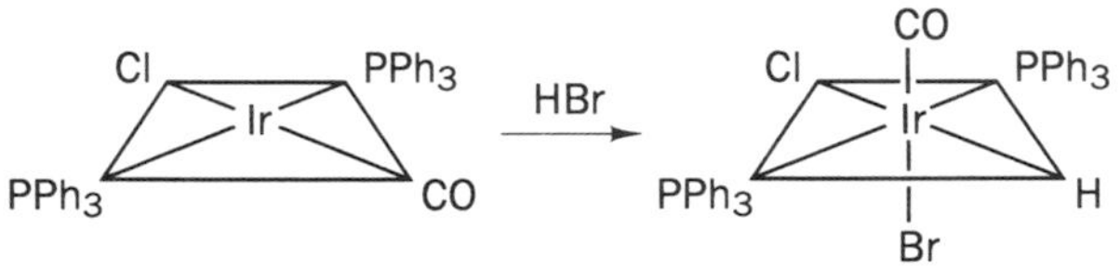

Fig. 4.2 Example of oxidative addition—to give Vaska's compound, *trans*-$[Ir(PPh_3)_2(CO)Cl]$.

Molecules such as H_2, O_2 and SO_2 can replace HBr in this reaction, to give complexes which readily lose these molecules again. The tendency to undergo oxidative addition reactions increases, roughly, from top right to bottom left in the d^8 series (Fig. 4.3).

Fig. 4.3 Series of d^8 configurations, showing the tendency to undergo oxidative addition.

[2] In polarography the potential between two electrodes in a solution is continuously varied and the consequent variation in current recorded. At certain potentials electrolytic reduction of species in solution occurs and the current rises. Because the cathode is very small (either a flow of mercury drops through a capillary or a thin platinum wire) the increase in current is diffusion-limited (and so concentration-dependent) and a characteristic step-like plot of current against voltage results, one step for each reduction process. The voltage corresponding to the mid-point of the riser part of a step is a characteristic of the reduction process and is termed the *half-wave potential*. The height of each step is a measure of the concentration of the corresponding species in solution.

Inevitably, oxidative addition reactions are not confined to these elements, these valence states or, indeed, to these geometries but it is for them that this type of reaction is most important.

Reactions which are, effectively, the reverse of oxidative addition also occur. Not surprisingly, they are known as *reductive elimination reactions*. The decomposition of complexes of Vaska's compound with molecules such as H_2 and O_2, mentioned above, provide obvious examples. Another is provided by elimination of CH_4 from some phosphite ester complexes of cobalt containing H and CH_3 as ligands, in which the cobalt may formally be regarded as changing from d^7, Co^{II} to d^9, Co^0:

$$cis\text{-}[H(CH_3)Co(P(OMe)_3)_4] \rightarrow CH_4 + [Co(P(OMe)_3)_4]$$

4.2.4 Thermal dissociation reactions

By controlled heating, some complexes can be degraded to others, a volatile compound being expelled. For example, when it is heated, ordinary blue copper(II) sulfate loses water in a stepwise manner until above ca 220 °C the anhydrous sulfate is left. The first water molecule to be lost comes from the lattice; subsequent ones come from the coordination sphere of the Cu^{II} and are replaced by sulfate oxygens:

$$CuSO_4 \cdot 5H_2O \xrightarrow{96.5\,°C} CuSO_4 \cdot 4H_2O \xrightarrow{102\,°C} CuSO_4 \cdot 3H_2O$$
$$\xrightarrow{115\,°C} CuSO_4 \cdot H_2O \xrightarrow{220\,°C} CuSO_4$$

Many other complexes behave similarly and heating (usually under vacuum) to a carefully controlled temperature is a useful preparative method. Hydrogen halide elimination, for example, is a reaction which occurs readily for almost any complex which has the electron-donor atom attached to a hydrogen (e.g. H_2O, ROH, NH_3, R_2NH) and the electron acceptor attached to a halogen (e.g. BF_3, $SnBr_4$, $FeCl_3$). So, in the preparation of a complex such as the first given in this chapter, $[BF_3(NH_3)]$, the product readily eliminates hydrogen fluoride to give a series of compounds and ultimately the polymeric solid BN. Another very common thermal reaction is the expulsion of one or more neutral ligands (as in the case of copper(II) sulfate), with a consequent reduction in the apparent coordination number of the central atom. In fact, quite often, some previously monodentate ligand (generally an anion) becomes either bidentate or a group that bridges two metal atoms (the acetate and halide anions, respectively, exemplify these two cases). Another possibility is that an anion, initially non-coordinated, becomes attached to the metal. The actual coordination number of the central atom is seldom reduced. Examples have already been given in Section 3.4.6.

Heating to a relatively high temperature can lead to the complete dissociation of the complex species. For example

$$K[BF_4] \rightarrow BF_3 + KF$$

and

$$BrF_2[RuF_6] \rightarrow BrF_3 + RuF_5$$

In the absence of some other suitable cationic species, complex fluorides containing the $[BrF_2]^+$ cation are formed when transition metals are dissolved in BrF_3. Thermal decomposition of these salts is a convenient way of making small quantities of many fluorides.

Another example of a complex prepared by a thermal dissociation reaction is the preparation of cis-$[Cr(en)_2Cl_2]^+$ by heating $[Cr(en)_3]Cl_3$:

$$[Cr(en)_3]Cl_3 \xrightarrow{210\,°C} cis\text{-}[Cr(en)_2Cl_2]Cl + en$$

The temperature of reaction has to be carefully controlled to achieve maximum yields; the reaction is complete in 2–3 h. In an analogous reaction trans-$[Cr(en)_2(SCN)_2]SCN$ is prepared by heating $[Cr(en)_3](SCN)_3$ to $130\,°C$.

Two experimental techniques have been much used to study the preparation of complexes by the thermal dissociation reactions.

Differential thermal analysis (DTA)

Two cells, one containing the complex under study and the other a similar amount of a thermally stable material, are slowly heated, each receiving an identical amount of thermal energy. The temperature difference between the two cells is measured and remains essentially zero until thermal dissociation of the complex occurs. Both the temperature (at a particular pressure, usually atmospheric) and enthalpy of dissociation may be obtained by this technique.

Thermogravimetric analysis (TGA)

The weight of a complex is measured as its temperature is raised. When thermal dissociation occurs the weight loss due to any volatile ligand expelled is measured. The empirical formula of the product may usually be deduced if the identity of the ligand expelled is known. Both dissociation pressure and temperature are recorded as they are, of course, interrelated.

4.2.5 Preparations in the absence of oxygen

It has been seen earlier that it may be necessary to work in the absence of water to prepare some complexes. An even wider range becomes accessible if we work in the absence of air—much of the field of organometallic chemistry, for example. Not that it is impossible to prepare organometallic compounds in the presence of air—Zeise's salt $K[PtCl_3(C_2H_4)]H_2O$, the first organometallic compound made, was prepared with no attempt to exclude air. However, it must be admitted that in this preparation the ethene itself serves to provide an inert atmosphere. Nowadays, it is prepared by bubbling ethene through a solution containing $[PtCl_4]^{2-}$ in strong hydrochloric acid (best with a trace of Sn^{II} as catalyst) for a few hours. Rather similar is the cyclooctene, C_8H_{14}, complex which serves as a precursor for many complexes of rhodium(I), $[RhCl(C_8H_{14})_2]_n$, made by allowing commercial hydrated $RhCl_3$ to stand for a week with cyclooctene in 2-propanol (which is oxidized to acetone in the reaction) in a flask filled with nitrogen.

Such one-step reactions are the exception rather than the rule. More commonly, it is necessary to carry out a series of successive reactions and procedures, such as refluxing, distilling, crystallization, filtration and washing in an inert atmosphere. The most evident way of doing this is to work in an inert-atmosphere filled glove box using conventional apparatus. This is both possible and common, but the volume of a typical box is such that it is rather difficult to reduce and maintain the oxygen concentration at an acceptable level. There exist sophisticated boxes with recirculation of the inert gas through oxygen-removing trains and entrance ports which can be thoroughly evacuated, but they are very expensive. At the other extreme, it has been pointed out that it is possible to work with a cheap transparent plastic glove, in which the fingers are used to store and mix reagents, nitrogen being passed in through the sealed-off wrist.

Many workers make use of so-called Schlenk tube techniques. This is the name given to a whole series of simple, but versatile, devices that enable reactions to be carried out in an essentially closed apparatus of low volume. Examples include the use of septum caps (those used to close the vials containing materials used for medical injections). Solvents can be taken in and out of the apparatus using hypodermic needles; nitrogen can be passed in through such a needle and allowed to escape through another; on removal of all needles the apparatus is automatically sealed. Filter sticks are used—glass tubes with a glass sinter sealed halfway down. If one is on top of the solution to be filtered, inversion of the apparatus (perhaps with gentle use of nitrogen gas pressure) leads to filtration. Solids can be placed in a limb of a tube which has a bend of ca. 90° in its middle and which is held with the angle at the top. When the solid is needed for reaction then, if the solid is in the left-hand limb, rotation by 180° about the right-hand limb causes the contents of the two limbs to mix. Tubes are interconnected by taps (often greaseless) so that alternative routes exist for gas and/or liquid flow and add to the versatility.

An example of a series of reactions involving Schlenk tubes is the preparation of $(CH_3)NGaH_3$. $GaCl_3$ is dissolved in diethylether and slowly added to a slurry of LiH in diethylether through a greaseless valve. After reaction, the product is filtered through a filter stick. To the filtrate ($Li[GaH_4]$ in diethylether) is added $[(CH_3)_3NH]Cl$ by the rotating arm technique. After vacuum removal of the solvent, the product is separated by vacuum sublimation from the reaction mixture.

Reactions involving metals, either bulk or, more commonly, finely divided, are an entry point for organometallic complexes of transition metals. Examples of direct reaction of a metal are

$$Ni + 4CO \rightarrow Ni(CO)_4$$

and

$$Fe + 5CO \rightarrow Fe(CO)_5$$

It should be noted that classical complexes of low-valence states can also be prepared from metals; thus, the simplest starting point for the preparation of pure complexes of chromium(II) is by the action of an oxygen-free acid on the pure metal in a inert atmosphere. More commonly, however, a reduction reaction is used in the preparation of organometallic complexes because in them the metal is usually in a low formal oxidation state. Examples of reduction reactions include those in which gaseous carbon monoxide is both reducing agent and reactant. For instance, when CO is passed over solid $RhCl_3$ at about 100° C, the chlorine-bridged dimer

$[Rh(CO)_2Cl]_2$ sublimes into the cold parts of the reaction vessel:

$$2RhCl_3 + 6CO \rightarrow [Rh(CO)_2Cl]_2 + 2COCl_2$$

This process can be carried further, in which the product reacts with carbon monoxide in hexane in the presence of a mild alkali:

$$2[Rh(CO)_2Cl]_2 + 6CO + 2H_2 \rightarrow Rh_4(CO)_{12} + 2CO_2 + 4HCl$$

The product is a black crystalline material. The molecules consist of a tetrahedron of rhodium atoms, each rhodium atom being bonded to the other three rhodiums and to three CO ligands, $[Rh(CO)_3]_4$ (this compound will be the subject of discussion in Chapter 15).

This same pattern was noted earlier, when describing the preparation of Cr^{III} complexes from chromates: that, if at all possible, a ligand is also used as reducing agent. Other examples are the reaction of rhodium trichloride with triphenylphosphine in hot ethanol to give Wilkinson's compound, $[RhCl(P(C_6H_5)_3)_3]$:

$$RhCl_3 + 4P(C_6H_5)_3 \rightarrow [RhCl(P(C_6H_5)_3)_3] + Cl_2P(C_6H_5)_3$$

This product is important in organic chemistry because it reversibly absorbs H_2 under normal laboratory conditions and catalyses the hydrogenation of alkenes and alkynes. Another example is the reduction of the $[OsCl_6]^{2-}$ anion, which is commercially available, with hydrazine hydrate, a liquid, without additional solvent, to give complexes with one and two coordinated N_2 molecules:

$$5(NH_4)_2[OsCl_6] + 13NH_2NH_2H_2O \rightarrow$$
$$4[Os(NH_3)_5(N_2)]Cl_2 + cis\text{-}[Os(NH_3)_4(N_2)_2]Cl_2 + 20HCl + 13H_2O$$

A very important compound in Ni^0 chemistry is bis(1,5-cyclooctadiene)-nickel(0) for the ligands are readily displaced. It is made by the reduction of $[Ni(acac)_2]$ by an aluminium alkyl, usually Al_2Et_6, in the presence of 1,5-cyclooctadiene using toluene as a solvent. The reaction occurs at room temperature or below, the product precipitating. An example of the use of this compound in synthesis is in the preparation of $Ni(PPh_3)_4$; the latter is formed at room temperature or below by the addition of triphenylphosphine in hexane.

As this last synthesis illustrates, the majority of reactions of transition metal organometallic compounds involve a starting material which is already in a low-valence state. As another example, hexacarbonylchromium reacts with arenes in a high-boiling solvent to give complexes with a π-bonded benzene ring:

$$Cr(CO)_6 + C_6H_6 \rightarrow [Cr(CO)_3(C_6H_6)] + 3CO$$

It is often a characteristic of organometallic reactions that they lead to a wide variety of products. For example, $Co_2(CO)_8$ reacts with CS_2 to give about 20 identified products; $Fe(CO)_5$ reacts with diphenylethyne, $(C_6H_5)C\equiv C(C_6H_5)$, to give about the same number. In such cases chromatographic methods are used to separate the individual component products.

Main group organometallics are usually prepared from the corresponding Grignard, lithium or mercury organic species,

$$[SnCl_4(Et_2O)_2] \xrightarrow[Et_2O]{EtMgBr} SnEt_4$$

$$2Al + 3HgEt_2 \rightarrow Al_2Et_6 + 3Hg$$

and similar methods are used for some transition metal compounds; the product in the example below is only stable at low temperatures:

$$3LiMe + WCl_6 \rightarrow WMe_6$$

(it is found to be important to use a 3:1 reactant ratio; the yield is only ca. 50%). Closely related are reactions in which a hydrocarbon, such as cyclopentadiene, C_5H_6, which has an acidic hydrogen (loss of a proton gives the $C_5H_5^-$ anion, with an aromatic π system), reacts either with a metal or strong alkali to give the anion. This is then commonly reacted with a metal halide or metal halide complex:

$$2C_5H_6 + 2K \xrightarrow{THF} 2KC_5H_5 + H_2$$

$$4KC_5C_5 + UCl_4 \rightarrow U(C_5H_5)_4 + 4KCl$$

Sometimes, the steps are contained within one reaction mixture, as in the preparation of ferrocene:

$$2KOH + 2C_5H_6 + FeCl_2 \cdot 4H_2O \rightarrow Fe(C_5H_5)_2 + 2KCl + 6H_2O$$

4.2.6 Reactions of coordinated ligands

It is almost a tautology to say that substitution reactions of inert complexes proceed slowly at room temperature; however, there are exceptions. For example, as has been mentioned, addition of acid to the $[Co(NH_3)_5CO_3]^+$ cation leads to the rapid evolution of carbon dioxide and formation of the complex ion $[Co(NH_3)_5OH_2]^{3+}$. Indeed, a whole series of facile interconversions exists between species containing Co–OH$_2$, Co–OH, Co–CO$_3$, Co–SO$_3$, Co–NO$_2$, and similar bonds. The explanation for this non-typical behaviour is that in none of these reactions is a Co–O bond broken (although, because they are commonly written as above, it is not immediately obvious that all of them contain Co–O bonds). Note that the NO_2^- anion may also bond through the nitrogen (see Section 3.4). It was mentioned earlier that in the reaction of $[Co(NH_3)_5OCO_2]^+$ with acids, it is the O–C bond that is broken, not the Co–O. Another reaction of the same type is:

$$[Cr(NH_3)_5(H_2O)]^{3+} \xrightarrow{HNO_2} [Cr(NH_3)_4(ONO)]^{2+}$$

and a rather unusual reaction which proceeds in aqueous solution is

$$[Co(NH_3)_5NCS]^{2+} \xrightarrow{H_2O_2} [Co(NH_3)_6]^{3+}$$

(the mechanism of this reaction does not seem to be known).

So far we have considered only reactions of an atom directly bonded to a transition metal. There has been much work on the more remote modification of a ligand. One relatively simple complex which has been the

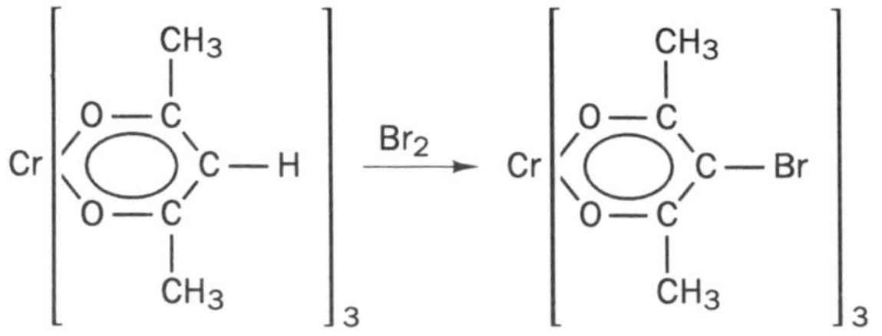

Fig. 4.4 Bromination of the ligand in trisacetylacetonatochromium(III).

subject of extensive study is trisacetylacetonatochromium(III). The action of bromine in acetic acid leads to bromination of the acetylacetone ring (Fig. 4.4).

It is reported that the N-halosuccinimides are the best agents for halogenating coordinated acetylacetonate rings. All the available evidence indicates that it is the coordinated ligand which reacts, and not any free ligand in equilibrium with it. Examples of other groups which have been used to replace the active hydrogen atoms in metal acetylacetonates are $-NO_2$, $-NH_2$, $-N_3^+$, $-CHO$, $-COCH_3$ and $-SCl$.

An important group of reactions of coordinated ligands which does involve the metal–ligand bond are the so-called *insertion reactions*. They are often of importance in the catalysis of organic reactions by transition metal complexes, although insertion reactions have a history dating back to the mid-19th century in the reaction

$$SbCl_5 + 2C_2H_2 \rightarrow Cl_3Sb(CHCHCl)_2$$

More relevant to the present topic is the reaction

$$CH_3Mn(CO)_5 + CO \rightarrow CH_3C(O)Mn(CO)_5$$

Isotopic and kinetic evidence have demonstrated unambiguously that the CO which enters the $Mn–CH_3$ bond is one of those already attached to the manganese; there is an equilibrium:

$$CH_3Mn(CO)_5 \rightleftharpoons CH_3–COMn(CO)_4$$

Another important, but often difficult to prove, example is the insertion of an alkene into a M–H bond to give an alkyl:

$$\begin{array}{c} CH_2\!\!=\!\!CH_2 \\ \downarrow \\ M–H \rightarrow M–CH_2–CH_3 \end{array}$$

Such reactions of coordinated ligands are important in the organometallic chemistry of transition elements; they may proceed in the opposite direction, an alkene hydride being formed from an alkyl. A trick widely adopted to prevent this reverse reaction is to replace the β hydrogens by alkyl groups or the β carbon by a silicon (with alkyl groups attached).

Another example is provided by the attack of alkoxide ions on coordinated CO to give $M–CO_2R$ groups:

$$[Ir(CO)_3(P(C_6H_5)_3)_2]^+ + CH_3O^- \rightarrow [Ir(CO)_2(CO_2CH_3)(P(C_6H_5)_3)_2]$$

Often, insertion reactions involve, formally, H^+ or H^- and may be difficult to distinguish from alkene insertion reactions of the type given above. Examples are given, firstly, by the reaction of the cobalticinium cation with (boro)-hydride,

$$[(\eta^5\text{-}C_5H_5)_2Co]^+ + H^- \rightarrow [(\eta^5\text{-}C_5H_5)Co(\eta^4\text{-}C_5H_6)]$$

and, secondly, the protonation of σ-allylic complexes (Fig. 4.5).

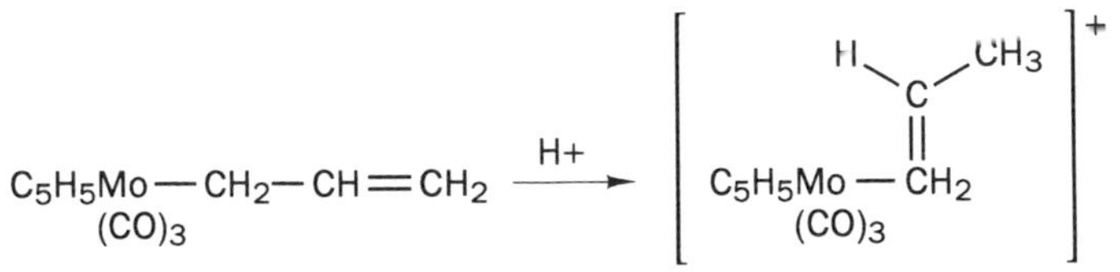

Fig. 4.5 Protonation of a σ-allylic complex.

$$\underset{(CO)_3}{C_5H_5Mo}-CH_2-CH=CH_2 \quad \xrightarrow{H+} \quad \left[\begin{array}{c} \overset{H}{\underset{}{\diagdown}} \underset{C}{\overset{}{\diagup}} \overset{CH_3}{} \\ \| \\ \underset{(CO)_3}{C_5H_5Mo}-CH_2 \end{array} \right]^+$$

A quite different, but very important, class of reaction of coordinated ligands is in the synthesis of new coordinated ligands. An excellent example is provided by the synthesis of cobalt sepulchrate (cf. Table 2.5) from $[Co(en)_3]^{3+}$ by the action of formaldehyde and ammonia in a basic medium (aqueous lithium carbonate), the reaction taking about 2 h to reach completion at room temperature (Fig. 4.6). The reaction occurs without rupture of the Co–N bonds. It seems that nucleophiles with the same C_{3v} local symmetry as ammonia may replace it in this synthesis—so CH_3–NO_2 leads to cages with $\equiv$C–NO_2 apices; the NO_2 group can be reduced, opening the

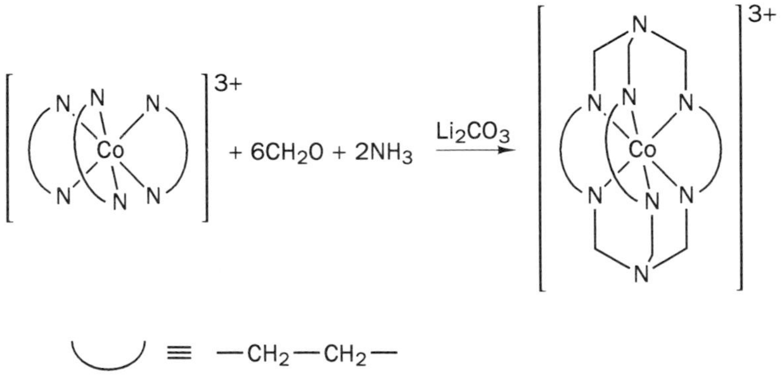

$$\left[\begin{array}{c} \\ \text{Co} \\ \\ \end{array} \right]^{3+} + 6CH_2O + 2NH_3 \quad \xrightarrow{Li_2CO_3} \quad \left[\begin{array}{c} \\ \text{Co} \\ \\ \end{array} \right]^{3+}$$

Fig. 4.6 Reaction of coordinated ligands to prepare cobalt supulchrate.

$$\smile \quad \equiv \quad -CH_2-CH_2-$$

way to an extensive study of the chemistry of one group of coordinated ligands. Syntheses of the type just described are often called *template reactions* for an obvious reason—the metal together with the ligands already in place form a template for the ligand to be created.

A rather important group of complexes which may also be made by template synthesis are those of imines. Imines are formed by condensation of an amine and a carbonyl (Fig. 4.7). The amine can be coordinated to a metal and the above reaction still proceeds, the amine (or, more usually, diamine) remaining coordinated. Typical is the reaction between bisethylene-diaminenickel(II) chloride and acetylacetone. An aqueous solution of the mixture, to which a few drops of pyridine have been added, is refluxed

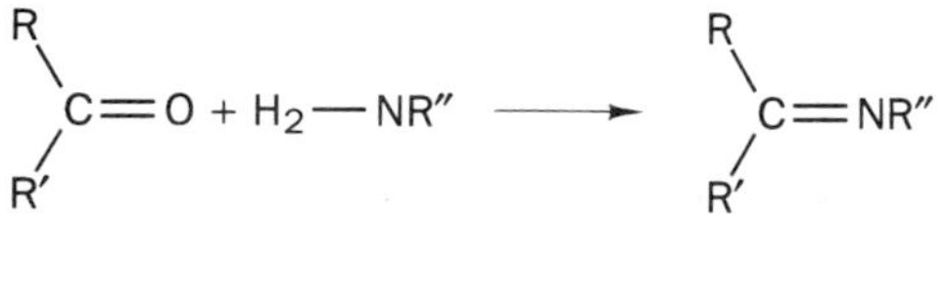

Fig. 4.7 Condensation of an amine and a carbonyl to form an imine.

$$\underset{R'}{\overset{R}{\diagdown}}C=O + H_2-NR'' \quad \longrightarrow \quad \underset{R'}{\overset{R}{\diagdown}}C=NR''$$

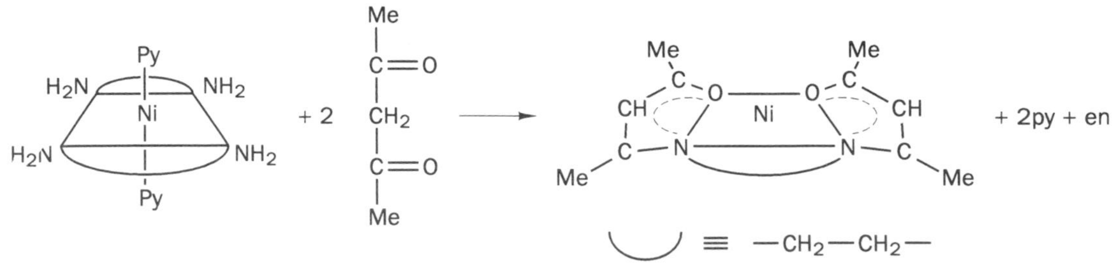

Fig. 4.8 Reaction between bisethylenediaminenickel(II) chloride and acetylacetone.

for 2 h to give the product. The reaction is probably between $[Ni(en)_2(py)_2]^{2+}$ and the acetylacetone (Fig. 4.8). Template syntheses are finding particular applicability to the synthesis of large, sometimes complicated, ligands. The improvement in yield of ligand, compared with an organic synthesis in the absence of metal, is sometimes quite spectacular.

Many reactions are known in which the presence of a metal ion influences the products of a reaction and the explanation for this may well lie in the different reactions of free and coordinated ligands. Peptide chemistry is one field in which this may prove to be of great importance.

4.2.7 The *trans* effect

The ligand arrangement around an atom after a substitution reaction may or may not be similar to that of the starting material, even for inert complexes. An example of such a change is provided by the formation of *trans*-$[Cr(en)_2(NCS)_2]^+$ on heating $[Cr(en)_3]^{3+}$ with solid ammonium thiocyanate at 130 °C. Similarly, *cis*-$[Co(NH_3)_4(H_2O)Cl]SO_4$ is converted into *trans*-$[Co(NH_3)_4Cl_2]HSO_4$ by the action of a mixture concentrated hydrochloric and sulfuric acids at room temperature.

The chemistry of platinum(II) and, to a lesser extent, those of Pd^{II} and Au^{III} (all three form d^8 square planar complexes) is therefore noteworthy in that the major product of a substitution reaction can be predicted with confidence. This is because the lability of a ligand bonded to platinum(II) is largely determined by the group which is *trans* to it and not by the nature of the ligand itself. Although this *trans* effect is not fully understood its operation is reasonably reliable and renders the synthesis of platinum(II) complexes, in particular, a class on its own. The stereochemistry of the products of reactions of platinum(II) complexes can often be varied by altering the order of reagent addition. An example of this is provided by the synthesis of *cis*- and *trans*-$[Pt(NH_3)(NO_2)Cl_2]^-$ starting from $[PtCl_4]^{2-}$ (Fig. 4.9).

Ligands can be arranged in a series depending on the relative magnitude of the *trans* effect which they exert. In view of the above discussion, it would be expected that different sources would agree on the sequence of ligands which corresponds to increasing *trans* effect. In fact, there is little unanimity, some lists appearing quite aberrant. The following sequence, however, is largely accepted:

$$F^- \approx OH^- \leq NH_3 \leq py < Cl^- < Br^- < I^- < -SCN^- \approx -NO_2^- < thiourea$$
$$< PR_3 \approx AsR_3 \approx H^- < CO \approx CN^- \approx C_2H_4$$

One can see from this list the difficulty of proposing a general explanation

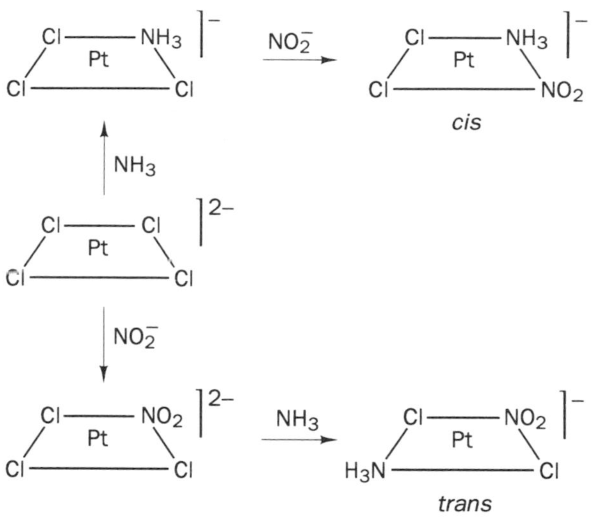

Fig. 4.9 The sequence of reagent addition can be altered to selectively provide *cis* and *trans* products from the same starting complex.

for the *trans* effect. Almost all those ligands exerting a strong *trans* effect are π bonding, but a π bonding explanation does not explain the position of H^- (a polarization model, based on the unique characteristics of the H atom is usually added to cover this). However, there seems no explanation of why the effect is largely confined to Pt^{II}, perhaps along with Pd^{II} and Au^{III}— certainly, neither the π bonding nor the polarization model is metal-specific. It could be that the notion that the *trans* effect is largely confined to Pt^{II} is incorrect; it is just that other elements have not been so extensively studied. Although this point has substance as far as Au^{III} and Pd^{II} are concerned, it does not seem to have general validity. Above all, it must be remembered that the *trans* effect is a kinetic effect, associated with bond breaking and formation. It could be more a phenomenon of the reaction pathway (an activated complex pathway, an activated complex or transition state) than the ground state. In Chapter 14 the kinetics of the reactions of square planar Pt^{II} complexes will be the subject of some discussion, a discussion that will include the *trans* effect.

4.2.8 Other methods of preparing coordination compounds

The title of this section rather overstates its contents. One could write endlessly on the subject. Rather, its purpose is to emphasize the variety of techniques available but which have not been mentioned elsewhere in this chapter. Ways of avoiding decomposition or, rather, reaction, in the presence of oxygen and water have been mentioned. No less important is the ability to avoid thermal decomposition. There are two related techniques available here. In both, reactants are cocondensed on a cold surface, cooled to anything from the temperature of liquid helium upwards. If the spectroscopic properties of the unstable species are the point of interest then the reactants are condensed along with an inert diluent, typically a noble gas such as argon, to give the product in a matrix of the noble gas—the so called *matrix-isolation* technique. If the preparation of large quantities is of more interest then no matrix is used. Typically, metal atoms are evaporated into the high vacuum chamber that holds the cooled surface, typically cooled with liquid nitrogen. Many heating techniques—resistive heating, induction heating, laser ablation are a few—may be used to generate the metal atoms. Along with the metal, but perhaps in alternate pulses, are condensed the chosen ligand(s). By such techniques tin carbonyls, compounds unstable at temperatures well below room temperature, have been prepared and characterized. Paradoxically enough, such metal atom synthesis methods can also provide a convenient high yield route to compounds which are stable at room temperature; commercial equipment is available for those wishing to work on a large scale. A simple version is also available for use in student laboratories. It is methods such as these which are being used to prepare fullerene, C_{60}, and related species. This soccer-ball-like molecule can encapsulate some ions—Sr^{2+} and La^{3+} are two examples—if suitable sources for them are present in the carbon rods from which the fullerene is prepared (they are evaporated by an electric arc struck between them to give a low yield of fullerenes). The bonding is these so-called *endohedral* molecules is discussed in Chapter 10.

Photochemical methods, usually irradiation with ultraviolet light from a mercury discharge lamp, but sometimes visible light too, have long been used in the preparation of coordination compounds. They depend on the fact that the chemistry of an electronically excited molecule is different from that of the same molecule in its electronic ground state. However, this can only be exploited synthetically if the excited state lives long enough for chemistry to be performed on it—and most excited states have very short lifetimes because excited molecules give up their extra energy to other molecules in a wide variety of processes. It is likely, therefore, that the success of photochemical methods depends, in part, upon there being an insulated step down the ladder of energy changes that lead to deactivation; a level of long lifetime, from which deactivation is slow. Such levels exist—lasers depend on them for their action—but it is difficult to predict them and so to predict whether photochemical methods will lead to new compounds or just to the destruction of those already present. One compound which is always made photochemically is the carbonyl $Fe_2(CO)_9$, by the reaction

$$2Fe(CO)_5 \xrightarrow{h\nu} Fe_2(CO)_9 + CO$$

In recent years it has become clear that there exist complexes which contain the H_2 molecule as a ligand; not two separated H atoms, but H_2. One method by which such complexes might be prepared is by the action of H_2 on suitable coordination compounds. Unfortunately, there is a problem—the solubility of H_2 in most solvents is rather low, so high pressures of H_2 are needed. Not surprisingly, most workers would prefer to work with low pressures of H_2 than with high. It is here that the ingenuity of experimentalists becomes apparent. Supercritical fluids have properties which in some ways resemble those of liquids. For example, they can act as solvents for coordination compounds, and in some ways they resemble gases in that they are miscible with gases, usually over the entire concentration range. Here, then, is a solution to the problem—study the reaction of coordination compounds with H_2 using supercritical fluids as solvents. The pressures needed to maintain supercriticality are often modest—a few tens of atmospheres for carbon dioxide or xenon, for instance. At the time of writing, this is an area in its infancy but it could lead to important new developments in the preparation of coordination compounds.[2]

Finally, the solid state. The preparation of solid state compounds by high temperature synthesis is a long established method. Unfortunately, the available techniques have been rather limited—grind the reactants together to a fine powder, heat, regrind, reheat and so on until uniformity is reached—is a typical procedure. But interest in the solid state is growing rapidly. For instance, some simple inorganic materials show long-range structural correlations which are difficult to understand or reproduce (for instance, despite the simple picture presented in most introductory inorganic textbooks, ZnS has been found to crystallize in several hundred different, but related, crystal structures, although there is no known method of controlling which form is produced). Again, the discovery of ceramic high-temperature superconductors, (a topic which is dealt with in Appendix 10) incorporating Cu^{III}, a rather unusual valence state, has sparked off a search for similar novel properties of solid state materials containing metal ions in unusual valence states.

There have been developments in synthetic methods. First, related to the 'heat and grind' method, gel/colloid methods of producing reaction precursors are giving much more control and more reproducibility of the final product. Secondly, hydrothermal methods are finding utility. In these, reactions are carried out at high temperature and pressure conditions in the presence of a solvent (not necessarily water, although this has been most commonly used). For instance, $Fe^{II}Fe_2^{III}F_8 \cdot H_2O$ was prepared in this way using liquid HF as a solvent.

[2] See M. Poliakoff, S. M. Howdle and S. G. Kazarian, *Angew. Chem., Int. Ed.* (1995), **34**, 1275.

Further reading

General

Whilst many textbooks of practical inorganic chemistry provide useful information, the most general source, used for much of the material in the present chapter, is the series Inorganic Syntheses that McGraw Hill started in 1939 and is still continuing, although now published by J. Wiley (over 20 volumes have been published).

Another source, almost as valuable, is *Handbook of Preparative Inorganic Chemistry*, G. Bauer (ed.) in two volumes, Academic Press, London 1963 and 1965.

A standard reference is D. F. Shriver, *The Manipulation of Air-Sensitive Compounds*, McGraw Hill, New York, 1969.

Other useful texts include:

- D. M. Adams and J.B. Raynor, *Advanced Practical Inorganic Chemistry*, Wiley, London, 1965.
- G. Pass and H. Sutcliffe, *Practical Inorganic Chemistry*, Chapman and Hall, London, 1974.
- R. J. Angelici, *Synthesis and Technique in Inorganic Chemistry*, Saunders, Philadelphia, 1977.

Specific

- *Reactions of Coordinated Ligands* (2 volumes), P. S. Braterman (ed.), Plenum, New York, 1987.

- E. C. Constable, *Metals and Ligand Reactivity*, E. Horwood, New York, 1990.

A recent discussion of template reactions is 'Template Syntheses', a long review by R. Hoss and F. Vögtle in *Angew. Chem., Int. Ed.* (1994), *33*, 375.

A useful source is Volume 1 of 'Comprehensive Coordination Chemistry', G. Wilkinson, R. D. Gillard and J. A. McCleverty (eds.), Pergamon Press, Oxford, 1987, Chapter 7.4 ('Reactions of Coordinated Ligands' by D. St. Black) and 7.5 ('Reactions in the Solid State' by H. E. LeMay Jr.).

Other useful texts include:

- F. Basolo and R. G. Pearson, 'The *trans* effect in Metal Complexes', *Prog. Inorg. Chem.* (1962) *4*, 381.
- J. R. Blackborow and D. Young, *Metal Vapour Synthesis in Organometallic Chemistry*, Springer-Verlag, Berlin, 1979.
- A. Rabenau, 'The Role of Hydrothermal Synthesis in Preparative Chemistry', *Angew. Chem., Int. Ed.* (1985) *24*, 1026.

Questions

4.1 The cation $[Mo(H_2O)_6]^{3+}$ can be prepared by dissolving $K_3[MoCl_6]$ in 0.5 M p-toluenesulphonic acid (which behaves similarly to the triflic acid discussed in the text) and allowing the solution to stand at room temperature in the absence of O_2, for a day. Suggest why this sequence is more successful than the action of H_2O on $K_3[MoCl_6]$.

4.2 Heating cis-$[Co(en)_2Cl_2]Cl$ in triflic acid, CF_3SO_3H (TH), at 100 °C for 3 h gives cis-$[Co(en)_2T_2]T$. Surprisingly, it was found that $trans$-$[Co(en)_2Cl_2]Cl$ also gives the cis product when heated with TH. Suggest a possible synthetic exploitation of this observation.

4.3 In the preparation of the complexes $Cr(CO)_5X$ (X = sulfur donor ligand) the recommended method is to UV-irradiate a solution of $Cr(CO)_6$ in THF (tetrahydrofuran) in an oxygen-free apparatus to give the complex $[Cr(CO)_5(THF)]$. This is followed by reaction with the ligand X. Direct reaction between $Cr(CO)_6$ and X generally leads to sulfur-bridged complexes. Explain the thinking behind this route to $Cr(CO)_5X$.

4.4 One procedure for the preparation of $Ir_4(CO)_{12}$ is to start with commercially available $Na_2[IrCl_6]$ in ethanol and reduce it to the $[IrCl_6]^{3-}$ ion with I^-. Gaseous CO is then passed into the solution, and together with the addition of solid K_2CO_3, leads to the product, a black solid. Give a retrospective rationale of this sequence.

5

Stability of coordination compounds

5.1 Introduction

The statement that a compound is stable is rather loose, for several different interpretations may be placed upon it. Used without qualification it means that the compound exists and, under suitable conditions, may be stored for a long period of time. However, a statement such as 'a compound is stable in water' may mean one of two things, either that there is no reaction with water which would lead to a lower free energy of the system (thermodynamic stability) or that, although a reaction would lead to a more stable system, there is no available mechanism by which the reaction can occur (kinetic stability). For example, there may not be enough energy available to break a strong bond, although once broken it could be replaced by an even stronger one. As we have seen, boron trifluoride forms a stable complex with trimethylamine, $[BF_3(N(CH_3)_3)]$. A similar complex is formed with trisilyl-amine, $[BF_3(N(SiH_3)_3)]$, which is thermodynamically unstable with respect to the reaction

$$[BF_3(N(SiH_3)_3)] \rightarrow (BF_2)N(SiH_3)_2 + SiH_3F$$

The complex $[BF_3(N(SiH_3)_3)]$ can be prepared and stored at low temperatures ($\approx -80\ ^\circ C$) since the decomposition then proceeds very slowly—at this temperature the complex is kinetically fairly stable. At room temperature the complex is kinetically unstable and the rate of decomposition is much greater. This is the key distinction made in Chapter 4 between kinetically inert and kinetically labile complexes. There it was pointed out that the species which crystallizes from a solution of a mixture of related labile complexes depends not only on the cation and ligand concentration but also on the solvent and crystallization temperature. Although it may be a relatively minor component in the solution, the least soluble complex is probably the one which crystallizes. In the solution there is a series of equilibria such that, if one component crystallizes, the concentrations of the others also change. This chapter is devoted to a discussion of the stability constants which characterize such equilibria.

5.2 Stability constants

When a complex is formed by the reaction[1]

$$M + L \rightleftharpoons ML$$

the equilibrium constant K_1 for the complex containing a single ligand will be

$$K_1 = \frac{[ML]}{[M][L]}$$

where, for the moment, activity coefficients of unity have been assumed. If ML adds a further molecule of L,

$$ML + L \rightleftharpoons ML_2$$

then the equilibrium constant for the complex containing two ligand molecules is

$$K_2 = \frac{[ML_2]}{[ML][L]}$$

In general, the equilibrium constant for the formation of the complex ML_n from ML_{n-1} will be

$$K_n = \frac{[ML_n]}{[ML_{n-1}][L]}$$

The equilibrium constants $K_1, K_2, \ldots, K_n$ are known as *stepwise formation constants*. Alternatively, one may consider the equilibrium constant for the overall reaction

$$M + nL \rightleftharpoons ML_n$$

as

$$\beta_n = \frac{[ML_n]}{[M][L]^n}$$

β_n is known as the nth *overall formation constant*. β_n is related to the stepwise formation constants $K_1, \ldots, K_n$ by

$$\beta_n = K_1 \times K_2 \times \cdots \times K_n$$

That is

$$\beta_n = \prod_{l=1}^{n} K_l$$

---

[1] Throughout this chapter we shall often not specify the charges on the species in reactions or equilibria. Square brackets are used to indicate both the concentrations of complex species and the species themselves. It will be clear from the context which is intended.

Table 5.1 Typical stability constant data for monodentate ligands. All values are logarithmic so, for the Sn^{2+}/Cl^- system, $\log K_1 = 1.51$

Metal ion	Ligand	Stability constants					
Sn^{2+}	Cl^-	$K_1 = 1.51$	$K_2 = 0.73$	$K_3 = -0.21$	$K_4 = -0.55$		
Pd^{2+}	Cl^-	$K_1 = 6.1$	$K_2 = 4.6$	$K_3 = 2.4$	$K_4 = 2.6$	$K_5 = -2.1$	$K_6 = -2.1$
Ni^{2+}	NH_3 (30 °C)	$K_1 = 2.67$	$K_2 = 2.12$	$K_3 = 1.61$	$K_4 = 1.07$	$K_5 = 0.63$	$K_6 - 0.09$ $\beta_6 = 8.01$
Cu^{2+}	NH_3 (30 °C)	$K_1 = 3.99$	$K_2 = 3.34$	$K_3 = 2.73$	$K_4 = 1.97$		
Mn^{2+}	F^-	$K_1 = 5.52$	$\beta_2 = 9.04$	$\beta_3 = 11.64$	$\beta_4 = 13.4$	$\beta_5 = 14.7$	$\beta_6 = 15.5$
Pb^{2+}	I^-	$K_1 = 1.98$	$\beta_2 = 3.15$	$\beta_3 = 3.81$	$\beta_4 = 3.75$	$\beta_5 = 3.81$	
Fe^{2+}	CN^-	$\beta_6 = 24$					
Fe^{3+}	CN^-	$\beta_6 = 31$					

These data refer to 25 °C unless otherwise stated and to zero ionic strength. As a comparison, the data for Ni^{2+}/NH_3 (30 °C) in 2 M NH_4NO_3 are:

$$K_1 = 2.78 \quad K_2 = 2.27 \quad K_3 = 1.65 \quad K_4 = 1.31 \quad K_5 = 0.65 \quad K_6 = 0.08$$

Notice, particularly, the effect of the change on K_6. The entries included in this table have been chosen to illustrate the variety of formats that are encountered and yet to be internally self-explanatory. As an example, the statement that $K_1 = 1.51$ for the Sn^{2+}/Cl^- system is to be interpreted as

$$\frac{[SnCl^+]}{[Sn^{2+}][Cl^-]} = 10^{1.51} = 32.6 \text{ L mol}^{-1}$$

There are the same number of overall formation constants as stepwise formation constants:

$$\beta_1 = K_1; \qquad \beta_2 = K_1 \times K_2; \qquad \beta_3 = K_1 \times K_2 \times K_3 \quad \text{etc.}$$

Some typical stability constants are given in Table 5.1. A point to remember is that when values of K_n, K_{n+1} etc. are similar then an equilibrium solution will contain mixtures of the complexes (the Pb^{2+}/I^- case in Table 5.1); when K_n, K_{n+1} values are very different then it is possible to obtain solutions containing, essentially, only a single complex. Generally speaking, $K_1 > K_2 > K_3$ etc. but, as Table 5.1 illustrates, exceptions occur. So, notice in this table the expression of the fact that the common anionic chloro complex of Pd^{2+} is $[PdCl_4]^{2-}$.

5.3 Determination of stability constants

In order to determine the values of n formation constants, $n + 2$ independent concentration measurements are needed. These can then be used to obtain the concentrations of the n species $ML, ML_2, \ldots, ML_n$ and also those of M and L. Two pieces of information are at once available; we (should!) know the quantities of M and L (or alternative starting materials) used in the measurement. This means that n additional pieces of information are needed. If it is certain that only one complex, of known empirical formula, is formed, then a measurement of the concentration of the uncomplexed M or L is sufficient to determine the formation constant. This measurement can be made in many ways: by polarographic or emf methods (if a suitable reversible electrode exists), by pH measurements (if the acid dissociation constant of HL is known) and by many other techniques, including the whole galaxy of spectroscopic methods. A recent source book for stability constant data (that by Connors, see Further Reading) distinguishes over 30 methods.

For the more general case where more than one formation constant is to be determined, the problem is usually more difficult. For inert complexes it may be possible to separate, and separately obtain the concentrations of, the various complex species. In this way Bjerrum was able to determine the six stability constants within the $[Cr(H_2O)_6]^{3+}$, $[Cr(H_2O)_5SCN]^{2+}$, ..., $[Cr(SCN)_6]^{3-}$ series. However, this is a potentially unreliable method and has been little used. Some methods of tackling the general problem will now be indicated. The variants are many for this is a field in which considerable ingenuity has been used in the design of experiments and in the analysis of experimental data. As an example consider a ligand for which the species H_2L, HL^- and L^{2-} all exist. The metal ion M and protons may be regarded as being in competition for L^{2-}. If we titrate H_2L against standard NaOH solution we obtain a pH–volume curve of the form shown dotted in the upper part of Fig. 5.1. Now add a known amount of M. Some H^+ will be displaced. If the mixture is again titrated with NaOH the curve shown dotted in the lower part of Figure 5.1 is obtained. Compare the titres $[NaOH]_1$ and $[NaOH]_2$ at the same pH (shown dashed in Fig. 5.1). Now, the H^+ liberated by the added M is dependent on the amount of ligand bound to M, ligand which was previously protonated, the average number of H^+ ions per ligand being n_H ($n_H = 2$ for H_2L). We have

$$[H^+ \text{ ion liberated}] = [\text{L bound to M}] \times n_H = [NaOH]_2 - [NaOH]_1$$

So,

$$[\text{L bound to M}] = \frac{[NaOH]_2 - [NaOH]_1}{n_H}$$

Define $\bar{n}$ as the average number of ligand molecules bound to M, so

$$\bar{n} = \frac{[\text{L bound to M}]}{[M]} = \frac{[NaOH]_2 - [NaOH]_1}{n_H[M]}$$

all the quantities in the right hand side expression are known.

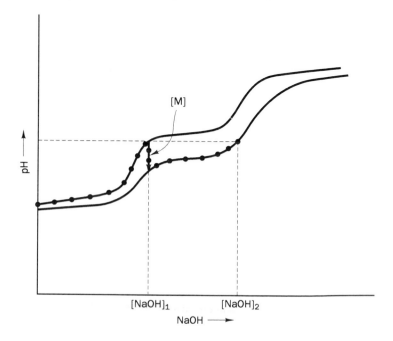

Fig. 5.1 The experimental sequence is shown by the solid circles on the titration curve. A protonated ligand is titrated against NaOH to a suitable pH. Metal ion is added (indicated by [M]) and the titration continued until the same pH is regained.

By such measurements the average number of ligand molecules complexed with each metal atom, $\bar{n}$, is obtained. Although it is always true that $\bar{n}$ indicates the most abundant species, i.e. $\bar{n} = 3$ means that the concentration of ML_3 is a maximum provided that equilibrium has been established, it does not immediately tell us anything about the other species. However, measurements on a variety of solutions with different $\bar{n}$ can. We first note that $\bar{n}$ is related to the composition of the mixture of complex ions by the equation

$$\bar{n} = \frac{1[ML] + 2[ML_2] + \cdots + N[ML_N]}{[M] + [ML] + [ML_2] + \cdots + [ML_N]} = \frac{\sum_{n=1}^{N} n[ML_n]}{[M]_t}$$

where the complex species are $ML, ML_2, \ldots, ML_N$ and $[M]_t$ is the total concentration of metal ion, complexed and uncomplexed, in the solution.

Now,

$$[M]_t = \sum_{n=0}^{N} [ML_n] = [M] + \sum_{n=1}^{N} [ML_n]$$

and

$$[ML_n] = \beta_n [M][L]^n$$

so, substituting for $[M]_t$ and $[ML_n]$ in the equation above,

$$\bar{n} = \frac{\sum_{n=0}^{N} n\beta_n [M][L]^n}{[M] + \sum_{n=0}^{N} \beta_n [M][L]^n} = \frac{\sum_{n=0}^{N} n\beta_n [L]^n}{1 + \sum_{n=0}^{N} \beta_n [L]^n}$$

If a series of solutions, with varying $\bar{n}$, is prepared and $[L]$ measured in each, then values of β_n, $n = 1, 2, \ldots, N$, must be chosen so that the above equality holds. This may be done, for example, by plotting $\bar{n}$ against $[L]$ and determining the βs by an iterative, best fit, method, most readily carried out using a computer. A large number of programs exist; a book detailing many of them is referenced at the end of this chapter. One further point: do not be deceived by the fact that there is no explicit dependence of $\bar{n}$ on $[M]$ in the above expression. $[L]$ is the *free* ligand concentration and its value *is* dependent on $[M]$.

If it is easier to measure $[M]$ than $[L]$, a different but related relationship is used. We know that

$$\frac{[M]_t}{[M]} = \frac{[M] + [ML] + [ML_2] + \cdots + [ML_N]}{[M]}$$

$$= \frac{1}{[M]} ([M] + \beta_1 [M][L] + \beta_2 [M][L]^2 + \cdots + \beta_N [M][L]^n)$$

$$= 1 + \beta_1 [L] + \beta_2 [L]^2 + \cdots + \beta_N [L]^N$$

that is,

$$\frac{[M]_t}{[M]} = 1 + \sum_{n=0}^{N} \beta_n [L]^n$$

If both [M] and [L] can be measured, keeping $[M]_t$ constant, but changing the total ligand concentration for each measurement, this relationship leads to a set of simultaneous equation in the βs (one equation for each measurement) which provide a quick way of determining β values. Alternatively, if [L] is made large, so that $[L] \gg [M]$, it is essentially constant. By varying $[M]_t$ and measuring [M], the βs may similarly be found.

A variety of optical methods is used to determine both complex formation and stability constants. Job's method of continuous variations is the best known. A wavelength is chosen at which the complex in question absorbs (usually in the visible or near-ultraviolet region). Several solutions are examined, for all of which $([M]_t + [L]_t)$ is a constant, C, although each has different values for $[M]_t$ and $[L]_t$. Here, $[L]_t$ is the total concentration of ligand, free and complexed. Some measure of the intensity of absorption (absorbance or optical density) is plotted against composition (usually against the ratio $[L]_t/([M]_t + [L]_t)$; that is, against $[L]_t/C$, a ratio we shall call α. As a simple example of the application of this method consider the case where only a single complex is formed. If a single complex is formed then we have an equilibrium

	M	+	nL	$\rightleftarrows$	ML_n
Initial concentration	$(1 - \alpha)C$		αC		0
Final concentration	$(1 - \alpha - \gamma)C$		$(\alpha - n_\gamma)C$		γC

where, as follows from the definition of α_1, αC is the initial concentration of the ligand L, γC is the final concentration of the complex species ML_n and n is a fixed (but unknown) number, although we expect it to be an integer. Explicitly, the equilibrium concentrations are

$$[M] = (1 - \alpha - \gamma)C$$
$$[L] = (\alpha - n\gamma)C$$
$$[ML_n] = \gamma C = \beta_n[M][L]^n$$

The values of [M], [L] and $[ML_n]$ all change with α; we wish to find the relationship that holds when $[ML_n]$ is a maximum. To do this we differentiate each of the above equations with respect to α. We obtain

$$\frac{d[M]}{d\alpha} = -C$$

$$\frac{d[L]}{d\alpha} = C$$

$$\frac{d[ML_n]}{d\alpha} = \beta_n[L]^{n-1}\left([L]\frac{d[M]}{d\alpha} + [M]\frac{n\,d[L]}{d\alpha}\right)$$

To find the maximum in $[ML_n]$ we set the last equation to zero and substitute the first two. The result is

$$[L](-C) + [M](nC) = 0$$

Introducing the explicit expressions for the concentrations [L] and [M] we obtain

$$-(\alpha_{max} - n\gamma) + n(1 - \alpha_{max} - \gamma) = 0$$

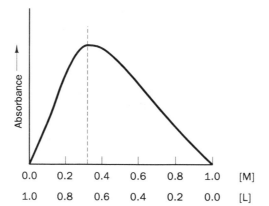

Fig. 5.2 The Job's plot obtained when only one complex species, ML_2, is formed between M and L.

that is,

$$\alpha_{max} = \frac{n}{1 + n}$$

It follows that since α is known (from the composition with which the solution was made up) then, if the visible or ultraviolet wavelength chosen corresponds to absorption by the complex, then n can be determined from the maximum in plot of absorbance against α. Conversely, if the absorption at the chosen value is primarily due to either M or L, the minimum of the absorbance versus α plot gives n. Figure 5.2 shows a typical Job's plot for a ML_2 complex ($\alpha_{max} = \frac{2}{3}$). The rounding at the peak is due to the fact that we have chosen to show the case of a not-very stable complex. The more stable the complex the less rounded the peak; indeed, in favourable cases the peak shape may be used to determine β_n. When more than one complex is formed there will usually be a corresponding number of peaks in a Job's plot, but stability constants may only be determined from such a plot under very special circumstances.

There is another method for the determination of stability constants which, although not much used for this purpose, is of some value in reaction kinetics as a method of estimating rate constants which cannot be measured readily (see Chapter 14). Suppose the reaction

$$ML_n + L \rightarrow ML_{n+1}$$

proceeds by a one-step process. The rate of the formation of ML_{n+1} is $k_f[ML_n][L]$, where k_f is the rate constant of this forward reaction. The corresponding backward reaction is

$$ML_{n+1} \rightarrow ML_n + L$$

and the rate of disappeance of ML_{n+1} is $k_b[ML_{n+1}]$, where k_b is the rate constant of the backward reaction. At equilibrium,

$$k_f[ML_n][L] = k_b[ML_{n+1}]$$

Therefore,

$$\frac{[ML_{n+1}]}{[ML_n][L]} = K_{n+1} = \frac{k_f}{k_b}$$

That is, if the rate constants k_f and k_b can be independently determined, the corresponding stepwise stability constant is given by their quotient.

So far, activity coefficients of unity have been assumed, but they seldom have this value. Consequently, the stability constants determined above are not constants at all, but are a function of concentration. The simplest way out of this difficulty is to determine the stability constants over a range of reactant concentrations and then extrapolate to zero concentration (where activity coefficients are unity). K_n and β_n at zero concentration are true thermodynamic equilibrium constants and are distinguished by the superscript T, thus ^TK_n and $^T\beta_n$ (although, to avoid complexity early in the chapter, they were not designated as such in Table 5.1). More commonly, activity coefficients are kept essentially constant by carrying out measurements in the presence of a backing electrolyte which keeps the ionic strength of the medium constant. Stability constants obtained in this way are proportional to the thermodynamic stability constants, the constant of proportionality being a function of the activity coefficients of the species involved in the complex. Such stability constants are referred to as *stoichiometric stability constants*. Strictly, it is the determination of these which we have discussed—they are given the symbols K_n and β_n which we have been using.

5.4 Stability correlations

Two schemes have been proposed which systematize the available stability constant data. In addition to data contained in the compilations given at the end of this chapter, more qualitative evidence, based for example on the results of displacement reactions, has been included in arriving at the generalizations. Historically, the first scheme is that due to Chatt and Ahrland who pointed out that electron acceptors may be placed in one of three classes. Class-a metals, the most numerous, form more stable complexes with ligands in which the coordinating atom is a first-row element (N, O, F) than with those of an analogous ligand in which the donor is a second-row element (P, S, Cl). Class-b has the relative stabilities reversed. It is not difficult to extend the stability relationships to include heavier donor atoms. Class-a behaviour is, then, typified by a stability order

$$F^- \gg Cl^- > Br^- > I^-$$
$$O \gg S > Se > Te$$

and

$$N \gg P > As > Sb > Bi$$

Class-b behaviour is rather more complicated and is typified by relative stability constants in the order

$$F^- \ll Cl^- < Br^- < I^-$$
$$O \ll S \approx Se \approx Te$$

Table 5.2 Classification of acceptor species (after Ahrland, Chatt and Davies)

Class-a behaviour
H, the alkali and alkaline earth metals, the elements Sc → Cr, Al → Cl, Zn → Br, In, Sb and I, the lanthanides and actinides

Class-b behaviour
Rh, Pd, Ag, Ir, Au and Hg

Borderline behaviour
The elements Mn → Cu, Tl → Po, Mo, Te, Ru, W, Re, Os and Cd

and

$$N \ll P > As > Sb > Bi$$

In addition, there is a third class of electron acceptor for which the stability constants do not display either class-a or -b behaviour uniquely. The class-a/class-b classification of some metal ions is given in Table 5.2 (normal valence states are assumed).

Although not included within the above classification, there are some other useful gradations which have been noted and are conveniently included at this point. For a given ligand, corresponding stability constants of complexes of bivalent ions of the first transition series are usually in the *natural order* (sometimes called the *Irving–Williams* order):

$$Mn^{II} < Fe^{II} < Co^{II} < Ni^{II} < Cu^{II} > Zn^{II}$$

Copper(II) does not coordinate a fifth and sixth ligand particularly strongly, and this order is incorrect for stability constants relating to CuL_5 and CuL_6 complexes. Complexes of chelating ligands also tend not to follow this order.

For non-transition metal ions complex stability decreases roughly in the order of *ionic potential* (or *polarizing power*), which is defined as (formal charge)/(ionic radius). Thus, corresponding stability constants decrease in the order

$$Li^+ > Na^+ > K^+ > Rb^+ > Cs^+$$
$$Mg^{2+} > Ca^{2+} > Sr^{2+} > Ba^{2+} > Ra^{2+}$$

and

$$Al^{3+} > Sc^{3+} > Y^{3+} > La^{3+}$$

provided that the ligand is not changed from one ion to the next and subject to the absence of size-matching conditions that will be discussed later in this chapter. Similarly, for approximately constant ionic radius, the stability constants are in the order of decreasing charge, thus,

$$Th^{4+} > Y^{3+} > Ca^{2+} > Na^+$$

and

$$La^{3+} > Sr^{2+} > K^+$$

Although the formal charge on an ion is a well-defined quantity, the actual charge (which is not so well defined) may differ significantly from it—we shall meet this point again in Section 7.5. Further, the concept of *ionic radius* is also somewhat elusive. If determined from crystal structures—so that, for example, the radius of Cl^- is given by the point of minimum electron density along the Na^+–Cl^- axis in NaCl—then it is not a quantity which is constant because it varies somewhat with the crystal structure chosen to measure it. In retrospect, it is perhaps surprising that the above inequalities hold as well as they do!

The second and more recent approach which has been used to classify metal ion–ligand interactions is based on the concept of *hard and soft acids and bases*, often denoted HSAB. Hard metal ions are those which parallel the proton in their attachment to ligands, are small, often of high charge and have no valence shell electrons that are easily distorted or removed. Soft metal ions are large, of low charge or have valence shell electrons which are easily distorted or removed. They bond strongly to highly polarizable ligands—which often have a very small proton affinity. Similarly, ligands are divided into those that are non-polarizable (hard) and those that are polarizable (soft). Remembering the Lewis definition of acids and bases as electron acceptors and electron donors respectively, the cations are classified as either soft or hard acids, whilst the ligands are classified as either soft or hard bases—hence, HSAB. An important empirical generalization is that the most stable complexes are those of soft acids with soft bases and of hard acids with hard bases.

This approach has the advantage that it is not restricted to complexes of transition metal ions. Indeed, it may be applied to a wide range of chemical equilibria and systemizes a great deal of chemical intuition. A particularly useful concept is that of *symbiosis*. A cation which is classified as a relatively hard acid (or, indeed, which is regarded as borderline) is made softer by the coordination of a soft ligand (or harder by the coordination of a hard ligand) and so it is more likely to add further soft (or hard) ligands. For the non-transition elements in particular, the principle of hard or soft acids and bases systematizes stability data in a useful way. In Table 5.3 some metal ions and ligands are classified as hard or soft. Generally, ligands in which the coordinating atom has a high electronegativity are hard bases; those in which it has a low electronegativity are soft.

Recent work[2] has made the connection between electronegativity and the hard–soft/acid–base concept clearer and has served to give the latter

Table 5.3 Classification of some (formally) ionic species as hard and soft acids and bases (after Pearson)

Hard acids	H^+	Li^+	Na^+	K^+					
	Be^{2+}	Mg^{2+}	Ca^{2+}	Sr^{2+}	Mn^{2+}				
	Al^{3+}	Sc^{3+}	Ga^{3+}	In^{3+}	La^{3+}				
	Cr^{3+}	Co^{3+}	Fe^{3+}	Ce^{3+}					
	Si^{4+}	Ti^{4+}	Zr^{4+}	Th^{4+}					
	VO^{2+}	VO_2^{2+}	MoO^{3+}						
Soft acids	Cu^+	Ag^+	Au^+	Tl^+	Hg^+				
	Cd^{2+}	Hg^{2+}	Pd^{2+}	Pt^{2+}					
	Tl^{3+}								
	Pt^{4+}								
Borderline	Zn^{2+}	Sn^{2+}	Pb^{2+}						
	Fe^{2+}	Co^{2+}	Ni^{2+}	Cu^{2+}	Ru^{2+}	Os^{2+}	Sb^{3+}	Bi^{3+}	Rh^{3+}	Ir^{3+}
Hard bases	H_2O	R_2O	ROH	NH_3	RNH_2				
	OH^-	OR^-	ClO_4^-	NO_3^-	CH_3^-	CO_3^-			
	SO_4^{2-}	CO_3^{2-}							
	PO_4^{3-}								
Soft bases	R_2S	RSH	P_3P	R_3As					
	RS^-	I^-	SCN^-	CN^-	H^-	R^-			
	$S_3O_3^{2-}$								
Borderline	Cl^-	Br^-	N_3^-	NO_2^-	SO_3^{2-}				

[2] R. G. Pearson, *J. Chem. Ed.* (1987) 64, 561 and *Inorg. Chem.* (1988) 27, 734.

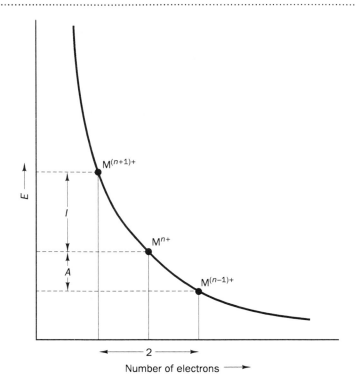

Fig. 5.3 The plot used to define the electronegativity and hardness of the species M^{n+} in terms of I and A. See the text for a full discussion.

concept a quantitative basis. This work starts from a rather different direction to that given above but, in the end, arrives at the same point. Consider Fig. 5.3. This figure shows a plot of the total energy of an atom (or molecule). If we start with M^{n+} (the central point on the graph) then the energy change to give $M^{(n+1)+}$ is the ionization potential of M^{n+}; this we will call I. Similarly, the energy change to give $M^{(n-1)+}$ is the electron affinity of M^{n+} (it is the negative of the ionization potential of $M^{(n-1)+}$). Call this A. Following Mulliken's definition of electronegativity, we equate $(I + A)/2$ with the electronegativity of M^{n+}. From Fig. 5.3 we see that if the curve joining $M^{(n+1)+}$ and $M^{(n-1)+}$ were a straight line then $(I + A)/2$ would be the slope of that straight line. To a first approximation, $(I + A)/2$, the electronegativity of M^{n+}, is the slope of the curve at M^{n+}. However, the curvature of the plot, the deviation from a straight line, is measured by the difference between I and A.

The approximate curvature at M^{n+}, $(I - A)/2$, may be taken as a measure of the hardness of M^{n+}. Physically, this definition relates hardness to a resistance to deformation of charge. Hard metal ions are those for which it costs a great deal to change electron densities. Unfortunately, this model cannot immediately be extended to all ligands; for anions, in particular, electron affinities are unknown. In such cases it seems that use of data for corresponding neutral species gives reasonable results. In Table 5.4 are given values of *absolute hardness* for typical metal and ligands.

At several points in this book there are mentioned the consequences of particular HOMO–LUMO (highest occupied molecular orbital–lowest unoccupied molecular orbital) energy separation patterns. The present discussion provides another. Hard ions or molecules have a large HOMO–

Table 5.4 Absolute hardness values for common metals and ligands

Species	Ionization potential, I	Electron affinity, A	Absolute hardness, $\frac{1}{2}(I - A)$
Cations			
Na^+	47.29	5.14	21.08
K^+	31.63	4.34	13.64
Cu^+	20.29	7.73	6.28
Cu^{2+}	36.83	20.29	8.27
Ag^+	21.49	7.58	6.96
Mg^{2+}	81.14	15.04	32.55
Ca^{2+}	50.91	11.87	19.52
Cr^{2+}	30.96	16.50	7.23
Cr^{3+}	49.1	30.96	9.1
Mn^{2+}	33.67	15.64	9.02
Mn^{3+}	51.2	33.67	8.8
Co^{2+}	51.3	17.06	8.22
Co^{3+}	51.3	33.50	8.9
Pd^{2+}	32.93	19.43	6.75
Pd^{2+}	31.94	18.76	8.46
Ligands			
CO	14.0	−1.8	7.9
PF_3	12.3	−1.0	6.7
C_2H_4	10.5	−1.8	6.7
C_5H_5N	9.3	−0.6	5.0
C_2C_2	11.4	−2.6	7.0
H_2O	12.6	−6.4	9.5
$(CH_3)_2S$	8.7	−3.3	6.0
NH_3	10.7	−5.6	8.2
F^a	17.42	3.40	7.01
Cl^a	13.01	3.62	4.70
Br^a	11.84	3.36	4.24

a Data for atoms; taken as models for the corresponding anions.

LUMO separation; soft ions or molecules have a small HOMO–LUMO difference—this is just another, orbital, interpretation of the quantity $(I - A)$. With these recent developments, it seems that the concept of hardness and softness will increasingly replace the older class-a/b distinctions. However, the latter nomenclature is widespread in the chemical literature and likely to remain so for some time. Indeed, the a/b distinction separates species that, from Table 5.3, show relatively small differences in hardness and so, perhaps, the notation will remain useful.

5.5 Statistical and chelate effects

The variations in stability constants which have been discussed so far in this chapter are typically rather large. Superimposed on them are some smaller but systematic variations which also merit attention.

Most stability constant measurements are made in aqueous solutions at a constant ionic strength. The species we have called M must, in reality, be a complex mixture of species. The majority will probably be $[M(H_2O)_n]^{x+}$, where n may well be four or six although larger numbers also occur. Other species present will include foreign anions—ones not involved in the reaction, but arising from the salt used to maintain a constant ionic strength—and some of these might well be coordinated to M. The addition reaction

$$ML_n + L \rightarrow ML_{n+1}$$

should more properly be written as a substitution reaction (L' is usually H_2O but in this section is to be generally taken to mean 'solvent molecule');

$$ML'_m L_n + I \rightarrow ML'_{m-1}L_{n+1} + L'$$

In this reaction it has been assumed that the number of ligands around M is constant, and equal to $(m + n)$. If we let this number (usually four or six) be N, then the reaction may be written as

$$ML'_{N-n}L_n + L \rightarrow ML'_{N-n-1}L_{n+1} + L'$$

As has been seen in Table 5.1, it is commonly found that for a given M and L there is a decrease in successive formation constants, $K_1, K_2 \ldots$ or $\beta_1, \beta_2 \ldots$ This is largely a statistical effect. It is easier to attach L to ML'_N than to $ML'_{N-1}L$ because there are N reaction sites on the former but only $N - 1$ on the latter (replacement of L in $ML'_{N-1}L$ by L gives an identical molecule).

Consider the equilibrium:

$$ML'_{N-n}L_n + L \rightleftarrows ML'_{N-n-1}L_{n+1} + L'$$

The rate of formation of $ML'_{N-n-1}L_{n+1}$ is, for a simple one-step reaction, $k_f[ML'_{N-n}L_n][L]$. If the statistical effect is the only one which varies as n changes, the forward rate constant will vary with n in proportion to the number of reaction sites, that is, the number of L' ligands in $ML'_{N-n}L_n$. We therefore rewrite the above rate as $k'_f(N-n)[ML'_{N-n}][L']$, where k'_f is expected to be constant for all values of $(N-n)$. Similarly, for the rate of disappearance of $ML'_{N-n-1}L_{n+1}$, $k_b[ML'_{N-n-1}L_{n+1}][L]$, the backward reaction rate constant, k_b will vary with n in proportion to the number of L ligands in $ML'_{N-n-1}L_{n+1}$. Hence, in the above cases $k_b = k'_b(n+1)$, where k'_b is constant for all values of $(n+1)$. It follows that

$$K_{n+1} = \frac{k_f}{k_b} = \frac{k'_f(N-n)}{k'_b(n+1)}$$

In a similar way it is found that

$$K_n = \frac{k'_f(N-n+1)}{k'_b n}$$

The ratio between successive formation constants

$$\frac{K_{n+1}}{K_n} = \frac{n(N-n)}{(n+1)(N-n+1)}$$

The value of this expression for $N = 6$ and $n = 1 \rightarrow 6$ are compared with the experimental data for the Ni^{2+}/NH_3 system (where L' is H_2O) in Table 5.5. The agreement is far from exact, but the predictions are of the correct order of magnitude. The experimental ratios are, as is commonly the case, rather smaller than those predicted statistically.

Table 5.5 Statistical predictions and experimental ratios for the equilibrium constants of the system

$$[Ni(H_2O)_{6-n}(NH_3)_n]^{2+} + NH_3 \rightleftharpoons [Ni(H_2O)_{5-n}(NH_3)_{n-1}]^{2+} + H_2O$$

Ratio	n	Experimental ratio (data from Table 5.1)	Statistical prediction[a]
K_2/K_1	1	0.28	0.42
K_3/K_2	2	0.31	0.53
K_4/K_3	3	0.29	0.56
K_5/K_4	4	0.36	0.53
K_6/K_5	5	0.19	0.42

[a] Calculated using $\dfrac{K_{n+1}}{K_n} = \dfrac{n(6-n)}{(n+1)(7-n)}$

Table 5.6 Comparative stability of the Ni^{2+}/NH_3 and Ni^{2+}/en systems

Equilibrium[a]	log K
Non-chelated complex	
$Ni^{2+} + 2NH_3 \rightleftharpoons [Ni(NH_3)_2]^{2+}$	5.05
$[Ni(NH_3)_2]^{2+} + 2NH_3 \rightleftharpoons [Ni(NH_3)_4]^{2+}$	2.96
$[Ni(NH_3)_4]^{2+} + 2NH_3 \rightleftharpoons [Ni(NH_3)_6]^{2+}$	0.73
Chelated complex	
$Ni^{2+} + en \rightleftharpoons [Ni(en)]^{2+}$	7.51
$[Ni(en)]^{2+} + en \rightleftharpoons [Ni(en)_2]^{2+}$	6.35
$[Ni(en)_2]^{2+} + en \rightleftharpoons [Ni(en)_3]^{2+}$	4.32

[a] Coordinated water is omitted for simplicity.

When a N–H bond in a ligand such as ammonia is replaced by a N–alkyl bond, the corresponding stability constants of complexes are usually lowered. This may be due partly to increased steric interaction in complexes of the substituted ligand, but the observation that the stability constants of complexes with sulfur-containing ligands are usually $H_2S < RSH < R_2S$— that is, in the opposite order—suggests that other factors also operate (for example, both the energy and the shape of the lone-pair-containing orbital(s) of the coordinating atom will change slightly on substitution). Even so, it might reasonably be expected that complexes of a bidentate ligand such as ethylenediamine ($NH_2–CH_2–CH_2–NH_2$) are less stable than the corresponding complex with two ammonia molecules. Quite the opposite is true. Complexes containing chelate rings are usually more stable than similar complexes without rings. This is termed the *chelate effect*, and is illustrated in Table 5.6; it was first met in Section 2.1. The origin of this effect has been the subject of some controversy. There have been those who have argued that the effect does not exist (despite the experimental observation that a ligand such as en seems invariably to displace NH_3; that bpy (2,2′-bipyridine) replaces py; that oxalate displaces acetate and so on). The problem is a fascinating one; it relates to the fact that quantities such as K_1 and K_2 etc. have units (mol^{-1} L) so that in comparing an equilibrium involving $2NH_3$ with one involving en, the corresponding βs have different units (mol^{-2} L^2

and mol^{-1} L, respectively). How does one compare quantities with different units? This is not the end. When we write $RT \ln K$ (as we shall do shortly) we have to recognize that we can only take the logarithm of a number. Where have the units of K gone? The answer here is that we really evaluate $RT \ln(K/1)$, where the 1 refers to a standard state which has the same units as K. This leads to a discussion of the characteristics of the standard state and this, in turn, to the recognition that there are three different concentration scales in use: molarity, molality and mole fraction. These concentration scales are *not* directly proportional to each other so that, for example, it is theoretically possible for a given solution to have activity coefficients of unity on one scale but not on the others. The chelate effect is not enormous and so it is not surprising that such fine points can assume undue prominence and that controversy and apparent misunderstanding exist! A very readable account of the general problem can be found in Chapter 2 of the book by Connors and in the articles given at the end of the chapter. Fortunately, we can side-step most of the pit-falls and learn more about the chelate effect from a more detailed analysis of stability constant data. To do this, we note the relationships

$$-RT \ln K = \Delta G^0 = \Delta H^0 - T\Delta S^0$$

which show that the chelate effect could originate in either the heat term, ΔH^0, the entropy term, ΔS^0, or both.

To proceed further, we could analyse the equilibrium constants in Table 5.6, in pairs, using these thermodynamic relationships. We can simplify matters by considering, instead, the equilibrium

$$[M(NH_3)_4]^{2+} + en \rightleftharpoons [M(NH_3)_2(en)]^{2+} + 2NH_3$$

Available data at 298 K for M = Ni^{II} and Zn^{II} (remembering that ΔS^0 is quoted in terms of J mol^{-1} K^{-1}, whilst ΔH^0 and ΔG^0 are given in kJ mol^{-1}) are

Ion	ΔG^0	ΔH^0	ΔS^0	$-T\Delta S^0$ (kJ mol^{-1})
Ni^{II}	−3.4	−2.0	4.8	−1.4
Zn^{II}	−1.5	0.1	5.3	−1.6

These results are qualitatively general: the chelate effect is largely an entropy effect; for non-transition metal ions the heat term is particularly small.

There are two important contributions to the ΔH^0 term when it makes a significant contribution. First, when two monodentate anionic ligands are brought together to occupy adjacent coordination sites in a complex, there will be an electrostatic repulsion between them against which work has to be done. The same is true for uncharged monodentate ligands because such ligands are always dipolar. For chelating ligands the coordinating centres do not have to be brought together and most of this repulsive energy is 'built in' (it makes a contribution to the enthalpy of formation of the ligand). That is, in the above equilibrium this contribution to the heat term would be expected to favour chelated species because the oriented ammonia molecules repel each other. The second, more variable, contribution to the ΔH^0 term comes from solvation energies. Each of the species in the equilibrium will be solvated (by hydrogen bonding, for example). Further, there will be anions closely associated with the cationic species. We shall see in Chapter 14 that there is kinetic evidence that counterions, although not directly bonded to the coordination centre—they are not in the first coordination sphere—are often present in an outer sphere (the so-called second coordination sphere). It does not seem possible to predict, in general, whether this term makes a positive or negative contribution to ΔH^0, but it appears to make a positive contribution when the coordination centre is a main group element.

The entropy effect is readily understood. An increase of randomness is associated with a concomitant increase in entropy. A complex molecule has a lower entropy than its separated and therefore independent components. In the equilibrium above there is a 50% increase in the number of independent molecules in the right-hand side compared with the left and, so, the right-hand side is favoured. Another aspect of the entropy effect is the following. When one end, and only one end, of an ethylenediamine molecule is coordinated, the effective concentration of the other end in the system, and the probability that it will coordinate, is high, because it is constrained to stay close to the cation. This means that it is easier to form a chelate ring than to coordinate two independent molecules because the two acts of coordination are related for the former, whilst for the latter they are entirely independent of each other. All these, then, are potential contributors to the chelate effect. They are difficult to relate to the kinetic data on the effect. These show that it occurs because the ring opening is slower than expected for the dissociation of the first of two independent ligands, not because the formation of the ring is faster (it is the ratio of these two that is the equilibrium constant K). It is understandable that the topic continues to be the subject of debate.

The chelate effect varies with the size of the ring formed on coordination. It is usually a maximum for five-membered rings and only slightly smaller for six-membered rings. It is not difficult to see one reason why this should be. For an octahedral complex, the angle subtended by a chelate ring at the metal, no matter the size of the ring, will be close to 90°. For simplicity, we fix it at this value. What will be the bond angles at the other atoms in the ring? Assuming planar rings, the average values will be:

Ring size	Average angle
4	90°
5	112°
6	126°
7	135°

So, if in a saturated ring system, such as that formed by en, the five-membered ring system will have angles that are closest to the natural, tetrahedral, value (109.5°). Similarly, for a ring that is conjugated, as in acac⁻, which may be drawn as shown in Fig. 5.4, the six-membered ring has an angle closest to the natural trigonal value of 120°. Of course, ring puckering can and does occur, but can only be of a large magnitude at a cost of either steric

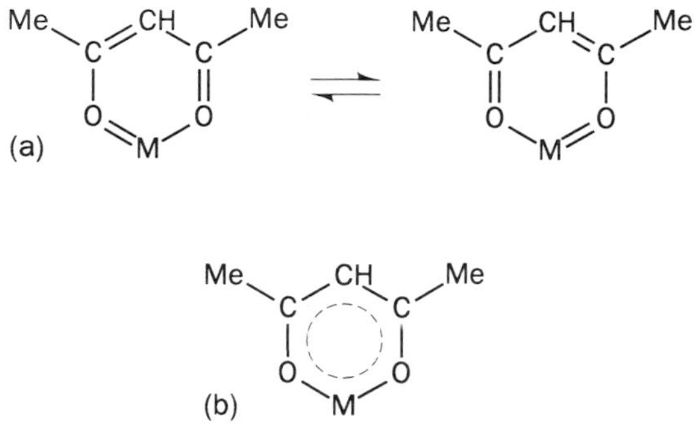

Fig. 5.4 (a) Two equivalent resonance hybrids (canonical forms) of coordinated acac. However, the conjugated nature of the system is more commonly represented as shown in (b).

interactions between rings or loss of conjugation within a ring. Further, the stabilization of six-membered chelate rings is not confined to conjugated systems, although the chelate stabilization of non-conjugated systems is usually greater for five-membered rings, where comparison is possible. Clearly, this discussion of the chelate effect could be elaborated considerably. What of polydentate ligands? This is a topic very relevant to the ligands encountered in the complexes of bioinorganic chemistry and discussion of it will be deferred until Chapter 16.

Although complexes with rings of size other than five or six members have been synthesized they show little sign of the chelate effect. This is partly because of the reduced effective concentration of the other end of the ligand for larger chelating molecules and because of the increase in work against electrostatic forces needed to bring the coordinating atoms together. Further, as has been mentioned, another potentially destabilizing influence for larger-membered rings is the relative difficulty of finding a sterically non-crowded ring configuration. The relationship between the geometry of free and coordinated multidentate ligands is currently the subject of research, particularly for the large ligands that occur in bioinorganic complexes (Chapter 16); it also finds a faint echo in Appendix 1. It is clear that ligand geometry is an important factor in determining the relative stabilities of complexes formed by such ligands. For instance, a coordinated ligand will normally be less flexible than the free ligand. This loss of motional freedom will be reflected in ΔS values and, thus, in the stability of the coordination compound. We shall return to this point in Chapter 16.

5.6 Solid complexes

In this chapter our concern has been with equilibria in solution. As has already been noted, it is sometimes possible to study a complex in solution but not to obtain it in the solid state or *vice versa*. Why should stability be related to phase in this way? It is because an additional energy factor is involved, the relative lattice energies of the crystals which might be formed on crystallization. Usually, the crystal form with the highest lattice energy will be obtained (this will be the least soluble complex). For an ionic complex species the lattice energy of a crystal will depend on the counterion with which it crystallizes. Lattice energies are a maximum when the cation and anion are of similar charge and size (and, usually, of the same hardness or softness). This suggests a method by which crystals containing an elusive complex may be obtained and, as we saw in Chapter 4, accounts for the widespread use in preparative complex chemistry of such species as $[PF_6]^-$, $[B(C_6H_5)_4]^-$, $[NEt_4]^+$, and $[As(C_6H_5)_4]^+$. Such symmetrical cations and anions have, however, one disadvantage. Because of their potentially high symmetry (O_h or T_h), they are somewhat spherical and have an unpleasant tendency to adopt a disordered arrangement in a crystal, making structure determination more difficult. This is one reason for the apparently perverse habit of preparing compounds in which the symmetry of such counterions is reduced by, for example, replacing one of the four phenyl groups in $[As(C_6H_5)_4]^+$ with an ethyl to give $[As(C_6H_5)_3(Et)]^+$.

5.7 Steric effects

As has been indicated above, steric effects can play a part in determining the stability of complexes. Indeed, this has become an area of much research. If a very bulky ligand coordinates it may well form a weak bond, in which case molecules containing it may well be rather reactive. If, on the other hand, despite its bulk it coordinates strongly, then it may well cause the bonding of other ligands to be weaker than normal. Again, an enhanced reactivity is likely. Enhanced reactivity, particularly in the field of catalysis, in which, often, a ligand dissociates from a metal ion at one point in the mechanistic sequences and recoordinates at another, is always interesting. A question at once arises: can we classify steric effect? The answer is that we can, to some extent. Before attempting this, however, there is an important point which has to be made. Consider a ligand such as ammonia but consider it as a free molecule. The energy required to make it rotate, to spin, about its threefold axis is low—it falls in the microwave region of the spectrum. When the ammonia molecule is coordinated to a metal this rotation becomes less free—it may well become a libration, in which it oscillates about its threefold axis as a solid unit but does not overcome the barrier preventing it reaching an equivalent position—but it is still of low energy. Everything depends on the thermal energy available compared to the barrier height. At room temperature the NH_3 groups in ammine complexes are commonly in a state of large-amplitude oscillation, if not almost free rotation. This is important because it means that the 'bumps' on the ammonia molecule—the hydrogen atoms—may well behave as rather smeared out as far as steric effects are concerned. It may not be too bad an approximation to regard each ligand as effectively acting as a circular cone, with its apex at the metal atom. This immediately leads us to some qualitative predictions. As far as steric effects are concerned, we would expect the linear M–N=C=S and M–N=C=Se systems to be more stable than their bent isomeric counterparts

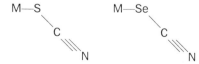

A case where this may well be evident is in the square planar complex formed between the tridentate ligand diethylenetriamine, $NH_2-CH_2-CH_2-NH-CH_2-CH_2-NH_2$, dien, and palladium(II), $[Pd(dien)(SeCN)]$. In this complex the $SeCN^-$ anion is coordinated through Se and is bent. When the terminal

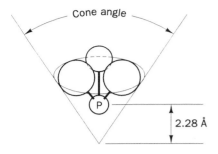

Fig. 5.5 The definition of the cone angle for a symmetrical phosphine, PR$_3$. It is necessary, of course, to define the cone angle with respect to the metal atom to which the phosphine is coordinated. It is assumed in the definition that this atom is 2.28 Å from the phosphorus. For a different coordinated atom a different metal–ligand separation would have to be assumed.

Table 5.7 Cone angles for some simple systems, see Fig. 5.5 (ligand charges are not shown

Ligand	Cone angle (degrees)	Ligand	Cone angle (degrees)
H	75	PH$_3$	87
Me	90	PF$_3$	104
F	92	P(OMe)$_3$	107[a]
CO, CN, N$_2$, NO	$\approx$95	P(OEt)$_3$	108[a]
Cl, Et	102	PMe$_3$	118
Br, C$_6$H$_5$	105	PCl$_3$	124
I	107	PEt$_3$	132
i-Pr	114	PPh$_3$	145
t-Bu	126	P(*i*-Pr)$_3$	160
C$_5$H$_5$	136	P(*t*-Bu)$_3$	182

[a] In obtaining these values, Tolman used a conformation which has yet to be observed.

NH$_2$ groups of the dien ligand are replaced by NEt$_2$ groups, the SeCN$^-$ anion is coordinated through nitrogen and is linear. Steric effects seem to be the factor causing the change. Can the effect be treated in a reasonably quantitative manner? The answer is yes. By a study both of crystal structures and of molecular models, Tolman[3] has compiled data for steric effects, described in terms of *cone angles*, defined as in Fig. 5.5, for a large number of ligands, particularly for those in which phosphorus is the donor atom. A selection of his data is reproduced in Table 5.7. There have been many variants in this field—the use of X-ray data to determine cone angles, the use of detailed theoretical calculations, methods of averaging when very different substituents are present, particularly on a phosphorus atom, and so on. Generally, these other approaches lead to cone angles which are somewhat smaller than those given by Tolman.

A rather different form of size effect occurs for those ligands which are cage-like and which largely envelop a cation at their centre. For these, it is the matching of the size of the central cavity of the ligand to the size of the cation which determines stability. In Chapter 3 the cryptand ligands were mentioned; these are shown again in Fig. 5.6. They are given shorthand names which simply list the number of oxygen atoms in each bridge between the two nitrogen atoms. Also in Fig. 5.6 are given the approximate radius of a hypothetical spherical cavity at the centre of each ligand. These values are to be compared with the approximate ionic radii of the alkali metal cations, as determined experimentally by X-ray data diffraction,[4] of Li$^+$ 0.76 Å, Na$^+$ 1.02 Å, K$^+$ 1.38 Å, Rb$^+$ 1.52 Å and Cs$^+$ 1.67 Å. The relative stabilities of the alkali metal cryptand complexes are

[211] Li$^+$ > Na$^+$
[221] Li$^+$ < Na$^+$ > K$^+$ > Rb$^+$
[222]   Na$^+$ < K$^+$ > Rb$^+$
[322]   Rb$^+$ < Cs$^+$

[3] C. A. Tolman, *Chem. Rev.* (1977) 77, 313.
[4] Data from R. D. Shannon, *Acta Cryst.* (1976) *A32*, 751.

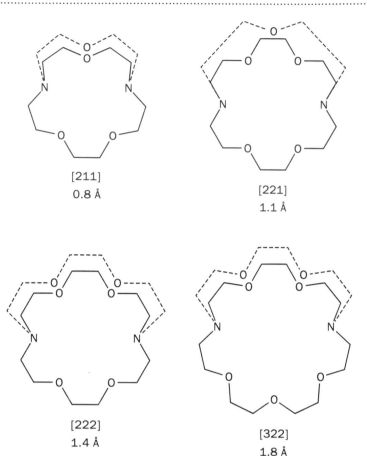

[211]
0.8 Å

[221]
1.1 Å

[222]
1.4 Å

[322]
1.8 Å

Fig. 5.6 Four cryptand ligands. The three numbers within the square brackets correspond to the three chains linking the two nitrogen atoms and give, in descending numerical order, the number of oxygen atoms in these chains. Below each ligand designation is given the approximate size of the cavity within that ligand.

The correspondence between the best cavity-size to cation-size match and maximum stability is perfect. Not surprisingly, this interpretation of stability constant data for encapsulating ligands is being widely applied, not least to some antibiotics and biological molecules which are involved in the transport of Na^+ and K^+ in living systems. It is also being applied to ligands which surround a molecule without totally encapsulating it—for instance, ligands that surround a metal, holding it in a square planar environment. Again, such species have biological analogues, to which we shall return in Chapter 16.

5.8 Conclusions

Although in this chapter many aspects of stability have been discussed, there are many others that have not. It is now accepted that for some ligands, metal–ligand double bonding occurs.[5] Just how are bond order and stability related? Is the distinction between inert and labile transition metal complexes related in any way to the d-electron configuration of the transition metal (we shall have more to say about this in Chapter 14)? In addition to the kinetic *trans* effect, discussed at the end of Chapter 4, there is a static *trans* effect, sometimes called the *trans influence*. In this, a metal–ligand bond

[5] See W. A. Nugent and J. M. Mayer, *Metal-Ligand multiple bonds*, Wiley, New York, 1988.

length, vibrational frequency or some other characteristic of the metal–ligand bond, is influenced by the ligand *trans* to it. So, in $[OsNCl_5]^{2-}$, in which the Os–N bond is short and usually regarded as a multiple bond, the Os–Cl bond *trans* to the nitrogen is over 10% longer than that *cis*. What is the consequence of the *trans* influence on stability? In the case quoted, it seems that the chlorine *trans* to nitrogen is kinetically the more labile. The observations and consequent questions are almost endless; for some, incomplete answers exist; for others, nothing. We note that although several aspects of the stability of coordination compounds have been discussed in this chapter we have avoided the most fundamental question of all: why is one particular complex more stable than another? What is the fundamental explanation for class-a and -b behaviour (or for the hard and soft distinction)? As has been seen earlier, an answer to this last question is emerging, but it is still incomplete. It is now generally accepted that the interaction of the orbitals of the two atoms to be bonded is the most important factor. This interaction depends on the matching of the orbitals—orbitals of similar energies often also have similar sizes—but it is not a simple matter because several orbitals or sets of orbitals of different energies and sizes on each atom are involved. Our discussion of LUMO–HOMO effects in the context of hard and soft ligands probably points the way ahead. Some aspects of this will be considered in more detail in Chapter 10. However, at the present time it does not seem possible to give more than a semiquantitative general answer to such questions.

Further reading

Stability constant data are to be found in some early, pioneering, compilations:

- *Stability Constants of metal–ion complexes*, Special Publication No. 17, The Chemical Society, Burlington House, London, 1964.
- *Stability Constants of metal–ion complexes, Supplement*, Special Publication No. 25, The Chemical Society, Burlington House, London, 1971.
- *Stability Constants of metal–ion complexes Part A (Inorganic Ligands)*, Pergamon Press, Oxford, 1982.
- *Stability Constants of metal–ion complexes Part B (Organic Ligands)*, Pergamon Press, Oxford, 1979.

All four of the above volumes are needed as data are not repeated unless new measurements are reported. The first two volumes give the more detailed and helpful description of the use of the tables.

A continuing compilation is that of A. E. Martell and R. M. Smith, *Critical Stability Constants*, Plenum, New York, Vols. 1–6 (1974, 1975, 1976, 1977, 1982 and 1989).

Related, and similarly arranged, thermodynamic data are to be found in J. J. Christensen and R. M. Izatt, *Handbook of Metal-Ligand Heats*, Marcel Dekker, New York, 1979.

A database is A. E. Martell and R. M. Smith, *Critical Stability Constants Database*, NIST, Gaithersburg, MD, USA, 1993.

Very useful is *Determination and Use of Stability Constants*, 2nd edn, VCH, New York, 1992.

Computer methods for the interpretation of experimental data are described in *Computational Methods for the Determination of Formation Constants*, D. J. Leggett (ed.), Plenum, New York, 1985.

With a content covering a much wider area than the subject of this chapter, but with much relevant material, particularly on measurement methods, is K. A. Connors, *Binding Constants*, Wiley, New York, 1987.

Graphical representations of data related to those discussed in this chapter will be found in J. Kragten, *Atlas of Metal–Ligand Equilibria in Aqueous Solution*, E. Horwood, Chichester, 1978.

An easy-to-read and enlightening article is 'A comparison of different experimental techniques for the determination of the stabilities of polyether, crown ether and cryptand complexes in solution', H.-J. Buschmann, *Inorg. Chim. Acta* (1992) *195*, 51.

Other relevant articles:

- 'Potentiometry revisited: The Determination of Thermodynamic Equilibria in Complex Multicomponent Systems', A. E. Martell and R. J. Motekaitis, *Coord. Chem. Rev.* (1990), *100*, 323.

- 'Steric Effects of Phosphorus Ligands', C. A. Tolman, *Chem. Rev.* (1977), *77*, 313.
- 'Ligand Interactions in Crowded Molecules', H. C. Clark and M. J. Hampden-Smith, *Coord. Chem. Rev.* (1987) *79*, 229.
- 'Ligand Steric Properties', T. L. Brown and K. J. Lee, *Coord. Chem. Rev.* (1993) *128*, 89.
- 'The Specification of Bonding Cavities in Macrocylic Ligands', K. Hendrick, P. A. Tasker and C. F. Linday, *Prog. Inorg. Chem.* (1985) *33*, 1.
- 'The Chelate Effect Redefined', J. J. R. Frausto da Silva, *J. Chem. Educ.* (1983) *60*, 390.
- 'Misunderstandings over the Chelate Effect', D. Munro, *Chem. in Britain* (1977) 100.

- 'Hard and Soft Acids and Bases—the Evolution of a Chemical Concept', R. G. Pearson, *Coord. Chem. Rev.* (1990), *100*, 403.
- 'Absolute Electronegativity and Hardness', R. G. Pearson, *Chem. in Britain* (1991) 444.
- 'Molecular Organization, Portal to Supramolecular Chemistry. Structural Analysis of the Factors Associated with Molecular Organization in Coordination and Inclusion Chemistry, including the Coordination Template Effect', D. H. Busch and N. A. Stephenson, *Coord. Chem. Rev.* (1990) *100*, 119.

Questions

5.1 In Fig. 3.27 was shown an example of a compound with a low barrier to internal rotation. There are many similar compounds, some of which spontaneously ignite when exposed to air. Comment on the applicability of the terms kinetic stability and thermodynamic stability to such species.

5.2 (Note that in this question square brackets indicate concentrations.) The consequences of the addition of base to aqueous Cr^{III} was followed for four years. $[Cr_2(OH)_2]^{4+}$ maximized after a few days whilst $[Cr_3(OH)_4]^{5+}$ increased throughout the period. $[Cr_4(OH)_6]^{6+}$ was constant after a few months. Comment on these observations in the light of the reported K_{ab} values:

$$K_{ab} = \frac{[Cr_a(OH)_b]^{n+}}{[Cr(OH)_2]^+[Cr_{a-1}(OH)_{b-2}]^{(n-1)+}}$$

$\log K_{22} = 5.1$; $\log K_{34} = 6.9$; $\log K_{46} = 5.2$

5.3 Use the data in Fig. 5.2 to show that, as claimed in the caption, it represents the formation of the ML_2 complex.

5.4 The experimental data used in Table 5.5 are those for zero ionic strength in Table 5.1. Calculate the corresponding experimental data for 2 M NH_4NO_3 (given in the caption to Table 5.1). Comment on the relative agreement of the two sets of experimental data with experiment.

5.5 Compare Tables 5.2 and 5.3 and thus comment on the question of whether the class-a and -b, and hard and soft classifications are simply restatement of the same experimental observations.

5.6 Use Table 5.4 to associate numbers with the broad divisions given in Table 5.3 (for instance, what is the range of absolute hardness values that corresponds to hard?).

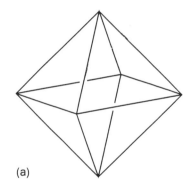

6

Molecular orbital theory of transition metal complexes

6.1 Introduction

The traditional approach to the electronic structure of transition metal complexes (which is the subject of the next chapter) is to assume that the only effect of the ligands is to produce an electrostatic field which relieves the degeneracy of the d orbitals of the central metal ion. The most serious defect of this model is that it does not recognize the existence of overlap, and hence the existence of specific bonding interactions, between the ligands and the metal orbitals. Yet calculations which assume reasonable sizes for the orbitals (together with a considerable body of physical evidence which will be reviewed in Chapter 12) point to the existence of overlap. How should this be taken account of? The simplest answer is to be found in the application of symmetry ideas to the problem, and this is the subject of this chapter. In it the reader will be assumed to have some familiarity with the basics of *group theory*. Appendix 3 gives an introduction to the subject; the following lines are intended to provide a brief overview of aspects needed to make a start on the present chapter.

It is convenient to focus the discussion on octahedral complexes (Fig. 6.1). These are molecules with high symmetry, possessing, for example, fourfold rotation axes, threefold rotation axes, twofold rotation axes, mirror planes, and a centre of symmetry (Fig. 6.2). Symbols such as a_{1g} ('aye one gee'), e_g and t_{1g} are used to distinguish the behaviour of different orbitals or sets of orbitals under the various operations associated with the rotation axes, mirror planes etc. of the octahedron. Note the shift from 'symmetry elements' (such as rotation axes, mirror planes) Fig. 6.2, to 'symmetry operations'—the *act* of rotating, reflecting etc. In group theory one is concerned with symmetry operations, rather than with symmetry elements. Lower case symbols are used to denote orbitals or, more generally, wavefunctions. So, a point to which we shall return, an s orbital of an atom at a centre of an octahedral

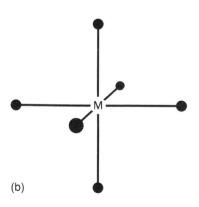

Fig. 6.1 (a) An octahedron. (b) An octahedral complex. The perspective adopted here is the same as that for (a), and was chosen to make (a) as easy as possible to visualize. The way that an octahedral complex is drawn will vary, the perspective adopted depending on the point under discussion.

Fig. 6.2 Some of the symmetry elements of an octahedron. The symbols used are those standard in chemical applications of group theory (see below). Where there is a number in brackets it indicates the total number of elements of that type.

C_4: fourfold rotation axis

C_3: threefold rotation axis

C_2: twofold rotation axis

i: centre of inversion

σ_h: (horizontal) mirror plane. Each of these planes contains two C_4 axes

σ_d: (dihedral) mirror plane. In the octahedron each of these planes contains two C_2 and two C_3 axes

Each C_4 and C_3 axis has two symmetry *operations* associated with them (the C_4 axis also has a coincident C_2). Not shown in the diagram is the 'leave alone, do nothing' operation. This is called the *identity operation* and is denoted by E. Although there is only one E (C_1) operation there is an infinite number of C_1 axes.

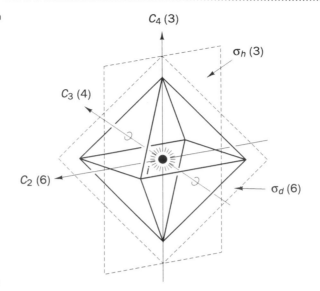

complex is referred to as an a_{1g} orbital. The symbol a, (and b, when it is met), with any suffixes or primes, indicates an orbital which is singly degenerate. Similarly e, with or without suffixes or primes, indicates a pair of degenerate orbitals.[1] Finally, t indicates a set of three degenerate orbitals. The subscripts g and u indicate behaviour under the operation of inversion in the centre of symmetry. An orbital of A_{1g} symmetry (an a_{1g} orbital) is turned into itself under inversion ($g = gerade$, German for even); something of A_{1u} symmetry (an a_{1u} orbital) is turned into the negative of itself, hence u (*ungerade*, odd). Other subscripts (and/or primes where these appear) serve only to distinguish general symmetry properties. So the symbols t_{1g} and t_{2g} represent two sets of triply degenerate orbitals, all centrosymmetric, but the members of one set behave differently from the members of the other under some symmetry operations. Although the same symbols may be applied to molecules of different symmetries there need be no obvious logical connection between them. Symmetry symbols and symmetry, reasonably, go together. A change of symmetry probably means a change of symbols. For example, although the s and three p orbitals of a metal atom at the centre of an octahedral complex are labelled a_{1g} and t_{1u}, in a tetrahedral complex the labels are a_1 and t_2.

In a non-linear polyatomic system one cannot, strictly, talk of σ, π and δ interactions. (Draw a σ bond involving the s and p orbitals of two atoms and then allow a third, non-colinear, atom to participate. The interactions between an orbital of this third atom with those forming the σ bond will not be purely σ unless the orbital on the third atom is a pure s orbital). For simplicity, in the discussion that follows, ligand–ligand overlap will be neglected (although some workers hold that this is of importance) and only ligand–metal interactions considered. This means that problems associated with the presence of third atoms can be neglected and a molecule is regarded

---

[1] The fact that some orbitals are members of degenerate sets will be taken for granted in the present chapter. In the following chapter it will be necessary to explore this degeneracy in much more detail and so a detailed discussion is deferred until then.

as held together by a network of diatomic interactions. From the point of view of the metal in a complex, each individual interaction may then be classified as either σ or π (when bonding is to another *metal* atom δ bonding may have to be included, as will be seen in Chapter 15). We start by considering octahedral complexes in some detail and, first of all, the metal–ligand σ bonding in them.

6.2 Octahedral complexes

6.2.1 Metal–ligand σ interactions

The interactions between the six ligand σ orbitals, one orbital on each ligand and pointing towards the metal, and the valence shell orbitals of the metal atom in an octahedral complex are the subject of this section. The case of a complex formed by a metal ion of the first transition series will be the one considered, so that the relevant valence shell orbitals of the metal are the 3d, the 4s, and the 4p. The ligand σ orbitals are all formally occupied by two electrons, which, in the simple picture of Chapter 1, one might regard as being donated to empty orbitals on the transition metal ion. Such a—*valence bond*—picture of the bonding is that presented by Pauling in his classic text *The Nature of the Chemical Bond* where the empty orbitals considered were d^2sp^3 metal orbital hybrids. Although there is increasing attention being paid to the valence bond model, we use a molecular orbital approach because this enables us more readily to exploit the octahedral symmetry of the molecule. Indeed, the context of the present chapter is largely symmetry-determined. The reader may easily see this by briefly comparing the content of the present chapter with that of Chapter 10. In Chapter 10 symmetry will be exploited as far as possible but that chapter also covers situations in which the symmetry is too low to be of any real assistance. The application of symmetry in the present chapter is at once a strength and a weakness; it enables an elegant and enlightening discussion but the study of real-life molecules often requires the content of Chapter 10.

The major impact of symmetry on the present discussion arises from the fact that only orbitals of the same symmetry type have non-zero overlap integrals (this is demonstrated in outline in Appendix 3). That is, if there are two sets of orbitals of E_g symmetry, (one on the ligands, the other on the metal), they will in general be non-orthogonal (i.e. have a non-zero overlap integral) and therefore interact. On the other hand both will have a zero overlap integral with all orbitals of A_{1g}, T_{1u}, T_{2g} and all other symmetry species, wherever these are located in the molecule, and will not interact with them. Not surprisingly, this fact enormously simplifies the problem. First then, we classify all the orbitals under consideration according to their symmetries. It is convenient to start by classifying the valence shell atomic orbitals of the central metal atom. It is an excellent exercise to demonstrate that the symmetry labels that follow correctly describe the transformation properties of the metal orbitals. This is an easy task for those with some experience of group theory; the inexperienced should immediately read Appendix 3.[2] Key is the character table of the octahedral group O, given in

---

[2] A detailed derivation is given in the literature in a paper that contains much other material relevant to the present chapter; see S.F.A. Kettle, *J. Chem. Educ.* (1966) *43*, 21.

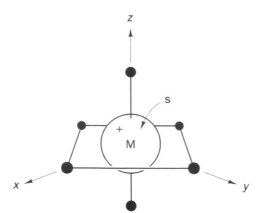

Fig. 6.3 The metal s orbital.

abbreviated form in Table 6.1 and also in Appendix 3 (Table A3.3). Because the 4s orbital (Fig. 6.3) of the central metal is turned into itself by all the operations of the group—multiplied by 1—its behaviour is described by the A_1 irreducible representation of Table 6.1. In the full octahedral group O_h (which, unlike the group O, contains the operation of inversion in a centre of symmetry), the label describing the behaviour of the 4s orbital is A_{1g} (note that an equivalent way of talking about an orbital labelled a_{1g} is to say that it 'is of A_{1g} symmetry' or that 'it transforms as A_{1g}'). In similar fashion, the set of three 4p orbitals (Fig. 6.4) have T_{1u} symmetry (T_1 in Table 6.1). Finally, the five 3d orbitals split into two sets, of E_g (Fig. 6.5) and T_{2g} (Fig. 6.6) symmetries (E and T_2 in Table 6.1). This latter splitting, which is denoted Δ, is of crucial importance to the understanding of the properties of transition metal complexes. One considerable attraction of crystal field theory, the subject of the next chapter, is that it gives a very simple physical explanation of this splitting. In the present context it is perhaps most easily regarded as a splitting which is group-theoretically allowed and so, presumably, may well exist. As the chapter develops we shall find good reasons for the splitting.

Evidently, the next step is to classify the ligand orbitals according to their symmetry types. Two problems arise. First, for a regular octahedral complex, all six ligand orbitals look alike—how then can they be classified differently? Second, a characteristic of a symmetry-classified set is that all of those symmetry operations which send an octahedron into itself send one member of the set into itself, another member or a mixture of members of the set. The reader who is not familiar with this pattern is asked to take it on trust for the moment. It will be demonstrated in the next chapter in showing the degeneracy of the metal $d_{x^2-y^2}$ and d_{z^2} orbitals in an octahedral crystal field (Section 7.3); alternatively, Appendix 4 is dedicated to the problem. As these symmetry operations send one ligand σ orbital into another, why do these orbitals not already constitute a set? The answer is, they do. However, this set can be broken down into smaller sets and it is to these latter that symmetry labels are attached.[3] In these sets the individuality of the ligand σ orbitals is lost. We talk of the wavefunctions of various sets of ligand orbitals rather than of the wavefunctions corresponding to individual ligand

Table 6.1 The O character table

O	E	$6C_4$	$3C_2$	$6C_2'$	$8C_3$
A_1	1	1	1	1	1
A_2	1	−1	1	−1	1
E	2	0	2	0	−1
T_1	3	1	−1	−1	0
T_2	3	−1	−1	1	0

[3] Appendix 6 shows how these subsets may be obtained.

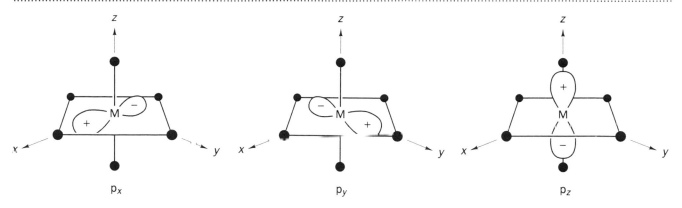

Fig. 6.4 The metal p_z, p_x and p_z orbitals.

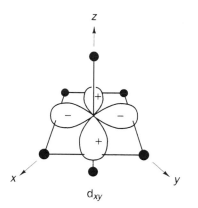

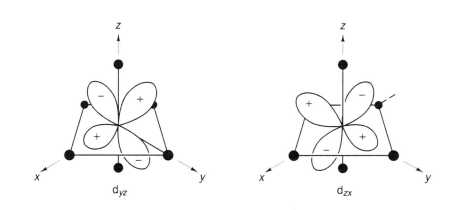

Fig. 6.5 The metal d_{xy}, d_{yz} and d_{zx} orbitals.

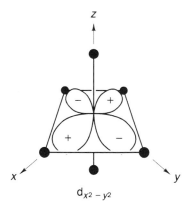

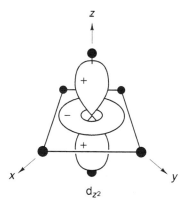

Fig. 6.6 The metal $d_{x^2-y^2}$ and d_{z^2} orbitals.

orbitals. The members of the symmetry-classifiable sets are linear combinations of the ligand σ orbitals and are often referred to either as 'ligand group orbitals' or 'symmetry-adapted combinations'. The step of moving from the individual ligand σ orbitals to their symmetry-adapted combinations involves the use of group theoretical procedures. The derivation is not difficult but it is lengthy and so would be out of place in the present section. It is given in Appendix 6, where several different ways of tackling the problem will be found. The explicit form of these ligand group orbitals is given in Table 6.2. The individual σ orbitals are identified by the labels $\sigma_1 \rightarrow \sigma_6$ as

Table 6.2 Ligand σ group orbitals of six octahedrally-orientated ligands

Symmetry	Ligand group orbitals
a_{1g}	$\dfrac{1}{\sqrt{6}}(\sigma_1 + \sigma_2 + \sigma_3 + \sigma_4 + \sigma_5 + \sigma_6)$
t_{1u}	$\begin{cases} \dfrac{1}{\sqrt{2}}(\sigma_1 - \sigma_6) \\[2mm] \dfrac{1}{\sqrt{2}}(\sigma_2 - \sigma_4) \\[2mm] \dfrac{1}{\sqrt{2}}(\sigma_3 - \sigma_5) \end{cases}$
e_g	$\begin{cases} \frac{1}{2}(\sigma_2 - \sigma_3 + \sigma_4 - \sigma_5) \\[2mm] \dfrac{1}{\sqrt{12}}(2\sigma_1 + 2\sigma_6 - \sigma_2 - \sigma_3 - \sigma_4 - \sigma_5) \end{cases}$

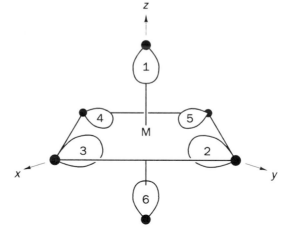

Fig. 6.7 The numbering system adopted for the ligand σ orbitals in an octahedral complex (cf. Fig. 3.4).

shown in Fig. 6.7, the labels being chosen in conformity with the convention given in Fig. 3.4. As mentioned earlier, in the explicit forms given in Table 6.2 it has been assumed that the ligand σ orbitals do not overlap each other, although non-zero ligand–ligand overlap integrals would only affect the normalization factors (the numerical coefficients at the front of each expression) given in the table, not the general form of the combinations.

The ligand σ group orbitals listed in Table 6.2 are of A_{1g}, E_g and T_{1u} symmetries and each set will overlap with metal orbitals of the same symmetries, so that the 4s(a_{1g}), 4p(t_{1u}) and 3d(e_g) orbitals of the metal will be involved in σ bonding. Of the valence shell orbitals of the metal atom only the 3d(t_{2g}) orbitals are not involved in this bonding. The interactions of A_{1g}, T_{1u} and E_g symmetries are shown pictorially in Figs. 6.8–6.10, where it can be seen that there is the expected close matching between ligand group orbitals and the corresponding metal orbitals. If symmetry had not been

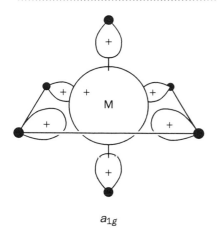

a_{1g}

Fig. 6.8 The metal–ligand σ bonding in an octahedral complex involving orbitals of A_{1g} symmetry. The metal s orbital is shown at the centre.

Fig. 6.9 The metal–ligand σ bonding in an octahedral complex involving orbitals of T_{1u} symmetry. In this figure the interaction involving the metal p_x orbital is shown; there are similar interactions involving p_y and p_z.

applied to the σ bonding problem then we would have had to consider the interaction between the six ligand σ orbitals and nine metal orbitals, with no means of knowing in advance that three of the 3d are not involved in this σ bonding. In contrast, not only does the symmetry break the problem up into subproblems—the A_{1g}, E_g and T_{1u}— but for each degenerate case, E_g and T_{1u}, only one member of each set need be considered (by symmetry, consideration of any other member must lead to the same result). That is, for each of the three subproblems only the interaction between a single metal orbital and a (delocalized) ligand orbital has to be treated. For simple models the calculations really can be done on the back of an envelope! Fig. 6.11 shows, schematically, the energy level pattern obtained as a result of these σ interactions. It includes metal and ligand electrons; the detailed problem of how to include those originating on the metal will be a major concern of Chapters 7 and 8. Note, particularly, the splitting Δ in Fig. 6.11. Many

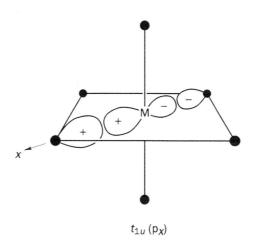

$t_{1u} (p_x)$

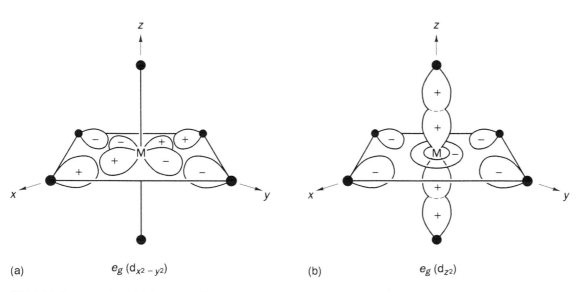

(a) $e_g (d_{x^2-y^2})$ (b) $e_g (d_{z^2})$

Fig. 6.10 The interaction (a) between metal $d_{x^2-y^2}$ and (b) d_{z^2} orbitals and the corresponding ligand group orbitals.

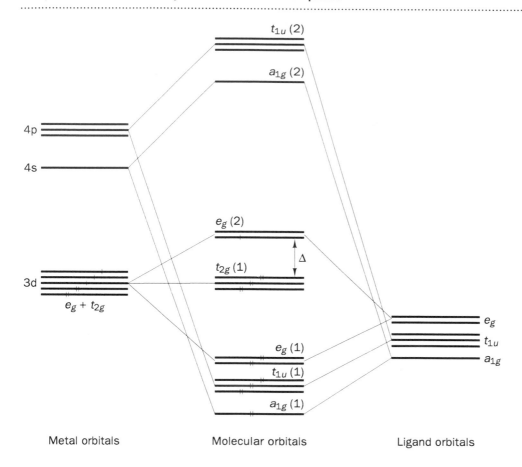

Metal orbitals   Molecular orbitals   Ligand orbitals

Fig. 6.11 A schematic molecular orbital energy level scheme for an octahedral complex of a first row transition metal ion (only σ interactions are included). The ligand orbitals are shown occupied; the metal d orbitals are partially filled. The example shown is appropriate to a d^7 ion in a strong-field (low-spin) complex. The full meaning of these latter terms will only become evident in Chapter 7. The numbers in parenthesis follow the established convention of counting from the bottom up. So, the bottom a_{1g} is $a_{1g}(1)$, the next $a_{1g}(2)$.

research groups have tried to carry out reasonably accurate calculations aimed, in part, at obtaining molecular orbital energy level schemes (such as that in Fig. 6.11) accurately, although this is a difficult task. It in this endeavour that many of the methods to be discussed in Section 10.3.2 were developed.

Figure 6.11 indicates that the idea that the ligands function as electron donors is correct. Electrons which, before the interactions were 'switched on', were in pure ligand orbitals are really in delocalized molecular orbitals which take them onto the metal atom—the electron density on the metal atom is increased as a result of the covalency. The consequences of this for the concept of formal valence states will be discussed later. The most important feature of Fig. 6.11 is the fact that the two lowest unoccupied orbital sets are, in order, t_{2g} and $e_g(2)$ (this latter being weakly σ antibonding). So far, no metal electrons have formally been included and so these unoccupied orbitals have to accommodate them. Of course, the number of electrons occupying these orbitals is the same as the number present in the valence shell orbitals of the metal before the metal–ligand interaction was switched on; the d electrons of the uncomplexed metal ion may be regarded as being distributed between the t_{2g} and $e_g(2)$ molecular orbitals. This is a situation which will be explored in the next chapter where care will be taken to state that Δ, the label given to the energy separation between a lower t_{2g} and an upper e_g set, is an experimental quantity. What the next chapter will not show is the way that the e_g set is antibonding. The present discussion also leads us to recognize that metal electrons in the $e_g(2)$ orbitals are to some extent delocalized over the ligands. In Chapter 12 experimental evidence will be adduced in support of this conclusion.

6.2.2 Metal–ligand π interactions

So far we have only considered σ bonding in an octahedral complex. What of π bonding? It appears to be generally true in chemistry that π is rather weaker than σ bonding. Thus, there is no compound known in which it has been established that, in the ground state, there is a π bond but no σ bond. On the other hand, the vast majority of molecules have at least one σ bond with, apparently, no associated π bond. Consequently, for our purposes, it can reasonably be assumed that the effects of π bonding will be to modify, but probably not drastically alter, Fig. 6.11.

Our basic approach to the treatment of π bonding is similar to that of σ bonding. It is convenient to assume that the ligands involved are simple, something like a halide or cyanide anion (cases which will be explicitly explored later). It follows that each ligand has two orbitals available for π bonding on each ligand—ligand p_x orbitals or (ligand) molecular orbitals with the same symmetry characteristics as the corresponding p_x orbitals. That is, they have their maximum amplitude perpendicular to the metal–ligand axis and this axis lies in a nodal plane of the ligand π orbitals (Fig. 6.12).

As there are two such orbitals on each ligand, there is a total of 12 in an octahedral ML_6 complex. Again, group theory can be applied to obtain the

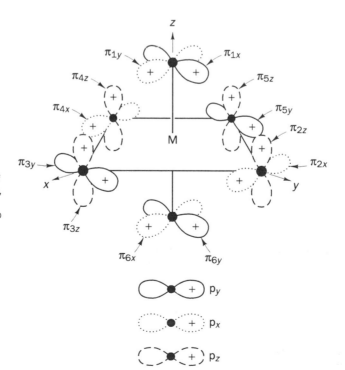

Fig. 6.12 Ligand π orbitals in an octahedral complex. The arrow heads represent the lobe of each orbital which has positive phase. Although, for generality, these orbitals are referred to as π, those actually drawn are pure p orbitals. All p orbitals are labelled according to octahedral axes using the outlining notation given at the foot of the diagram.

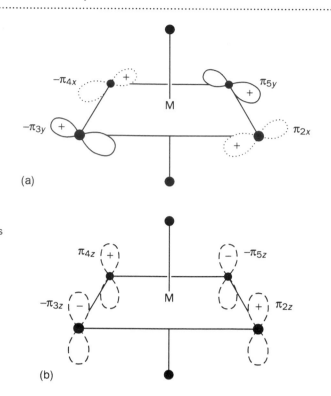

Fig. 6.13 (a) One of the three ligand π group orbitals of T_{1g} symmetry. Notice that of the 12 π ligand orbitals only four are involved in this particular combination. The remaining eight similarly divide into two sets of four, with no member common to any two sets. A similar pattern holds for all the other π combinations shown in the following figures. (b) One of the three ligand π group orbitals of T_{2u} symmetry. As for all other π combinations in adjacent figures, the combination shown is the first listed in the appropriate section of Table 6.3.

Table 6.3 π ligand group orbitals for the six ligands in octahedral complexes (a definition of the labels is given in Fig. 6.12)

Symmetry	Ligand group orbital
t_{1u}	$\frac{1}{2}(\pi_{2z} + \pi_{3z} + \pi_{4z} + \pi_{5z})$
	$\frac{1}{2}(\pi_{1x} + \pi_{2x} + \pi_{6x} + \pi_{4x})$
	$\frac{1}{2}(\pi_{1y} + \pi_{3y} + \pi_{6y} + \pi_{5y})$
t_{1g}	$\frac{1}{2}(\pi_{2x} - \pi_{3y} - \pi_{4x} + \pi_{5y})$
	$\frac{1}{2}(\pi_{1x} - \pi_{2z} - \pi_{6y} + \pi_{4z})$
	$\frac{1}{2}(\pi_{1y} - \pi_{3z} - \pi_{6x} + \pi_{5z})$
t_{2g}	$\frac{1}{2}(\pi_{2x} + \pi_{3y} - \pi_{4x} - \pi_{5y})$
	$\frac{1}{2}(\pi_{1y} + \pi_{3z} - \pi_{6x} - \pi_{4x})$
	$\frac{1}{2}(\pi_{1x} + \pi_{2z} - \pi_{6y} - \pi_{5z})$
t_{2u}	$\frac{1}{2}(\pi_{2z} - \pi_{3z} + \pi_{4z} - \pi_{5z})$
	$\frac{1}{2}(\pi_{1x} - \pi_{2x} + \pi_{6x} - \pi_{4x})$
	$\frac{1}{2}(\pi_{1y} - \pi_{3y} + \pi_{6y} - \pi_{5y})$

symmetry-adapted combinations. The symmetry-adapted combinations consist of four sets with three ligand group orbitals in each set. The sets have T_{1g}, T_{1u}, T_{2g} and T_{2u} symmetries; explicit expressions for them are given in Table 6.3 using the ligand π orbitals labelled as shown in Fig. 6.12. The derivation of the explicit forms of these functions is somewhat lengthy and is not given in this book; the procedure follows that given in Appendix 6; a detailed but simple treatment is given elsewhere.[4] There are no metal orbitals of T_{1g} and T_{2u} symmetries within our chosen valence set (although, just for the record, one set of metal f orbitals transforms as T_{2u} and one set of metal g orbitals as T_{1g}). The ligand π group orbitals of T_{1g} and T_{2u} symmetries are therefore carried over, unmodified, into the full molecular orbital description. A representative example of each is shown in Fig. 6.13. We are left with T_{2g} and T_{1u} sets. The T_{1u} set will interact with the metal p orbitals (also of T_{1u} symmetry), but it is simpler to think of their effect on the occupied T_{1u} σ molecular orbitals in Fig. 6.11 and to consider two cases.

Case 1: The ligand π orbitals are occupied

In this case, illustrated in Figs. 6.14 and 6.15(c), it follows that the T_{1u} symmetry ligand π orbitals are also occupied. Interaction with the $t_{1u}(1)$ molecular orbitals of Fig. 6.11 will raise or lower this latter set depending on whether its energy is higher or lower than that of the ligand $\pi(T_{1u})$ set. As long as we retain two occupied t_{1u} sets the orbital occupancy is unaffected

[4] Appendix 4 in *Symmetry and Structure*, 2nd edn., by S. F. A. Kettle, Wiley (Chichester and New York) 1995. Note that the atom labelling pattern used in this text differs from that in the present.

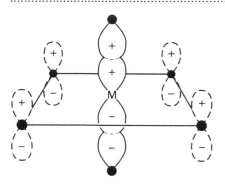

Fig. 6.14 A ligand π group orbital of T_{1u} symmetry and the molecular orbital of Fig. 6.9 with which it interacts (that involving the metal p_z orbital).

by this interaction and, because the $t_{2g} - e_g(2)$ separation is our concern, for this case we may forget the t_{1u} orbitals and their interactions.

Case 2: The ligand π orbitals are unoccupied

In this case it follows that the ligand π combinations of T_{1u} symmetry are also unoccupied. Interaction between the ligand π and (occupied) $t_{1u}(1)$ sets of Fig. 6.11 will result in a repulsion between the two and the $t_{1u}(1)$ set will be stabilized somewhat but the orbital occupancy will remain unchanged. This is illustrated in Figure 6.15(b). We could, as a separate case, consider the interaction between the $t_{1u}(2)$ molecular orbital set of Fig. 6.11 and the ligand $\pi(t_{1u})$ set. Again, this would only be necessary if the interaction leads to one of the orbital sets becoming occupied. In fact, there is no recognized case in which the ligand $\pi(t_{1u})$ set, whether occupied or not, needs to be considered.

We are left with the ligand π set of T_{2g} symmetry. This set may interact with the metal t_{2g} orbitals (d_{xy}, d_{yz}, d_{zx}) which, so far, have been non-bonding because there is no ligand σ combination of T_{2g} symmetry. The consequences

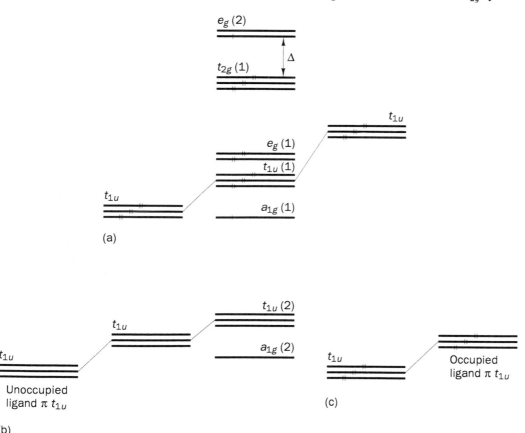

(a)

(b)

Unoccupied ligand π t_{1u}

(c)

Occupied ligand π t_{1u}

Fig. 6.15 The interaction between the lowest t_{1u} orbitals of Fig. 6.11(a) with (b) an unoccupied ligand $\pi(t_{1u})$ set and (c) an occupied ligand $\pi(t_{1u})$ set. Largely ignored in the text is the $t_{1u}(2)$ set of Fig. 6.11; corresponding arguments can be developed, with similar conclusions, for this set.

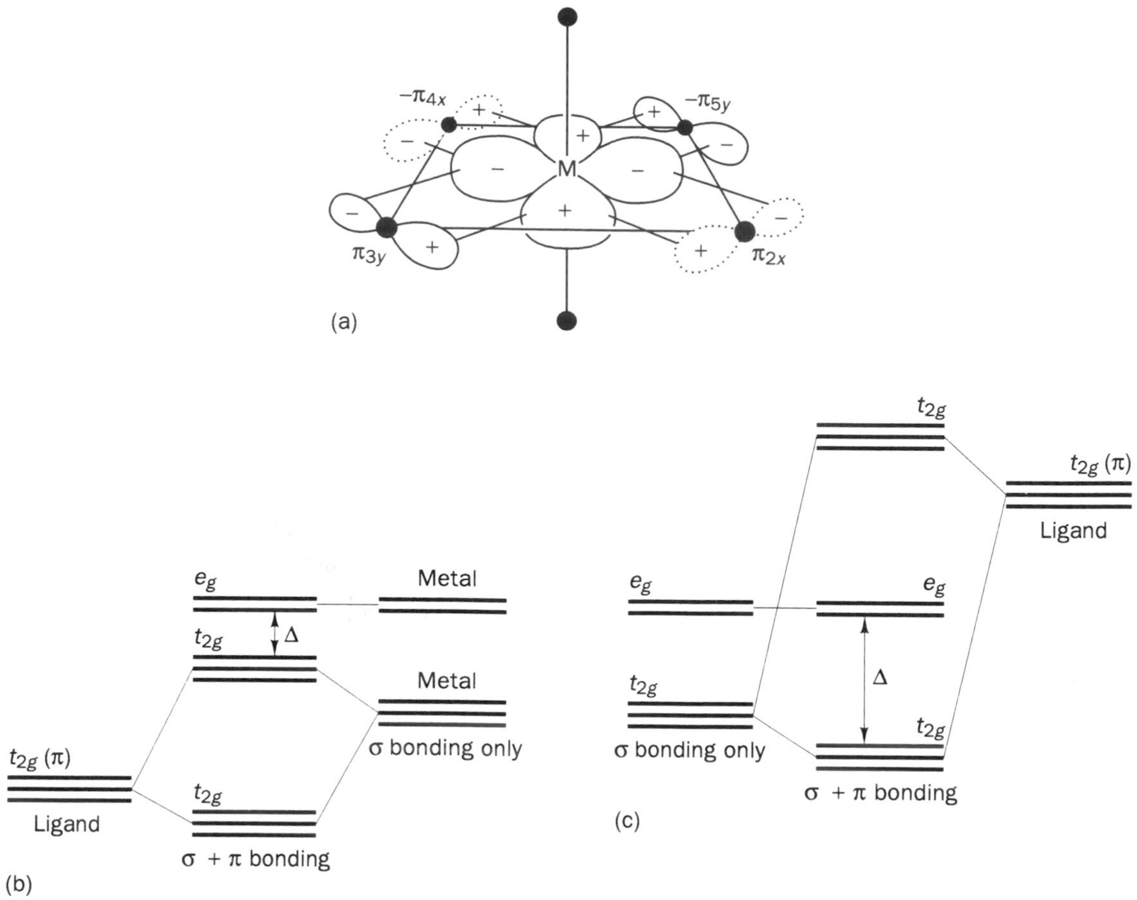

(a)

(b)

(c)

Fig. 6.16 (a) The interaction between a metal t_{2g} orbital (d_{xy}) and the corresponding ligand π combinations. Overlapping lobes are joined by lines. (b) The consequences when the ligand $\pi(t_{2g})$ are low lying (and occupied); Δ is decreased. (c) The consequences when the ligand $\pi(t_{2g})$ are high lying (and empty); Δ is increased.

of interaction between the ligand $\pi(t_{2g})$ orbitals and the metal t_{2g} orbitals shown in Fig. 6.11 depend upon which of the two t_{2g} sets is higher in energy. The alternative situations are shown in Fig. 6.16. If the ligand π set is the higher in energy then the metal t_{2g} set is pushed down; if the ligand π set is the lower, the metal set is raised in energy. As Fig. 6.16 shows, these movements have a direct effect on the $t_{2g} - e_g(2)$ splitting. That is, the magnitude of Δ depends, in part, on π bonding. It is believed that the halide anions provide examples of the situation shown in Fig. 6.16(b). For these ligands, two of the filled p orbitals in their valence shell will be the orbitals involved in π bonding and interaction with them will raise the energy of the metal t_{2g} orbitals making the latter weakly antibonding. This explains, partially at least, why complexes containing halide anions as ligands have relatively small values of Δ (see Table 7.1). In contrast, the cyanide anion is associated with large values of Δ. The cyanide anion possesses two sets of π orbitals both of which will interact with the metal orbitals. The effect of the occupied $C\equiv N^-$ π bonding orbitals will be similar to that of the occupied π orbitals of the halides. However, these π orbitals have energies well removed from those of the metal d orbitals and so their effect is small. Much more important are the empty $C\equiv N^-$ π antibonding orbitals, which behave as shown in Fig. 6.16(c). Interaction with the metal t_{2g} orbitals stabilizes the latter (which become weakly π bonding) and so increases the metal

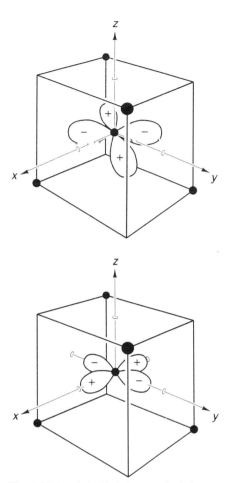

Fig. 6.17 Metal d orbitals in a tetrahedral complex.

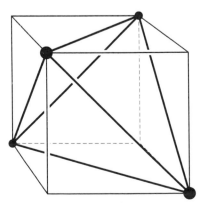

Fig. 6.18 The relationship between a cube and a tetrahedron.

$t_{2g} - e_g(2)$ separation of Fig. 6.11. This behaviour is clearly consistent with the observation that the cyanide anion gives rise to large values of Δ; it exerts a very large ligand field. In Chapter 10 we will see that the carbon monoxide ligand behaves similarly.

Finally, a cautionary note. There is theoretical evidence, despite all that has been said in this chapter, that in a complex such as $[Cr(CN)_6]^{3-}$ the d electrons are *not* in the highest occupied orbitals. Rather, that all the CN^- π occupied and, indeed, the CN^- σ bonding orbitals, all occupied, have higher energies.[5] The reason for this apparently ridiculous behaviour is to be found in a phenomenon which so often wrecks our simple pictures—electron repulsion within the molecule. How can there be filled orbitals above partially occupied ones? The increase in electron repulsion energy in taking an electron from a (delocalized and therefore diffuse) π occupied or σ bonding orbital and putting it into a (largely localized and a therefore concentrated) metal d orbital 'costs' more than the energy gained from moving the electron into what is a more stable orbital. The holes in the d orbitals are protected! Fortunately, this complication turns out to be unimportant for the discussion in this chapter. It doesn't really matter too much where the d orbitals are relative to the ligand orbitals. As has been commented 'if one is only interested in the energy pattern, ligand field theory remains a reliable guide'. Since virtually all of the measurements made on transition metal complexes are concerned with the detailed energy pattern, all is well. The reader who is either unhappy with this situation or is so interested in it that they wish to learn more, should turn to Section 10.7 where it is encountered again, in the context of ferrocene, and discussed in more detail. In Section 12.7 measurements that go beyond energy patterns will be described and compared with theory, whereupon the relative energies of the ligand orbitals will become of importance.

6.3 Tetrahedral complexes

In this and the next section the bonding in complexes with other-than-octahedral stereochemistries are considered. That the d orbitals split into t_2 and e sets in tetrahedral complexes may be seen from Fig. 6.17. The key is to recognize that a tetrahedron is closely related to a cube. If, starting from one vertex, lines are drawn across the face diagonals of a cube to connect with the opposite vertices and the process continued, then a tetrahedron results (Fig. 6.18). If Cartesian axes are drawn as in Fig. 6.17 then $x \equiv y \equiv z$ so that d_{xy}, d_{yx} and d_{zx} must be degenerate. The degeneracy of $d_{x^2-y^2}$ and d_{z^2} follows just as it did for octahedral complexes (for a detailed justification, see Section 7.3 and/or Appendix 4). The other metal orbitals in the valence shell are $4s(a_1)$ and $4p_x, 4p_y, 4p_z(t_2)$, again considering a first-row transition element. It is a simple exercise in group theory to show that the four ligand σ orbitals give A_1 and T_2 symmetry-adapted combinations. The ligand group orbitals may be obtained by the methods of Appendix 6 (for the third method described there, also use the operations

[5] There are many papers in the literature which indicate this. See, for example: Vanquicken-borne et al., *Inorg. Chem.* (1984) *23*, 1677; *Excited States and Reactive Intermediates*, ACS Symposium Series 307 (1986) 2; and *Inorg. Chem.* (1991) *30*, 2978.

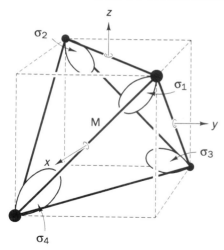

Fig. 6.19 The labels used in the text for the ligand σ orbitals of a tetrahedral complex.

Table 6.4 Ligand group orbitals of σ symmetry in a tetrahedral complex; the labels used are those of Fig. 6.19

Symmetry	Ligand group orbital
a_1	$\frac{1}{2}(\sigma_1 + \sigma_2 + \sigma_3 + \sigma_4)$
t_2	$\begin{cases} \frac{1}{2}(\sigma_1 + \sigma_2 - \sigma_3 - \sigma_4) \\ \frac{1}{2}(\sigma_1 - \sigma_2 - \sigma_3 + \sigma_4) \\ \frac{1}{2}(\sigma_1 - \sigma_2 + \sigma_3 - \sigma_4) \end{cases}$

E, $8C_3$ and $3C_2$). Using the orbital labels shown in Fig. 6.19, the A_1 and T_2 ligand group orbitals are given in Table 6.4 and pictured in Fig. 6.20, a figure which also shows the metal orbitals of the same symmetries.

As before, only orbitals of the same symmetry species may interact with each other. The A_1 interactions are straightforward but the T_2 more complicated than the corresponding octahedral case because both metal sets $4p_x$, $4p_y$, $4p_z$ and $3d_{xy}$, $3d_{yz}$, $3d_{zx}$ have T_2 symmetry and so interact with the ligand orbitals of this symmetry. It does not appear possible to make a general statement about the details of the outcome of the T_2 interactions but that given in Fig. 6.21 is about as close as one can get. Figure 6.21 gives a schematic σ bonding-only molecular orbital diagram for a tetrahedral complex. As we shall see in the next chapter, just as in octahedral complexes, the more stable d orbital set according to crystal-field theory, that of e symmetry, is not involved in σ bonding but the less stable, the t_2, is involved.

Many tetrahedral complexes involve the oxide anion as a ligand (e.g. the MO_4^{n-} anions commonly known as permanganate, chromate, and ferrate) and contain a metal atom in a high formal valence state. Since oxygen is usually regarded as forming two covalent bonds and because a high metal charge will favour ligand-to-metal charge migration, it can be anticipated that π bonding may well be of potential importance for such tetrahedral complexes. Unfortunately, the consequences of π bonding are not as clear-cut as for octahedral complexes. An important difference between octahedral and tetrahedral complexes is that the latter do not have a centre of symmetry. Such a centre of symmetry separates the p orbitals from the d orbitals on the metal (the ps are *ungerade* and the ds are *gerade*) and no centrosymmetric ligand field can mix them. A tetrahedral ligand field *can* mix ps and ds on the metal and there is good evidence that such mixing occurs—it will be met in Chapter 8.

The π bonding problem in tetrahedral molecules starts difficult and remains so throughout. It is not a trivial task to demonstrate that the ligand π orbital symmetry-adapted combinations are of $T_1 + T_2 + E$ symmetries. To show this it is important to chose the orientation of the ligand π orbitals carefully if the task is to be made (relatively) easy. Guidance is given in Fig. 6.22. Appendix 4, and the character of -1 described at the end, may well be needed also. The next step, that of the generation of the ligand group orbitals, is not trivial.[6] The explicit expressions for the ligand group orbitals

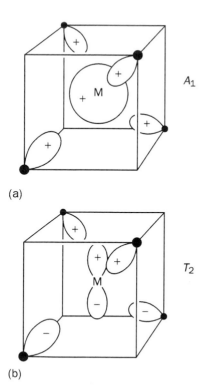

(a)

(b)

Fig. 6.20 (a) Interaction of the ligand a_1 group orbital of Table 6.3 with the metal s orbital. (b) Interaction of the first ligand t_2 σ group orbital listed in Table 6.4 with the corresponding metal p orbital.

[6] A fairly simple derivation is given in detail in S. F. A. Kettle, *Symmetry and Structure*, 2nd edn., Wiley, Chichester, 1995, Appendix 4.

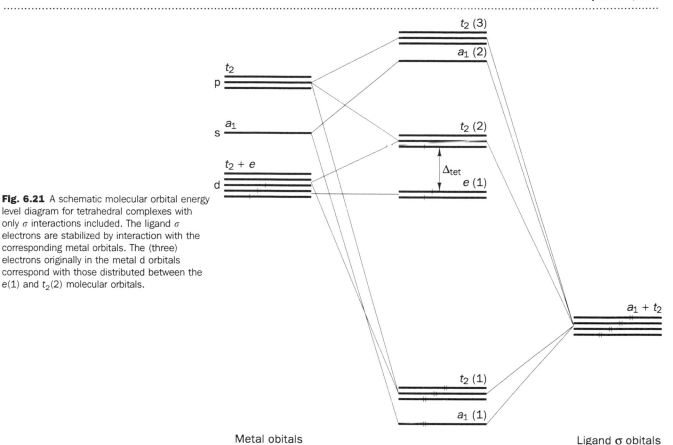

Fig. 6.21 A schematic molecular orbital energy level diagram for tetrahedral complexes with only σ interactions included. The ligand σ electrons are stabilized by interaction with the corresponding metal orbitals. The (three) electrons originally in the metal d orbitals correspond with those distributed between the $e(1)$ and $t_2(2)$ molecular orbitals.

Metal obitals

Ligand σ obitals

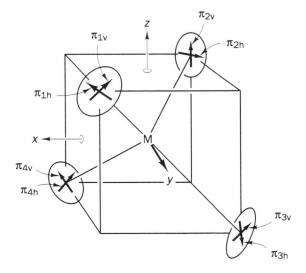

Fig. 6.22 A symbolic representation of the ligand π orbitals in a tetrahedral complex. On each ligand there are two π orbitals and these may be regarded as lying in a plane perpendicular to the local ligand–metal axis (at each ligand this plane is represented by a circle which, however, may appear as an ellipse because of the perspective). At each ligand a π orbital is labelled π_v if it lies in a plane containing the z axis and π_h if it is perpendicular to this plane.

are given in Table 6.5 using the notation of Fig. 6.22 and illustrated in Figs. 6.23–6.25, the relevant metal orbitals being included in the e and t_2 cases. For tetrahedral complexes π bonding involves all of the metal d orbitals, not just one set (as in the octahedral case), and it is not possible to give a simple diagram analogous to Fig. 6.16. Both the $t_2(2)$ and e levels of Fig. 6.21 will

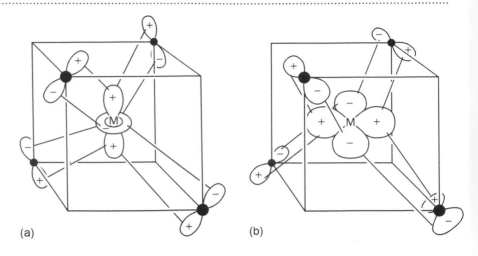

Fig. 6.23 Ligand π group orbitals of E symmetry: (a) that which overlaps with the metal d_{z^2}; (b) that which overlaps with the metal $d_{x^2-y^2}$. The pattern of overlaps in this diagram is worthy of careful study.

(a)   (b)

Table 6.5 π ligand group orbitals of a tetrahedral complex. Figure 6.22 gives a definition of the orbitals

Symmetry	Ligand group orbital
e	$\begin{cases} \frac{1}{2}(\pi_{1v} + \pi_{2v} - \pi_{3v} - \pi_{4v}) \\ \frac{1}{2}(\pi_{1h} + \pi_{2h} - \pi_{3h} - \pi_{4h}) \end{cases}$
t_1	$\begin{cases} \frac{1}{2}(\pi_{1h} + \pi_{2h} + \pi_{3h} + \pi_{4h}) \\ \frac{1}{2\sqrt{2}}(\pi_{3h} - \pi_{4h} + \sqrt{3}(\pi_{1v} - \pi_{2v})) \\ \frac{1}{2\sqrt{2}}(\pi_{1h} - \pi_{2h} + \sqrt{3}(\pi_{3v} - \pi_{4v})) \end{cases}$
t_2	$\begin{cases} \frac{1}{2}(\pi_{1v} + \pi_{2v} + \pi_{3v} + \pi_{4v}) \\ \frac{1}{2\sqrt{2}}(\sqrt{3}(\pi_{1h} - \pi_{2h}) - \pi_{3v} + \pi_{4v}) \\ \frac{1}{2\sqrt{2}}(\sqrt{3}(\pi_{3h} - \pi_{4h}) - \pi_{1v} + \pi_{2v}) \end{cases}$

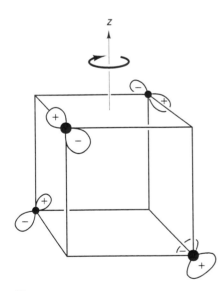

Fig. 6.24 A ligand π group orbital of T_1 symmetry. This orbital has the same symmetry properties as a rotation about the z axis. It is the fact that the ligand π orbitals have been chosen to be oriented with respect to the z axes (see caption to Fig. 6.22) that makes this orbital, mathematically and pictorially, particularly simple. Its two partners in Table 6.5 have the same symmetry properties as rotations about the x and y axes, respectively.

change with π bonding and it is not possible to discuss the behaviour of their separation, Δ_{tet}, in general terms although it is quite clear that Δ_{tet} will be sensitive to π bonding. However, both e and t_2 π interactions are likely to be either both bonding or both antibonding so that the difference between the e and $t_2(2)$ orbital energy changes may be small and this may mean that π bonding has a relatively small effect on Δ_{tet}.

A schematic molecular orbital pattern for a tetrahedral complex with a significant π bonding contribution is shown in Fig. 6.26. It is emphasized that the relative energies shown for the molecular orbitals in this figure are to be regarded as highly flexible.

6.4 Complexes of other geometries

Apart from one example, other geometries will not be discussed in detail because there are none for which we could arrive at firm general conclusions. Instead, in Table 6.6 are listed the symmetries of the σ and π ligand group

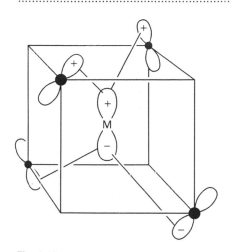

Fig. 6.25 A ligand π group orbital of T_2 symmetry together with the metal p_z with which it overlaps. The dominant orbital overlaps are indicated. The form of this ligand group orbital is particularly simple for the reason given in the caption to Fig. 6.24. Note carefully the phases of the overlap—they are not incorrect!

Fig. 6.26 A schematic molecular orbital energy level diagram for a tetrahedral transition metal complex in which both σ and π interactions are important.

orbitals for common geometries and compared with the symmetries of the orbitals comprising the valence set of the transition metal ion. This table is group theoretical in origin and is designed to enable the reader to develop a qualitative bonding argument for almost any complex with up to eight ligands with any significant symmetry (if a molecule is only slightly distorted from a high symmetry it often pays to pretend that there is no distortion at all, as a first approximation). The reader who is unhappy about the plethora of labels in Table 6.6 need not be too concerned—these labels can be regarded as simply indicating what interactions can occur—labels have to be identical for an interaction to be possible; there is no vital need to enquire into their deeper (group theoretical) significance.

As an example of the use of Table 6.6 its application to a trigonal bipyramidal complex of D_{3h} symmetry (Fig. 6.27) will be outlined. First, it has to be recognized that our discussion of octahedral (O_h) and tetrahedral (T_d) complexes will be of little direct use in discussing ML_5 (D_{3h}) complexes. Had we been concerned with complexes with either six or four ligands, however, it might well have been a good idea to start from the appropriate high symmetry case, as mentioned above. As Table 6.6 shows, in a ML_5 complex the metal d orbitals split up into three sets, d_{z^2} has A_1' symmetry, $d_{x^2-y^2}$ and d_{xy}, together, are of E' symmetry whilst d_{yz} and d_{zx}, together, have E'' symmetry. The somewhat surprising fact that $d_{x^2-y^2}$ pairs with d_{xy}

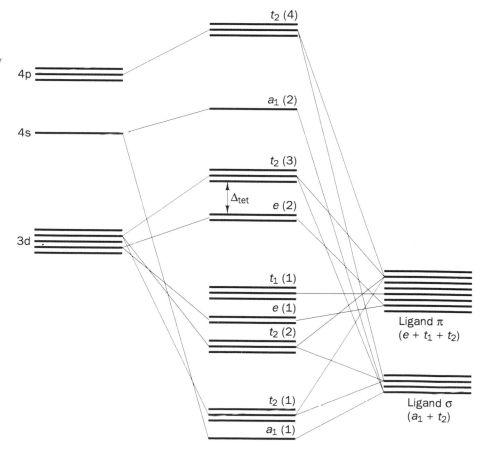

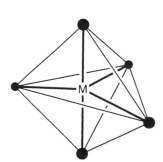

Fig. 6.27 A trigonal bipyramidal ML_5 complex.

Table 6.6 The symmetries of ligand σ, π and metal orbitals for common geometries

Molecular geometry	Symmetry symbol	Symmetries of metal orbitals						
		s	p_z	p_x	p_y	d_{z^2}	$d_{x^2-y^2}$	d_{xy}
Octahedral, ML_6	O_h	a_{1g}		t_{1u}		e_g		
Tetrahedral, ML_4	T_d	a_1		t_2		e		
trans-Octahedral, ML_4M_2'	D_{4h}	a_{1g}	a_{2u}	e_u		a_{1g}	b_{1g}	b_{2g}
Square planar, ML_4	D_{4h}	a_{1g}	a_{2u}	e_u		a_{1g}	b_{1g}	b_{2g}
Octahedral with a bidentate chelating ligand, $M(L_2)_3$	D_3	a_1	a_2	e		a_1	e	
Trigonal bipyramidal, ML_5	D_{3h}	a_1'	a_2''	e'		a_1'	e'	
All cis-octahedral, ML_3L_3'	C_{3v}	a_1	a_1	e		a_1	a_1	a_2
Tetrahedral ML_3L'	C_{3v}	a_1	a_1	e		a_1	a_1	a_2
One face centred octahedral, $ML_3L_3'L''$	C_{3v}	a_1	a_1	e		a_1	a_1	a_2
Square based pyramidal, ML_4L'	C_{4v}	a_1	a_1	e		a_1	b_1	b_2
Octahedral, ML_5L' or trans-$ML_4L'L''$	C_{4v}	a_1	a_1	e		a_1	b_1	b_2
cis-Octahedral	C_{2v}	a_1	a_1	b_1	b_2	a_1	a_1	a_2
Tetrahedral ML_2L_2'	C_{2v}	a_1	a_1	b_1	b_2	a_1	a_1	a_2
Square face-centred trigonal prism, $ML_4L_2'L''$	C_{2v}	a_1	a_1	b_1	b_2	a_1	a_1	a_2
Dodecahedral, ML_8	D_{2d}	a_1	b_2	e_1		a_1	b_1	b_2
Square antiprism, ML_8	D_{4d}	a_1	b_2	e_1		a_1	e_2	

and not d_{z^2} is easily explained—it is a general rule that the axis of highest symmetry is (almost[7]) always chosen as the z axis, so here we choose the threefold rotation axis; for an octahedron the z axis was a C_4 and for a tetrahedron an S_4.

The next step is to include, qualitatively, σ bonding in the picture. The five ligand group σ orbitals are of $2A_1' + A_2'' + E'$ symmetries. They can easily be obtained by the methods of Appendix 6 (treat non-equivalent ligands separately) and are pictured in Fig. 6.28. From this figure it is evident that

[7] When working with an icosahedral molecule such as $[B_{12}H_{12}]^{2-}$ or C_{60} life is much easier if a C_2 (rather than a C_5) axis is chosen as z because x and y can then also be orientated along C_2s and the three coordinate axes are symmetry-related. With z chosen to lie along a C_5 axis the three coordinate axes are not symmetry-related.

d_{yz}	d_{zx}	Symmetries of the ligand orbitals		Comments
		σ	π	
		$a_{1g} + e_g + t_{1u}$	$t_{1g} + t_{1u} + t_{2g} + t_{2u}$	
		$a_1 + t_2$	$e + t_1 + t_2$	b_{1g}/b_{2g}; b_{1u}/b_{2u} possibly exchanged
e_g		$2a_{1g} + a_{2u} + b_{1g} + e_u$	$a_{2g} + b_{2g} + a_{2u} + b_{2u} + 2e_g + 2e_u$	
e_g		$a_{1g} + b_{2g} + e_u$	$a_{2g} + b_{2g} + a_{2u} + b_{2u} + e_g + e_u$	
e		$a_1 + b_2 + 2e$	$2a_1 + 2a_2 + 4e$	
e''		$2a'_1 + a'_2 + e'$	$a'_2 + a''_2 + 2e' + 2e''$	
e		$2a_1 + 2e$	$2a_1 + 2a_2 + 4e$	the threefold axis is the z axis
e		$2a_1 + e$	$a_1 + a_2 + 3e$	
e		$3a_1 + 2e$	$2a_1 + 2a_2 + 5e$	
e		$2a_1 + b_1 + e$	$a_1 + b_1 + a_2 + b_2 + 3e$	
e		$3a_1 + b_1 + e$	$a_1 + b_1 + a_2 + b_2 + 4e$	
b_2	b_1	$3a_1 + b_1 + a_2 + b_2$	$2a_1 + 4b_1 + 2a_2 + 4b_2$	
b_2	b_1	$2a_1 + b_1 + b_2$	$2a_1 + 2b_1 + 2a_2 + 2b_2$	b_1 and b_2 may be interchanged by some authors
b_2	b_1	$3a_1 + 2b_1 + a_2 + b_2$	$3a_1 + 4b_1 + 3a_2 + 4b_2$	
e		$2a_1 + 2b_2 + 2e$	$2a_1 + 2b_1 + 2a_2 + 2b_2 + 4e$	
e_3		$a_1 + b_2 + e_1 + e_2 + e_3$	$a_1 + b_1 + a_2 + b_2 + 2e_1 + 2e_2 + 2e_3$	

both of the a'_1 ligand orbitals may interact with the same metal a'_1 orbital. The relative importance of the two interactions will depend on the geometry of the complex—which of the (two) axial and the (three) equatorial ligands are the closer to the metal? Usually, the equatorial. The problem is made more complicated, of course, by the presence of two metal orbitals which are A'_1, because, in addition to d_{z^2}, the metal s orbital has this symmetry. Although by no means always justified by detailed calculations, it is often convenient to take sum and difference of the d_{z^2} and s orbitals. This leads to one metal A'_1 s–d mixed orbital largely, if not exclusively, interacting with the axial ligands and one with the equatorial, thereby simplifying the problem. The metal p_z orbital has A''_2 symmetry and so interacts uniquely with the ligand group orbital of this symmetry.

The two ligand σ orbitals of A_1' symmetry:

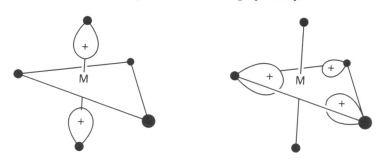

may each interact with either or both of the metal orbitals of A_1' symmetry:

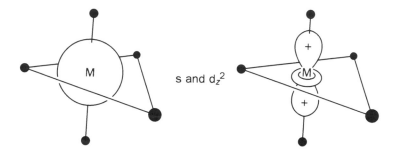

s and d_{z^2}

Similarly, the ligand A_2'' σ orbital interacts with the metal p_z orbital

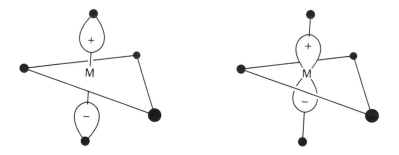

and the ligand e' orbitals interact with metal p_x and p_y orbitals:

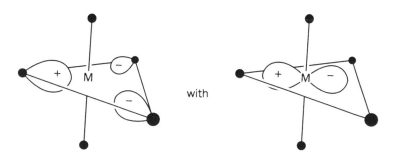

with

Fig. 6.28 σ bonding in a D_{3h} complex.

and

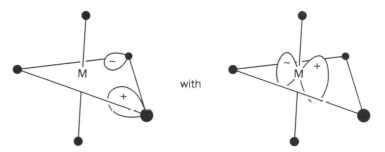

with

Fig. 6.28 *continued*.

Finally, there are both d orbital and p orbital sets of E' symmetry (the d we have discussed above, the p are p_x and p_y). Again, the general outcome is not clear. Nonetheless, we can make an educated guess. Although there is no required relationship, we can actually relate the trigonal bipyramidal problem to that of the octahedron, discussed earlier. If, for each, we separate out two axial ligands, then in the trigonal bipyramidal they are partnered by three equivalent coplanar ligands, in the octahedron by four. Perhaps the outcome in the two cases is not too dissimilar. If this is so, it would be concluded that the a_1' (d_{z^2}) and e' ($d_{x^2-y^2}$ and d_{xy}) orbitals are actually σ antibonding. The metal d orbitals of E'' symmetry, d_{yz} and d_{zz}, are non-bonding. When π bonding is included, the ligand orbitals a_2', a_2'', $2e'$ and $2e''$ are added to the problem (Table 6.6). They will not be discussed in detail; we merely note that the a_1' orbitals are only involved in σ bonding, the e'' only in π bonding and the e' in both σ and π bonding. The appearance of two orbitals of a symmetry species ($2a_1'$, $2e_1'$, $2e''$) in the above discussion need cause no concern. They arise because the axial and equatorial ligands are not symmetry related and so contribute additively. The way they were handled for the $2a_1'$ case illustrates the general approach—include them all together, the actual number of them is not important.

6.5 Formal oxidation states

In crystal field theory a complex ion is assumed to be composed of a cation, M^{n+}, surrounded by, but not overlapping with, a number of ligands. What is the effect of covalency on the electronic nature of the cation? Let us consider two octahedral complex ions, one of Fe^{3+} (d^5), the other of Fe^{2+} (d^6), both assumed to have as many unpaired electrons as is reasonably possible, and with identical ligands. Figure 6.29 gives schematic molecular orbital energy level diagrams for the two complexes. The electrons in the $a_{1g}(1)$, $t_{1u}(1)$, and lower $e_g(1)$ orbitals originated on the ligands but in the complex ions, occupy molecular orbitals and are therefore delocalized onto the cation to some extent.

Because of its higher charge and smaller size, the polarizing power of the Fe^{3+} cation would be expected to be greater than that of Fe^{2+} so that the transfer of electron density from the ligands to the cation would be expected to be greater for Fe^{3+} than for Fe^{2+}. Let us suppose that an effective transfer

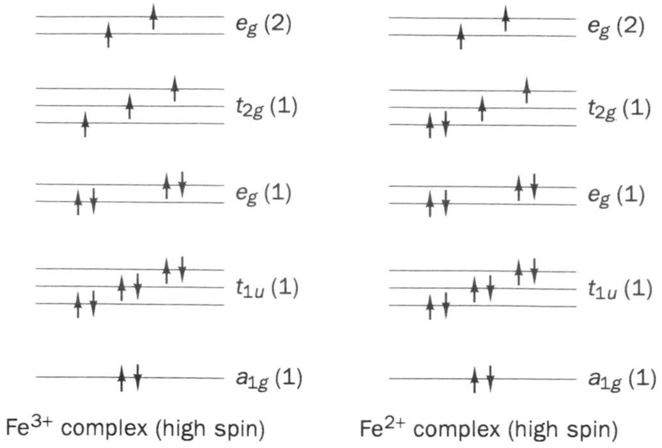

Fig. 6.29 Schematic molecular orbital energy level diagrams for high spin Fe^{3+} (d^5) and Fe^{2+} (d^6) octahedral complexes.

of two electrons occurs for the former and one for the latter (we choose integers for simplicity and ignore the rather difficult problem of how to calculate the number of electrons transferred). The resultant charges on the two cations are therefore both $+1$ (($+3-2$) and ($+2-1$)) although we started with one Fe^{3+} and one Fe^{2+} complex. The charges we assumed, $3+$ and $2+$, are free-ion charges. One expects that the magnitude of an actual charge will always be less than the absolute magnitude of a free-ion charge, for both ligands and cations. This, of course, is a restatement of Pauling's electroneutrality principle. Contemporary calculations indicate that whilst the actual charge of a Fe^{3+} ion is likely to be rather greater than that on a Fe^{2+} ion if the ligands are identical, the difference between them is only of the order of one-third of an electron. The actual charges themselves would be of the order of unity (positive). It is not surprising, therefore, that the use of free-ion charges and valence states, will sometimes prove difficult. In a transition metal complex containing H as a ligand, should this be regarded as H^+ or H^-? The charge on the metal depends on which we choose. In practice, the problem could either be sidestepped or worked the other way round—a formal charge first assigned to the metal from which the, equally formal, charge to be allocated to the H ligand would be deduced. Alternatively, the problem could be resolved by appeal to chemistry—does the H ligand behave more like H^+ or H^- in its reactions? In transition metal chemistry the answer is quite often H^- because transition metal hydrides are more characteristically reducing agents than acids. Although the charges indicated by the symbols Fe^{2+} and Fe^{3+} are misleading, this representation is far from valueless. In particular, these charges lead to a correct count of the number of electrons in the t_{2g} and upper e_g orbitals such as in Fig. 6.11. It is essential to get this number right if we are to be able to correctly interpret the physical and chemical properties of a complex. The usual compromise adopted is to refer to iron(III) and iron(II), as has usually been done throughout this book, rather than Fe^{3+} and Fe^{2+}, thereby avoiding the difficult problem of the actual charge distribution within the molecule. At some points in the text, usually for emphasis or to facilitate electron counting, there has been a reversion to the Fe^{3+} and Fe^{2+} convention.

It is common practice to assign a charge to species such as $[Fe(H_2O)_6]^{2+}$. This assumes that there is no covalent interaction between the complex ion and surrounding molecules. For example, it is assumed that in aqueous solution $[Fe(H_2O)_6]^{2+}$ either does not hydrogen bond with the solvent or that any hydrogen bonding does not affect the electron distribution within the complex ion. When attention is focused on the metal ion in these species, this approximation introduces no difficulties. For quantitative work on the interaction of a complex ion with its environment—and such an interaction can be of vital importance in bioinorganic chemistry—it might become sensible to further modify the nomenclature and to call this species $[Fe(H_2O)_6]^{II}$ or something similar.

6.6 Experimental

The content of this chapter has been theoretical and so an experimental section seems somewhat out of place. Yet it is not, for experimental evidence from a variety of sources is becoming available which helps pin-point the strengths and weaknesses of the discussion. The focus is on X-ray and neutron diffraction data. Although in X-ray crystallography it is usual to assume that atoms are spherical (so that it is assumed that on each atom in a crystal there is a spherical distribution of electron density and it is this which is responsible for diffracting X-rays), the precision of current apparatus and techniques, at their best, is such that deviations from spherical can be measured. In Fig. 6.30 are shown these deviations for the cobalt ion and four coplanar NO_2^- ligands in the complex ion $[Co(NO_2)_6]^{3-}$. In this complex the cobalt(III) has a $t_{2g}^6 e_g^0$ configuration and the consequences of this are evident in Fig. 6.30—there is a depletion of electron density in the 'e_g region', i.e. along metal–ligand bonds, and buildup in the 't_{2g} regions', i.e. between metal–ligand bonds, just as expected. The reader may also note a buildup of electron density near the N atom of the Co–N bonds; even the N–O bonding electrons are visible. This result is typical of X-ray electron density difference measurements. They support the general picture presented both in this chapter and in Chapter 7.

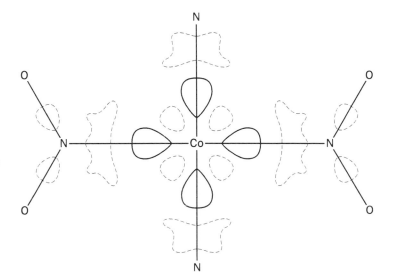

Fig. 6.30 Electron density difference map for the $[Co(NO_2)_6]^{3-}$ ion in the CoN_4 plane. Within a solid contour the electron density has been depleted and within a dotted contour it has increased. Adapted and reproduced with permission from S. Ohba, K. Toriumi, S. Sato and Y. Saito, *Acta Cryst.* (1978) *B34*, 3535.

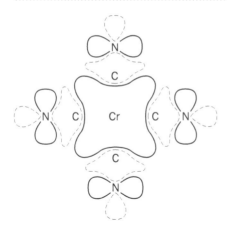

Fig. 6.31 Spin density distribution in the $[Cr(CN)_6]^{3-}$ anion in the CrC_4 plane. The solid contour defines a region of spin-up electron density and the dotted contour of spin-down. This diagram is from a model of the experimental data which over-sharpens some features so that fine detail is not reliable. Adapted and reproduced with permission from B. B. Figgis, S. B. Forsyth and P. A. Reynolds, *Inorg. Chem.* (1987) *26*, 101.

In most neutron diffraction experiments, it is (loosely put) the consequences of neutrons bouncing off nuclei that are measured. Electrons are not involved. However, this last statement is not always true. Neutrons, like electrons, have a spin of $\frac{1}{2}$ and so, like electrons, behave a bit like tiny bar magnets, an analogy that will be pursued in Chapters 8 and 11. This means that, experimentally, one can sort, and work with, neutrons either with spin up or with spin down. If such a beam of polarized neutrons impinges on a crystal containing unpaired electrons then the associated electron magnets will interact with the neutron magnets. If the electron spins, the electron magnets, are at least partially orientated by application of a magnetic field the electron–neutron interaction will depend on whether the neutron magnets are up or down. The crystals of many transition metal complexes contain unpaired electrons and diffraction measurements made with polarized neutrons enable unpaired electron density distributions to be measured. The $[Cr(CN)_6]^{3-}$ anion has a $^4A_{2g}(t_{2g}^3)$ ground state, is of high symmetry (the detailed geometrical structure of many ligands is such that their complexes can never be truly octahedral), and is of particular interest because of the expected involvement of chromium to CN^- π bonding. Figure 6.31 shows the experimental spin density distribution in this anion. Most obviously and surprisingly, it contains both spin up and down densities. Ignoring this problem for the moment it is noticeable that the spin up density on the chromium is essentially where we expect it to be—in the t_{2g} orbitals, between the metal–ligand axes. Further, some of this electron density has been transferred to a π orbital on the nitrogen, in accord with our π bonding model. So far, so good, but what of the negative spin density? It is energetically favoured for the three electrons in the t_{2g} shell to have their spins parallel—as will be seen in the next chapter, we have a $^4A_{2g}$ ground state. One talks of 'the (electron) exchange stabilization'. Why should this privilege of exchange between parallel spins be restricted to the chromium t_{2g} electrons? In fact, other electrons can participate. In particular, in the lone pair on the carbon of CN^-, the two electrons are not uniformly distributed in space. The one with spin up tends to be closer to the chromium t_{2g} electrons with spin up—this is exchange-preferred. There are two consequences. First, these ligand electrons appear somewhat merged with the chromium t_{2g} electrons in Fig. 6.31. Second, they leave behind a balancing down spin density in a σ orbital on the carbon atom, thus accounting for the experimental result.

Where does all this leave the model developed in this chapter? Clearly, basically correct (if one is prepared to accept that the wrong placement of ligand orbital energy levels is unimportant) but unable to account for fine details such as unpaired spin distributions (the total uneven spin density on each carbon in $[Cr(CN)_6]^{3-}$ amounts to about 0.1 electron). How can the model be improved? The explanation given for the observation of an uneven spin density depended on, loosely, how one electron behaved consequent on the behaviour of another. The model presented in this chapter is a one-electron model; electrons were talked of as individuals. In order to explain fine details this one-electron model is inadequate, two-electron correlations have to be included in our treatment, the second time in this chapter that this conclusion has been deduced. This is not the last time that we shall find a need to explicitly include electron correlation. At this point all that needs to be added is that theoretical models which include electron correlation not only predict, qualitatively, the presence of uneven spin densities in $[Cr(CN)_6]^{3-}$ but also give quantitative data that are not too far from the experimental observations. However, the routine and accurate inclusion of correlation effects in calculations on transition metal compounds remains an elusive goal. It is the subject of much current theoretical research. It highlights the difference in difficulty presented by accurate calculations on transition metal compounds and organic molecules—for the latter, the inclusion of correlation is essentially a solved problem. As seen earlier, at the end of Section 6.2, inclusion of electron correlation in calculations on transition metal complexes may mean orbital energy level patterns rather different from those given by calculations which exclude it.

Further reading

Most contemporary texts in inorganic chemistry include a treatment of the material in this chapter, although in less detail and depth. They may be useful, however, in painting a broad-brush picture. Other treatments tend to be rather mathematical and to give at least as much emphasis to crystal field theory as to molecular orbital (or to their combination, ligand field theory); some older works remain the easiest to follow. Examples are:

1. A classic text, *Introduction to Ligand Field Theory* C. J. Ballhausen, McGraw Hill, New York, 1962; the reader of the present book may well find it easiest to skip the first few chapters of Ballhausen at a first reading.

2. Other texts or compilations—again, some selectivity will be needed.
(**a**) *Introduction to Ligand Fields*, B. J. Figgis, Wiley, New York, 1966.
(**b**) *Some Aspects of Crystal Field Theory*, T. M. Dunn, R. S. McClure and R. G. Pearson, Harper and Row, New York, 1966.

(**c**) *Modern Coordination Chemistry*, J. Lewis and R. G. Wilkins, Interscience, New York, 1960.

Recent theoretical work is, largely readably, illustrated in *The Challenge of d and f Electrons*, D. R. Salahub and M. C. Zerner (eds.), Symposium Series 394, American Chemical Society, 1989.

Electron density distributions in inorganic compounds are reviewed by K. Toriumi and Y. Saito in *Adv. Inorg. Chem. Radiochem.* (1984) 27, 27.

A good introduction to spin density distributions—which concludes that long range effects may well be important—is 'The Magnetization Density of Hexacyanoferrate(III) Ion Measured by Polarized Neutron Diffraction in $Cs_2KFe(CN)_6$' by C. A. Daul, P. Day, B. N. Figgis, H. U. Güdel, F. Herrin, A. Ludi and P. A. Reynolds, *Proc. Roy. Soc. Lond.* (1988) A419, 205.

A very worthwhile article which emphasizes that main group and transition metal ions have much in common is 'The Roles of d Electrons in Transition Metal Chemistry: a New Emphasis' by M. Gerloch, *Coord. Chem. Rev.* (1990) 99, 117.

Questions

6.1 Use Table 6.1 and Fig. 6.2 to show that the following metal orbitals have the symmetries indicated

$$s \; : \; a_{1(g)}$$
$$p \; : \; t_{1(u)}$$
$$d \; : \; e_{1(g)} + t_{2(g)}$$

Table 6.1 does not enable the g and u suffixes to be determined. The answer to this problem is worked out in detail in the reference at the foot of page 97; the solution in this reference includes the suffixes.

6.2 Rehearse the arguments which lead to the conclusion that only interactions of T_{2g} symmetry need to be considered when the problem of π bonding in octahedral complexes is studied.

6.3 Although both octahedral and tetrahedral complexes are cubic ($x \equiv y \equiv z$), it is possible to be much more specific about the effects of metal–ligand bonding in the former. Why?

6.4 Show that the five M–L σ bonding orbitals in a ML_5 trigonal bipyramidal complex have $2A_1'' + A_2'' + E'$ symmetries. For this problem the character table of the D_{3h} point group is needed:

D_{3h}	E	$2C_3$	$3C_2$	σ_h	$2S_3$	$3\sigma_d$
A_1'	1	1	1	1	1	1
A_2'	1	1	-1	1	1	-1
E'	2	-1	0	2	-1	0
A_1''	1	1	1	-1	-1	-1
A_2''	1	1	-1	-1	-1	1
E''	2	-1	0	-2	1	0

6.5 As indicated in the text, Fig. 6.30 shows a depletion of electron density along the Co–N axis; it is to be noted that this depletion is close to the Co (as expected). However, close to the N there is a buildup of electron density. Suggest reasons for this buildup.

7

Crystal field theory of transition metal complexes

7.1 Introduction

Although little use is made now of the theory presented in this chapter, it contains the basis of all of those that are used. It provides the foundation, particularly for the understanding of spectral and magnetic properties; all else is elaboration and refinement. A knowledge of simple crystal field theory is therefore essential to an understanding of the key properties of transition metal complexes and particularly those covered in Chapters 8 and 9. This chapter deals exclusively with transition metal complexes. In one or more of their valence states, the ions of transition metals have their d orbitals incompletely filled with electrons. As a result, their complexes have characteristics not shared by complexes of the main group elements. It is the details of the description of these incompletely filled shells which is our present concern; this is in contrast to the discussion of the previous chapter where the topic was scarcely addressed. Ions of the lanthanides and actinides elements have incompletely filled f orbitals and so necessitate a separate discussion which will be given in Chapter 11.

In 1929 Bethe published a paper in which he considered the effect of taking an isolated cation, such as Na^+, and placing it in the lattice of an ionic crystal, such as NaCl. In particular, he was interested in what happens to the energy levels of the free ion when it is placed in the electrostatic field existing within the crystal, the so-called *crystal field*. The energy levels of a free ion show a considerable degeneracy, particularly if one is prepared to ignore effects which cause only small splittings. That is, in the free ion there exist sets of wavefunctions, each member of any set being quite independent of all other wavefunctions (i.e. orthogonal to them), yet all members of any one set correspond to the same energy. What happens to these ions when placed in an ionic crystal? Do the wavefunctions which in the free ion had the same energy still all have the same energy in the crystal? Bethe showed that in some cases the free ion degeneracy is retained and in others it is lost, the crucial factors being the geometry of the crystalline environment and the term (^1S, ^3P, ^2D, ^1F etc.) of the wavefunctions of the free ion.

Two years later, in 1931, Garrick demonstrated that a simple ionic model gives heats of formation for transition metal complexes which are in remarkably good agreement with the experimental values. That is, these complexes behave as if the bonding between the central metal ion and the surrounding ligands is purely electrostatic, just as in a simple picture of the bonding in NaCl. If this is so, then Bethe's work may be applied to complexes as well as to ionic crystals and the energy levels of the central metal ion related to those of the same ion in the gaseous state. All that is needed is a suitable quantitative calculation to obtain the energy level splittings due to the crystal field. This approach to the electronic structure of transition metal complexes is known as *crystal field theory* and it is the subject of the present chapter.

7.2 Symmetry and crystal field theory

Almost all the material of the present chapter arises from the symmetry of the molecules considered. This symmetry finds expression in the subject of group theory and, as in the previous chapter, to follow the arguments we use the reader will need to be reasonably conversant with group-theoretical jargon. The reader who needs a refresher course on the subject, or perhaps somewhere in the depths of this chapter becomes uncertain of their command, is reminded that there is a brief review of the essentials in Appendix 3, although this is not to be regarded as a substitute for a proper study.[1] Some of the language that will be used is reviewed in the next few paragraphs and, as in the previous chapter, the discussion will be confined to octahedral complexes.

In the present chapter the symbols A_{1g}, A_{2g}, E_g, T_{1g} and T_{2g} will be encountered with a spin state designated, for example, ^2E_g ('doublet ee gee') and $^3T_{1g}$. The lower-case symbols used in Chapter 6 are also of frequent use, so, a_{1g}, e_g, t_{1g} etc. These symbols may also have superscripts. We shall refer to t_{2g}^2, ('tee two gee two'), for example when talking about two electrons in a set of t_{2g} orbitals (other books may refer to this as $(t_{2g})^2$). Symbols such as t_{2g}^3 or $(t_{2g})^3$ indicate that the set of three orbitals labelled t_{2g} are occupied by three electrons. This usage follows the similar use of s, p and, very important for this chapter, d to indicate an orbital occupancy. One talks of sp($\equiv$ s^{1}p^1), d^2, d^3, d^{3}p^2 and so on as *configurations* (appropriate to an isolated atom). In just the same way, symbols such as t_{2g}^2, t_{2g}^3, and $t_{2g}^2e_g^1$ refer to electron configurations in (for the examples given) octahedral (O_h) symmetry.

It often happens that, for isolated atoms, one wishes to discuss all the electrons—or, certainly, the outermost ones—collectively rather than as individuals. One then refers to *states* or *terms* and uses uppercase symbols.

---

[1] The present author has written a, hopefully, easy to read and follow, non-mathematical text on group theory called *Symmetry and Structure*, 2nd edn., published by Wiley, Chichester, 1995.

The difference between these names is rather subtle and different usages are met. The most common is that in which the detailed mathematical specification decreases in the order state > level > term. So, in most of our discussion we will talk of terms, although the repeated use of this word may give rise to some ugly language and although in some papers and texts the word state is used almost interchangeably with it. After a phenomenon known as spin–orbit coupling has been included we will talk of levels and if we wish to be even more detailed and talk about an individual wavefunction, we will use the name state. Care must be taken to avoid confusing this use of state with phrases such as 'the ground state' and 'an excited state'. For our present purposes the important thing to note is that under the rotational operations of an octahedron[2] any S term has the same symmetry as an s orbital, a P term the same symmetry properties as a set of three p orbitals and so on. Key, however, is the fact that whilst a solitary electron in a set of p orbitals gives a P term, it transpires that two electrons in the d orbitals can give one also, as can three d electrons. We talk of 'the P term arising from the p^1 configuration', or 'the P term arising from the d^2 configuration' and so on. A crucial factor turns out to be the number of unpaired electrons, n, associated with each term and this is indicated by the number $(n + 1)$ as a superscript thus: ^2P, ^3D and so on. If the choice of $(n + 1)$ seems odd—perhaps n seems more sensible—reflect on the fact when $n = 1$, that is, when there is just one unpaired electron, this electron can have spin up or down—there are *two*, $(n + 1)$, different spin possibilities. Although it is usual to arrive at symbols such as ^3P by feeding electrons into orbitals, the absence of any specific reference to these orbitals in the final symbol suggests that they are merely a convenient vehicle, and not essential. This is the case. Symbols such as P and ^3P, like their lower case counterparts, are a consequence of the (rotational) symmetry of a sphere. Indeed, the symbols S, P, D, etc. are the labels of irreducible representations of the relevant spherical group.

Although an octahedron has a much lower symmetry than a sphere it would be reasonable to expect that many-electron wavefunctions would be handled similarly. This is so—symbols such as $^3T_{1g}$, ^2E_g and $^1A_{1g}$, like t_{1g}, e_g and a_{1g} orbitals, imply, respectively, triple, double and single orbital degeneracy. In each case they are associated with a spin degeneracy which, in each of these three examples, is identical to the spatial degeneracy. However the two vary independently and so symbols such as $^2T_{1g}$, ^3E_g and $^6A_{1g}$ are perfectly reasonable.

7.3 Crystal field splittings

In crystal field theory a complex is regarded as consisting of a central metal cation surrounded by ionic or dipolar ligands which are electrostatically attracted to the cation. The bonding within the complex arises from the electrostatic attraction between the nucleus of the metal cation and the electrons of the ligands. The interaction between the electrons of the cation and those of the ligands is entirely repulsive. These repulsions will be central

---

[2] In the previous chapter, the extension to include operations such as reflection and inversion was indicated; a similar discussion will appear later in this chapter.

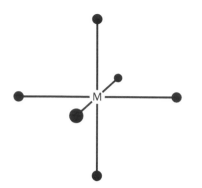

Fig. 7.1 An octahedral complex. In this chapter the way that an octahedral complex is drawn will vary, the perspective adopted depending on the point under discussion. Frequently, lines will be drawn which represent the edges of the octahedron rather than chemical bonds.

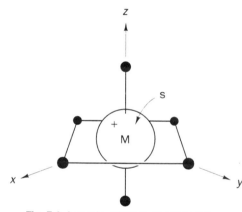

Fig. 7.2 A metal s orbital in an octahedral complex.

to the content of this chapter, for they are largely responsible for the energy level splittings which were the subject of Bethe's paper. As we will see, group theory tells us to expect these splittings (this is what Bethe showed). However, group theory tells us nothing about the magnitude or even the sign of the splittings. We need a specific model to do this and the simplest is the crystal field model. Actually, if we try to get numbers out of the model which can then be compared with experiment we find that the model is not a very good one. This is one aspect which has led to the development of the more realistic model of Chapter 6. However, historically, virtually the entire detail of the theory of transition metal ions was developed using the crystal field model. The trick is never to use the model to get numbers. Rather, it is used to focus our attention on energy differences and the relationships between them. Experimental data are then used to obtain the numbers! It is this, together with the fact that (as we shall see) the results are largely symmetry-determined that lead to the utilization of the method. Consider the octahedral complex shown in Fig. 7.1. What will be the effect of the crystal field on a single s electron of the central metal ion (Fig. 7.2)? The ligand–metal electron repulsion which we have associated with the crystal field will raise the energy of the s electron (or an S term), but as there is no orbital degeneracy, no orbital splitting can result. Next, what will be the effect of the crystal field on a single p electron (or a P term) of the metal ion? As is evident from Fig. 7.3 all the p orbitals are equally affected by the crystal field and so, no matter which of them the p electron occupies, the repulsion is the same. That is, the p orbitals (or the components of a P term) remain triply degenerate in an octahedral crystalline field.

The case of a single d electron (or a D term) is both more difficult and more interesting. All five d orbitals are not spatially equivalent. Three, d_{xy}, d_{yz} and d_{zx}, are evidently equivalent for they are equivalently situated with respect to the ligands (Fig. 7.4) and may be interchanged by simply interchanging the labelling of the Cartesian coordinate axes. The other two d orbitals, $d_{x^2-y^2}$ and d_{z^2}, are not equivalent, although they both have their maximum amplitudes along the Cartesian coordinates axes (Fig. 7.5). Interchange of the labels associated with each of these axes has the effect not of interchanging the orbitals but of generating new orbitals. So, starting with the coordinate system of Fig. 7.5 the interchange $x \to z \to y \to x$ gives us

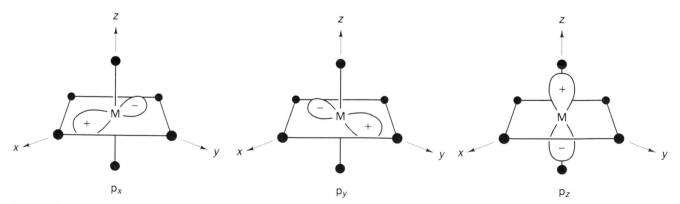

p_x ⬚ p_y ⬚ p_z

Fig. 7.3 A set of metal p orbitals in an octahedral crystal field. Because the coordinate axes of the octahedron are equivalent so too are the p orbitals. They remain triply degenerate.

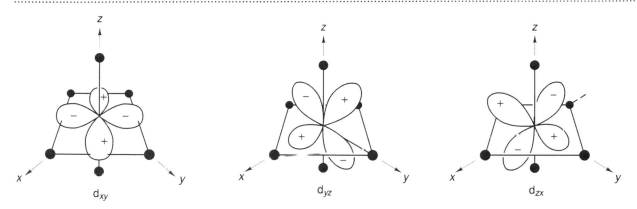

Fig. 7.4 The d_{xy}, d_{yz} and d_{zx} metal orbitals in an octahedral crystal field. Because they are all equivalent (they can be interconverted by rotating around the threefold axis approximately perpendicular to the plane of the paper) they remain triply degenerate.

Fig. 7.5 The $d_{x^2-y^2}$ and d_{z^2} orbitals in an octahedral crystal field. Although they look very different, the fact that they can be mixed by an axis relabelling shows that they are a degenerate pair (see the text and Fig. 7.6).

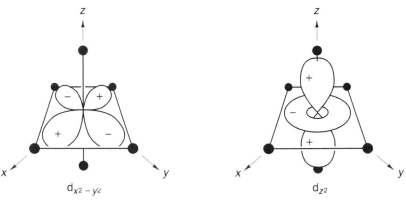

Fig. 7.6 When the octahedral axes of Fig. 7.5 are relabelled

$$z \to x$$
$$\nwarrow \ \swarrow$$
$$y$$

then the orbitals d_{x^2} and $d_{y^2-z^2}$ are obtained. It may help to see this if both of the octahedra drawn are mentally rotated by 120° anticlockwise so that the axes return to the positions they have in Fig. 7.5. Since nothing else has been added or removed, the orbitals d_{x^2} and $d_{y^2-z^2}$ must be mixtures of the original d_{x^2} and $d_{y^2-x^2}$.

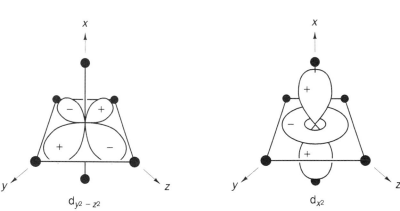

Fig. 7.6 in which these same d orbitals are now labelled d_{x^2} and $d_{y^2-z^2}$. We have not invented something new; the 'new' orbitals are simply mixtures of the 'old'. The fact that it is possible to mix two orbitals by such a trivial operation as relabelling the axes shows that the two orbitals are degenerate. If they were not, the mixed orbitals would have different energies from those of the starting pair, and it is obviously ridiculous that the energies should be a function of the labelling of the axis system. In Appendix 4 it is shown in a more formal way that the new d orbitals are simply mixtures of the old ones, d_{z^2} and $d_{x^2-y^2}$. We conclude that the d orbitals (or a D term) split into

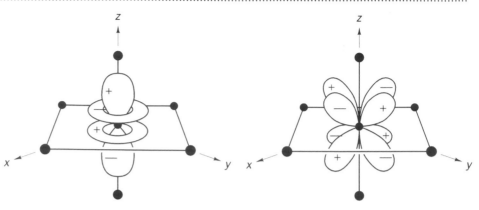

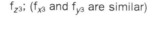

f_{z^3}; (f_{x^3} and f_{y^3} are similar)

$f_{z(x^2-y^2)}$; ($f_{x(y^2-z^2)}$ and $f_{y(z^2-x^2)}$ are similar)

Fig. 7.7 In an octahedral crystal field the f orbitals of a metal atom split into two sets which are triply degenerate and one orbital which is singly degenerate.

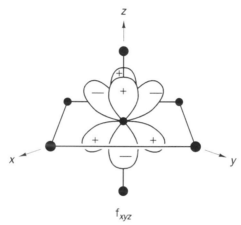

f_{xyz}

two sets, a set of three degenerate orbitals and a set of two degenerate orbitals. The relative energies of these two sets will be discussed shortly.

Finally, we consider the effect of an octahedral crystal field on a single f electron (or an *F* term). Fig. 7.7 pictures the seven f orbitals in an octahedral environment. The lobes of three, f_{x^3}, f_{y^3} and f_{z^3} point along axes. Three, $f_{z(x^2-y^2)}$, $f_{x(y^2-z^2)}$ and $f_{y(z^2-x^2)}$ have lobes located in coordinate planes (for the $f_{z(x^2-y^2)}$ orbital shown in Fig. 7.7 these are the *zx* and *yz* planes). The last, f_{xyz}, has lobes pointing between all coordinate axes. Clearly in an octahedral crystal field the f orbital sevenfold degeneracy is lost to give two sets of triply degenerate orbitals and one singly degenerate orbital.

So far it has been shown that sets of d and f orbitals (and therefore *D* and *F* terms) split into subsets in an octahedral crystal field but nothing has been said about the relative energies of these subsets. For the moment, the discussion will be restricted to orbitals because it is easy to give pictures of them. In subsequent sections the discussion will be extended to the corresponding terms. In preparation for this extension, it would be helpful if the reader has some idea of their derivation so that he or she is fully aware, for example, that an *F* state means seven spatial (as opposed to spin) functions, just like a set of f orbitals. One of the simplest ways of appreciating this is through the Russell–Saunders coupling scheme, that which is adopted to obtain the explicit functions themselves. This scheme is outlined in

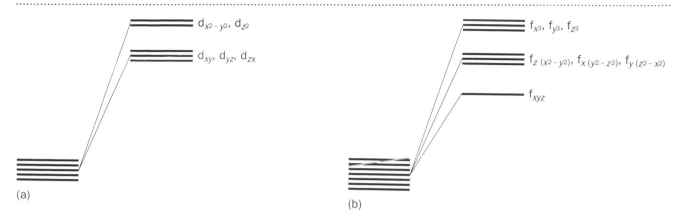

$d_{x^2-y^2}, d_{z^2}$

d_{xy}, d_{yz}, d_{zx}

$f_{x^3}, f_{y^3}, f_{z^3}$

$f_{z(x^2-y^2)}, f_{x(y^2-z^2)}, f_{y(z^2-x^2)}$

f_{xyz}

(a)

(b)

Fig. 7.8 The combined repulsive and splitting effect of an octahedral crystal field on the energies of (a) a set of metal d orbitals and (b) a set of f orbitals.

Appendix 5; part of it is also given in Section 11.3 which contains an example worked through in outline. If difficulty is encountered as the argument develops over the following pages, these resources should provide help.

Consider the d orbitals shown in Figs. 7.4 and 7.5. The more stable set is that in which an electron experiences least repulsion—destabilization—from the electrons on the ligands. Evidently, this set is that composed of d_{xy}, d_{yz} and d_{zx}, because this set keeps the d electrons away from the ligand electrons, at least when the ligands are represented as point charges. This set has T_{2g} symmetry and we shall refer to these orbitals as 'the t_{2g} orbitals' or 'the t_{2g} set'. The less stable set, the 'e_g orbitals' or 'e_g set', consists of d_{z^2} and $d_{x^2-y^2}$.

In a similar way it can be seen that the relative stabilities of the f orbitals (Fig. 7.7) is: f_{xyz} (a_{2u}), most stable; $f_{x(y^2-z^2)}$, $f_{y(z^2-x^2)}$ and $f_{z(x^2-y^2)}$ (t_{2u}), intermediate stability; f_{x^3}, f_{y^3} and f_{z^3} (t_{1u}), least stable. The splitting patterns for d and f orbitals are shown in Fig. 7.8[3] We shall return to the f orbital splitting but for the moment confine ourselves to the d orbital case. The vast majority of experimental data on transition metal complexes gives information on the splitting between the d orbitals but not on their absolute displacements from the free ion energy. It is therefore convenient to delete this unknown quantity from the diagrams and to regard the free ion energy as lying at the centre of gravity of the energies of the split orbitals (Fig. 7.9). The splitting between the t_{2g} and e_g sets of d electrons we shall call Δ (some authors prefer to call it 10 Dq). For elements of the first transition series Δ has a value of around 10 000 cm^{-1}; for dispositive transition metal ions its value is usually 5000–15 000 cm^{-1}; and for tripositive ions 10 000–30 000 cm^{-1}. Its value increases roughly in proportion to the cation charge, depends markedly on the ligands and, to a smaller extent on the metal (within any one transition series). A complex of the second or third transition series has a value of Δ which is up to twice that of the corresponding first row complex.

The modified d orbital splitting pattern showing only the crystal field effect given in Fig. 7.9 defines the splitting which is crucial to our discussion. However, the argument we used to derive the splitting used plausibility in place of mathematics so we cannot be absolutely confident of the results. Ultimately, the justification for the splitting in Fig. 7.9 is experimental so

---

[3] Beware the assumption that this diagram, although correct, may be used to explain the properties of ions containing unpaired f electrons. The situation is more complicated, as will become evident in Chapter 11.

Fig. 7.9 The d orbital splitting in an octahedral crystal field after the repulsion term common to both sets has been deleted. The unsplit d orbitals are at the centre of gravity of the split sets.

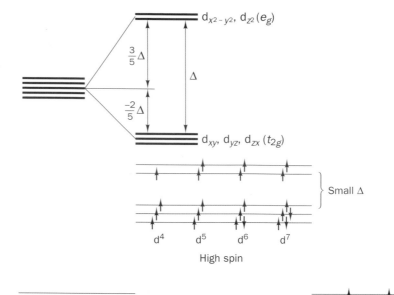

Fig. 7.10 The high spin (small Δ) and low spin (large Δ) possibilities for d^4–d^7 octahedral complexes.

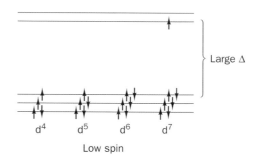

that Δ is to be regarded as an experimental quantity. This is why we prefer Δ to 10 Dq to describe the overall splitting—the D and q in 10 Dq relate to specific mathematical functions in crystal field theory.

One immediate consequence of the splitting of d orbitals into t_{2g} and e_g sets must be recognized. When there are between four and seven d electrons present there exist two quite different low-energy ways of allocating these electrons to the t_{2g} and e_g orbitals. These are shown in Fig. 7.10. There is a competition between the tendency of the electrons to stay as far apart as possible—they repel each other—and the preference to occupy the lowest-energy empty orbitals (remember that there is no covalency in our model—pairing electrons brings no bonding stabilization). The *high spin* arrangement is the one in which the interelectron repulsion between the d electrons is smallest—the electrons are spatially less concentrated—and the interelectron exchange stabilization is greatest (which is also why the 'maximum number of spins parallel' arrangement is of lowest energy). There also exists the *low spin* arrangement which differs from the high spin in that one or more electrons have been transferred from the less stable e_g to the more stable t_{2g} orbitals. Each electron so transferred contributes a crystal field stabilization of Δ to the system, but only at the cost of an increase in the electron repulsion destabilization and a decrease in the exchange stabilization (the latter, together, are often called the *pairing energy*). For any one transition metal ion with from four to seven d electrons the vital factor determining whether a particular complex is of the high or low spin type is the magnitude of Δ. The change from one type to the other is discontinuous; for other d^n configurations there may be related, but less dramatic, continuous changes associated with the transition from high to low type. We may note at this point that the two types of complex display quite different spectral and magnetic properties. For example, the $[Fe(CN)_6]^{4-}$ ion, which is a low spin complex of Fe^{II}, a d^6 ion, is yellow and has no unpaired electrons. The $[Fe(H_2O)_6]^{2+}$ ion, a high spin complex of Fe^{II} is pale blue[4] and is paramagnetic, with four unpaired electrons. It is easy to show the difference in the number of unpaired electrons. If finely ground crystals of $K_4Fe(CN)_6$ and then of $FeSO_4 \cdot 7H_2O$ are separately dropped onto the poles of a strong magnet the latter show a definite tendency to stick, whereas the former do not (neither stick to unmagnetized steel). This different behaviour leads to an interest in the magnetic properties of transition metal complexes and this forms the subject of Chapter 9.

In the crystal field model, the ligands are approximated by point charges or dipoles, the value of Δ for a particular complex depending on the magnitude of both this charge and that on the metal. This suggests that it should be possible to place ligands in order of increasing effective charge and, therefore, of increasing Δ. Further, this order should be the same for all metals. Such an order of ligands was discovered by Tsuchida before the advent of crystal field theory, and is called the *spectrochemical series*. As its name implies the series was discovered as a result of a study of the (visible region) spectra of transition metal complexes. It is therefore particularly relevant to the content of Chapter 8, which deals with these spectra. An abbreviated spectrochemical series, in order of increasing Δ, is:

$$I^- < Br^- < SCN^- \text{ (S-bonded)} < Cl^- < F^- < OH^- < H_2O < SCN^- \text{ (N-bonded)}$$

$$< NH_3 \approx py < SO_3^{2-} < bpy < NO_2^- \text{ (N-bonded)} < CN^-$$

A similar series exists for the variation with metal ion in which it is seen, as already mentioned, that Δ increases with the formal charge on the ion and down the periodic table:

$$Mn^{II} < Ni^{II} < Co^{II} < Fe^{II} < V^{II} < Fe^{III} < Cr^{III} < V^{III} < Co^{III} < Mn^{IV} < Rh^{III}$$

$$< Pd^{IV} < Ir^{III} < Pt^{IV}$$

Despite the above argument, the former series should not be regarded as a series in which the charge on the ligand increases from left to right. Other factors are involved, as was evident from the discussion in Chapter 6.

---

[4] It commonly appears green, but this is caused by contamination with a small amount of the yellow $[Fe(H_2O)_6]^{3+}$.

Table 7.1 M and L_l values for octahedral complexes

Metal	M	Ligand	L_l
Mn^{III}	0.80	Br$^-$	1.27
Ni^{II}	0.89	Cl$^-$	1.33
Co^{II}	0.93	F$^-$	1.50
Fe^{II}	1.00	OH$^-$	1.57
Cu^{II}	1.20	H$_2$O	1.67
V^{II}	1.23	SCN$^-$	1.72
Fe^{II}	1.40	NH$_3$	2.08
Cr^{III}	1.74	bpy	2.38[a]
V^{III}	1.86	NO$_2^-$	2.50
Co^{III}	1.90	CN$^-$	2.83
Ti^{III}	2.03		
Mn^{III}	2.10		
Mn^{IV}	2.30		
Rh^{III}	2.70		
Ir^{III}	3.20		
Pt^{IV}	3.60		

[a] For each donor atom; multiply by 2 for the bidentate ligand.

These two series can be brought together for octahedral complexes, including those which contain a mixture of ligands (here we anticipate the 'rule of average environment', to be described in Section 8.9). Using Table 7.1, one simply forms the product

$$M \sum_l n_l L_l \times 10^3$$

to predict a value for Δ (cm^{-1}). Here M and L_l are taken from Table 7.1 and n_l is the number of ligands of type l associated with L_l in the complex (for genuine octahedral complexes $n_l = 6$). This procedure has been the subject of some controversy in the literature, but the fact is that it works remarkably well.

We now return to the problem we met at the beginning of the chapter. What are the energy levels of a transition metal ion in an octahedral crystal field? There are alternative approaches to this problem depending on whether one is interested in a high or a low spin complex. Before embarking on this discussion, a limitation in our definition must be recognized. We have talked of high and low spin complexes, but recognized that this distinction is only applicable to ions with between four and seven d electrons. Clearly, the magnitude of Δ can vary in a similar way for the other d electron configurations, it is just that its variation does not have such evident consequences. When we wish to talk of the entire set of d electron cases, d^1–d^9, we shall talk, instead, of *weak field* (small Δ) and *strong field* (large Δ) complexes although, in fact, these terms are not even synonymous with high spin and low spin for the d^4–d^7 cases (the relationship will be made clear later). The two cases will be considered separately and followed by a discussion of the real-life case, in which the crystal field is of a strength intermediate between those appropriate to the two extremes.

7.4 Weak field complexes

Weak field complexes are those for which the crystal splitting Δ, is smaller than the electron-repulsion and -exchange energies. This at once indicates a suitable theoretical approach to a discussion of their electronic structure. In a general discussion of the energy levels of a free atom or ion one usually considers the various interactions in order of importance. The most important is the attraction between an electron and the nucleus. Next, the effects of interelectron repulsion (this includes the exchange energy) are considered, then the coupling between the spin and orbital motion of the electron (spin–orbit coupling) and so on.

There would be no need to preserve this pecking order if each step in the calculation were carried out exactly for both the ground and all excited states of the atom—and this is what one would hope to achieve if a computer were used to attack the problem. However, a deeper understanding is obtained by following an algebraic approach. Here, each step usually involves approximations and is only carried out for the ground and a few low-lying excited states. Consequently, the step-wise procedure becomes necessary to ensure that the properties of the ground state and terms immediately above it in energy are described with fair accuracy.

In crystal field theory the central atom of a complex is regarded as a free ion subject to an additional perturbation due to its environment. Evidently, this additional perturbation must be introduced at the correct point in a calculation and, for weak field complexes, this is after the effects of electron repulsion have been dealt with. That is, when the calculation has reached the stage of the classification of terms, such as ^2D, ^3F, ^1S, and so on. At this point a very important characteristic has emerged. The most stable electron arrangement in the free ion, which will be preserved in weak field complexes, is one in which electron–electron repulsion is a minimum. The electrons are as far apart as is possible. This will be when they occupy different orbitals as much as is possible and this, in turn, means that the term of highest spatial (orbital) and spin multiplicity will be the ground state. In this section we shall therefore principally be interested in such terms. However, the way also has to be prepared for the strong field case and this will necessitate consideration of other, excited state, terms. Most important will be those of the same spin degeneracy as the ground state, for some of the physical observables associated with transition metal complexes—colour, for instance —can only be understood if they are included.

As has been pointed out, the spatial degeneracy implied by labels such as P, D, F, G, ... (3, 5, 7 and 9, respectively) arise from the high rotational symmetry of a sphere. In the O_h symmetry of an octahedral complex, the central metal atom, instead of an environment with spherical symmetry, is in one with only 24 rotational operations (cf. Fig. 6.2). This reduction in symmetry means that there will also be a reduction in spatial degeneracy. What we must now do, then, is determine how, for example, the 21 wavefunctions of the term which is the ground state of the free ion d^2 configuration, ^3F, (three spin wavefunctions each combined with any one of seven spatial wave functions) split under the influence of the crystal field. This, of course, is where we came in at the beginning of the chapter and is the topic which was the subject of Bethe's paper. We ignore the spin degeneracy (because the effects of a crystal field are the same whether an electron spin is up or down; the crystal field may only have a small indirect effect on the spin degeneracy through the coupling which exists between the spin and orbital motions of an electron). We are left with the problem of the F term and, as has already been pointed out, the splitting of an F term parallels the splittings of f orbitals so that we conclude that the F term is split into three substates, two of which are triply degenerate. It has already been seen that a set of f orbitals split up into t_{1u}, t_{2u} and a_{2u} subsets. Similarly, an F term splits up into either T_{1u}, T_{2u}, and A_{2u} or T_{1g}, T_{2g} and A_{2g} subsets. Which of these is correct is determined by the g or u nature of the configuration from which the F term is derived. Because f orbitals are u in character the ^2F term corresponding to an f^1 configuration splits up into $^2T_{1u}$, $^2T_{2u}$, and $^2A_{2u}$ components. Similarly, the ^3F term derived from the d^2 configuration splits into $^3T_{1g}$, $^3T_{2g}$ and $^2A_{2g}$ components because the d orbitals are g in character. Note a clever, but potentially misleading, trick here: the group of all rotations of a sphere does not include the operation of inversion in a centre of symmetry, i, and so the labels, S, P, D, ... tell us nothing about behaviour under this operation. As a result, we are able to apply them equally to g and u functions but as soon as a centre of symmetry is encountered (as in O_h) we have to check whether we are dealing with g

Table 7.2 Crystal field components of the ground and some excited terms of d^n ($n = 1$–9) configurations

Configuration	Free ion ground term	Crystal field subterms	Important excited term	Crystal field term
d^1	^2D	$^2T_{2g} + {}^2E_g$		
d^2	^3F	$^3T_{1g} + {}^3T_{2g} + {}^3A_{2g}$	^3P	$^3T_{1g}$
d^3	^4F	$^4T_{1g} + {}^4T_{2g} + {}^4A_{2g}$	^4P	$^4T_{1g}$
d^4	^5D	$^5T_{2g} + {}^5E_g$		
d^5	^6S	$^6A_{1g}$		
d^6	^5D	$^5T_{2g} + {}^5E_g$		
d^7	^4F	$^4T_{1g} + {}^4T_{2g} + {}^4A_{2g}$	^4P	$^3T_{1g}$
d^8	^3F	$^3T_{1g} + {}^3T_{2g} + {}^3A_{2g}$	^3P	$^3T_{1g}$
d^9	^2D	$^2T_{2g} + {}^2E_g$		

or u functions. In this chapter our concern is only with the crystal field splitting of terms derived from d^n configurations and because all d orbitals are centrosymmetric, we shall encounter only g suffixes.

Table 7.2 gives the behaviour for all of the transition metal ions (d^1–d^9 configurations); it lists ground terms and, where the information is needed later, the behaviour of the lowest excited state. Two points should be noted in connection with this table. First, that only S, P, D, and F terms occur; for each the splitting is similar to that of the corresponding orbitals. Secondly, the table has some symmetry. Apart from the first column, the bottom half is the mirror image of the top half. These two features combine to simplify the remainder of our discussion of weak field complexes. The ground and excited terms of transition metal ions have been introduced in Table 7.2 without any justification apart from that in Appendix 5. Such a justification will be needed for the f electron systems in Chapter 11. The reader who is unhappy with the present *ex cathedra* presentation should read Section 11.6 where the procedure is detailed; they may be fortified by the knowledge that d electron systems are easier than their f electron counterparts! The crystal field splittings given in Table 7.2 follow from the discussion earlier in this chapter.

We turn now to the problem of the relative energies of the crystal-field components listed in Table 7.2. Consider the D terms, which give T_{2g} and E_g components. Which component is the more stable and by how much? It will be taken for granted that it is an experimental fact that the five d orbitals split as shown in Fig. 7.9 the splitting being denoted by Δ. Consider the d^1 case, Fig. 7.11. Here, the ground state will be that in which the electron occupies the lowest, t_{2g}, orbitals. Just as a d^1 configuration gives rise to a ^2D term so, too, t_{2g}^1 configuration (which is what we have here) gives rise to a $^2T_{2g}$ term. Similarly, the (excited) e_g^1 configuration gives rise to a ^2E_g term. We conclude that, for the d^1 case, the $^2T_{2g}$ term is the more stable because it means that the solitary d electron is in the t_{2g} orbitals. The (excited) ^2E_g term is generated from it by excitation of an electron from the t_{2g} orbitals to the e_g orbitals. This, by definition, requires an energy Δ, so we conclude that the $^2T_{2g}$ and ^2E_g terms are also separated by the energy Δ.

Two points should be noted. First, a detailed argument is required to relate the splitting of orbital energies to the splitting of term energies.

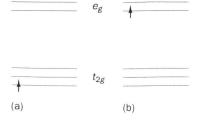

Fig. 7.11 (a) Ground and (b) excited states derived from the d^1 configuration in an octahedral crystal field.

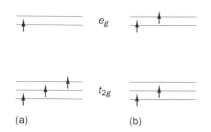

(a)   (b)

Fig. 7.12 (a) Ground and (b) an excited state with the same spin multiplicity, derived from the d^4 configuration in an octahedral crystal field. In the ground state there is a hole in the e_g orbitals; in the excited state the hole is in the t_2g orbitals.

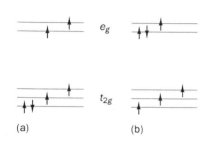

(a)   (b)

Fig. 7.13 (a) Ground and (b) lowest excited state of the same spin multiplicity, derived from the d^6 configuration in an octahedral crystal field.

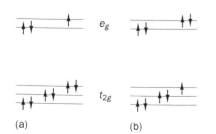

(a)   (b)

Fig. 7.14 (a) Ground and (b) excited state derived from the d^9 configuration in an octahedral crystal field.

This caution is necessary because, as will be seen in the following paragraph, the fact that t_{2g} orbitals are more stable then e_g does not imply that T_{2g} terms are necessarily more stable than E_g. Second, nine more electrons could be accommodated in the orbitals of Fig. 7.11. The other positions are vacant or, as it is more usually put, occupied by holes. A filled shell of electrons has spherical symmetry; so too does a half-filled shell, provided that no orbital of the shell is doubly occupied. The complement of a shell half-filled with electrons is a half shell of holes which, therefore, also has spherical symmetry. As has been indicated, our discussion is symmetry-derived and something of spherical symmetry (more precisely, something which is totally symmetric) never changes symmetry arguments and makes a constant contribution to energies (this is the reason that inner-shell electrons can be ignored). It follows that when we have a half shell of electrons or, equivalently, holes, this half shell may be neglected. This is illustrated in the next paragraph, which develops the electron–hole relationship.

We consider the d^1 case in which the t_{2g} orbital is the one occupied. There are two ways of describing the situation in which the t_{2g} orbitals are occupied by a single electron. We may say that the situation differs from spherical symmetry either by the presence of a t_{2g} electron or by the presence of two holes in the e_g set and two holes in the t_{2g} set—that is, a hole is missing in the t_{2g} set. Obviously, it makes no sense to use the hole description for $d^1(t_{2g}^1)$ case but sometimes it *is* useful to talk in terms of holes. Such a case is provided by the ground state ^5D term of the d^4 configuration. The quintet spin state means that all four of the d electrons have parallel spin, as shown in Fig. 7.12. The electron distribution differs from spherical symmetry by the presence of four electrons, or, what is equivalent, by the presence of one hole. Just as it was sensible to discuss the d^1 case in terms of one electron rather that four holes, so it is sensible to discuss the d^4 case in terms of one hole rather than four electrons. The most stable situation is that in which the hole is in the e_g orbitals. The ground term is therefore ^5E_g. At an energy Δ above the ground state is the $^5T_{2g}$ term, in which the hole is in the t_{2g} orbitals. In the d^4 case, therefore, the splitting of the E_g and T_{2g} levels is the inverse of that in the d^1 case. It is worthwhile emphasizing again that, as has just been seen, the fact that t_{2g} orbitals are more stable than e_g does *not* mean that T_{2g} terms are automatically more stable that E_g.

In the ^5D term arising from the d^6 configuration the orbital occupation is as shown in Fig. 7.13. It differs from spherical symmetry by the presence of a single electron, which is most stable when in the t_{2g} orbitals. The splitting therefore follows the d^1 case, the $^5T_{2g}$ term being more stable than the ^5E_g term by an energy Δ. Similarly, the ^2D term arising from the d^9 configuration differs from spherical symmetry by the presence of a hole in the e_g orbitals in the ground state, as shown in Fig. 7.14. The ground state is therefore of ^2E_g symmetry and the excited state, at an energy of Δ above, is of $^2T_{2g}$ symmetry.

Why is the hole formalism so useful? When a set of t_{2g} orbitals contains a single electron a T_{2g} term results. What if it contains two electrons; does this also result in a T_{2g} term? The answer, as we shall see later, is that it does, but other possibilities also exist (in the same way that a d^2 configuration gives rise to other than D terms). Similarly, if we evaluate the terms arising from the $t_{2g}^3 e_g^1$ configuration (using methods to be described later) we find a multitude of them. Some terms correspond to singlet spin terms, others to triplets and one to a quintuplet. If we are interested in the splitting of the ^5D term (the ground state of the free ion d^4 configuration) we are only interested in the spin quintuplet, and work is obviously involved in sorting it out from the others. The hole formalism does this work for us. Similarly, if we wish to write down an explicit expression for the wavefunctions of the ^5D term arising from the d^4 configuration it is much easier to write down a wavefunction appropriate to the single hole than to write down a (product) wave function appropriate to the four electrons. However, care is needed in the use of these pseudowavefunctions because holes behave differently from electrons. For example, it was argued earlier that metal electrons are repelled by the ligand field. It follows that, in contrast, holes are attracted by a ligand field.

The lowest term arising from the d^2 configuration of a free metal ion is ^3F. It has already been shown that this gives $^3T_{1g}$, $^3T_{2g}$ and $^3A_{2g}$ components in a ligand field. What is their relative ordering? Following our discussion of the splitting of f orbitals in a crystalline field we might anticipate that the T_{2g} term would be of intermediate stability and, following our discussion of the splitting of D terms, expect that T_{1g} and A_{2g} would alternate as the ground state for the d^4, d^7 and d^8 weak field cases. Detailed calculations confirm both of these predictions. Although it is beyond the scope of this book to give the detailed calculations,[5] this ordering of terms will be justified, and values for the relative energies of the components obtained, in Section 7.5. Having done this, the subject becomes easier because the splittings of all other F terms follow from the fact that for the ^3F term derived from the d^2 configuration the $^3T_{1g}$ term is lowest and $^3A_{2g}$ the highest. This will now be done fairly briefly; the task will be made easier by anticipating the results we shall later derive for the orbital occupancies associated with each term. This derivation will also help to deepen the understanding of the following results. At this point, remember that all starts from the d^2 case and follows either directly (d^7), or by the hole–electron analogy (d^3 and d^8). For d^7 and d^8, the spherical symmetry of a half-filled shell is also involved.

In the ($^3T_{1g}$) ground term derived from the ^3F term of the d^2 configuration we will find that there are two electrons in the t_{2g} orbitals.[6] For the ^4F term arising from the d^3 case, the $^4T_{1g}$ term derived from it in an octahedral crystal field corresponds to the presence of two *holes* in the t_{2g} orbitals and so is an excited state (Figs. 7.15 and 7.16); the $^4A_{2g}$ state is the ground state in this case. Note the switch from electrons to holes in the last two sentences and the consequent change of ground state into excited state when referring to the $^4T_{1g}$ term. For the ^4F term of the d^7 configuration the ground state is $^4T_{1g}$; the configuration differs from spherical symmetry by the presence of two electrons in the t_{2g} orbitals (Fig. 7.17) and so the splitting parallels the d^2 case (the spin states differ, of course, but this is irrelevant). The ^3F term

---

[5] A reasonably simple treatment is given in *Valence Theory* by J. N. Murrell, S. F. A. Kettle and J. Tedder, Wiley, London (1965) Chapter 13.

[6] This statement is not quite correct, as we shall see at the end of Section 7.5. Similarly, errors are contained in the statements on the d^3, d^7 and d^8 configurations. These errors are made in the interest of linguistic and conceptual simplicity and in no way invalidate the general argument.

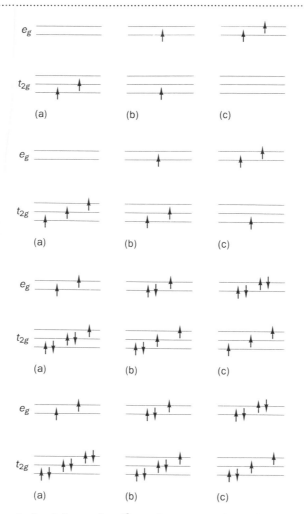

Fig. 7.15 (a) Ground and (b, c) two excited state configurations of a d^2 ion in an octahedral crystal field. The triple orbital degeneracy of the $^3T_{1g}$ ground state may be associated with the three possible orbital sites for the hole in (a).

Fig. 7.16 (a) Ground and (b, c) two excited state configurations of a d^3 ion in an octahedral crystal field. Note that there is only one distinguishable way of arranging the electrons in the three orbitals in the ground state—and so the ground state is orbitally non-degenerate.

Fig. 7.17 (a) Ground and (b, c) two excited state configurations of a d^7 ion in an octahedral crystal field.

Fig. 7.18 (a) Ground and (b, c) two excited state configurations of a d^8 ion in an octahedral crystal field.

derived from the d^8 configuration gives rise to an excited $^3T_{1g}$ term (Fig. 7.18), two holes occupying the t_{2g} orbitals, and a $^3A_{2g}$ ground state, paralleling the d^3 case—and so the opposite splitting pattern to d^2—remember that holes and electrons behave in opposite ways in crystal fields.

The only other weak-field case which remains to be discussed is that of the ^6S ground state of the d^5 configuration. This ground state is orbitally non-degenerate and so no crystal field splitting can occur. The behaviour of this ^6S term parallels that of an s orbital and becomes $^6A_{1g}$ in an octahedral crystal field.

So far the P excited states of the d^2, d^3, d^7, and d^8 configuration which were given in Table 7.2 have not been mentioned. These excited states give rise to a T_{1g} term which is of the same spin multiplicity as the T_{1g} term derived from the corresponding free-ion F ground term, the energy of which has just been discussed. These two T_{1g} terms, being of the same spin multiplicity and orbital symmetry, may interact under the influence of the crystal field. In the limit of a very weak crystal field—and such a field is our present concern—this interaction is correspondingly weak and may be ignored. Attention is drawn to it at this point as it is referred to in Sections 7.5 and 7.6.

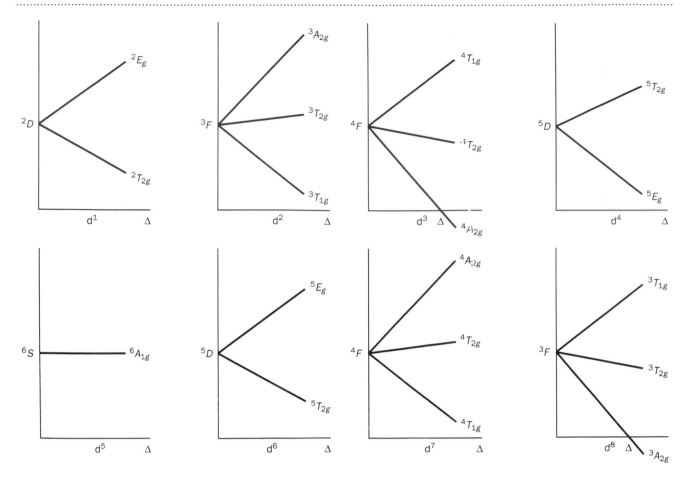

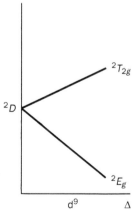

Fig. 7.19 The splitting of the weak field d^n ($n = 1$–9) ground state terms in an octahedral crystal field.

The arguments of this section are summarized by showing in Fig. 7.19 the splittings, in the weak field limit, of the ground state free-ion terms listed in Table 7.2. The behaviour shown is simple—the energies are a linear function of Δ (introduction of the additional T_{1g} levels mentioned above, will, when we include them, introduce a curvature into the T_{1g} behaviour). So far, the actual energies for the components of split D terms are the only ones to have been discussed; although F terms have been considered in outline, the energies of their components in an octahedral crystal field have not been obtained—this will be done later.

7.5 Strong field complexes

Strong field complexes are distinguished from weak field complexes because their crystal field splitting energy, Δ, is greater than the energies associated with electron pairing. In an actual calculation this means that instead of finding the terms arising from a d^n configuration—otherwise the first step in a step-wise, pecking order, calculation—one first applies the crystal field perturbation. This divides the d orbitals into e_g and t_{2g} sets. A d^n configuration splits up into sets which differ in the occupancy of the t_{2g} and e_g orbitals. So, for example, the d^2 configuration splits into three sets t_{2g}^2, $t_{2g}^1 e_g^1$ and e_g^2. A

Table 7.3 Strong-field substates for d^n configurations

d^n	Strong field configurations
d^1	t_{2g}^1, e_g^1
d^2	t_{2g}^2, $t_{2g}^1 e_g^1$, e_g^2
d^3	t_{2g}^3, $t_{2g}^2 e_g^1$, $t_{2g}^1 e_g^2$, e_g^3
d^4	t_{2g}^4, $t_{2g}^3 e_g^1$, $t_{2g}^2 e_g^2$, $t_{2g}^1 e_g^3$, e_g^4
d^5	t_{2g}^5, $t_{2g}^4 e_g^1$, $t_{2g}^3 e_g^2$, $t_{2g}^2 e_g^3$, $t_{2g}^1 e_g^4$
d^6	t_{2g}^6, $t_{2g}^5 e_g^1$, $t_{2g}^4 e_g^2$, $t_{2g}^3 e_g^3$, $t_{2g}^2 e_g^4$
d^7	$t_{2g}^6 e_g^1$, $t_{2g}^5 e_g^2$, $t_{2g}^4 e_g^3$, $t_{2g}^3 e_g^4$
d^8	$t_{2g}^6 e_g^2$, $t_{2g}^5 e_g^3$, $t_{2g}^4 e_g^4$
d^9	$t_{2g}^6 e_g^3$, $t_{2g}^5 e_g^4$

complete list of these substates is given in Table 7.3 which possesses the same sort of symmetry as Table 7.2. The number of substates listed for a d^n configuration is the same as that for a d^{10-n} configuration although inspection shows clear differences in orbital occupancies. These differences are only apparent. The reader should easily be able to show that, if the hole formalism is used for one of them, the difference between them disappears. For the d^8 configuration, for example, the hole formalism leads to t_{2g}^2, $t_{2g}^1 e_g^1$ and e_g^2 hole configurations, the same as the electron configurations listed for the d^2 configuration in Table 7.3. The relative energies of the terms given in Table 7.3 are readily evaluated; in evaluating them the advantage of arbitrarily placing the energy of the d orbitals of the free ion at the centre of gravity of those of the complexed ion will be seen. With this convention, it is obvious (Fig. 7.9) that the t_{2g} orbitals are stabilized by $\frac{2}{5}\Delta$ and the e_g orbitals destabilized by $\frac{3}{5}\Delta$. Following the usual sign convention, the energy of the t_{2g} orbitals is $-\frac{2}{5}\Delta$ and that of the e_g orbitals $\frac{3}{5}\Delta$. That is, each electron in a t_{2g} orbital contributes $-\frac{2}{5}\Delta$, and each electron in an e_g orbital $\frac{3}{5}\Delta$, to the total d orbital energy. As an example of the application of this, consider the t_{2g}^1, $t_{2g}^1 e_g^1$ and e_g^2 configurations derived from d^2. These have energies, respectively, of $2(-\frac{2}{5}\Delta) = -\frac{4}{5}\Delta$, $(-\frac{2}{5}\Delta + \frac{3}{5}\Delta) = \frac{1}{5}\Delta$, and $2(\frac{3}{5}\Delta) = \frac{6}{5}\Delta$; three levels, each Δ away from its neighbour(s).

This is a convenient point at which to begin to introduce some diagrams to which our discussion is leading. The beginnings of the diagram relevant to the d^2 configuration is given in Fig. 7.20. The weak field limit ($\Delta = 0$) is represented by the left hand vertical line on which will be seen two of the labels which featured in the discussion of the weak field case, F and P. At the right hand edge of the diagram (Δ infinite) will be seen—a bit buried— the energies of the three strong-field configurations arising from the d^2 configuration in the octahedral case (t_{2g}^2 at $-\frac{4}{5}\Delta$, $t_{2g}^1 e_g^1$ at $\frac{1}{5}\Delta$ and e_g^2 at $\frac{6}{5}\Delta$). The second of these configurations differs from the first in that it has one fewer t_{2g} electrons and one more e_g. The third differs from the second in the same way. Now the energy required to move an electron from a t_{2g} orbital to an e_g is Δ (this is the definition of Δ). It follows that in Fig. 7.20 the three configurations are equally spaced (by Δ) along the right-hand vertical axis. Along the horizontal axis Δ runs from 0 to ∞, as in Fig. 7.20, and this entire range is included in the figure. It follows that the Δ scale must be nonlinear. Nonetheless, it will be assumed to be linear near the weak field and strong field axes and that the non-linearity is somehow accommodated in-between. As Fig. 7.20 shows, the discussion so far has left us in a rather difficult position. At the weak field limit we know the symmetries of the crystal field terms but only for one, the $^3T_{1g}$ arising from the free ion ^3P term (usually denoted $^3T_{1g}(P)$) do we know an energy. Because the ^3P term does not split in a crystal field, the energy of the $^3T_{1g}(P)$ term is independent of Δ. The energy is therefore shown as a horizontal line at the weak field limit. In the strong field limit we know energies; all terms arising from the e_g^2 configuration have an energy of $\frac{6}{5}\Delta$, all arising from $t_{2g}^1 e_g^1$ have an energy of $\frac{1}{5}\Delta$ and all arising from t_{2g}^2 have an energy of $-\frac{4}{5}\Delta$. In Fig. 7.20 lines of the appropriate slope have been drawn at the strong field limit but we neither know the symmetry labels to be attached to these lines nor, indeed, whether each line represents more than one energy level. Our next task is to tackle this latter problem; once this is done we shall be able to complete Fig. 7.20.

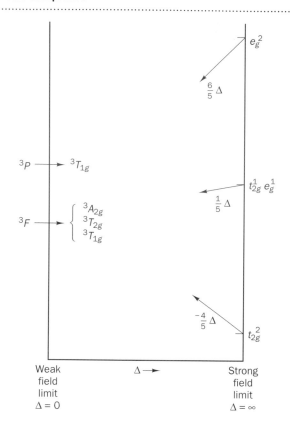

Fig. 7.20 The weak and strong field limits for the d^2 configuration in an octahedral crystal field.

By the end of the discussion of weak field complexes the effects of both the (weak) crystal field and electron repulsion had been included. To be consistent we must consider the effects of electron repulsion in strong field complexes, and this we now do. Electron repulsion causes some arrangements of electrons within an unfilled shell to be more stable than others—those arrangements in which unpaired electrons are kept farthest apart and have parallel spins will be the most stable (Hund's rules). That is, in a free atom or ion, electron repulsion causes the terms arising from a configuration to have different energies, as we have seen. Similarly, in a crystal field, the terms arising from a configuration like t_{2g}^2 will, in general, have different energies because of electron repulsion. How does one determine the terms arising from such a configuration? What are the relative energies of these terms? We shall answer these questions by looking at the group theory of the problem. Results that are qualitatively correct will be obtained, with no need to evaluate a single integral.

Table 7.4 is a table of *direct products*; its derivation is included in Appendix 3. This table is important, for it is used whenever one is simultaneously interested in two similar quantities associated with an octahedral molecule; for example, if we are interested in the symmetry properties of two electrons as a pair rather than as individuals. Similarly, this table will be used to discuss spectra, for which the ground and excited states of a molecule have to be considered simultaneously. Table 7.4 does not give g and u suffixes; they could have been included, but this would have made the table four times larger with no increase in real content. The way that these suffixes may

Table 7.4 The O_h direct product table ($g \times g = u \times u = g$; $g \times u = u$)

O_h	A_1	A_2	E	T_1	T_2
A_1	A_1	A_2	E	T_1	T_2
A_2	A_2	A_1	E	T_2	T_1
E	E	E	$A_1 + A_2 + E$	$T_1 + T_2$	$T_1 + T_2$
T_1	T_1	T_2	$T_1 + T_2$	$A_1 + E + T_1 + T_2$	$A_2 + E + T_1 + T_2$
T_2	T_2	T_1	$T_1 + T_2$	$A_2 + E + T_1 + T_2$	$A_1 + E + T_1 + T_2$

be added will be shown shortly but first the meaning of the entries in Table 7.4 must be explained.

As our immediate aim is to complete Fig. 7.20 our discussion will be confined to the d^2 case. Consider first the $t_{2g}^1 e_g^1$ configuration. The first electron may be fed into any one of three t_{2g} orbitals and the second into any one of two e_g. That is, there are $3 \times 2 = 6$ ways of feeding the two electrons in; there are six orbitally-different wavefunctions. Table 7.4 shows that the direct product of t_2 (extreme left-hand column) with e (top row) written $T_2 \times E$, is equal to $T_1 + T_2$ or, including g suffixes in an obviously sensible way, $T_{2g} \times E_g = T_{1g} + T_{2g}$ (note that $T_{2g} \times E_u$ or $T_{2u} \times E_g$ would have given $T_{1u} + T_{2u}$). The other possibility, $T_{2u} \times E_u$, would also have given $T_{1g} + T_{2g}$ ($g \times g = u \times u = g$; $u \times g = g \times u = u$). The sum of the degeneracies implied in T_{1g} and T_{2g} ($3 + 3 = 6$) is the same as the number of orbital wavefunctions arising from the $t_{2g}^1 e_g^1$ configuration ($3 \times 2 = 6$). It will not surprise the reader to learn that these six $t_{2g}^1 e_g^1$ two-electron wavefunctions divide into two sets of three each, one set of T_{1g} symmetry and one set of T_{2g} symmetry. That is, the configuration $t_2^1 e_g^1$ gives rise to T_{1g} and T_{2g} terms. This is no accident—it is a group theoretical requirement and would be just as valid in a different context (molecular vibrations, for instance, if we simultaneously excite T_{2g} and E_g molecular vibrations of an octahedral complex the molecule could end up in either a—vibrational—T_{1g} or T_{2g} state). So far, the spin of the electrons has not been mentioned. Because in the $t_{2g}^1 e_g^1$ configuration the two electrons always occupy different orbitals, there are no constraints and all paired (singlet) and parallel (triplet) spin arrangements are compatible with all the orbital symmetries that arise. We conclude that the $t_{2g}^1 e_g^1$ configuration gives rise to $^3T_{1g}$, $^3T_{2g}$, $^1T_{1g}$ and $^1T_{2g}$ terms. This means that the line at the strong field limit in Fig. 7.20 with slope $\frac{1}{5}\Delta$ is a superposition of lines corresponding to these four terms. However, because Fig. 7.20 is only concerned with triplet spin terms, only the labels $^3T_{1g}$ and $^3T_{2g}$ have to be added to this line in Fig. 7.20. As a check on the answer that has been obtained it is helpful to count wavefunctions. Previously, we only counted orbital functions; what if we include spin? Now, there are six different ways of putting an electron into the t_{2g} orbitals (three orbitals, two spins) and four ways of putting one into the e_g, a total of 24 different ways of putting in the two electrons; that is, there are 24 different wavefunctions. This is just the number implied by $^3T_{1g} + ^3T_{2g} + ^1T_{1g} + ^1T_{2g}$ ($9 + 9 + 3 + 3$). The counts agree, as they have to.

We now consider the e_g^2 configuration, associated with the strong field line of slope $\frac{6}{5}\Delta$ in Fig. 7.20. The first electron can be fed into the

e_g orbitals in any one of four ways (two orbitals, two spins). The second cannot be in the same orbital with the same spin as the first and so the number of distinct two-electron wavefunctions is $\frac{1}{2}(4 \times 3) = 6$. The $\frac{1}{2}$ arises from the word distinct: for instance, the 4×3 includes both ($\uparrow\downarrow$) and ($\downarrow\uparrow$) yet these are not distinct; we have counted everything twice. As Table 7.4 shows, the direct product $E_g \times E_g = A_{1g} + A_{2g} + E_g$. We know that the e_g^2 configuration must give rise to at least one spin triplet (because we can put one electron into each e_g orbital and they can be put in with parallel spin). But, as Fig. 7.20 shows, in the weak field limit there is no $^3A_{1g}$ or ^3E_g term, only $^3A_{2g}$. We conclude that the e_g^2 configuration gives rise to the terms $^1A_{1g} + {}^3A_{2g} + {}^1E_g$. This conclusion can be checked by counting the number of two-electron wavefunctions implicit in these terms. It is $1 + 3 + 2 = 6$, in agreement with the number obtained at the beginning of this paragraph. Since we are only interested in triplet spin terms, the label $^3A_{2g}$ may now be added to the line of slope $\frac{6}{5}\Delta$ in the strong field limit of Fig. 7.20. Notice an important distinction between the treatment of the $t_{2g}^1 e_g^1$ and the e_g^2 (and, in the next paragraph, the t_g^2) configurations. For the $t_{2g}^1 e_g^1$ case the spin and orbital motions of the electrons were treated independently—a term of given orbital symmetry appeared both as spin singlet and as spin triplet. When the two electrons are in the same orbital set this independence disappears; a term of given orbital symmetry appears as *either* a spin singlet *or* as a spin triplet, never both. This apparently arbitrary distinction actually arises from the fact that the (spin plus orbital) wavefunctions have to be antisymmetric (go into themselves multiplied by a factor of -1) on interchange of two electrons in the same orbital set. The connection between the requirement and the consequence is not immediately obvious but can be expressed group theoretically by an extension of the concept of the direct product beyond that used in this chapter.

We now turn to the t_{2g}^2 configuration but we already know what we expect to find—those terms which appear at the weak field side of Fig. 7.20 and which have not yet been obtained. That is, we are looking for a $^3T_{1g}$ term, the only spin triplet unaccounted for. As Table 7.4 shows, the direct product $T_{1g} \times T_{1g} = A_{1g} + E_g + T_{1g} + T_{2g}$. We expect that the addition of spin labels will lead to $^1A_{1g} + {}^1E_g + {}^3T_{1g} + {}^1T_{2g}$. Is this correct? A partial check can be made by counting two-electron wavefunctions. The configuration t_{2g}^2 implies $\frac{1}{2}(6 \times 5) = 15$ two-electron wavefunctions. This is the number implicit in $^1A_{1g} + {}^1E_g + {}^3T_{1g} + {}^1T_{2g}$ $(1 + 2 + 9 + 3 = 15)$. It is reasonable then, to add the label $^3T_{1g}$ to the line of slope $-\frac{4}{5}\Delta$ at the strong field limit of Fig. 7.20.

All that we need to complete Fig. 7.20 is a knowledge of the relative energies (i.e. the Δ dependence) of the $^3A_{2g} + {}^3T_{1g} + {}^3T_{2g}$ components of the ^3F term in the weak field limit. This is easy. We know the energies of the $^3A_{2g}$ and $^3T_{2g}$ states in the strong field limit; they are $\frac{6}{5}\Delta$ and $\frac{1}{5}\Delta$, respectively. There is no reason why these Δ dependencies should change as Δ decreases and so we conclude that these values are also their energies in the weak field limit. All that remains is to obtain the energy of the $^3T_{1g}$ term arising from the ^3F term (usually denoted $^3T_{1g}(F)$). We might be tempted to simply look at the $^3T_{1g}$ term arising from the t_{2g}^2 strong field configuration and conclude that the answer is $-\frac{4}{5}\Delta$. However, a little thought will indicate that caution is needed. The $^3T_{1g}(P)$ term does not depend on Δ in the weak field limit but has to correlate with the $^3T_{1g}$ term arising from the $t_{2g}^1 e_g^1$ configuration, a term which has an energy of $\frac{1}{5}\Delta$ (this has to correlate because if it did not the non-crossing rule—the rule that the energy levels of terms of the same symmetry and spin degeneracy do not cross—would be violated). So, the term $^3T_{1g}(P)$ changes its dependence on Δ as Δ itself changes—and this is caused by the presence of the second $^3T_{1g}$ term, with which it interacts—and it is this latter that we are interested in. The moral is: be careful when there is more than one term of a given symmetry and spin multiplicity. Actually, it is not difficult to calculate the energy of the $^3T_{1g}(F)$ term. Note that it is the *splitting* of the ^3F weak field term that gives $^3A_{2g} + {}^3T_{1g} + {}^3T_{2g}$ terms. The word splitting implies that the energies of the components sum to that of the ^3F term; their Δ dependencies sum to zero. Denoting the energy of the $^3T_{1g}(F)$ term by $\mathscr{E}(T_{1g})$, taking care to weight each energy by the number of wavefunctions with this energy, we have

$$(3 \times \tfrac{6}{5}\Delta) + (9 \times \mathscr{E}(T_{1g}) + (9 \times \tfrac{1}{5}\Delta) = 0$$

That is, $\mathscr{E}(T_{1g}) = -\tfrac{3}{5}\Delta$; our caution was justified—in the weak field limit the $^3T_{1g}(F)$ term does not have the same energy as in the strong field limit. The two energies differ by $-\tfrac{1}{5}\Delta$, equal and magnitude but opposite in sign to the difference between the energies of the $^3T_{1g}(P)$ term in the same limits. The crystal field has caused the two $^3T_{1g}$ terms to interact; they 'push' each other apart by equal and opposite amounts. Figure 7.20 can now be completed. The final version is shown in Fig. 7.21 which embodies all the content of the above discussion. One final point. As befits the above discussion, in this figure straight lines are drawn representing $^3A_{2g}$ and $^3T_{2g}$ terms connecting the weak and strong field limits. Given what was said earlier about the non-linear scale of the Δ axis in this figure, such straight lines cannot be justified. The moral is clear: Fig. 7.21, and related ones which will be presented shortly, are of pedagogical use only. However, they serve as an excellent introduction to related diagrams (Tanabe–Sugano diagrams) which have well-defined energy scales.

The discussion so far almost, but not quite, enables an extension to all d^n systems either by the electron–hole parallel, by neglect of half-filled shells, or both. The omissions can be dealt with by arguments paralleling that which follows for the t_{2g}^3 configuration, a configuration which poses a problem. The most obvious way of obtaining the orbital terms arising from this configuration is to simply take each of the orbital terms arising from the t_{2g}^2 configuration and combine each with a further t_{2g} orbital function. That is, to form the triple direct product $T_{2g} \times T_{2g} \times T_{2g}$ or, equivalently, consider the sum of direct products

$$(A_{1g} \times T_{2g}) + (E_g \times T_{2g}) + (T_{1g} \times T_{2g}) + (T_{2g} \times T_{2g})$$

This would be wrong. Some of the spin singlet wavefunctions of the T_{2g}^2 configuration represent two electrons occupying the same orbital. To form direct products blindly would, for some of the three-electron wavefunctions, be to allocate all three electrons to one orbital! The simplest way to avoid this problem and obtain the correct answer is as follows. In the $^3T_{1g}$ term of the t_{2g}^2 configuration we know that the electrons have parallel spins and so *must* occupy different orbitals (because of the Pauli exclusion principle). Adding a third t_{2g} electron can never give us three electrons in one orbital.

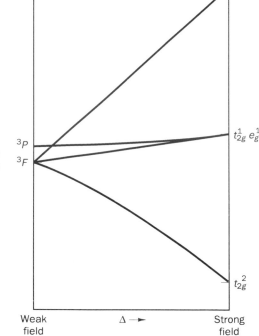

Fig. 7.21 The correlation between weak and strong field limits for the d^2 configuration in an octahedral crystal field.

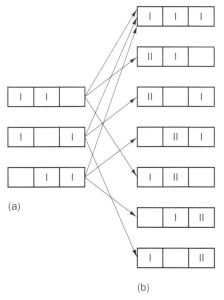

(a)

(b)

Fig. 7.22 (a) The three orbital arrangements associated with the triple orbital degeneracy of the $^3T_{2g}$ ground state of the d^2 (t_{2g}^2) configuration (cf. Fig. 7.15). (b) The seven different orbital arrangements corresponding to the t_{2g}^3 configuration may be derived from those of the t_{2g}^2 as indicated. It will be noticed that there are three arrows connecting with the top line of (b). One spin arrangement associated with this top line is 'all spins parallel'—$^4A_{2g}$. The other two must be spin doublets, which, together with the six below, give a total of eight spin doublet functions, corresponding to the $^2E_g + {}^2T_{1g} + {}^2T_{2g}$ terms (cf. Table 7.5).

Figure 7.22 gives diagrammatic representations of the $^3T_{1g}$ orbital wavefunctions (spin is not specified). Also in Fig. 7.22 are shown all the possible ways of orbitally allocating electrons in t_{2g}^3 configuration, again without specifying spin. It is easy to see that all of the t_{2g}^3 arrangements may be obtained from those of the orbital components of the $^3T_{1g}$ term. This suggests that the direct product $T_{1g} \times T_{2g}$ will give us the symmetries of all the sets of t_{2g}^3 three-electron wavefunctions. This is so; from Table 7.4 it is seen that they are A_{2g}, E_g, T_{1g} and T_{2g}. The spin multiplicities of these terms now have to be added. It is easy to see that only one spin quartet term can exist and that this is orbitally singly degenerate—there is only one way of allocating three electrons with α spin to the three t_{2g} orbitals. The other terms must therefore be doublets; that is, we have $^4A_{2g}, {}^2E_g, {}^2T_{1g}$ and $^2T_{2g}$. Again, the total degeneracy $(4 + 4 + 6 + 6 = 20)$ equals the number of distinguishable and allowed ways of feeding three electrons into the t_{2g} orbitals:

$$\frac{(6 \times 5 \times 4)}{(1 \times 2 \times 3)} = 20$$

Similar arguments may be applied to all the other configurations which arise, although when either or both of the t_{2g} and e_g shells are more than half full it is convenient to work in terms of holes. The results, of course, are the same as those for the similar electron configurations, both for space and spin. Results are collected together in Table 7.5 which may appear frightening in its complexity. Fortunately, Hund's rule applies to each configuration and we shall usually be interested only in the most stable terms, those of highest spin multiplicity, of each configuration.

Table 7.5 Terms arising from $t_{2g}^m e_g^n$ configurations

Number of d electrons	Configuration		Terms arising
1, 9	t_{2g}	$t_{2g}^5 e_g^4$	$^2T_{1g}$
	e_g	$t_{2g}^6 e_g^3$	^2E_g
2, 8	t_{2g}^2	$t_{2g}^4 e_g^4$	$^3T_{1g}$, $^1A_{1g}$, ^1E_g, $^1T_{2g}$
	$t_{2g}e_g$	$t_{2g}^5 e_g^3$	$^3T_{1g}$, $^3T_{2g}$, $^1T_{1g}$, $^1T_{2g}$
	e_g^2	$t_{2g}^6 e_g^2$	$^3A_{2g}$, $^1A_{1g}$, ^1E_g
3, 7	t_{2g}^3	$t_{2g}^3 e_g^4$	$^4A_{2g}$, ^2E_g, $^2T_{1g}$, $^2T_{2g}$
	$t_{2g}^2 e_g$	$t_{2g}^4 e_g^3$	$^4T_{1g}$, $^4T_{2g}$, $^2A_{1g}$, $^2A_{2g}$, 2^2E_g, 2^2T_{1g}, 2^2T_{2g}
	$t_{2g}e_g^2$	$t_{2g}^5 e_g^2$	$^4T_{1g}$, 2^2T_{1g}, 2^2T_{2g}
	e_g^3	$t_{2g}^6 e_g$	^2E_g
4, 6	t_{2g}^4	$t_{2g}^2 e_g^4$	$^3T_{1g}$, $^1A_{1g}$, ^1E_g, $^1T_{1g}$
	$t_{2g}^3 e_g$	$t_{2g}^3 e_g^3$	^5E_g, $^3A_{1g}$, $^3A_{2g}$, 2^3E_g, 2^3T_{2g}, 2^3T_{1g}, $^1A_{1g}$, $^1A_{2g}$, ^1E_g, 2^1T_{1g}, 2^1T_{2g}
	$t_{2g}^2 e_g^2$	$t_{2g}^4 e_g^2$	$^5T_{2g}$, $^3A_{2g}$, ^3E_g, 3^3T_{1g}, 2^3T_{2g}, 2^1A_{1g}, $^1A_{2g}$, 3^1E_g, $^1T_{1g}$, 3^1T_{2g}
	$t_{2g}e_g^3$	$t_{2g}^5 e_g$	$^3T_{1g}$, $^3T_{2g}$, $^1T_{1g}$, $^1T_{2g}$
	e_g^4	t_{2g}^6	$^1A_{1g}$
5	t_{2g}^5	$t_{2g}e_g^4$	$^2T_{2g}$
	$t_{2g}^4 e_g$	$t_{2g}^2 e_g^3$	$^4T_{1g}$, $^4T_{2g}$, $^2A_{1g}$, $^2A_{2g}$, 2^2E_g, 2^2T_{1g}, 2^2T_{2g}
	$t_{2g}^3 e_g^2$		$^6A_{1g}$, $^4A_{1g}$, $^4T_{2g}$, 2^4E_g, $^4T_{1g}$, $^4T_{2g}$, 2^2A_{1g}, $^2A_{2g}$, 3^2E_g, 4^2T_{1g}, 4^2T_{2g}

In the preceding paragraphs, our concern has largely been with the d^2 case. The results obtained for this configuration may readily be extended to other cases; this extension is given in the next section. The present section is concluded by enquiring into the d orbital occupancy when a weak field d^2 ion is in its ground state (the d^7 case is similar). The energy of the $^3T_{1g}$ ground term, $-\frac{3}{5}\Delta$, must correspond to an electron distribution of $\frac{9}{5}$ electrons in the t_{2g} orbitals and $\frac{1}{5}$ in the e_g orbitals:

$$(\tfrac{9}{5} \times -\tfrac{2}{5}\Delta) + (\tfrac{1}{5} \times \tfrac{3}{5}\Delta) = -\tfrac{3}{5}\Delta$$

If talking in terms of fractions of electrons seems strange, remember that these fractions refer to the probability distribution of the electron density. Physically, because electron repulsion is larger than the crystal field in weak field complexes, this repulsion forces some electron density into the e_g orbitals.

7.6 Intermediate field complexes

In the vast majority of transition metal complexes the energies associated with electron repulsion and the crystal field are of the same order of magnitude. This means that neither the weak field limit (electron repulsion $\gg$ crystal field) nor the strong field limit (crystal field $\gg$ electron repulsion) are met with in practice. There is no separate theory for the intermediate, real life, region. It is approached from either the strong or the weak field end, whichever seems the more appropriate. For d^4–d^7 complexes the choice is dictated by the distinction used when introducing high and low spin complexes—by the number of unpaired electrons present on the metal ion.

The qualitative behaviour of the energy levels in the intermediate-field region can readily be obtained from a knowledge of the energy levels in the weak and strong field limits. When discussing strong field complexes every possible term arising from every possible configuration was considered. However, for weak field complexes only the terms arising from the lowest free ion term were included in the discussion. To give a complete treatment of the intermediate field region the weak field treatment must be extended to include the terms arising from all of the other free ion terms. This information is given in Table 7.6, which also includes the splittings of those terms which we have already considered in detail. The terms arising from d^n configurations are listed in Table 7.7.

The relative energies of the components of free-ion terms in a crystal field may be obtained by the methods which have already been described or by others which are related to them. Two problems arise. First, as has already been noted, for a given d^n configuration there is usually more than one term of a given spin multiplicity and symmetry. These different terms interact, the interaction between them becoming increasingly important as the field increases. Second, if we wish to show the transition from weak to strong fields diagrammatically, the diagrams will be rather complicated (and so, too, would the calculations—this is where a computer comes in handy). For the d^4 and d^6 configurations there are 43 energy levels the behaviour of

Table 7.6 Crystal field splitting of Russell–Saunders terms arising from d^n configurations in an octahedral crystal field. The spin multiplicity, not included in this table, is the same for the crystal field terms as for the parent Russell–Saunders term

Russell–Saunders term	Crystal field components
S	A_{1g}
P	T_{1g}
D	$E_g + T_{2g}$
F	$A_{2g} + T_{1g} + T_{2g}$
G	$A_{1g} + E_g + T_{1g} + T_{2g}$
H	$E_g + 2T_{1g} + T_{2g}$
I	$A_{1g} + A_{2g} + E_g + T_{1g} + 2T_{2g}$

Table 7.7 Terms arising from d^n configurations

Configuration	Terms[a]
d^1, d^9	^2D
d^2, d^8	^3F, ^3P, ^1G, ^1D, ^1S
d^3, d^7	^4F, ^4P, ^2H, ^2G, ^2F, 2^2D, ^2P
d^4, d^6	^5D, ^3H, ^3G, 2^3F, ^3D, 2^3P, ^1I, 2^1G, ^1F, 2^1D, 2^1S
d^5	^6S, ^4G, ^4F, ^4D, ^4P, ^2I, ^2H, 2^2G, 2^2F, 3^2D, ^2P, ^2S

[a] 2^2D means that there are two distinct ^2D terms.

which would have to shown in a diagram (combine the relevant data in Tables 7.6 and 7.7).

In order to use diagrammatic representations of the behaviour of these energy levels it is necessary to make simplifications. We know which are the more stable terms in the weak and strong field limits and we will show only these. If we do not wish to solve the problem of the interactions between terms exactly, the non-crossing rule gives a qualitative idea of what happens. It is convenient first to consider the behaviour of the ground and low-lying terms obtained in the weak-field limit as the crystal field increases, and then to extend the discussion to include the strong field limit. This is also a convenient point at which to broaden our discussion to include all d^n configurations.

The behaviour of the T_{2g} and E_g states derived from the D term of highest spin multiplicity for the d^1, d^4, d^6 and d^9 configurations is given in Fig. 7.23. This figure summarizes the relevant discussion of the two previous sections and, in particular, the relevant parts of Fig. 7.19. Figure 7.23 is a modified form of a type of diagram first introduced by Orgel and usually called *Orgel diagrams*. It will be noted that some of the g suffixes that have been used in the discussion so far have been dropped in Fig. 7.23. The reason for this is that, as will be seen in the next section, by so doing, tetrahedral complexes can be included in Fig. 7.23. As expected, in the weak field (and, in this case, also in the strong field), limit the $E_{(g)}$ terms in Fig. 7.23 have a $\pm\frac{3}{5}\Delta$ dependence on Δ (+ when we work in terms of holes, and this means the d^4 and d^9 cases, and − when we work with electrons, the d^1 and d^6 cases). The $T_{2(g)}$ terms have a $\pm\frac{2}{5}\Delta$ dependence (− for holes, + for electrons). Because d^4 and d^6 configurations give rise to quintet spin terms but d^1 and d^9 to doublet, there is no mention of spin terms in Fig. 7.23 so that all of these cases can be included in the same diagram.

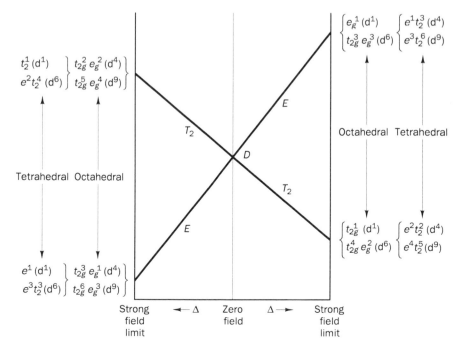

Fig. 7.23 A modified Orgel diagram for weak field complexes with free ion D terms as ground state (d^1, d^4, d^6 and d^9).

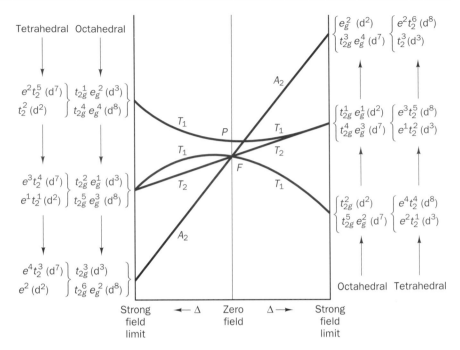

Fig. 7.24 A modified Orgel diagram for weak field complexes with free ion F terms as ground state (d^2, d^3, d^7 and d^8).

The ground and low-lying weak field terms of the d^2, d^3, d^7 and d^8 configurations are given in Fig. 7.24. Again, this figure summarizes the relevant discussion in the previous sections. In particular, the interaction between the $T_{1(g)}(P)$ and $T_{1(g)}(F)$ terms has been included and is responsible for the appearance of curvature on the lines associated with these terms in the intermediate field region. Just as for the splitting of D terms, so the splitting of F terms is inverted for holes (d^3 and d^8) compared with electrons (d^2 and d^7). In the former cases this brings the $T_{1(g)}(F)$ and $T_{1(g)}(P)$ terms close together and, although the magnitude of the interaction between these two terms is the same as for the d^2 and d^7 cases, the *effect* of the interaction is much greater and leads to considerable curvature in the Orgel diagram in the intermediate field region.

The only case not so far included, the d^5 case, is trivial and a diagram is not given for it (a more complicated one will be given later). A diagram given at this point would consist of a single horizontal straight line joining the ^6S free ion term to the $t_2^3 e^2$ strong field configuration (again, the g suffixes are omitted).

The next step is to include in these diagrams additional terms appropriate to strong field complexes for those cases where these have different ground-state spin multiplicities from the corresponding weak field complexes. Figures 7.25–7.27 include the additions, the last named showing the d^5 case. It will be noticed that the g suffixes, appropriate to octahedral complexes, have been reinstated in these figures—and this is because they cannot be extended to tetrahedral complexes. Figure 7.26 is particularly interesting because, whereas in the d^3 case a spin quartet ($^4A_{2g}$) remains the ground state for all values of Δ, for the d^7 case the $^4T_{1g}$ weak field ground state is supplanted by a ^2E_g as the crystal field increases. The reasons for this difference are either contained in or implicit in our earlier discussion and the

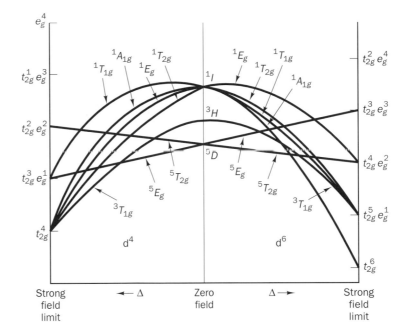

Fig. 7.25 A modified Tanabe–Sugano diagram for d^4 and d^6 octahedral complexes.

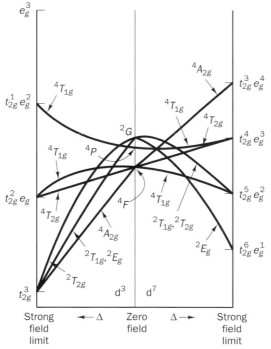

Fig. 7.26 A modified Tanabe–Sugano diagram for d^3 and d^7 octahedral complexes.

reader will find that explaining it provides a useful check on his or her understanding.

Figures 7.25–7.27 show in a qualitative fashion the crystal field energy levels changes which lead to the experimental distinctions between high and low spin complexes; the close relationship between d^n and d^{10-n} configurations being brought out by including them in the same diagram.

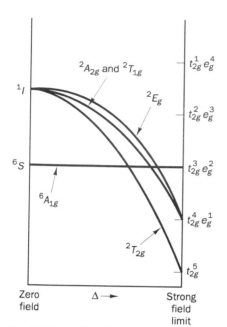

Fig. 7.27 A modified Tanabe–Sugano diagram for d^5 octahedral complexes.

For the diagrams relevant to the d^4–d^7 configurations at some point the lowest level at the weak field limit is replaced as ground state by a term of lower spin multiplicity with increase in crystal field. This is the point of transition from a high spin to a low spin complex, although it would not be correct to regard it also as a point of transition from weak to strong field (although it is often convenient to loosely relate the two transitions). As has been seen, the high spin–low spin transition does not occur at a single point for a given d^n configuration—it is a function of the ligands in a complex. For a given complex, the crossover point will be determined by an appropriate balance of crystal field and electron repulsion energies. Mathematically, in the crossover region both high and low spin terms would be treated using an intermediate field approach (which is why the transition is not also between strong and weak fields). The intermediate field approach is applicable whenever curvature is present in the lines shown in Figs. 7.24–7.27. The fact that curvature, always in one sense, is present on the lines corresponding to low spin terms in these figures indicates that they have been drawn in a selective way—other lines, at higher energies, with opposite curvature have been omitted. Further, as has been seen, Figs. 7.23–7.27 have a limitation. The scale along the horizontal axis cannot be properly defined. To interpret spectra, in particular, it is convenient to have diagrams in which the Δ scale is accurately linear. In Chapter 8 such plots will be introduced—they are called *Tanabe–Sugano diagrams*. Those lines which are shown meeting at $\Delta = \infty$ in Figs. 7.23–7.27 appear as parallel lines in the accurate Tanabe–Sugano diagrams because there is no attempt to include infinity in them!

7.7 Non-octahedral complexes

So far, this chapter has been concerned entirely with octahedral complexes. Whilst the majority of complexes are octahedral, almost all of them display some slight deviation from the ideal geometry. Other complexes have quite different geometries, as was seen in Chapter 3. For the moment the discussion will be extended in a more limited way, so as to include only tetrahedral and square planar complexes.

7.8 Tetrahedral complexes

In tetrahedral complexes the five d orbitals of a transition metal ion are again split into a set of three (d$_{xy}$, d$_{yz}$, and d$_{zx}$), denoted t_2, and a set of two (d$_{x^2-y^2}$ and d$_{z^2}$) denoted e. These labels are those obtained by dropping the g suffix from the labels given to the same orbitals in the octahedral group, but the reader must be wary, for this pattern does not always hold. As was seen in Chapter 6 the label t_{1u} is used when discussing, for example, the metal p orbitals in octahedral complexes. This label becomes t_2, not t_1, when carried over to tetrahedral complexes.

That the d orbitals split into t_2 and e sets in tetrahedral complexes may be seen from Fig. 7.28. As in Chapter 6, the key is to recognize that a tetrahedron is closely related to a cube (Fig. 7.29). If Cartesian axes are

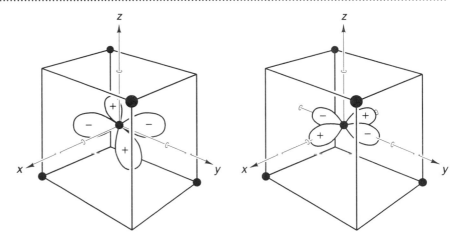

Fig. 7.28 Metal d orbitals in a tetrahedral ligand field.

drawn as in Fig. 7.28 then $x \equiv y \equiv z$ so that d_{xy}, d_{yz} and d_{zx} must be degenerate (Section 6.3 and Appendix 4). The relative ordering of these two sets energetically may also be seen from Fig. 7.28. For example, one may say, loosely, that each lobe of the d_{xy} orbital is half a cube edge away from each ligand (regarded as a point), but each lobe of the $d_{x^2-y^2}$ orbital is half a cube diagonal away. As for the octahedral case, this argument must be regarded as indicative only. The conclusion that the e set is less destabilized by the ligand field than the t_2 set (Fig. 7.30)—the opposite sense of splitting to that which occurs in an octahedral field—is confirmed by experiment and by detailed calculations. Although the agreement between experiment and the crystal field calculations is no better than for the octahedral case, these calculations also show that, for a given metal and ligand and constant metal–ligand distance, the magnitude of the splitting in a tetrahedral complex is $\frac{4}{9}$ of that in an octahedral. That is,

$$\Delta_{\text{tet}} = -\tfrac{4}{9}\,\Delta_{\text{oct}}$$

Experimentally, this relationship has been found to hold remarkably well, a fact which is perhaps not too surprising in crystal field theory because the $\frac{4}{9}$ factor arises from the geometric relationship of a tetrahedral complex to an octahedral—it is the squared ratio of the number of ligands: $\left(\frac{4}{6}\right)^2$. It becomes more surprising when, as in the previous chapter, the π bonding in octahedral and tetrahedral complexes is introduced and compared.

The next step is to obtain crystal field splitting diagrams of the type that were given in the last section. Surprisingly, we already have them! Remember that when discussing, for example, the d^8 weak field octahedral case the energy levels were obtained from those of the d^2 case by talking in terms of holes rather than electrons. Because holes are attracted by the ligands, whereas electrons are repelled, for holes the crystal field splitting of the d orbitals is, effectively, $-\Delta$.[7] That is, the d^8 case is obtained from the d^2 by replacing Δ by $-\Delta$. Now, the crystal field splitting (for electrons) in tetrahedral complexes is $-\frac{4}{9}\Delta_{\text{oct}}$. That is, the splitting diagram for a d^2 ion in a tetrahedral crystal field is qualitatively the same as that for a d^8 ion in

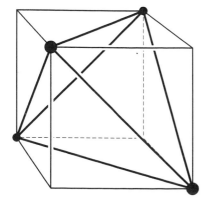

Fig. 7.29 The relationship between a cube and a tetrahedron.

[7] As in previous sections we shall use Δ without any suffix to denote the octahedral case. Other cases (and the octahedral also, when in the interest of clarity) carry suffixes.

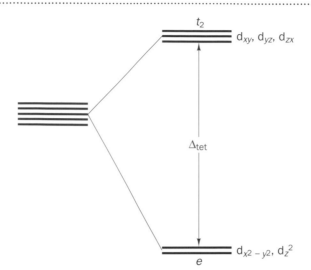

Fig. 7.30 d orbital splitting in a tetrahedral ligand field. As in Fig. 7.9, a common repulsion term has been omitted.

an octahedral field and accurately so if we include a scale factor of $\frac{4}{9}$. It was for this reason that we dropped some of the g suffixes in Figs. 7.23 and 7.24. Fig. 7.23 shows the energy levels of d^1, d^4, d^6 and d^9 weak field complexes in both octahedral and tetrahedral ligand fields. Similarly, Fig. 7.24 shows, qualitatively, the energy levels of d^2, d^3, d^7 and d^8 weak field complexes in both of these fields. The reader should easily be able to show that d^3–d^6 ions would be expected to form weak and strong field tetrahedral complexes, differing in the spin multiplicity of their ground states. Strong field tetrahedral complexes are very, very rare so they will not be discussed further. In particular, no attempt has been made to include tetrahedral complexes in Figs. 7.24–7.27. A distinction between Orgel and Tanabe–Sugano diagrams is that the former include both tetrahedral and octahedral cases whilst the latter refer to octahedral complexes only.

7.9   Square planar complexes

There are two conceptually different approaches to square planar complexes. One may regard the d orbital splitting as being that obtained when two *trans* ligands are simultaneously removed from an octahedral complex (conventionally, along the z axis). Alternatively, one may use the same approach as that used for octahedral and tetrahedral complexes and consider the splitting of the free ion d orbitals by a square planar ligand field. Since, in a square planar complex, $x \equiv y\ (\neq z)$ (Fig. 7.31) with the ligands in the xy plane, the d_{zx} and d_{yz} orbitals will be more stable than d_{xy}, leading to the qualitative d orbital splitting pattern shown in Fig. 7.32. If, as shown in Fig. 7.32, one regards a square planar complex as derived from an octahedral, then there is no evident reason to expect any significant change in energy levels of d orbitals largely localized in the xy plane. That is, d_{xy} and $d_{x^2-y^2}$ are essentially unchanged in energy. This is what is shown in Fig. 7.32 but a disadvantage must be recognized. This is that the centre of gravity of the split energy levels no longer coincides with the free-ion level. This defect is corrected in Fig. 7.33, but Fig. 7.33 has its own problems: it might easily lead one to

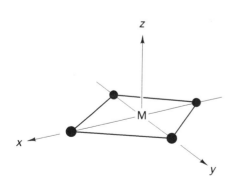

Fig. 7.31 The usual choice of coordinate axes in a square planar complex.

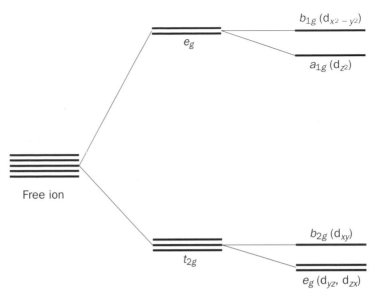

Fig. 7.32 Correlation of the d orbital splitting in a square planar ligand field with those in an octahedral ligand field.

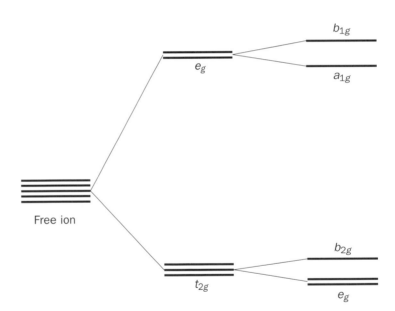

Fig. 7.33 Modified correlation between octahedral and square planar crystal field splittings.

conclude that in square planar complexes d_{xy} and $d_{x^2-y^2}$ suffer an extra repulsion. This is but a minor disadvantage and Fig. 7.33 is the one generally encountered. Note that whilst Δ may be regarded as the energy difference between the centres of gravity of the upper pair and lower trio of levels, two additional splitting energies are needed to give a complete specification of the energy level diagram for square planar complexes. These are the splittings of the b_{1g}–a_{1g} orbitals and of the b_{2g}–e_g. There is no evident connection between these energies because the consequence of removing the ligands on the z axis must be expected to be different for the d_{z^2} and the d_{xz} and d_{yz}

Table 7.8 Correlation between symmetry labels used for octahedral and square planar geometries. Although g and u suffixes are not given and have to be added, the expressions in brackets indicate how the various d orbitals transform. Thus, d_{z^2} has A_{1g} symmetry in a square planar complex

Octahedral	Square planar
A_1	A_1
A_2	B_1
E	$A_1(z^2) + B_1(x^2 - y^2)$
T_1	$A_2 + E$
T_2	$B_2(xy) + E(zx, yz)$

orbitals. Not surprisingly, theoretical efforts have been made to relate the two splittings and thus make the problem of interpreting the d–d electronic spectra of square planar complexes more tractable.

Note that the symmetry labels used for square planar complexes are different from those used for octahedral complexes. The correlation between the two sets is shown in Table 7.8. In this table are included some symmetry labels which have not yet been used in the discussion, anticipating a need for them in later chapters. The g and u suffixes have been omitted in both geometries in order to keep the table compact. (The rule, of course, is $g \rightarrow g$ and $u \rightarrow u$.) There is some freedom about the choice of the B_1 and B_2 labels in square planar complexes in that they may be interchanged. What one author called B_1 another may call B_2. In this book the choice shown in Table 7.8 is used. Table 7.8 may be used for either orbitals or terms. For example, the $^2T_{2g}$ term of a d^1 configuration in an octahedral ligand field splits into two, $^2B_{2g}$ and ^2E_g, if the ligand field is reduced to square planar. Of these, the ^2E_g is more stable because it corresponds to the single d electron occupying the d_{zx} and d_{yz} orbitals (which are more stable than d_{xy}, cf. Fig. 7.33).

7.10 Other stereochemistries

The d orbital splitting patterns in low-symmetry geometries other than those which have been discussed are most readily obtained in the way outlined for square planar complexes. The results for two important geometries are given in Table 7.9. However, care must be exercised in using this table. First, as for square planar complexes, some choice exists in the allocation of symmetry labels for digonally distorted octahedral complexes (Fig. 7.34). Secondly, the choice of z axis for an octahedral complex (along a fourfold rotation axis) is not consistent with the usual choice for trigonal distorted octahedral complexes, for which the threefold axis is chosen as z axis (Fig. 7.35). This means that, for example, the d_{z^2} orbital in the latter geometry is *not* the same d_{z^2} as that in the corresponding octahedral complex, but turns out to be a mixture of the octahedral d_{xy}, d_{yz} and d_{zx} orbitals. This perhaps seems rather strange but, in fact, illustrates a rather general point which will be more fully dealt with in Chapter 11. There it will be found that two complete sets of f orbitals exist, the shapes of the orbitals in one set not all being matched in

Table 7.9 Correlation between symmetry labels used for true octahedral, trigonally distorted octahedral and digonally distorted octahedral geometries, As for Table 7.8 g and u suffixes are omitted but must be added. So, $A_1(xy)$ and $A(z^2)$ mean, respectively, that d_{xy} transforms as A_{1g} (trigonally distorted octahedron) and d_{z^2} transforms as A_g (digonal distortion). For digonally distorted complexes the labels B_1, B_2 and B_3 (with or without g or u suffixes) may be used differently by different workers, although those with g suffixes are never interchanged with those with u suffixes

Octahedral	Trigonally distorted octahedral field	Digonally distorted octahedral field
A_1	A_1	A
A_2	A_2	A
E	E	$A(z^2) + A(x^2 - y^2)$
T_1	$A_2 + E$	$B_1(xy) + B_2(yz) + B_3(zx)$

the other. Which set is used depends on the molecular geometry, just as different d orbital sets are used for octahedral and trigonally distorted octahedral geometries.

One final point: just as we found that the number of unpaired electrons in an octahedral complex can vary with the crystal field, so, too, in other geometries. However the effects tend to be more subtle and difficult to disentangle. Thus, for five-coordinate complexes they are interwoven with the trigonal bipyramidal and square pyramidal structural possibilities.[8]

7.11 Ligand field theory

This chapter is concluded by returning again to octahedral transition metal complexes. In Chapter 6 the molecular orbital method was found to give an account of the bonding in these complexes which gave good general agreement with the observations except at points where the independent electron model is deficient. In the present chapter crystal field theory had rather little to say about bonding but showed considerable value in modelling the behaviour of d electrons in complexes. In that they focus on rather different aspects, could they not be combined into a single, unified, theory? They can and have been. This theory, called *ligand field theory*, differs from simple crystal field theory in that allowance is made for the existence of covalency. This is done by dropping a requirement inherent in simple crystal field theory: that when a quantity is available from spectral data on free ions—an electron repulsion energy, for instance—it should used in crystal field calculations. Covalency means that the orbitals occupied by the d electrons in a complex must be expected to differ from those of the free ion and so, too, therefore should the value of electron repulsion and other energies. In practice, this means that such quantities become parameters in the theory, given those values which produce the best agreement with experiment. These additional parameters will be met in the next two chapters. The magnitudes of the additional parameters are determined experimentally,

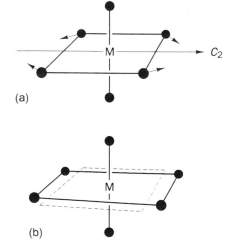

Fig. 7.34 A digonally distorted octahedral complex: (a) an octahedral complex and a distortion which is symmetric with respect to the C_2 axis indicated; (d) the distorted molecule with the original octahedron indicated dotted.

[8] See R. R. Holmes, *J. Amer. Chem. Soc.* (1984) *106*, 3745, for a detailed discussion.

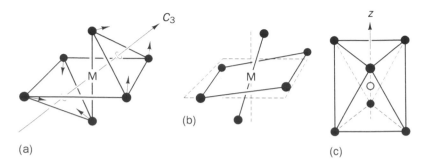

Fig. 7.35 A trigonally distorted octahedral complex: (a) an octahedral complex and a distortion which is symmetric with respect to the C_3 axis indicated; (b) the distorted molecule with the original octahedron shown dotted; (c) an alternative view of the trigonally distorted octahedron.

but the number of independent observables is usually less than the number of parameters. It cannot, therefore, be asserted that ligand field theory, is in general, proven. One may say only that it provides a consistent explanation for experimental data using parameters which, almost invariably, have physically reasonable values. Not surprisingly, there have been attempts to devise cunning experiments which provide additional information so that we end up with more data than parameters. For instance, instead of studying complexes randomly orientated in solution one might look at molecules fixed in a single crystal of known structure. The orientation of the crystal, and with it any inherent molecular anisotropy (such as a tetragonal distortion), could be varied relative to the direction of polarization of polarized incident light, for example, and so give additional data. However, octahedral and tetrahedral complexes are isotropic ($x \equiv y \equiv z$) so this method is only applicable to lower symmetry molecules requiring even more parameters to start with, as was seen in Section 6.9. Suitable cases for study are already bad ones! Again, instead of working at one temperature, we could study a property— the magnetic properties of a complex for instance—as a function of temperature. The varying thermal population of low-lying energy levels (a magnetic field frequently induces suitable small splittings, as we shall see in Chapter 9) then gives information on the energy separation between these levels. Unfortunately, at the very low temperatures that really lead to the depopulation of levels that are only slightly split apart, and so offer the greatest potential for additional useful information, other weak phenomena—such as the freezing out of lattice vibrations—have to be considered and this requires yet more parameters! We seem to be hitting a brick wall at every turn. This situation has inspired a search for theoretical models which might reveal interrelationships between the various parameters of ligand field theory whilst themselves involving relatively few, if any, additional parameters. Many such models have been investigated, including those that we will meet in Chapter 10. Entire books have been written on the subject.[9] Suffice to say that the search continues. Although clear progress has been made, no single model is currently accepted as providing a generally applicable method.

We conclude by reconsidering what we must now call the ligand field splitting parameter of an octahedral transition metal complex, Δ, since the factors affecting its magnitude are evidently more complicated than we at first supposed. We have encountered three such factors:

[9] See, for example, M. Gerloch and R. S. Slade *Ligand-Field Parameters*, Cambridge University Press, Cambridge, 1973.

- First, the magnitude of Δ depends on the electrostatic field generated by the ligands—the crystal field.
- Secondly, it depends on ligand–metal σ bonding because the energy of the metal e_g orbitals depends on this.
- Thirdly, it depends on the ligand–metal π bonding, since this affects energy of the metal t_{2g} set.

This does not exhaust the list of factors influencing the magnitude of Δ, but there are believed to be no others of comparable importance. A recitation of most of the evidence that supports the ligand field model in preference to the crystal field model is deferred until Section 12.1.

Further reading

In practice it is not possible to separate the further reading relevant to this chapter from that appropriate to Chapter 6. The contents of the two chapters go together. In addition to the references given at the end of Chapter 6, two others which follow the pattern of the present chapter but with a more mathematical approach are Chapter 13 of *Valence Theory* by J. N. Murrell, S. F. A. Kettle and J. Tedder, Wiley, London,

1970 and, more simply, in Chapter 12 of *The Chemical Bond* by the same authors, J. Wiley, Chichester, 1985.

An excellent introduction is given in 'Ligand Field Theory' by J. S. Griffiths and L. E. Orgel, *Quarterly Rev., Chem. Soc. (London)* (1958) *11*, 381. This particular article was perhaps more responsible than any other for the introduction of the ideas of crystal and ligand field theories into inorganic chemistry.

Questions

7.1 Outline the difference(s) between members of the following pairs:

1. symmetry elements and symmetry operations;
2. t_{2g}^2 and t_{2g}^3;
3. a p orbital and a *P* term;
4. T_{1g} and $^3T_{1g}$.

7.2 Draw diagrams showing the splitting of the following orbital sets in an octahedral crystal field: s, p, d and f. Make an educated guess at the likely pattern of splitting of a set of g orbitals (the answers to all of these may be deduced from Table 7.6).

7.3 Write brief notes on:

1. the d^5 high and low spin configurations in an octahedral ligand field;
2. pairing energy;
3. the spectrochemical series.

7.4 Starting with (a) the weak field and then (b) the strong field octahedral d^5 configurations, write down the excited state configurations which may be derived from them. Comment on the consequences of the differences between the two lists (these

differences will be reflected in diagrams such as that in Fig. 7.27).

7.5 With as little reference to the text as possible, rehearse the arguments that lead to Fig. 7.20.

7.6 With as little reference to the text as possible, rehearse the arguments that lead from Fig. 7.20 to Fig. 7.21.

7.7 A tetrahedron and a cube are closely related. Indeed, as Fig. 7.29 shows, a cube may be regarded as composed of two tetrahedra. Suggest, qualitatively, the form of the relationship that exists between Δ_{tet} and Δ_{cubic} for complexes involving the same metal and ligands. The present chapter contains data which suggest a probable quantitative relationship. What is it?

7.8 You have just astonished the scientific world by your discovery of a class of monodentate ligands that form cubic complexes, $[ML_8]^{2+}$, with all of the ions of the first transition series. It is clear that, depending on the choice of L, some metals may form either high spin or low spin complexes. For which ions of the first transition series does the possibility of high and low spin cubic complexes occur?

7.9 Enumerate and briefly explain those factors which are believed to make a significant contribution to Δ, the ligand field splitting parameter of an octahedral complex.

Electronic spectra of transition metal complexes

8.1 Introduction

In this chapter our concern will be with the electronic spectra of transition metal complexes, particularly those of the first transition series. Although nowadays there are fewer studies of these spectra, per se, than previously, their study is essential if the electronic excited states of complexes are to be understood. The energy required for the promotion of an electron from one orbital to another or, more precisely, the excitation of a molecule from its electronic ground state to an electronic excited state, corresponds to absorption of light in the near-infrared, visible or ultraviolet regions of the spectrum. For transition metal complexes the absorption bands in the first two of these regions are relatively weak and are associated with transitions largely localized on the metal atom. The ultraviolet bands are intense. They are associated with the transfer of an electron from one atom to another and so are called *charge-transfer* bands. These bands are responsible for the colour changes associated with indicators for inorganic cations, such as the thiocyanate test for Fe^{III} or the indicators used in EDTA (compleximetric) titrations, but in these cases the charge-transfer bands fall in the visible region of the spectrum. The intense bands will be dealt with in the latter part of this chapter; for the moment our concern will be with the weaker bands. Whilst these are most simply explained by crystal field theory, a detailed comparison of the data with the theoretical predictions will show that ligand field theory provides a more appropriate explanation.

When spectral bands are weak there is a reason. Often it means that they are bands which are forbidden but—obviously—not totally forbidden. The weak bands with which we are concerned are of this type. Electronic transitions which correspond to strong bands are electric dipole in type. Classically, the electric vector associated with the incident light beam behaves like a pair of alternating + and − charges across the molecule. These oscillating charges induce an oscillating dipole in the molecule; when the

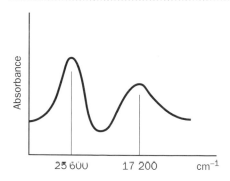

Fig. 8.1 The d–d spectrum of the [V(H$_2$O)$_6$]$^{3+}$ cation (d^2).

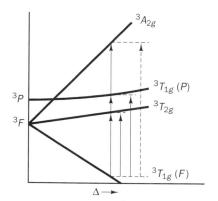

Fig. 8.2 The part of the d^2 Tanabe–Sugano diagram needed for the interpretation of the spectrum in Fig. 8.1.

frequency of the molecular oscillating dipole corresponds to a natural frequency of the molecule, resonance occurs and the molecule acquires energy at the expense of the light wave. Clearly, at the heart of this explanation is an electric dipole—hence the phrase *electric dipole allowed*. In an atom, the electric dipole-allowed transitions are those associated with transitions between the orbitals s ↔ p, p ↔ d and d ↔ f (where the double-headed arrows indicate that the transition is allowed in either direction).[1] Forbidden are s ↔ d, p ↔ f, s ↔ s, p ↔ p, d ↔ d etc. The weak bands observed in the visible region for transition metal complexes are d ↔ d in type, and so, formally forbidden but made slightly allowed by mechanisms which will be discussed later. First, we need an explanation for the number of bands and their relative energies. We start with the crystal field model and later consider the changes needed to accommodate the ligand field approach.

As an example, consider the pale violet coloured ion [V(H$_2$O)$_6$]$^{3+}$ which shows two weak absorptions in the visible. One is in the blue at ca. 3900 Å and the other in the yellow at ca. 5800 Å. The spectrum of this d^2 ion is shown in Fig. 8.1. The explanation provided by crystal field theory for the weak transitions is presented in diagrams such as Fig. 8.2, which shows the effect of an octahedral crystal field on the energy levels of a d^2 ion such as V^{III}. This energy level diagram was developed in a qualitative fashion in Chapter 7 (see Fig. 7.21) but for the purposes of the present chapter, a form in which the energy scale is well defined is needed and obtaining this is our first task. The ground state is $^3T_{1g}(F)$ and spin-allowed transitions— those in which the spin is the same in both ground and excited states—may occur to the $^3T_{2g}$, $^3T_{1g}(P)$ and $^3A_{2g}$ excited states, this being the usual order of increasing energy. The origin of these predictions is shown by the arrows in Fig. 8.2 (the triple-headed arrow is that relevant; those to the right explain it). As is evident from Fig. 8.1, experimentally two transitions are observed. Calculations (see below) indicate that the third, high energy, band is usually obscured by an intense charge-transfer band in this region. Later, it will transpire that this third band should, in any case, be particularly weak and broad. It thus appears that we can account, qualitatively, for the spectrum. Whether we can do so quantitatively is discussed in the next section.

8.2 Electronic spectra of V^{III} and Ni^{II} complexes

Before a quantitative assessment of the applicability of the crystal field model is possible, explicit expressions for the energies of the excited states of the V^{III} ion relative to that of the ground state are needed. These energies depend on three quantities: the crystal field splitting parameter Δ, the magnitude of the ^3F–^3P separation in the free ion, and the magnitude of interaction between the two $^3T_{1g}$ terms. The second of these quantities is usually denoted $15B$ where B is a sum of electron repulsion integrals which it is convenient not to evaluate explicitly and which is referred to as a *Racah parameter*. Other Racah parameters, also composed of electron repulsion integrals, which the reader may encounter are denoted A and C. The third of the quantities mentioned above, the energy of interaction of the two $^3T_{1g}$ terms, is, fortunately, a function of Δ and B only, so that these two quantities are

[1] Strictly, of course, we should work with terms rather than orbitals.

all that is needed to interpret the spectra of octahedral V^{III} complexes. $15B$ is the separation of the ^3F and ^3P free ion terms, so it should be possible to obtain the value of B from atomic spectral data. This corresponds to the pure crystal field approach, but, significantly, agreement between experiment and theory for complexes is only generally obtained if B is treated as a parameter, the value of which may be varied. It turns out that B has to be given a value which is rather smaller than that obtained from atomic spectra. This parameterization of B is part of the ligand field method and is consistent with our ideas of the consequences of covalency in transition metal complexes. Covalency implies

1. that the metal electrons will be partially delocalized onto the ligands;

2. that the effective positive charge on the transition metal will be smaller than in the free ion.

Both of these effects mean that the metal electron cloud will be more diffuse in the complex than in the free ion, and repulsion between the electrons making up this cloud is therefore reduced. This conclusion is confirmed by a more detailed analysis.

The energies of the crystal field spin triplet terms of the d^2 configuration, relative to the ^3F free-ion term as zero, are given explicitly below. It is beyond the scope of the present text to derive them so we give these expressions without proof;[2] however, the reader may readily check that they reduce to those given in Section 6.5 for the weak field limit[3] $\Delta \to 0$ (expand the square root, using the binomial theorem in the form $(y^2 + xy)^{1/2} \approx y + x/2$). These expressions also give the strong field energies; note that the $15B$ term was not included in the energies given in Section 6.5, not that this is important because in the strong field limit the effects of electron repulsion are negligible in comparison with the ligand field. In Chapter 7 this situation was accommodated by making the ligand field infinite but in the present context it is simpler to set $B = 0$. The square root expressions are those which allow for the mixing of the $^3T_{1g}(F)$ and $^3T_{1g}(P)$ terms by the crystal field. Indeed, it should be evident from the form of these two energy level expressions that they are obtained as the roots of a quadratic equation. The energies of the $^3T_{1g}$ and $^3A_{2g}$ terms are those which were obtained in Chapter 7.

Term	Energy
$^3T_{1g}(F)$	$\frac{1}{2}[15B - \frac{3}{5}\Delta - (225B^2 + 18B\Delta + \Delta^2)^{1/2}]$
$^3T_{2g}$	$\frac{1}{5}\Delta$
$^3T_{1g}(P)$	$\frac{1}{2}[15B - \frac{3}{5}\Delta + (225B^2 + 18B\Delta + \Delta^2)^{1/2}]$
$^3A_{2g}$	$\frac{6}{5}\Delta$

It follows that the two observed bands, assigned to the $^3T_{2g} \leftarrow \ ^3T_{1g}(F)$[4]

---

[2] A derivation is given in J. N. Murrell, S. F. A. Kettle and J. M. Tedder, *Valence Theory*, Wiley, London and New York, 1956, Chapter 13. A simpler derivation is given by the same authors in *The Chemical Bond*, Wiley, Chichester and New York, 1985, Chapter 12.

[3] Strictly, of course, the weak field limit is $\Delta = 0$. However, substituting this value simply gives the $^3F-^3P$ separation in the free ion, $15B$.

[4] Here we follow the usual spectroscopic practice of giving the excited state first.

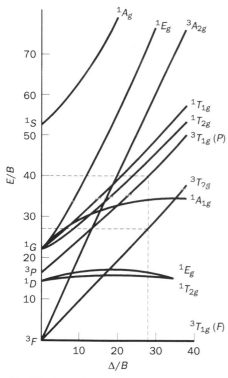

Fig. 8.3 A complete Tanabe–Sugano diagram for octahedral d^2 complexes.

and $^3T_{1g}(P) \leftarrow {}^3T_{1g}(F)$ transitions, have energies given by the differences between the two relevant expressions above:

Transition	Energy
$^3T_{2g} \leftarrow {}^3T_{1g}(F)$	$\frac{1}{2}[\Delta - 15B + (225B^2 + 18B\Delta + \Delta^2)^{1/2}]$
$^3T_{1g}(P) \leftarrow {}^3T_{1g}(F)$	$(225B^2 + 18B\,\Delta + \Delta^2)^{1/2}$

We have experimental transition energies to fit these expressions, so we can now obtain values for Δ and B. This task is not difficult. Divide the two expressions above by B. We then have

$$\frac{E(^3T_{2g} \leftarrow {}^3T_{1g}(F))}{B} = \frac{1}{2}\left[\frac{\Delta}{B} - 15 + \left(225 + 18\frac{\Delta}{B} + \frac{\Delta^2}{B^2}\right)^{1/2}\right] \quad (8.1)$$

$$\frac{E(^3T_{1g}(P) \leftarrow {}^3T_{1g}(F))}{B} = \left(225 + 18\frac{\Delta}{B} + \frac{\Delta^2}{B^2}\right)^{1/2} \quad (8.2)$$

The right-hand side expressions are functions of Δ/B. Similarly, the energies of the terms themselves, divided by B, may be expressed as functions of Δ/B. Tanabe and Sugano have published diagrams which show these relationships; they take as their energy zero the energy of the ground state, so that, for the d^2 case, they effectively plot the functions

$$\frac{E(^3T_{2g} \leftarrow {}^3T_{1g}(F))}{B}$$

and

$$\frac{E(^3T_{1g}(P) \leftarrow {}^3T_{1g}(F))}{B}$$

against Δ/B. It is to be emphasized that the ground state is coincident with the abscissa in their diagrams. If a change in ground state occurs—and this can happen for d^4–d^7 ions, where a high spin/low spin distinction exists—there is a discontinuity in the diagram. Figure 8.3 is a Tanabe–Sugano diagram for the octahedral d^2 case. A complete set of Tanabe–Sugano diagrams is given in Appendix 7; similar sets appear in many textbooks of inorganic chemistry. It may well be found helpful to compare these, real, Tanabe–Sugano diagrams with the modified versions introduced in Chapter 7. We shall return to the Tanabe–Sugano diagrams later in this chapter and the discussion there should explain any difficulties encountered when making the comparison just suggested.

Consider as a specific example the $[V(H_2O)_6]^{3+}$ ion, which has the spectrum shown in Fig. 8.1; it has weak peaks at ca. 17 200 cm^{-1} (ca. 5800 Å) and 25 600 cm^{-1} (ca. 3900 Å). We conclude that $E(^3T_{1g}(P) \leftarrow {}^3T_{1g}(F)) = 17\,200$ cm^{-1} and that $E(^3T_{1g}(P) \leftarrow {}^3T_{1g}(F)) = 25\,600$ cm^{-1}. Equations 8.1 and 8.2—and therefore Fig. 8.1—can be used provided that the Bs appearing on the left-hand side of each equation are eliminated. This can be done if

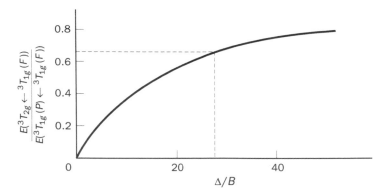

Fig. 8.4 A plot of eqn 8.3, useful in the interpretation of the spectra of octahedral d^2 complexes.

eqn 8.1 is divided by eqn 8.2 to give

$$\frac{E(^3T_{2g} \leftarrow 3T_{1g}(F))}{E(^3T_{1g}(P) \leftarrow {}^3T_{1g}(F))} = \frac{\dfrac{\Delta}{B} - 15 + \left(225 + 18\dfrac{\Delta}{B} + \dfrac{\Delta^2}{B^2}\right)^{1/2}}{2\left(225 + 18\dfrac{\Delta}{B} + \dfrac{\Delta^2}{B^2}\right)^{1/2}} \qquad (8.3)$$

In the case of $[V(H_2O)_6]^{3+}$ the value of this quotient is $17\,200/25\,600 = 0.67$. We can now proceed in one of two ways—and a third will immediately occur to the computer enthusiast; it is not difficult to write a suitable program to do the work. We could apply a trial and error process to Fig. 8.3 until we find the correct value of Δ/B (in our case $\Delta/B = 28$, shown dotted in Fig. 8.3). Alternatively, we could plot the right-hand side of eqn 8.3 against Δ/B. Such plots are useful when a large number of data have to be analysed. That appropriate to the d^2 configuration is shown in Fig. 8.4 where, again, the $[V(H_2O)_6]^{3+}$ case is indicated by dotted lines. Having determined Δ/B from Fig. 8.4 (again, of course, obtaining a value of 28), we return to Fig. 8.3 and use this ratio to obtain $E(^3T_{1g}(P) \leftarrow {}^3T_{1g}(F))/B$ and $E(^3T_{1g}(P) \leftarrow {}^3T_{1g}(F))/B$. Alternatively, and more accurately, we could substitute for Δ/B in eqns 8.1 and 8.2. The right-hand sides of these equations are then found to have values of 25.9 and 38.6, respectively so, using the experimental transition energies, both give $B = 665 \text{ cm}^{-1}$. Since $\Delta/B = 28$ it follows that $\Delta = 18\,600 \text{ cm}^{-1}$. This analysis follows the ligand field approach, with B regarded as a parameter.

In the crystal field approach we use the free-ion value of B, 860 cm^{-1}. It follows that $E(^3T_{1g}(P) \leftarrow {}^3T_{1g}(F))/B = 17\,200/860 = 20$ and $E(^3T_{1g}(P) \leftarrow {}^3T_{1g}(F))/B = 25\,600/860 = 29.8$. Using Fig. 8.3 these lead to values of Δ/B of 22.5 and 18.0 respectively; that is, Δ values of $19\,400$ and $15\,500 \text{ cm}^{-1}$. This internal inconsistency does not occur in ligand field theory. However, it is only removed by introducing an additional parameter, and, consequently, there are no additional data with which to test the theory, for we have used two experimental quantities (transition energies) to define two parameters Δ and B. Only if the $^3T_{2g} \leftarrow {}^3T_{1g}(F)$ were to be observed could we test the theory. Using the ligand field values for Δ and B obtained above, together with the expression for the transition energy obtained from the term energies given earlier, this transition is predicted to be at ca. $36\,000 \text{ cm}^{-1}$, where it is obscured by strong charge-transfer bands. However, a weak band has been observed in this region in closely related complexes.

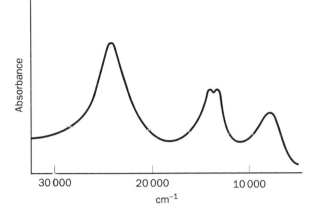

Fig. 8.5 The d–d spectrum of the [Ni(H$_2$O)$_6$]$^{2+}$ cation (d^8).

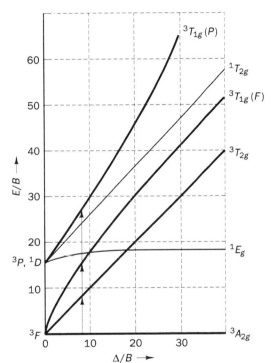

Fig. 8.6 The part of the d^8 Tanabe–Sugano diagram believed to be relevant for the interpretation of the spectrum in Fig. 8.5.

As a further example of the interpretation of d-d spectra consider the case of the d–d ion Ni^{II} in the complex [Ni(H$_2$O)$_6$]$^{2+}$, the spectrum of which is shown in Fig. 8.5. The three bands, at 8500 cm^{-1}, ca. 14 100 cm^{-1} and 25 300 cm^{-1} are assigned, respectively, to the $^3T_{2g} \leftarrow {}^3T_{2g}$, $^3T_{1g}(F) \leftarrow {}^3A_{2g}$ and $^3T_{1g}(P) \leftarrow {}^3A_{2g}$ transitions (see the simplified Tanabe–Sugano diagram in Fig. 8.6). Of these, the central band shows some structure which will be discussed later in this chapter. The energy level expressions are derived immediately from those given earlier for d^2, except that now the $^3A_{2g}$ level is the ground state (which means that we simply have to change the sign of Δ). It follows that the lowest energy transition has an energy Δ. So, we take $\Delta = 8500$ cm^{-1}. The next transition, to the $^3T_{1g}(F)$ term, has an energy of

$$\tfrac{1}{2}[15B + 3\Delta - (225B^2 - 18B\Delta + \Delta^2)^{1/2}]$$

(remember, the sign of Δ has to be changed, here we are dealing with two holes). In the crystal field model the free-ion value of B, 1082 cm^{-1} is inserted, to obtain a second value of Δ. This can be done by dividing throughout by B, to obtain

$$\frac{14\,100}{1082} = \frac{1}{2}\left[15 + 3\frac{\Delta}{B} - \left(225 - 18\frac{\Delta}{B} + \frac{\Delta^2}{B^2}\right)^{1/2}\right]$$

and solving for Δ/B, thus obtaining $\Delta/B = 7.71$. With the free-ion value of B (the crystal field model) this gives $\Delta = 8340$, a bit different from the previous value. In the ligand field model we insert $\Delta = 8500$ and solve for B (the B^2 terms disappear) to obtain $B = 900$ cm^{-1}.

In this particular example there was a third band to consider. Really, this should be set on a par with the other two to obtain values for Δ and B, particularly on the ligand field model. Instead, we will use our calculated ligand field values of Δ and B to predict its position—for this will provide a test of the ligand field model. The energy of the $^3T_{1g}(P) \leftarrow {}^3A_{2g}$ transition is

$$\tfrac{1}{2}[15B + 3\Delta + (225B^2 - 18B\Delta + \Delta^2)^{1/2}]$$

which, inserting $B = 900$ cm^{-1} and $\Delta = 8500$ cm^{-1} predicts that the third band will be at 24 900 cm^{-1}. In contrast, the crystal field model ($B = 1082$ cm^{-1}, $\Delta = 8500$ cm^{-1}) predicts 27 400 cm^{-1} or, with $\Delta = 8340$ cm^{-1}, the transition is predicted to be at 27 100 cm^{-1}. Clearly, the ligand field theory is to be preferred.

As this example shows, those cases in which there are sufficient data to provide a test of ligand field theory show that it leads to reasonable agreement between observed and calculated band positions. It will be noted that the discussion has been confined to spin-allowed bands (both ground and excited states were spin triplets). Other, extremely weak bands occur and have been assigned to spin-forbidden transitions (for example $^1T_{2g} \leftarrow {}^3T_{1g}(F)$).

The spin-allowed spectral bands of all other cases, weak field and strong field alike, may be analysed in a similar way to that given above. Tanabe and Sugano have given diagrams for all other d^n cases (except d^1 and d^9 which are trivial, having only one excited state at an energy Δ above the ground state) and a complete set is given in Appendix 7. Alongside each diagram in this Appendix is an example of a spectrum to which it refers. The reader who glances forward to this Appendix at this point will find that there are quite often complicating features which have not so far been discussed. Most will be covered in later in this chapter. As has been demonstrated, both Δ and B may be obtained from Tanabe–Sugano diagrams with sufficient accuracy for most purposes. The physical reasonableness of the results obtained may be assessed by reference to the spectrochemical series (for Δ) and the free-ion value of B. The latter are given in Table 8.1 for ions of the first-row transition elements. It is assumed that an incorrect assignment of the spectral bands will lead either to an incompatibility with the Tanabe–Sugano diagram or to physically unreasonable values of Δ and B.[5]

---

[5] Of particular interest to the computer enthusiast are general expressions for Δ and B. See Y. Do, *J. Chem. Ed.* (1990) 67 134.

Table 8.1 Free-ion values of the electron repulsion parameter B for first row transition metal elements with d^n configurations, measured in cm^{-1}. A blank indicates that the value is not known; a dash indicates that the configuration has either one or no electrons outside a closed shell

Metal atom charge   **Metal atom**

	Ti	V	Cr	Mn	Fe	Co	Ni	Cu
0	560	579	790	720	805	789	1025	–
+1	681	660	710	872	870	879	1038	1218
+2	719	765	830	960	1059	1117	1082	1239
+3	–	860	1030	1140				
+4	–	–	1040		1144			

It is convenient at this point to complete our discussion of the reduction in the free-ion value of the Racah parameter B which occurs on complex formation. This, it will be recalled, appears to be a consequence of the metal electrons being delocalized over a larger volume of space in the complex than in the free ion. It has proved possible to arrange ligands in a series such that, for a given metal ion, the B value required to fit the spectra of the $[ML_6]^{n+}$ ions decrease down the series

$$F^- > H_2O > NH_3 > en > -NCS^- > Cl^- \approx CN^- > Br^- > S^{2-} \approx I^-$$

This series has been called the *nephelauxetic series* (nephelauxetic = cloud expanding). There are exceptions, but this order holds quite well. It should be noted that what are probably the most polarizable ligands give the lowest B values and vice versa. Another way of looking at the list is to recognize that hard bases are to the left and that soft are to the right (see Section 5.4). A similar list exists for metal ions but it is difficult to attach any clear interpretation to it, although some regularities may be teased out. Such a list, with B decreasing left to right relative to the free ion values, is

$$Mn^{II} \approx V^{II} > Ni^{II} \approx Co^{II} > Mo^{III} > Cr^{III} > Fe^{III} > Rh^{III} \approx Ir^{III} > Co^{III}$$

$$> Mn^{IV} > Pt^{IV} > Pt^{VI}$$

8.3 Spin-forbidden transitions

As mentioned above, extremely weak bands, assigned to spin-forbidden transitions, may sometimes be observed in the spectra of transition metal ions. Indeed, for the d^5 high spin case (Mn^{II} and some Fe^{III} complexes), the ground state is the only sextet spin term ($^6A_{1g}$) and so all the observed d–d transitions are spin-forbidden. Evidently, the spin selection rule, like so many others, is not absolute. Similarly, most of this chapter is concerned with transitions which, in large measure, are d–d transitions and therefore forbidden (they are often said to be *Laporte forbidden*). Later we shall return to the problem of why these forbidden transitions occur.

The problem of accounting quantitatively for the spin-forbidden bands is rather more difficult than carrying out calculations for their spin-allowed counterparts. Not surprisingly, a new parameter has to be introduced. This is the Racah parameter C, another electron repulsion parameter, which like B, is expected to be numerically smaller than its free-ion value. If the treatment given above for the d^2 case is followed, C appears in the final expressions for (transition energy)/B as the quotient C/B. Before a Tanabe–Sugano diagram which includes terms of all multiplicities can be constructed it is necessary to give a numerical value to C/B (or, as it is sometimes called, γ). For different values of C/B the Tanabe–Sugano diagram will be different. However, this parameter only affects the interscaling between terms of different spin multiplicities—so that, for instance, the relative energies of the $^3A_{2g}$, $^3T_{1g}(P)$, $^3T_{2g}$ and $^3T_{1g}(F)$ terms of the d^2 configuration are independent of C/B.[6] The Tanabe–Sugano diagrams which are commonly encountered are plotted for $C/B \approx 4.5$. This value is rather larger than the free-ion ratio which, for the first row transition elements, has an average value of ca 4.0 (but which varies from 3.2 to 4.8). Before the availability of suitable computer programs it was not an uncommon practice to assign spin-forbidden bands using whatever Tanabe–Sugano diagram was to hand. When the job has been done properly it has been found that it is possible to obtain excellent agreement between theory and experiment. In the computer-based approach it is not generally possible to work with relatively simple equations of the type that were used above for the d^2 and d^8 cases. It is necessary to go back to the beginning and to set up the problem in a way that is rather different from the presentation given in the present text.

8.4 Effect of spin–orbit coupling

The spin-allowed d–d bands which dominate the visible spectrum of complexes of many transition metal ions are rather broad, with half-widths of ca. 3000 cm^{-1}. This means that if there is some interaction within the system which has been overlooked, its presence will not be apparent from the electronic spectrum unless it causes splittings in the energy levels of perhaps 1000 cm^{-1}. In fact, two such effects have been omitted—spin–orbit coupling and the Jahn–Teller effect.

When developing crystal and ligand field theories, it was assumed that the orbital properties of the electrons are something quite separate from their spin. This separation is reflected in the symbolism; in the symbol ^3F for example, the 3 refers to the spin multiplicity and the F to the orbital multiplicity. Now, the electron has an intrinsic magnetic moment—that is, it behaves like a tiny bar magnet—the *spin* of the electron describes the same phenomenon. These tiny bar magnets, spins, orientate themselves in a magnetic field, a phenomenon which makes relevant the magnetic measurements described in the next chapter. Similarly, there is an orbital-derived magnetic moment for all but S terms; this orbital moment may be likened to the magnetic moment of a solenoid, the circulation of the electron around the nucleus being akin to the circulation of a current through the turns of a

---

[6] The general development in this and the preceding chapter follow the traditional pattern and are based on the so-called Slater–Condon–Shortley model for the description of the energy differences between terms arising from the same configuration. Fundamental objections have recently been raised to certain features of this model but the general picture that it presents survives, although, for example, some detailed features that it attributes to electron repulsion may actually arise from nuclear attraction!

solenoid. We see that the separation of the spin and orbital properties of the electrons is equivalent to assuming that two magnets, spin-derived and orbital-derived, do not interact. Of course, they interact, i.e. couple, to some extent and this is reflected in the phenomenon of spin-orbit coupling. The formal theory of spin–orbit coupling treats the phenomenon as the coupling of two angular momenta rather than bar magnets. We shall have to move in this more formal direction in our discussion of the properties of f electron systems (Chapter 11), although we shall there attempt to compensate by introducing pictorial representations of spin–orbit functions. The consequences of spin–orbit coupling for d electron systems are discussed in more detail in the next chapter. At this point it is sufficient to simply state that spin–orbit coupling causes splitting of some of the degeneracies implicit in the orbital energy level diagrams encountered so far. For example, a $^4T_{1g}$ term is 12-fold degenerate (4×3). This splits into three sublevels as a consequence of spin–orbit coupling because, loosely, some arrangements of spin and orbital magnets are energetically more stable than others. The effective number of spin–orbit components of crystal field terms depends only on their spin and orbital multiplicities and these are shown in Table 8.2. The actual magnitude of the splitting caused by spin–orbit coupling may be determined from the electronic spectra of gaseous transition metal atoms and ions (excited, for instance, in an electric discharge). For these it depends on the metal. Although, in theory, the dependence should vary with the atomic number Z as Z^4, the actual experimental dependence on Z is not so simple, as Table 8.3, which gives spin–orbit splitting constants for some typical ions, shows.

For Ti^{III} the free-ion value of the spin–orbit splitting constant, denoted by ζ (zeta), is 158 cm^{-1}. This is the quantity which determines the actual spectral splittings observed and which varies from one ion to another. So, given band half-widths of ca. 3000 cm^{-1}, fine structure caused by spin–orbit coupling is most unlikely to be seen in the spectra of Ti^{III} complexes. For Ni^{II} the value is 703 cm^{-1}, which means that small splittings might be caused by this effect, although none have been unambiguously found. Spin–orbit coupling is more likely to be of importance in the assignment of the very weak bands associated with spin-forbidden transitions (which, as we will see, largely owe their intensity to spin–orbit coupling). These are frequently much narrower than those corresponding to spin-allowed transitions and, consequently, much smaller splittings will be visible. For Rh^{III} and Ir^{III}, both of which form stable complexes, ζ has values of ca. 1400 and 4400 cm^{-1}, respectively so, for complexes of these ions spin–orbit splitting of d–d bands is to be expected. Indeed, spin–orbit coupling is of potential importance for all ions of the second and third transition series.

Table 8.2 Spin–orbit levels arising from crystal field terms

Crystal field term	$^1A_{1g}$	$^6A_{1g}$	$^2A_{2g}$	$^3A_{2g}$	$^4A_{2g}$	^2E_g	^5E_g
Spin–orbit levels	1	1	1	1	1	1	1
Crystal field term	$^3T_{1g}$	$^4T_{1g}$	$^5T_{1g}$	$^2T_{2g}$	$^3T_{2g}$	$^4T_{2g}$	$^5T_{2g}$
Spin–orbit levels	3	3	3	2	3	3	3

Table 8.3 Atomic spin–orbit coupling parameters (cm^{-1})

Spin orbit coupling is apparent as splittings in the spectra of many-electron atoms. However, it is not difficult to derive one-electron quantities from these splittings and these one-electron quantities enable comparisons to be made between different electron configurations. It is these one-electron parameters that are listed in this table. All the configurations covered are of the form d^n with the exception of the uncharged metal atoms, all of which are d^{n}s^2, and some highly charged ions of Ti and V, all which are p^n. At the foot of each column of data for the first row transition metals is given a number, m, which indicates the m dependence on Z, the atomic number, of the data above in the form Z^m. An asterisk indicates a closed shell species so that no spin–orbit splitting is seen in the spectrum.

Metal	Charge on metal atom						
	0	1+	2+	3+	4+	5+	6+
Ti	(123)	104	131	158	*	(3716)	(3973)
V	(179)	154	187	220	253	*	(4900)
Cr	(248)	219	256	296	337	378	*
Mn	(334)	294	343	338	436	486	536
Fe	(431)	388	441	499	554	612	673
Co	(550)	500	561	625	695	760	830
Ni	(691)	634	703	775	851	934	1012
Cu	(857)	*	870	931	1037	1127	1224
m	7	7.5	7	6.5	6	6	5.5
Ru	(1042)	968	1082	1201	1319	1441	1564
Rh	(1259)	1177	1299	1426	1567	1689	1825
Os	(3381)	3174	3531	3898	4259		
Ir	(3909)	3690	4056	4430	4814		

Data are adapted from *Handbook of Atomic Data* by S. Fraga, J. Karawowski and K. M. S. Saxena, Elsevier, Amsterdam, 1976.

A more recent compilation is given by J. Bendix, M. Brorson and C. E. Schäffer, *Inorg. Chem.* (1993) *32*, 2838 but the definitions that they adopt, although with merit, are not those generally encountered in the literature.

This is why our discussion of crystal and ligand theories has been exemplified by complexes formed by elements of the first transition series. Had we included spin–orbit coupling within the crystal field model (and in a more complete treatment this would have been done), within the crystal field model it would have appeared with its free-ion value. Not surprisingly, in ligand field theory it becomes a parameter which, characteristically, is found to have a value somewhat lower than that found for the free ion.

8.5 Jahn–Teller effect

The second effect which has been neglected is that due to low-symmetry crystal fields. As the discussion in this chapter has been confined to octahedral ML$_6$ complexes one might assume that low-symmetry ligand fields could safely be ignored, but this is not so. A theorem due to Jahn and Teller states that *any* non-linear ion or molecule which is in an orbitally degenerate term will distort to relieve this degeneracy. This means that all E_g, T_{1g} and T_{2g} terms of d^n configurations, in principle, are unstable with respect to some distortion which reduces the symmetry. Of course, as

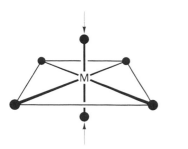

Fig. 8.7 An octahedral complex, compressed along one fourfold axis (taken to be the z axis).

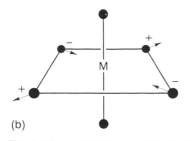

Fig. 8.8 Two metal–ligand bond-length change vibrations of an octahedral complex. Those shown, together, have E_g symmetry; (a) has a displacement pattern of a d_{z^2} orbital, (b) resembles $d_{x^2-y^2}$.

has been discussed above, some of the orbital degeneracy may be lost because of spin–orbit coupling. If this mechanism were to totally relieve the orbital degeneracy then there would be no need to consider the Jahn–Teller theorem.

The Jahn–Teller theorem (which is group-theoretical in origin) tells us that a regular octahedral complex will often be unstable with respect to a distortion, but it says nothing at all about the magnitude of the distortion. An exceedingly small distortion, small enough to escape detection by the most sensitive technique, could in principle satisfy the Jahn–Teller requirement. Indeed, there is often argument over the crystallographic evidence cited in support of the phenomenon, although the importance of the Jahn–Teller effect is generally accepted and it is frequently invoked in explanations. The most convincing evidence for the operation of the Jahn–Teller effect in transition metal complexes is found in studies on Cu^{II} complexes, although its importance is rather more clearly established in other areas, notably some parts of solid-state physics.

Physically, the Jahn–Teller effect may be regarded as operating as follows. Suppose that an octahedral complex is momentarily distorted as a result of a molecular vibration and has the shape shown in Fig. 8.7. This vibration will cause d_{z^2} and $d_{x^2-y^2}$ to lose their degeneracy (the distorted molecule has the symmetry of the *trans* octahedral molecule in Table 6.6). Suppose too that in the undistorted complex the e_g orbitals were occupied by a single electron. In the distorted complex this electron could be in either d_{z^2} or $d_{x^2-y^2}$. Consider both possibilities. The distortion shown in Fig. 8.7 results from a metal–ligand stretching mode which is of the form shown in Fig. 8.8(a); from this figure two things should be evident. First, that the caption to Fig. 8.7 is incomplete—not only do the axial ligands move in, but the equatorial ligands move out, although by a smaller amount, actually one half. Secondly, the general nodal pattern and amplitudes in Fig. 8.8(a) follow those characteristic of a d_{z^2} orbital. We know that, in an octahedron, d_{z^2} is partnered by $d_{x^2-y^2}$. So, the vibration in Fig. 8.8(a) is partnered by that in Fig. 8.8(b). The two, as a pair, have e_g symmetry. We could discuss either member but the discussions would differ in detail. We shall consider only the d_{z^2}-like vibration, that shown in Fig. 8.8(a.) This is the one for which, if led to a static distortion, simple crystal-field arguments would lead us to expect that the solitary e_g electron would be more stable in the $d_{x^2-y^2}$ orbital. If the arrows in this diagram are reversed, so that the distorted complex is one in which the four in-plane ligands move in and the axial ligands move out compared with their positions in the undistorted molecule, then we would expect the e_g electron to be more stable in d_{z^2}. Consider now Fig. 8.9, which shows a diatomic potential energy diagram which we apply, separately, to each bond in our complex. We compare the cost in energy terms of the (two short, four long) and (two long, four short) alternatives. In doing this we are just talking about 'cost' and so ignoring the Jahn–Teller stabilization which is the driving force for the distortion— but, as said earlier, the argument is indicative only, it certainly is not complete.

Remembering that the amplitude associated with the two ligands is twice that associated with the four, it can be seen that the inherent asymmetry in the potential energy surface indicates that the cost of (two short, four long) will be greater than that of the (two long, four short). This conclusion is in

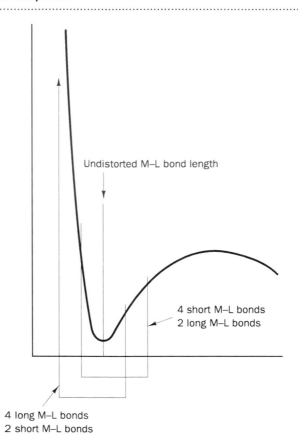

Fig. 8.9 A M–L (diatomic) potential energy diagram which enables an assessment of energy changes consequent of the alternative distortions shown in Fig. 8.8.

Undistorted M–L bond length

4 short M–L bonds
2 long M–L bonds

4 long M–L bonds
2 short M–L bonds

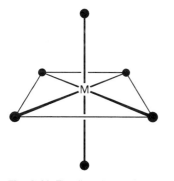

Fig. 8.10 The distortion commonly found in octahedral complexes of Cu^{II}. In real life the picture is often more complicated because the two long-bonded ligands differ from the other four.

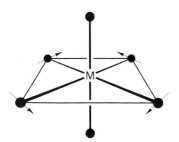

Fig. 8.11 A vibrational mode of T_{2g} symmetry of an octahedral complex. These vibrations change bond angles, not bond lengths.

accord with the distortions that have been observed in crystal structure determinations and other techniques. Octahedral complexes of Cu^{II} are almost invariably distorted, commonly in the way shown in Fig. 8.10. This is entirely consistent with the operation of the Jahn–Teller effect because the ground-state electron configuration of a Cu^{II} ion in an octahedral field is $t_{2g}^6 e_g^3$, ^2E_g. It has an odd number of electrons in the e_g orbitals.

There is one result of the general theory which should be mentioned. It can be shown that in a regular octahedron the Jahn–Teller effect only operates by way of vibrations of e_g symmetry when the electronic state is ^nE_g (the value of n is irrelevant because the Jahn–Teller theorem applies only to space functions, not spin). For electronic states of either $^nT_{1g}$ or $^nT_{2g}$ symmetries then the Jahn–Teller effect operates through vibrations of either e_g or t_{2g} symmetries; a vibration of the latter symmetry is shown in Fig. 8.11. Of course, the orbital degeneracy in an octahedral complex may be relieved by distortions other than those shown in Figs. 8.8 and 8.11. However, in such cases we may conclude that whatever is responsible for the distortion it is not the Jahn–Teller effect. In particular, all of the Jahn–Teller-active vibrations of an octahedron carry the g suffix and this means that they cannot give rise to a distortion which destroys the centre of symmetry of an octahedron. Distorted octahedral complexes which lack a centre of symmetry cannot owe their distortion to the operation of the Jahn–Teller effect. This account of the Jahn–Teller effect indicates why it is of little importance when the t_{2g} orbitals are unequally occupied. Occupation of these orbitals

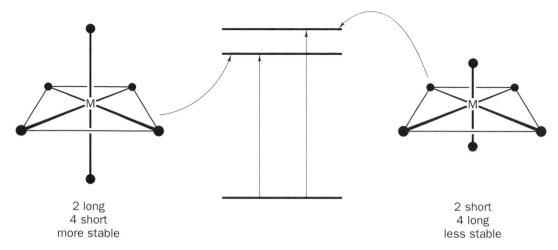

	2 long	2 short
	4 short	4 long
	more stable	less stable

Fig. 8.12 The effect of the distortion shown in Fig. 8.8(a) on an excited state in which the Jahn–Teller effect is operative. It follows that the excited state must be orbitally degenerate; this figure demonstrates the loss of that degeneracy consequent on the distortion.

involves less destabilization than does occupation of the e_g orbitals—they are, to a first approximation, non-bonding.

The Jahn–Teller effect is of importance in the spectra of octahedral transition metal complexes because the transitions observed usually involve the excitation of a single electron from a t_{2g} to an e_g orbital. It follows that there must be an odd number of e_g electrons either before or after the transition. That is, if the ground state is not subject to a Jahn–Teller distortion, the excited state is, and vice versa. To explore the effect of a Jahn–Teller distortion on the electronic spectrum of a complex we shall consider a simplified model. Suppose that the Jahn–Teller effect is operative in the excited state and that molecular distortions have transiently distorted the molecule in its electronic ground state so that it has the shape shown in Fig. 8.10. Suppose too that the stable molecular geometry in the excited state is also that shown in Fig. 8.10—that is, it is that briefly assumed by the ground state. Then there also exists a less stable excited state for which the molecular geometry is as shown in Fig. 8.12. The two excited states differ in that in the one the e_g electron is in $d_{x^2-y^2}$ and in the other it is in d_{z^2}. If whilst the ground state has the molecular geometry of Fig. 8.10 it absorbs light and assumes an electronically excited state then we have to consider the two possibilities shown in Fig. 8.12. First, the excited state could be the more stable one, so that the molecule finds itself in its vibrational ground state. Secondly, the molecule could assume the less stable excited state but would then find itself well away from its equilibrium geometry—it is vibrationally excited, possibly by several quanta. These quanta may each have an energy of ca. $300 \, \text{cm}^{-1}$ (assuming no great change in frequency between ground and excited state). Remembering the intrinsic difference in energy between the two excited levels in their vibrational ground states, we conclude that such excited state Jahn–Teller distortions may well be spectrally apparent. For example, this is believed to be the explanation of the asymmetry in the $^2E_g \leftarrow {}^2T_{2g}$ d–d transition of the $[\text{Ti}(\text{H}_2\text{O})_6]^{3+}$ ion shown in Fig. 8.13.

If the Jahn–Teller effect operates in the electronic ground state of a complex it is unlikely to be apparent in the electronic spectrum. If the two ground states are split sufficiently far apart for transitions to the excited state

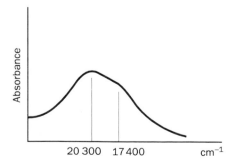

Fig. 8.13 The asymmetry in the d–d spectrum of the $[\text{Ti}(\text{H}_2\text{O})_6]^{3+}$ ion (d^1), believed to result from the operation of the Jahn–Teller effect in the excited state.

to be resolvable, the upper of the two ground states will not be sufficiently thermally populated for transitions from it to be seen. On the other hand, in this situation there will be an additional transition in the infrared, although probably difficult to locate amongst the forest of vibrational bands in this spectral region. This additional band corresponds to a simultaneous electronic and vibrational excitation from the lower of the ground states to the upper.

A Jahn–Teller distortion may be either static or dynamic and so far only the former has been considered. There are, in fact, theoretical reasons for expecting static Jahn–Teller distortions to be atypical, which is perhaps why it is so difficult to point to firm evidence for the effect in ground states. The dynamic case occurs when the potential barrier separating, for example, the three equivalent distortions of the type shown in Fig. 8.10, an elongation along x, y or z axes, is of the order of kT. The distortion then rapidly alternates between the possibilities. In this way a small Jahn–Teller effect may be unobserved—the time average is an undistorted structure (as will become apparent towards the end of Chapter 9, small distortions may be detected by magnetic measurements). Because the time taken to jump from one configuration to another is of the same order as the time taken for a simple vibration, a measurement taking much longer than this to perform will give only an average. A measurement of 10^{-10} s duration takes too long! The dynamic Jahn–Teller effect does not affect the visible appearance of the electronic spectra of transition metal complexes but this is because of the energetics of the effect, not because of the timescale.

8.6 Band contours

The d–d bands of transition metal complexes are weak, sometimes very weak. Some of the bands are sharp, some broad and some so broad that it is difficult to be certain that they exist. In the next section we shall consider the problem of intensities; in this section, the problem of band shapes. The two are related. The very weak peaks are usually sharp. If they were not, they would escape detection and no doubt many broad, very weak, peaks do pass undetected.

When discussing the Jahn–Teller effect a refinement was introduced into our model of a transition metal complex: it was allowed to vibrate. This is an important refinement for it enables us to understand both band intensities and shapes. Consider an octahedral metal complex in which the totally symmetric metal–ligand stretching mode (the breathing mode) is excited. That is, the complex retains its octahedral geometry because all of the M–L bonds are contracting and elongating in phase. The essential point is that the repulsion between the ligands and metal d electrons and, therefore, the crystal field splitting parameter, Δ, will vary with the metal–ligand distance. On the molecular orbital model, the overlap between the ligand and metal orbitals will be modulated by the vibration. The conclusion is that Δ is not a fixed quantity but rather one that varies with metal–ligand separation, so that at any instant it varies from one molecule to another. A convenient way of thinking about this is to regard a collection of molecules as being represented, not by a line in a Tanabe–Sugano diagram (as in Fig. 8.3) but by a band, covering a range of values of Δ/B. An absorption peak will therefore not consist of a narrow line but, usually, a broad one, a superposition of a multitude of sharp ones. The relative breadths of the spectral lines will be determined by the relative slopes of the lines which, in a Tanabe–Sugano diagram, represent the excited states. For example, Fig. 8.3 leads us to predict that transitions to the $^3T_{1g}(P)$ and $^3T_{2g}$ terms of the d^2 configuration will give bands of similar widths. As Fig. 8.1 shows, this prediction is roughly confirmed by experiment. We further predict that the transition to the $^3A_{2g}$ excited state will give a band which is about twice as broad as the other two. This, together with its inherent low intensity (see below), is no doubt why it is difficult to observe. However, as the Ni^{II} example given earlier in this chapter shows, it is possible to see such transitions in favourable cases.

On the other hand, some of the lines in a Tanabe–Sugano diagram are almost parallel to that representing the ground state. These lines are invariably amongst those derived from the same strong field configuration as the ground state. For example, as Fig. 8.3 shows, in the d^2 configuration the Δ dependence of the $^1A_{1g}$, ^1E_g and $^1T_{2g}$ terms is roughly parallel to that of the $^3T_{1g}(F)$ ground state. These four terms are those which arise from the t_{2g}^2 strong field configuration. That is, a transition from the ground to another of these terms corresponds to a rearrangement of electrons within the t_{2g} orbitals and so does not depend on the t_{2g}–e_g separation, Δ. This statement is rigorously true in the strong field limit but is only approximately true in weak fields. In the latter case, mixing of the $^1A_{1g}$, ^1E_g and $^1T_{2g}$ excited terms derived from the $t_{2g}^1 e_g^1$ and e_g^2 configurations, into those derived from the t_{2g}^2 configuration, makes the lowest $^1A_{1g}$, ^1E_g and $^1T_{2g}$ term each have a different dependence on Δ. However, in intermediate and strong ligand fields it is true that the energies of transitions to these terms are scarcely modulated by the Δ variations caused by ligand vibrations. It follows that transitions to these terms will appear as relatively sharp lines in the spectrum. However, they are very weak because they are spin forbidden (a spin triplet to a spin singlet) and for this reason are not easy to detect. The existence of these transitions is exploited in those lasers which involve transition metal ions, for instance in the ruby laser. Ruby is an Al_2O_3 crystal containing a small amount of Cr^{III} as impurity, the chromium being approximately octahedrally surrounded by oxygen atoms. It is on this Cr^{III}, a d^3 ion, that we focus our attention. As Fig. 8.14 shows, the energies of the transitions from the ground term to the lowest ^2E_g, $^2T_{1g}$ and $^2T_{2g}$ terms are essentially independent of Δ and so the bands are sharp. Emission of light from Cr^{III} ions in the ^2E_g excited term as they fall back to the ground state gives rise to the essentially monochromatic red light emitted by the ruby laser.

In the above section only the effect of the totally symmetric breathing vibration on the value of Δ was considered. Similar conclusions follow if vibrations of other symmetries are considered in detail. As will be seen in the next section, some of these other vibrations are responsible for the appreciable intensity in formally forbidden d–d transitions.

8.7 Band intensities

The fact that very weak d–d transitions may be observed in atomic spectroscopy indicates the approximate nature of the rule that only s ↔ p, p ↔ d, d ↔ f etc. orbital transitions are allowed. The spin-allowed d–d transitions in an octahedral metal complex are of much higher intensity than

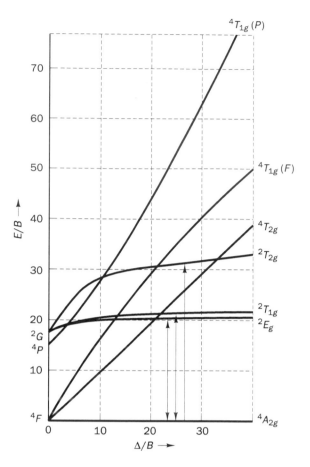

Fig. 8.14 A simplified Tanabe–Sugano diagram for the d^3 configuration (appropriate to Cr^{III}), showing the excited terms which have a dependence on Δ/B which is almost the same as that of the ground state. Such parallel behaviour only occurs when the ground and excited terms correlate with the same strong field configuration. In the present case this is t_{2g}^3 (see Table 7.5).

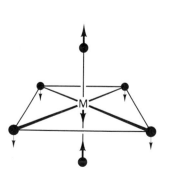

Fig. 8.15 An M–L bond stretch vibration of T_{1u} symmetry of an octahedral complex.

their atomic counterparts so that some new intensity-generating mechanism exists in the complex which is not present in the free ion. Not surprisingly, it involves metal–ligand vibrations. Consider, for example, an octahedral complex which has momentarily been distorted by the vibration of T_{1u} symmetry shown in Fig. 8.15. This vibration destroys all of the threefold and all but one of the fourfold axes (that along which the vibration takes place). The molecular symmetry is, in fact, the same as that of a *trans*-$ML_4L'L''$ complex, C_{4v}. As Table 6.6 shows, in this symmetry, p_z, s, and d_{z^2} carry the same symmetry label (a_1) and so may mix. Similarly, p_x and p_y together have the same symmetry (e) as d_{zx} and d_{yz} and so these pairs of orbitals may also mix. In the distorted molecule of Fig. 8.15 what we have called d_{z^2} becomes a mixture of d_{z^2} with a little of p_z and of s. The contamination by these other orbitals disappears when the molecule becomes accurately octahedral again. Similarly, what we have called d_{zx} is contaminated by p_x and d_{yz} by p_y. This means that if an electronic excitation occurs whilst the molecule is distorted, a transition in which an electron jumps formally from d_{zx} to d_{z^2} also contains a component in which the electron jumps from d_{zx} to p_z and from p_x to d_{z^2}, both of which are allowed by the selection rules. The d–d transition gains intensity, therefore, because it is not entirely d–d. The intensity depends on the extent of d–p mixing and, in turn, this depends on the nature of the distortion at the moment of excitation. Any distortion which fails to remove the centre of symmetry of the molecule will not introduce any intensity, for the essential requirement is that the vibration mixes atomic orbitals of g and u types (and, so, in our example, d and p orbitals, respectively). This topic is explored in some detail in Appendix 8, an appendix that shows that there will be perhaps ten vibrational subpeaks, each with a Δ modulation and Jahn–Teller broadening, leading to the relatively weak, broad peaks observed.

The above discussion is associated with a breakdown in the Born–Oppenheimer approximation. This states that electronic and vibrational energy levels may be treated separately, so that under this approximation an electronic transition is something quite separate from a vibrational transition. However, in our discussion we have found that the one depended on the other; that there is a coupling between vibrational and electronic states. This is called *vibronic* coupling. The intensity introduced into a d–d transition by vibronic coupling cannot come from nowhere. As is evident from our discussion, it is in fact 'stolen' from what in a regular octahedron is an allowed transition—for example, the equivalent in the complex of the free ion d → p transition, a transition which will lie in the ultraviolet. A more detailed analysis shows that the magnitude of the stolen intensity is expected to be an inverse function of the energy separation between the d–d band and the allowed band. This means that the highest energy d–d band, that with the smallest separation from the allowed band, will generally be the most intense, other things being equal. This perhaps explains why the $^3T_{1g}(P) \leftarrow {}^3A_{2g}$ transition in $[\text{Ni}(\text{H}_2\text{O})_6]^{2+}$—a two-electron promotion (see Fig. 7.18)—is seen at all.

We now turn to the origin of the intensities of the spin-forbidden d–d transitions. One possible source is a magnetic dipole mechanism. In electron paramagnetic (spin) and nuclear magnetic resonance spectroscopies (EPR and NMR), electromagnetic radiation is used to 'turn over' the spin of either an electron or nucleus in a molecule (this simple view of the process is adequate for our present purposes). The electromagnetic radiation does this by virtue of its associated magnetic vector and the process is said to be *magnetic dipole allowed*. Classically, the alternating N–S magnetic dipole associated with the light wave causes similar magnetic dipoles in the molecule to alternate in direction in phase with it. When the alternation coincides with a resonant frequency, a transfer of energy from the light wave to the molecule occurs. Although a similar process undoubtedly contributes to the intensity of the spin-forbidden d–d bands—such transitions are magnetic dipole allowed—detailed calculations indicate that its contribution is small, although, as we shall see in Chapter 11, this mechanism is of importance for f electron systems. Much more important is spin–orbit coupling. Our discussion in Section 8.4 demonstrated that the spin and orbital properties of an electron are not completely independent of each other. Earlier in this section we described a vibrational mechanism by which an orbital transition becomes weakly allowed, so, in principle at least, spin–orbit coupling may transfer some of this allowedness into what is, formally, a spin-forbidden transition. Let us consider a specific case.

As we have seen, the ground state of an octahedral d^8 complex is $^3A_{2g}$. A low-lying spin-forbidden transition is to a ^1E_g term and a spin-allowed transition is to the $^3T_{1g}$ term (see the modified Tanabe–Sugano diagram shown in Fig. 8.16). Now, spin–orbit coupling has the effect of contaminating

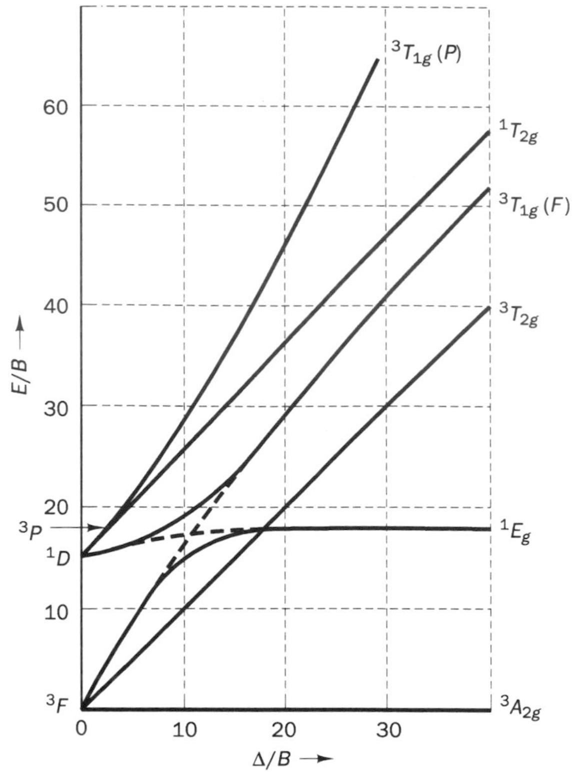

Fig. 8.16 Spin–orbit coupling causes an interaction between the ^1E_g low-lying excited state of an octahedral d^8 complex with two of the wavefunctions of a $^3T_{1g}$. This interaction leads to mutual repulsion and a non-crossing of the energy levels.

two of the nine wavefunctions of the $^3T_{1g}$ term with those of the ^1E_g term (of which, of course, there are also two) and vice versa. When the energy separation between these two terms is small (and this occurs at $\Delta/B \approx 11$ in Fig. 8.16) the contamination becomes gross pollution! The result is that for these four coupled wavefunctions the two lines on a Tanabe–Sugano diagram do not cross but repel each other and behave as shown in Fig. 8.16 so that those two wavefunctions which in a weak field belonged to $^3T_{1g}$ become ^1E_g in a strong field and vice versa. This means that some of the intensity of what was the $^3T_{1g} \leftarrow ^3A_{2g}$ transition (before we included spin–orbit coupling) is transferred to the $^1E_g \leftarrow ^3A_{2g}$ transition once spin–orbit coupling is included. All other spin-forbidden transitions are believed to gain intensity by similar mechanisms. One explanation which has been suggested for the double-headed peak in the $[Ni(H_2O_6)]^{2+}$ spectrum of Fig. 8.5 is that it originates in this mixing, the d–d intensity being shared by the two components. Against this, it has been argued that a detailed analysis, including all interactions, leads to a diagram sufficiently different from Fig. 8.16 for it to be unlikely that this explanation holds for this particular example—it is not true that the complex falls at the crossover point. An alternative explanation for the double-headed band is that the splitting originates in the Jahn–Teller effect associated with the $t_{2g}^5 e_g^3$ strong field configuration with which the $^3T_{1g}(F)$ excited term correlates. But in this case why is the lower band, to $^3T_{2g}$, which also correlates with $t_{2g}^5 e_g^3$ not also split—and, if this is the explanation why is the band-splitting not found in most, if not all, octahedral Ni^{II} complexes? At the present time there is no agreed explanation for the splitting.

Two points should be mentioned. First, although the $^1E_g \leftarrow {}^3A_{2g}$ transition of the above example would, by the arguments of the previous section, be expected to be sharp, mixing with the $^3T_{1g} \leftarrow {}^3A_{2g}$ transition will broaden it. In general, the lower the intensity of a spin-forbidden band the narrower it is likely to be. Secondly, the greater the spin–orbit coupling, the more important the spin-forbidden transitions are likely to be in the spectra. We shall have more to say about this when we discuss f electron systems in Chapter 11. The relative magnitudes of spin–orbit coupling constants has been discussed in Section 8.4. We conclude this section by considering transitions such as $^3A_{2g} \leftarrow {}^3T_{1g}(F)$ of the d^2 case. In the strong field limit this corresponds to a transition from the t_{2g}^2 to the e_g^2 configuration. Now, except in a very special circumstances, the effect of electromagnetic radiation is to excite one electron at a time, not two. In the strong field limit the above transition is therefore forbidden (and rigorously so, within the general approximation scheme that we are using). In the intermediate and weak field regions the $^3T_{1g}(F)$ level is a mixture of the t_{2g}^2 and $t_{2g}^1e_g^1$ configurations (this was discussed in Section 6.5) and so the $^3A_{2g} \leftarrow {}^3T_{1g}(F)$ transition has a $t_{2g}^1e_g^1 \leftarrow e_g^2$ component which, of course, corresponds to a one-electron jump. This component can acquire intensity by a vibronic mechanism. It follows that the most favourable situation for observing the $^3A_{2g} \leftarrow {}^3T_{1g}(F)$ transition in d^2 complexes will be with very weak field complexes. In this situation it will lie at a relatively long wavelength and, hopefully, will not be obscured by the charge-transfer bands which, as we will see later, commonly occur in the near ultraviolet. In contrast to the conclusion that we have just reached, we note that, of all the transitions in the $[Ni(H_2O)_6]^{2+}$ spectrum, the highest is the most, not the least, intense. This clearly indicates a weakness in our model; most probably the fact that we have ignored any mixing between d–d and the intense charge transfer transitions that will be discussed in Section 8.10; a mixing that was also mentioned earlier.

8.8 Tetrahedral complexes

So far in this chapter we have only discussed octahedral transition metal complexes. However, all of what we have said may be carried over to the tetrahedral case, provided that we remember that Δ_{tet} is of opposite sign to Δ_{oct} (in a tetrahedral complex d_{xy}, d_{yz} and d_{zx} are above $d_{x^2-y^2}$ and d_{z^2}, a point developed in some detail in Section 7.7). This means, for example, that the spectrum of a tetrahedral d^2 complex can be interpreted using a Tanabe–Sugano diagram for the octahedral d^8 case because, as was discussed in Section 7.4, the splitting pattern for d^8 is the opposite to that for d^2. It is to facilitate this application that Tanabe–Sugano diagrams quite often do not contain any g or u suffixes in their term labels. There is only one major point of difference that has to be discussed and this concerns the intensities of d–d bands of tetrahedral complexes. As is evident from Fig. 8.17, where the spectrum of the octahedral $[Co(H_2O)_6]^{2+}$ ion is compared with that of the tetrahedral $[CoCl_4]^{2-}$ anion, these intensities are at least an order of magnitude greater in tetrahedral complexes than in their octahedral counterparts. Evidently, some new intensity-generating mechanism is available in tetrahedral complexes.

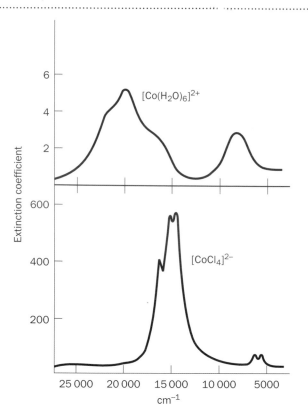

Fig. 8.17 The spectra of the d^7 species $[Co(H_2O)_6]^{2+}$ (which is weak field) and $[CoCl_4]^{2-}$. The latter is two orders of magnitude more intense.

One important aspect of the electronic structure of tetrahedral complex ions which was not discussed in detail in Chapter 7—although it was mentioned in Chapter 6 and is contained in Table 6.6—is that, unlike the octahedral case, in a tetrahedral complex the metal p orbitals have the same symmetry (t_2) as do the metal d_{xy}, d_{yz} and d_{zx} orbitals. Because they are of the same symmetry, the t_2 set of d orbitals will be mixed by the crystal field with the metal p orbitals. This in turn means that an $e \rightarrow t_2$ transition contains some d $\rightarrow$ p component and this component is an allowed transition. Evidently, the intensity of the $e \rightarrow t_2$ transition is related to the extent of d–p mixing so that this may be worked backwards and the extent of mixing assessed, at least qualitatively, from the intensities of d–d bands in the spectra. Notice both the similarity and the difference between the intensity-generating mechanisms for d–d transitions in octahedral and tetrahedral complexes. Both depend on d–p mixing but only for the tetrahedral case does this mixing occur for the non-distorted molecule.

8.9 Complexes of other geometries

As the ligand-field splitting of the d orbitals in lower symmetry complexes lead to lower degeneracies and so a need for more parameters to describe them (see Table 6.6), so the d–d transitions of complexes of other than octahedral and tetrahedral geometries can only be discussed by introducing these parameters. For square planar complexes, for instance, two additional parameters must be introduced.

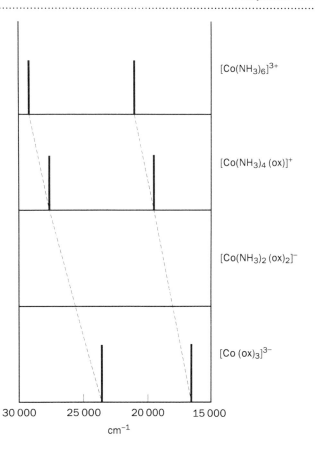

Fig. 8.18 The spectrum of $[Co(NH_3)_4(ox)]^+$ (second down, indicated by vertical bars) to a first rather good approximation is a weighted interpolation of those of $[Co(NH_3)_6]^{3+}$ and $[Co(ox)_3]^{3-}$. This is an example of an application of the rule of average environment.

The number of observed spectral bands is seldom sufficient to allow all of the parameters to be simultaneously determined. Commonly, therefore, one has to be content either with tentative band assignments or to extract data appropriate to an oversimplified model—assuming that deviations from an ideal geometry may be ignored, for example. A useful first approximation to the assignment of the spectra of coordination compounds in which the symmetry is low by virtue of the coordination of several different ligands is provided by the *rule of average environment*. This states that, for example, the value of Δ for the complex $[ML_3L_3']$ will be approximately the mean of the Δ values for $[ML_6]$ and $[ML_6']$; similarly, for $[ML_4L_2']$ the observed Δ will be approximately

$$\tfrac{1}{3}(2\Delta[ML_6] + \Delta[M_6'])$$

As mentioned in Section 7.3, the rule of average environment may be used with the data in Table 7.1. The ligand non-equivalence may manifest itself by a splitting of the bands expected for an octahedral complex with the value of Δ predicted by the rule of average environment. An example of the application of this rule to spectra is contained in Fig. 8.18, where it is shown that the spectrum of $[Co(NH_3)_4(ox)]^+$ may be predicted rather well from those of $[Co(NH_3)_6]^{3+}$ and $[Co(ox)_3]^{3-}$.

8.10   Charge-transfer spectra

As is evident from the simple description given in Section 8.1, if the absorption of light is to cause an electronic transition within an atom or molecule, it is essential that the absorption results in a movement of charge density. This displacement may be localized on one atom—as it is in the d–d spectra which have been the subject of this chapter up to this point—or it may be the displacement of charge from one atom to another. Electronic transitions which can be ascribed to the latter process are termed charge-transfer transitions. These are not always simple processes, as shown by the fact that change of solvent or of ions present in solution can influence the position of a charge-transfer band, particularly when the complex is asymmetric (and so has some inherent polarity). For the moment, such complications will be ignored. Three main categories of charge-transfer processes may then be distinguished in coordination compounds.

Class 1—Intraligand transitions

A ligand such as SCN^- has internal charge-transfer transitions, usually located in the ultraviolet region of the spectrum. Corresponding transitions occur in the coordinated ligand but can usually be identified by comparison with the spectrum of the free ligand. Any metal ion to which the ligand is complexed is not significantly involved in the transition. This class will therefore not be discussed further.

Class 2—Ligand to metal (reduction) charge-transfer transitions (LMCT)

These are a common type of transition, in which a ligand electron is transferred to a metal orbital and the charge separation within the complex thereby reduced. The process can lead to a permanent transfer of charge so that the compound involved is photolytically changed. This offers not only a method of measuring the intensity of a light beam (actinometry) but in recent years has been the subject of much research in an attempt to couple such a process into the reaction

$$2H_2O \rightarrow 2H_2 + O_2$$

If this could be cheaply achieved it would have considerable economic consequences. In Chapter 14 the topic will be briefly explored. In the present section we are not concerned with such permanent charge transfer but, rather, the process in which electron transfer from one orbital to another is caused by the absorpion of light (the return of the molecule to its ground state subsequently occurring by some energy degradation mechanism which is not our present concern). One problem associated with this class of charge-transfer transitions is the lack of certainty about which orbital the electron comes from. Thus, one is tempted to discuss such a transition in terms of molecular orbital energy level diagrams such as those of Figs. 6.11 or Fig. 6.26. Figure 6.26 is more complete than is Fig. 6.11 because the former covers both ligand σ and π electrons, a total of six on each ligand. But a monatomic ligand such as F^- or O^{2-} has eight valence electrons. The two which have not been included are in a σ orbital which points away from the metal atom. These σ orbitals give rise to ligand group orbitals which are probably higher in energy than most of the occupied ligand orbitals of

Figs. 6.11 and 6.26 because they are not stabilized by interactions with the metal orbitals. They might, then, be expected to give the lower energy charge transfer bands (as, indeed might some of the other ligand orbitals mentioned above, for the reason indicated at the end of Section 6.2). This is probably the case but, as the intensity of charge-transfer bands is a function of the overlap of the orbitals between which the overlap occurs, the corresponding intensities are small. Clearly, there is a danger that these bands may be confused with d–d transitions. This is in contrast to most charge-transfer bands, which are intense, and so easy to distinguish.

Class 3—Metal to ligand (oxidation) charge-transfer transitions (MLCT)

In this type of transition a metal electron is transferred to an orbital largely located on the ligands. It is therefore, in a sense, the opposite to type of charge-transfer transition described in the prevous section. So, it corresponds to an oxidation of the metal and reduction of the ligand. An example is provided by the $[M(H_2O)_6]^{2+}$ ions of the first transition series, for which Dainton has shown that the energy of the first charge-transfer band is linearly related to the redox potential of the system

$$M^{2+}(aq) \rightarrow M^{3+}(aq) + e^-(aq)$$

Much of what has been said about ligand metal charge-transfer bands applies to this class also—they, too, can be used as a basis for actinometry; it would be attractive to combine both types in a catalysed photolytic decomposition of water.

It is not always easy to decide whether category 2 or 3 is involved in a particular transition. If it is 2, then, for an allowed transition in an octahedral complex, the excitation must be from a ground-state orbital of u symmetry (and, therefore, t_{1u} or t_{2u} as there are no other occupied u orbitals) and the electron must be excited into either a t_{2g} or e_g orbital. If both t_{2g} and e_g sets are incompletely filled then both excitations may occur and two bands (or families of bands) separated by Δ are to be expected. If the t_{2g} set is full then only excitation to the e_g set will occur. This probably explains why there is a long wavelength charge-transfer band in $[Fe(CN)_6]^{3-}$ (d^5) which has no counterpart in $[Fe(CN)_6]^{4-}$ (d^6). A similar pattern is observed in the anions $[IrBr_6]^{2-}$ (d^5) and $[IrBr_6]^{3-}$ (d^6)—the latter pair are illustrated in Fig. 8.19. In both cases the band in the d^5 species which disappears is about Δ away from the one which remains (the Δ being that of the higher valence ion, of course), as the above explanation requires. If we regard charge-transfer spectra as involving an electron moving towards or away from a cation then the fine details of the cation's electronic structure assume a lesser importance, although it may influence the number of bands, as we have seen. Adopting this viewpoint, then in cases such as those just discussed in which there are similar complexes differing by a single formal charge, we would expect ligand $\rightarrow$ metal charge-transfer to become easier—and so of lower energy—as the formal charge on the metal ion increases. Conversely, metal $\rightarrow$ ligand charge-transfer would be expected to become more difficult and so of higher energy. These, then, are criteria by which the type of charge transfer transition involved in a particular band can tentatively be identified.

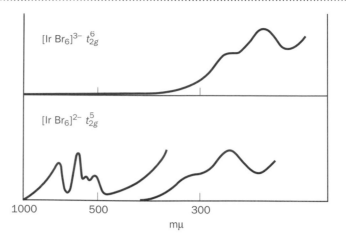

Fig. 8.19 The visible and near-ultraviolet spectra of the related octahedral complexes $[IrBr_6]^{2-}$ and $[IrBr_6]^{3-}$, d^5 and d^6, respectively. There is a complicated long-wavelength band pattern in the d^5 case which is absent in the d^6. The complexity suggests that orbitally degenerate states are involved, consistent with a charge-transfer transition in which t_{2g}^5 (ground state) becomes a t_{2g}^6 in the excited state.

Because the charge-transfer process involves electron transfer the spectra are sometimes called *redox spectra* and, certainly, band positions can correlate with the ease of oxidation or reduction of the metal ion or ligand—we have already met one example of this. A second, rather different, example is given in Fig. 8.20. This figure shows the charge-transfer spectra of $[Co(NH_3)_5X]^{2+}$, $X = Cl^-$, Br^- and I^-. As our discussion leads us to expect, the transition energies decrease in the order $Cl^- > Br^- > I^-$. Usually, in such series the charge-transfer process does not lead to a permanent oxidation or reduction. However, along the series $[CuCl_4]^{2-}$, $[CuBr_4]^{2-}$ and $[CuI_4]^{2-}$ charge-transfer absorption occurs at longer and longer wavelength until, for the latter, the process becomes irreversible and decomposition into Cu^I and I_2 becomes a spontaneous process. Similarly, copper(II) chloride is green, copper(II) bromide is black and, of course, copper(II) iodide is not stable. These sequences provide an excellent demonstration that the particular charge-transfer process involved is ligand-to-metal in nature. Similarly, $[FeF_6]^{3-}$ is virtually colourless, but replacement of the fluoride ion with the heavier halogens leads first to yellow solutions (Cl^-), then brown (Br^-) and finally to the spontaneous reduction of Fe^{III} by I^-, at least in aqueous media.

It has been found possible to calculate the positions of charge-transfer bands by assuming that they depend on an electronegativity difference between the ligand and the metal ion. The values of these so-called *optical electronegativities* indicate, as observed, that the first charge-transfer band in the spectrum moves to lower energies in the order:

$$Pt^{IV} > Rh^{III} > Cu^{II} > Co^{III} > Ir^{III} > Ni^{II}$$

Finally, as already mentioned, it seems clear that the solvent molecules can be involved in the charge-transfer process. Detailed discussions of charge-transfer spectra in transition metal complexes quite often label the corresponding bands quite separately, giving them the label CTTS—*charge transfer to solvent*. So, the fact that FeI_3 has recently been prepared in non-aqueous media suggests that the solvent—water—is not always the mere spectator that it was implicitly assumed to be above.

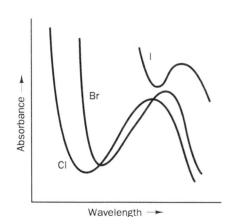

Fig. 8.20 The lowest energy charge-transfer bands of the complexes $[Co(NH_3)_5X]^{2+}$, $X = Cl^-$, Br^-, I^-. Qualitatively, the less electronegative the X, the longer the wavelength (the easier the transition).

8.11 Intervalence charge-transfer bands

So far in this chapter we have followed the assumption implicit in the previous two chapters—that we could treat transition metal ions as individuals, isolated along with their ligands. It will become apparent in the next chapter that there is magnetic evidence that this assumption is not always valid and so it is appropriate to recognize that similar spectroscopic evidence also exists. The evidence is not confined to transition metal ions and so we shall broaden this section to cover the entire periodic table (or almost!), although referring solely to 'metals' for linguistic simplicity. It quite often happens that when a metal appears in two different valence states in a compound that the colour of the compound is not that corresponding to the sum of the colours of its components. Examples are I_3^-, which may be regarded as derived from I^0—I_2 has a purple vapour—and I^-, the I^- anion is colourless; soluble Prussian blue, the deep blue almost black $KFe[Fe(CN)_6] \cdot H_2O$, prepared from Fe^{II}, pale blue, and $[Fe(CN)_6]^{3-}$, yellow.

These colours arise because there is an overlap between the orbitals on the two or more metal atoms with different valence states. This overlap can be direct, as in I_3^-, or indirect, through mutual overlap with an intervening ligand, as in $KFe[Fe(CN)_6] \cdot H_2O$, where the CN^- anion is involved. This overlap involves both ground and excited electronic states. The intense colours arise because if the electron density (or just 'electron') that is excited is in a ground state orbital largely located on one atom, then in the excited state it is largely located on the other. This pattern is easy to see for the case of just two interacting atoms. The effect of the overlap between two orbitals is to give sum and difference combinations. If the sum combination is, in fact, largely located on one atom the difference combination has no choice but to be largely located on the other. Movement of electron density from ground to the excited state will be associated with a large electric dipole change and, hence, absorption. These are called *intervalence-transfer* bands, to distinguish them from the charge-transfer absorptions discussed in the previous section. Commonly, but not always, they involve identical atoms, differing only in their formal valence states—iodine and iron in the two examples that have been cited. The ground and excited states of such atoms are likely to have a compatibility seldom shared by dissimilar atoms.

Not all mixed valence compounds show intervalence bands. Those that show not the slightest evidence for them are often called class 1. In class 2 intervalence compounds the intervalence-transfer band may well dominate the visible spectrum, swamping the spectra of the individual ions. However, these may be seen; the important point is that the different ions retain their chemical individuality. Soluble Prussian blue is a case in point. This contains Fe^{II} and Fe^{III} bridged by CN^- ligands. Isotopic tracer measurements show that there is no doubt—the Fe^{III} is bonded to the N atom of the CN^- ligand and the Fe^{II} to the C. If radioactive Fe is used in the preparation and the compound subsequently decomposed, the activity remains in the iron of the same valence state in which it was incorporated. A class 3 also exists, exemplified by the I_3^- anion, in which it is not possible to associate a unique valence state to individual metal ions; typically, but as our example shows, not always, they are structurally indistinguishable. When they form an

Table 8.4 Examples of intervalence-transfer compounds (after Robins and Day)

Class 1—no detectable interaction

$Cr^ICr^{III}F_5$	Cr_2F_5
$Fe^{II}Fe_2^{III}F_8 \cdot 10H_2O$	$Fe_3F_8 \cdot 10H_2O$
$Pb_2^{II}Pb^{IV}O_4$	Pb_3O_4 (red lead)
$Ti^ITl^{III}(CN)_4$	$Tl_2(CN)_4$

Class 2—clearly interaction, but chemical identity retained

$KFe^{II}[Fe^{III}(CN)_6]$	soluble Prussian blue
$Fe_4^{III}[Fe^{II}(CN)_6]_3$	insoluble Prussian blue
$[Cu^ICu^{II}Cl_6]^{3-}$	$[Cu_2Cl_6]^{3-}$
$[Au^IAu^{III}Cl_6]^{2-}$	$[Au_2Cl_6]^{2-}$
$[Sb^{III}Sb^{IV}Br_{12}]^{4-}$	$[Sb_2Br_{12}]^{4-}$
$[Pb^{II}Pb^{IV}Cl_{12}]^{6-}$	$[Pb_2Cl_{12}]^{6-}$

Class 3—clearly interaction, chemically distinct species cannot be identified

$[Cu_4S_3]^-$
$[Cu_4S_4]^-$
$[Nb_6Cl_{12}]^{2+}$
$[Nb_6Cl_{14}]^{4+}$
$[Tc_2Cl_8]^{3+}$
AgF_2

extended lattice such compounds typically have a metallic appearance and even metal-like electrical conductivity. Examples of all three classes are given in Table 8.4.

Finally, we return to transition metal ions, but in a slightly different context. The class 1, 2 and 3 behaviours which have just been detailed refer to interactions between two or more ions in different valence states. In the next chapter we will meet interactions which can be between ions in the same valence state, interactions which are evident in the magnetic properties of the ions. Magnetically coupled (or *exchange coupled*, as it is usually put) ions also show spectroscopic evidence for the coupling—what in an isolated ion are spin-forbidden transitions appear with enhanced intensity (often temperature dependent, the intensity rising with temperature) and additional fine structure. In addition, new absorptions appear with energies approximating the sum of energies of two single excitations. Overall, the colour may appear somewhat different. Thus, copper(II) acetate, an exchange-coupled dimer, is green, whereas the common colour for copper(II) salts is blue.

8.12 Conclusions

The discussion in this Chapter has, at times, been somewhat complicated, the complications centring around such topics as the extraction of quantitative data from spectra or the details of how a forbidden band obtains intensity. The assignment of the band pattern has, tacitly, been assumed to be straightforward. This is by no means always so. For first row transition metal complexes there is seldom a problem but for the others the larger values of Δ mean that there is more overlap between d–d and charge-transfer bands.

Further, the larger values of the spin–orbit coupling constants for these elements make intensity and band-width criteria for distinguishing between spin-allowed and spin-forbidden transitions less reliable. Even more difficult is the case of lower-than-octahedral symmetry. It is probably true that the electronic spectrum of no transition metal complex has been studied in as much detail as has that of the square-planar $[PtCl_4]^{2-}$ ion. Despite all of this work, it is only relatively recently that some general agreement on the assignment of the spectrum has been achieved and few would regard as completely improbable a revision of some aspect of the current interpretation.

Finally, apart from a brief reference to the ruby laser, the content of this chapter has been concerned with the absorption of light. Some complexes emit light also, giving rise to the phenomenon of fluorescence (the excited, emitting state has the same spin multiplicity as has the ground state) and phosphorescence (the excited state has a different spin multiplicity from the ground state). Such processes are currently exciting much interest but as the identity of the actual excited states involved is seldom agreed they will not be discussed in this chapter. They will be referred to in Chapter 14 because a related topic, that of the reactions of electronically excited states, is better defined.

Further reading

An old but still very useful article that not only covers the material indicated in the title but which also explains very simply how spectral data may be interpreted and also lists spectral data for many compounds is 'The Nephalauxetic Effect. The Calculation and Accuracy of the Interelectronic Repulsion parameters for cubic high spin d^3, d^3, d^7 and d^8 systems' by E. König in *Struct. Bonding* (1971) 2, 175.

Much of the basic work in this subject was done in the 1960s and so many of the more useful references tend to date from this period. Examples are:

- C. J. Ballhausen 'Intensities of Spectral Bands in Transition Metal Complexes' *Prog. Inorg. Chem.* (1960) 2, 251.

- J. Ferguson 'Spectroscopy of 3d Complexes' *Prog. Inorg. Chem.* (1979) 12, 158.

- C. J. Jørgensen *Absorption Spectra and Chemical Bonding in Complexes* Pergamon, Oxford, 1962. This book is a useful source of data on individual species.

- C. J. Ballhausen *Introduction to Ligand Field Theory* McGraw-Hill, New York, 1962. This book is something of a bible in the field; Chapter 10 is relevant to the present chapter.

- A useful source is Volume 1 of *Comprehensive Coordination Chemistry* G. Wilkinson, R. D. Gillard and J. A. McCleverty (eds.), Pergamon Press, Oxford, 1987, Chapter 6, 'Ligand Field Theory' by B. N. Figgis.

- T. M. Dunn 'The Visible and Ultraviolet Spectra of Complex Compounds', Chapter 4 in *Modern Coordination Chemistry* J. Lewis and R. Wilkins (eds.), Interscience, New York, 1960.

- C. K. Jørgensen 'The Nephalauxetic Series' *Prog. Inorg. Chem.* (1962) 4, 73

- M. B. Robin and P. Day 'Mixed Valence Chemistry—a Survey and Classification' *Adv. Inorg. Chem. Radiochem.* (1967) 10, 247.

- N. S. Hush 'Intervalence-Transfer Absorption' part 1 (with G. C. Allen) and part 2, both in *Prog. Inorg. Chem.* (1967) 8, 357 and 391, respectively.

- A useful reference, which although devoted to crystal spectra is also of value for solution work, is N. S. Hush and R. J. M. Hobbs, *Prog. Inorg. Chem.* (1968) 10, 259.

- A more recent, more mathematical, discussion is by K. Y. Wong and P. N. Schaty 'A Dynamic Model for Mixed-Valence Compounds' *Prog. Inorg. Chem.* (1981) 28, 369.

- P. J. McCarthy and H. U. Güdel 'Optical Spectroscopy of Exchange-Coupled Transition Metal Complexes' *Coord. Chem. Rev.* (1988) 88, 69.

Probably the most useful current reference, one that contains not only theory but also data on a large number of species is A. B. P. Lever 'Inorganic Electronic Spectroscopy' Elsevier, Amsterdam, 1984.

A much earlier article by the same author gives a short, simple but useful review of 'Charge Transfer Spectra of Transition Metal Complexes' in *J. Chem. Educ.* (1974) 51, 612.

Questions

8.1 An octahedral vanadium(III) complex has d–d bands at 20 000 and 30 000 cm^{-1}. Interpret these data in terms of (a) the crystal field and (b) the ligand field models.

8.2 Whereas the central d–d band in the spectrum of $[Ni(H_2O)_6]^{2+}$ (Fig. 8.5) shows a splitting, the corresponding band of $[Ni(NH_3)_6]^{2+}$, at ca. 17 300 cm^{-1}, shows, at most, a slight asymmetry. That in $[Ni(en)_3]^{2+}$, at 18 400 cm^{-1}, is quite symmetrical. Do these two latter observations enable a distinction to be made between the two explanations offered in the text for the splitting observed for $[Ni(H_2O)_6]^{2+}$?

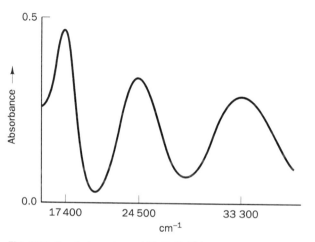

Fig. 8.21 The d–d spectrum of $[Cr(H_2O)_6]^{3+}$.

8.3 Figure 8.21 shows the d–d spectrum of $[Cr(H_2O)_6]^{3+}$. Provide an explanation for the bands observed. Note: this question can be answered at several levels. Simplest is to use the appropriate Tanabe–Sugano diagram in a qualitative fashion. Deeper, is to use the d^2 discussion of the text together with the hole–electron analogy.

8.4 The following are the positions of the d–d bands of some Ni^{II} complexes. The assignments correspond to those of $[Ni(H_2O)_6]^{2+}$ (Fig. 8.5). To what extent do these data conform to those predicted by the rule of average environment? Data are in 10^3 cm^{-1}.

$[Ni(en)_3)]^{2+}$	11.2	18.4	29.0
$[Ni(en)_2(gly)]^+$	10.8	17.9	28.6
$[Ni(en)(gly)_2]$	10.5	17.3	28.1
$[Ni(gly)_3]^-$	10.1	16.6	27.6

8.5 Outline the spectral consequences of (a) the Jahn–Teller effect and (b) spin-orbit coupling for the first row transition metal ions. How would corresponding second and third row ions be expected to differ?

8.6 Use the Tanabe–Sugano diagrams given in Appendix 7 to suggest transitions which might be candidates to provide monochromatic emitted (laser) light. What is the feature common to all such transitions?

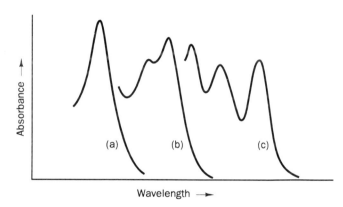

Fig. 8.22 The charge-transfer spectra of (a) $[CoCl_4]^{2-}$, (b) $[CoBr_4]^{2-}$ and (c) $[CoI_4]^{2-}$.

8.7 Figure 8.22 shows the charge transfer spectra of the ions $[CoCl_4]^{2-}$, $[CoBr_4]^{2-}$ and $[CoI_4]^{2-}$. What is the correspondence between spectra and species? Give reasons for your choice.

Magnetic properties of transition metal complexes

9.1 Introduction

In the last chapter, in Section 8.4, it was suggested that the spin of an electron is best thought of as meaning that the electron behaves like a tiny bar magnet. When there are several unpaired electrons the spin degeneracies can be thought of as resulting from the variety of ways of arranging bar magnets side-by-side. Similarly, the orbital motion of the electron, the circulation of charge around the nucleus, can be thought of as leading to a solenoid-like magnet. These spin and orbital magnets will interact with an applied magnetic field, so that when an atom or molecule is placed in a magnetic field any spin degeneracy may be removed—the different resultant magnets behave differently. Thus, a level which is orbitally non-degenerate but is a spin doublet has this spin degeneracy split into two levels with slightly different energies in a magnetic field (corresponding to the N–S and S–N arrangements). If there is orbital degeneracy this too may be removed by a magnetic field. As we shall see, it is such splittings which determine the magnetic properties of a complex. Note particularly that either or both of spin and orbital degeneracies give rise to the magnetic effects. Sometimes statements such as 'the number of unpaired electrons in a complex may be determined from magnetic susceptibility measurements' are encountered. Whilst true within their own context, they should not be read as meaning that in electron spin lies the sole source of magnetic effects.

The splittings produced by magnetic fields are very small, about 1 cm^{-1} for a field of 0.5 T, and, for the majority of cases, are proportional to the magnetic field. Because the splittings are so small, any particular atom or molecule may be in any one of the several closely spaced states resulting from the splitting. For a macroscopic sample, however, there will be a Boltzmann distribution between the levels; at room temperature kT is about 200 cm^{-1} so ample energy is available. Clearly if we are to be able to interpret experimental results, we must consider both the splittings and the Boltzmann distribution over them. However, because the effect of the magnetic field is so small we must consider any other interaction which involves energies corresponding to more than about 1 cm^{-1}. This is because such interactions will play a part in determining the ground state of the complex before the magnetic field is switched on—this field will only cause the subsequent splitting. Two such interactions—spin–orbit coupling and the presence of low-symmetry components in what is otherwise an octahedral crystal field—have already been discussed in outline. The latter effect can seldom be neglected. One might think, for example, that an isolated $[Co(NH_3)_6]^{3+}$ ion would be accurately octahedral. This is not so, for there is an incompatibility between the threefold axis of each NH_3 and the coincident fourfold axis of the CoN_6 octahedron. Add to this the effect of the environment (and this means 'do not consider an isolated molecule; in solution it will be surrounded by a jumble of solvent molecules and in a crystal by anions—and these will seldom respect its octahedral symmetry'), recall the Jahn–Teller effect, remember that some metal–ligand vibrations will be thermally excited down to quite low temperatures and that lattice vibrations will persist to an even lower temperature, and it becomes evident that from the point of view of magnetism, a regular octahedral environment is a rare, if not extinct, species. For basically octahedral molecules, however, an octahedral model is a good first approximation and may be refined to take account of the above effects. For the majority of this chapter we shall confine ourselves largely to such octahedral complexes.

Before we can proceed, some of the vocabulary of magnetism has to be introduced. Closed shells of electrons have neither spin nor orbital degeneracy and are represented by a single wavefunction. A magnetic field therefore produces no splitting. It does, however, distort the electron density slightly, in a manner akin to that predicted by Lenz's law in classical electrodynamics. That is, effectively, a small circulating current is produced, the magnetic effect of which opposes the applied magnetic field. Because there is no resistive damping, the current remains until the magnetic field is removed. Molecules with closed shells are therefore repelled by a magnetic field and are said to exhibit diamagnetism or to be *diamagnetic*.

Suitably oriented magnets are attracted towards the magnetic field of a stronger magnet and the same is true for any orbital magnet and the intrinsic (spin) magnet associated with an unpaired electron. Molecules with unpaired electrons are therefore attracted into a magnetic field[1] and are said to exhibit paramagnetism or to be *paramagnetic*. For any transition metal ion which has both closed shells and unpaired electrons the diamagnetism of the former and the paramagnetism of the latter are opposed. All that can be measured is their resultant. Fortunately, the effect of paramagnetism is about 100 times as great as that of diamagnetism so that it takes a great deal of the latter to swamp the former. Fortunately, too, it is found that the effects of diamagnetism are approximately additive—each atom makes a known, and approximately constant, contribution to the diamagnetism of a molecule and so diamagnetism may be fairly accurately allowed for once the empirical formula of a complex is known. This means that it is possible to deduce the

---

[1] Exceptions to this statement exist at very low temperatures.

intrinsic paramagnetism of a complex to an acceptable level of precision and to compare it with theoretical predictions. Finally, a discussion in terms of an isolated molecule requires that the (para)magnetic units do not interact; that is, these units must be spatially well separated. The jargon is to say that the sample must be magnetically dilute. The detailed theory of magnetically non-dilute substances is beyond the scope of this book, although some general comments on the topic are contained in Section 9.12. Actually, measurements at very low temperatures have shown that some of the most classic of coordination compounds exhibit a tiny amount of magnetic coupling between the transition metal ions and so are slightly non-dilute. Examples include $K_3[Fe(CN)_6]$, $[Ni(en)_3](NO_3)_2$, $K_2[Cu(H_2O)_6](SO_4)_2$ and $[Ni(H_2O)_6]Cl_2$. At temperatures much above those of liquid helium such compounds behave as magnetically dilute.

9.2 Classical magnetism

A theory of magnetism was developed long before the advent of quantum mechanics. This section contains an outline of this theory and defines some of the experimental quantities which may well be encountered.

When a substance is placed in a magnetic field, H, the total magnetic induction within the substance, B, is proportional to the sum of H and M, where M is the magnetization of the substance:

$$B = \mu_0(H + M)$$

where μ_0 is the vacuum permeability (in the CGS system of units μ_0 is unity, in SI it is defined to be exactly $4\pi \times 10^{-7}\,N\,A^{-2}$; note that $N\,A^{-2}$ is the same as $H\,m^{-1}$—either may be encountered). This equation may be written as

$$\frac{B}{H} = \mu_0\left(1 + \frac{M}{H}\right) = \mu_0(1 + \kappa)$$

where κ is the volume susceptibility (the ratio of the induced magnetization to the applied magnetic field). A more useful quantity is the susceptibility per gram, χ (chi), called the specific susceptibility, which is given by

$$\chi \times \frac{\kappa}{\rho}$$

where ρ is the density of the substance. Alternative forms of χ which are encountered are χ_A and χ_M, the atomic and molar susceptibilities, respectively. These latter are obtained by multiplying χ by the atomic and molecular weights, respectively, of the magnetically active atom or molecule. They are the susceptibilities per (gram) atom and mole, respectively. The former is no longer used but will be met in the older literature. Because our interest is in paramagnetism, some correction has to be made for the underlying diamagnetism, a topic dealt with in more length in Appendix 9. If such a correction has been made, it is indicated by a prime, thus χ_A' and χ_M'. Now,

$$\chi_M = \frac{\kappa}{\rho} \times (MW) \times \frac{MV}{H}$$

where (MW) is the molecular weight and V is the molar volume. Therefore,

$$\chi_M = \frac{\text{total magnetization per mole}}{H}$$

$$= \frac{\text{average magnetization per mole} \times N_A}{H}$$

that is

$$\chi_M = \frac{\overline{m} N_A}{H} \tag{9.1}$$

where $\overline{m}$ is the average moment per molecule and N_A is Avogadro's number.

If we have a collection of identical molecules, each of magnetic moment μ and free to orient itself in a magnetic field, then in such a field there will be some alignment but this will be opposed by the thermal motion of the molecules. That is, the measured moment decreases with increasing temperature although μ itself is a constant. Langevin showed that in this situation the average (measured) magnetic moment $\overline{m}$, and the actual moment μ, are related by:

$$\overline{m} = \frac{\mu^2 H}{3kT} \tag{9.2}$$

a derivation of which will be found in almost every text on magnetism. His derivation contains the assumption that $B = \mu_0 H$, so the theory can only be expected to hold for gaseous molecules, although, as is commonly done, we shall ignore this limitation. Combining the last two equations we have

$$\chi_M = \frac{N_A \mu^2}{3kT} = \frac{N_A^2 \mu^2}{3RT} \equiv \frac{C}{T} \tag{9.3}$$

where C, the Curie constant, is equal to $N_A^2 \mu^2 / 3R$. The equation $\chi = C/T$ is known as the Curie law—susceptibility is inversely proportional to the absolute temperature. Surprisingly, this law is obeyed rather well by many liquids and solids, and in particular by complexes of the first row transition elements. The origin of this general agreement is not clear, except under rather special limiting conditions. The detailed quantum mechanical treatment does not at all readily lead to a prediction of a C/T type of behaviour. For complexes of the first row transition elements it turns out that the agreement is because the spin–orbit coupling constants are comparable in magnitude to kT; at low temperatures they do not follow the Curie law so well. Rearranging the above equation we find:

$$\mu = \frac{(3RT\chi_M)^{1/2}}{N_A} \tag{9.4}$$

It is convenient to express μ in units of Bohr magnetons,[2] β, and so we write $\mu = \mu_{\text{eff}}\beta$. Combining the last two equations we find that

$$\mu_{\text{eff}} = \frac{(3RT\chi_{\text{M}})^{1/2}}{N_{\text{A}}\beta} \tag{9.5}$$

Notice that, defined in this way, μ_{eff} is a pure number, the Bohr magneton number, which refers to a single molecule. Although often met, it is not strictly correct to call it the effective magnetic moment, and it certainly is incorrect to express it in units of Bohr magnetons, although this is a usage met all too often.

9.3 Orbital contribution to a magnetic moment

We now turn to the quantum mechanical approach to the phenomenon of paramagnetism and discuss first the orbital contribution to a magnetic moment. It is convenient at this point to make a few general statements which will be explained by the subsequent discussion.

The phrase orbital degeneracy was used in Section 9.1, implying that for individual cases one may or may not exist. If in the free ion—in the absence of a magnetic field—there is orbital degeneracy (that is, if we have anything other than an S state ion) the free ion has an *orbital magnet*. If the orbital degeneracy is lost in a real environment—by chemical bonding or crystal field effects—the orbital contribution to the total magnetic moment is said to be *quenched*. If the orbital degeneracy is merely reduced, the orbital contribution is partially or incompletely quenched. Orbital degeneracy, however, although a necessary condition for an orbital moment, is not a sufficient condition.

It may be recalled that in Section 8.4 the orbital moment was likened to the magnetic effect produced by a current in a solenoid. An even better analogy would be with the magnetic moment associated with a current flowing around a circular ring of superconducting material. If the super-conducting ring is rotated by 90°—or any other angle—about its unique axis, one is left with a physically identical situation (for a solenoid the ends of the wire coil would be in a different position after the rotation). The requirement for a non-zero orbital-derived moment is similar. An orbital

---

[2] The Bohr magneton is a fundamental quantity in the quantum theory of magnetism. Bohr's explanation of atomic structure was based on the assumption that the angular momentum of an electron circulating about the nucleus of an atom is quantized and equal to $nh/2\pi$, where n is an integer and h is Planck's constant. That is, $nh/2\pi = ma^2\omega$, where m is the mass of the electron, a the radius of its orbit and ω its angular velocity in radians. The area of the circular orbit is πa^2 and the current to which the electron circulation is equivalent is $e \times (\omega/2\pi)$. From the theory of a current flowing through a circular loop of wire, the magnetic moment associated with the circulating electron is equal to the product

$$\text{current} \times \text{area} = \frac{e\omega}{2\pi} \times \pi a^2 = \frac{ea^2\omega}{2}$$

From Bohr's postulate, this equals

$$\frac{ea^2}{2} \times \frac{nh}{2\pi ma^2} = n\frac{he}{4\pi m} = n\beta, \quad \text{where } \beta = \frac{he}{4\pi m}$$

That is, the magnetic moment is an integer times β, where β is the Bohr magneton. It transpires that β is also the fundamental quantity in the modern quantum mechanical treatment.

degeneracy is needed—and this only occurs when orbitals are unequally occupied—but this orbital degeneracy must be such that there exist two or more degenerate orbitals which can be interconverted by rotation about a suitable axis. Consider the d_{zx} and d_{yz} orbitals in an octahedral complex, shown in Fig. 9.1. Rotation by 90° about the z axis interconverts these two orbitals, so that if an electron were initially in the d_{zx} orbital it could circulate about the z axis by jumping between the d_{zx} and d_{yz} orbitals alternately. This circulation is equivalent to a current flowing and so it produces a magnetic effect. The circulation may take place in either direction, clockwise or anticlockwise, and in the absence of a magnetic field the two possibilities are degenerate (this is implicit in the orbital degeneracy). On application of a magnetic field the two directions of circulation have different energies, and this is associated with the loss of orbital degeneracy in a magnetic field.

Four comments are relevant at this point. First, the d_{xy} and d_{zx} orbitals are interconverted by a rotation about the x axis and the d_{xy} and the d_{yz} by a rotation about the y axis, and so the discussion above is relevant to these pairs also. Second, a more detailed analysis shows that an electron in the d_{xy} orbital does not have to jump 90° to get into the d_{yz} orbital because these two orbitals overlap each other. Note that this statement is not incompatible with the fact that these two orbitals have a zero overlap *integral*. A continuous range of rotations is allowed, so that the electron cloud experiences no barrier to free rotation about the z axis. Third, the Jahn–Teller theorem requires that the orbital degeneracy which has just been invoked, actually, never exists! However, because they are largely non-bonding, it is usually assumed that any Jahn–Teller splitting of the d_{xy}, d_{yz} and d_{zx} degeneracy is small and that the orbital contribution to the moment is only slightly quenched. Physically, this means that an electron circulating about the z axis in the d_{zx} and d_{yz} orbitals does experience a barrier to free rotation but that this barrier is usually small relative to thermal energies. Finally, the d_{xy} and $d_{x^2-y^2}$ orbitals are interconverted by a 45° rotation about the z axis. However, in an octahedral complex they are not degenerate and so give rise to no magnetic effect. In a free atom, on the other hand, they *are* degenerate and so contribute to the orbital magnetic moment. Summarizing, it turns out that for octahedral transition metal ions, only ground states of T_{1g} and T_{2g} orbital symmetries give rise to orbital-derived magnetic moments. E_g ground states have no such moment.

A question which at once arises is 'can a complex which has no unpaired electrons be paramagnetic just because of its orbital magnetism'? The answer is yes; two examples are provided by the permanganate ion MnO_4^- and by the ion $[Co(NH_3)_6]^{3+}$. This seems strange. There is no case in which the ground state of a transition metal ion has orbital degeneracy without also having a spin degeneracy. So, both of the two examples just given have orbitally non-degenerate ground states and, of course, no spin degeneracy. We would expect them to be diamagnetic. Without orbital degeneracy, how can an orbital magnetism exist? The answer lies in the existence of an excited state in each case which *does* have the required orbital degeneracy and yet is a spin singlet. The magnetic field mixes some of the excited state into the ground state (or, equivalently, pushes a tiny bit of electron density into the excited state) and so the ground state assumes some of the properties of the excited. In particular, measurement shows that such a complex has

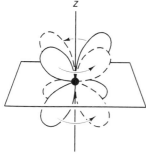

Fig. 9.1 The circulation of electron density about the z coordinate axis in an incompletely filled t_{2g} shell of an octahedrally coordinated transition metal ion. The electron density may be thought of as jumping from d_{zx} to d_{yz} to d_{zx} and so on, but see the text for a more accurate description.

a—small—orbital paramagnetism, which is sometimes large enough to cancel out the inherent diamagnetism. Because this paramagnetism does not depend on the thermal population of levels—unlike simple paramagnetism—it is called temperature independent paramagnetism, or TIP for short.[3] TIP is no mere academic curiosity. For diamagnetic transition metal ions it can be very important in determining the position of NMR resonances (the Co^{III} case is particularly well studied and is cited in the 'Further reading' at the end of Chapter 12). In Co^{II} octahedral complexes the magnitude of the TIP is often very comparable to the correction that has to be made for the underlying diamagnetism. In such cases, it is better to make no correction, and rely on the approximate cancellation of the two effects, than to correct for one but not the other.

9.4 Spin contribution to a magnetic moment

As has been mentioned earlier in this chapter, the phenomenon of electron spin is best thought of as the electron behaving like a tiny bar magnet. This bar magnet is essentially insensitive to its environment because the latter imposes an electrical, rather than a magnetic, field. A single bar magnet can be thought of as having two orientations with respect to an applied magnetic field, parallel and antiparallel, and these have different energies; with a pair of bar magnets the possibilities increase to three (both parallel, one parallel one antiparallel, both antiparallel), again each of the three arrangements have different energies. Higher spin multiplicities can similarly be given simple physical descriptions. It is possible to derive a simple equation which holds when the spin is the sole cause of the magnetic properties of a complex, and this will be done in Section 9.10. Despite its simplicity, we cannot immediately move to it because it is important to be aware of the circumstances in which it is applicable. These are when other, complicating, effects are absent. First, these complications must be explored. In the general case, both orbital and spin motions will contribute to the magnetism displayed by a complex; their effects may be additive or opposed. Further, the spin and orbital magnets may interact with each other, the phenomenon of spin–orbit coupling. This will be covered in the next section.

9.5 Spin–orbit coupling

Spin–orbit coupling has to be included in a discussion of the magnetic properties of transition metal complexes whenever a spin-derived magnetic moment and an orbitally-derived magnetic moment coexist. Spin–orbit

---

[3] It is instructive to look at the Co^{III} case in more detail. The d electron configuration of Co^{III} is $t_{2g}^6 e_g^0$ and so the ground state is $^1A_{1g}$. Earlier in the section it was found helpful to talk in terms of the magnetic properties of a current flowing in a ring of superconducting material. This analogy is helpful in the present context. In the point group O_h the set of three rotations R_x, R_y and R_z transform as T_{1g}. The analogy suggests that the components of the magnetic field also transform as T_{1g}. This is, indeed, the case. Now, if an excited state—call its symmetry species Γ—is to be mixed with the ground state, $^1A_{1g}$, by the magnetic field, then the direct product $T_{1g} \times \Gamma$ must contain $^1A_{1g}$, the symmetry of the ground state. This only occurs when $\Gamma = T_{1g}$ and so it has to be an excited $^1T_{1g}$ state that gives rise to the TIP of Co^{III} species. Reference to the d^6 Tanabe–Sugano diagram shows that such a low-lying excited state does, indeed, exist.

coupling splits terms with both spin and orbital degeneracy into a number of sublevels. This phenomenon is relevant to the spectroscopic properties of transition metal complexes and so in Table 8.2 were listed the number of components obtained as a result of spin–orbit coupling.

There is a simple method by which one may discover whether spin–orbit coupling mixes two d orbitals. For each d orbital, note the number of its nodal planes which also contain the z axis (i.e. those in which the z axis lies on a nodal plane). If two d orbitals differ by either one or zero in these numbers then the two orbitals may be mixed by spin–orbit coupling. Using this approach it follows that spin–orbit coupling mixes d orbitals as indicated below. Numbers in brackets indicate the number of nodal planes which, for that orbital, contain the z axis:

The origin of this pattern will become more evident from the detailed discussion that will be given Sections 11.4 and 11.5.

The magnitude of the spin–orbit coupling for a particular ion is usually given in terms of one of two different so-called spin–orbit coupling constants, ζ (zeta) and λ. The former is the one-electron spin–orbit coupling constant and is useful when comparing the relative magnitudes of spin–orbit coupling for different ions. In practice, for many-electron ions, what is measured is a resultant spin–orbit coupling between the resultant spin magnetic moment and the resultant orbital moment. It is this latter spin–orbit coupling constant which is called λ. For ground states the two constants are simply related:

$$\lambda = \pm \frac{\zeta}{2S}$$

where S is the spin multiplicity of the ion and the plus sign refers to d^1–d^4 ions and the minus sign to d^6–d^9. For these latter, it is simplest to think of holes circulating, the holes having the opposite charge to electrons. This point is detailed in appendix 12, which contains a specific example, although this particular appendix is specifically addressed to a problem that will be encountered in Chapter 11. Values for λ (and thence ζ) are obtained for free ions from atomic spectral data and for complex ions from magnetic measurements of the type discussed in this chapter.

9.6 Low-symmetry ligand fields

Because they can, and usually do, remove orbital degeneracies—and thus reduce the orbital contribution to the magnetism—low-symmetry ligand fields cannot be ignored in a study of the magnetic properties of transition metal complexes. Unfortunately, it is no easy matter to determine the splitting effects of low symmetry fields; usually it is necessary to work with two splitting parameters. Because the effects of low symmetry are so much more important in magnetism than in, say, spectroscopy, the magnitude of the parameters is best determined magnetically. However, the values of the distortions may be small and so impossible to detect by any other method; the distortion may not even be evident in a structure determination, for instance. In this situation the magnitude of the parameter becomes something of a 'fudge factor'—it is given the value that produces best agreement between experiment and theory.

There is an important theorem due to Kramer, which states that when the ground-state configuration has an odd number of unpaired electrons there exists a degeneracy which a low symmetry ligand field cannot remove. This degeneracy, usually known as Kramer's degeneracy, arises from the fact that an orbital may be occupied by an electron in two ways—the spin may be up or it may be down. In the absence of a magnetic field these two orientations have the same energy. The application of a magnetic field causes the two to differ in energy—by about $1\,\text{cm}^{-1}$—and it is the greater occupation of the more stable in a macroscopic sample which causes the sample to be attracted into a magnetic field. When there is an even number of unpaired electrons a low-symmetry ligand field *can* relieve degeneracies, but the application of a magnetic field then causes the lowest state to become even more stable.

9.7 Experimental results

If there were no orbital contribution to the magnetic moment of a complex ion, one would have to worry a great deal less about the effects of spin–orbit coupling and low-symmetry fields. Table 9.1 lists the moments that would be expected for ions of the first transition series if there were no orbital contribution (spin-only moments) and compares these with room temperature experimental data. Also indicated in Table 9.1 is whether a significant orbital contribution is to be expected (that is, whether there is a ground state T_{1g} or T_{2g} term). In Section 9.10 the spin-only equation used to predict the moments given in Table 9.1 will be derived. On the whole, the agreement in this table is not at all bad and if all one is interested in is whether a complex is high or low spin then, if the oxidation state is known, a simple room temperature magnetic susceptibility measurement[4] on a tetrahedral or octahedral complex of a first-row element may readily be interpreted using the data in this table. A detailed analysis shows that the reasonable agreement between spin-only and experimental moments shown in Table 9.1 is somewhat fortuitous. For d^1–d^4 ions, spin–orbit coupling has the effect of reducing the observed moment and, roughly, cancels any orbital contribution. For d^6–d^9 ions, spin–orbit coupling increases the observed moment and adds to the orbital contribution. This, together with the increased magnitude of the spin–orbit coupling constants for these ions explains the few gross disagreements between spin-only and experimental values. Because of the very large spin–orbit coupling constants of elements of the second and third transition series, their complexes usually have moments much

[4] Appendix 8 contains both a description of how such measurements are made and the treatment of the data obtained; how the diamagnetic corrections are made, for instance.

Table 9.1 Comparison of calculated spin-only moments and experimental data for magnetic moments of ions of the first transition series

Ion	Configuration[a]	Orbital contribution expected?	Theoetical spin-only value	Range of experimental values found at room temperature
Octahedral complexes				
Ti^{3+}	d^1	Yes	1.73	1.6–1.75
V^{4+}	d^1	Yes	1.73	1.7–1.8
V^{3+}	d^2	Yes	2.83	2.7–2.9
Cr^{4+}	d^2	Yes	2.83	ca. 2.8
V^{2+}	d^3	No	3.88	3.8–3.9
Cr^{3+}	d^3	No	3.88	3.7–3.9
Mn^{4+}	d^3	No	3.88	3.8–4.0
Cr^{2+}	d^4 hs	No	4.90	4.7–4.9
Cr^{2+}	d^4 ls	Yes	2.83	3.2–3.3
Mn^{3+}	d^4 hs	No	4.90	4.9–5.0
Mn^{3+}	d^4 ls	Yes	2.83	ca. 3.2
Mn^{2+}	d^5 hs	No	5.92	5.6–6.1
Mn^{2+}	d^5 ls	Yes	1.73	1.8–2.1
Fe^{3+}	d^5 hs	No	5.92	5.7–6.0
Fe^{3+}	d^5 ls	Yes	1.73	2.0–2.5
Fe^{2+}	d^6 hs	No	4.90	5.1–5.7
Co^{2+}	d^7 hs	Yes	3.88	4.3–5.2
Co^{2+}	d^7 ls	No	1.73	1.8
Ni^{3+}	d^7 ls	No	1.73	1.8–2.0
Ni^{2+}	d^8	No	2.83	2.8–3.5
Cu^{2+}	d^9	No	1.73	1.7–2.2
Tetrahedral complexes[b]				
Cr^{5+}	d^1	No	1.73	1.7–1.8
Mn^{6+}	d^1	No	1.73	1.7–1.8
Cr^{4+}	d^2	No	2.83	2.8
Mn^{5+}	d^2	No	2.83	2.6–2.8
Fe^{5+}	d^3 hs	Yes	3.88	3.6–3.7
unknown	d^4 hs	Yes	4.90	
Mn^{2+}	d^5 hs	No	5.92	5.9–6.2
Fe^{2+}	d^6 hs	No	4.90	5.3–5.5
Co^{2+}	d^7	No	3.88	4.2–4.8
Ni^{2+}	d^8	Yes	2.83	3.5–4.0

[a] hs = high spin, ls = low spin.
[b] Note that low-spin tetrahedral complexes are very rare—if any exist at all—and are not included in this table.

smaller than the spin-only values. We shall see why this is so in Section 9.9. Lest it be thought that the sole effect of spin–orbit coupling is that of making life more complicated, it should be mentioned that a large spin–orbit coupling tends to reduce the sensitivity of the magnetic moment to low-symmetry fields, leading to a more octahedral-like behaviour—provided that the complex is approximately octahedral to start with.

9.8 Orbital contribution reduction factor

So far in this chapter an essentially crystal field approach has been followed. How has it to be modified to take account of covalency? One way is to let the spin–orbit coupling constant vary, to become a parameter, rather than giving it its free-ion value. Because a molecular orbital differs from an atomic orbital, we would expect them to have different magnetic moments and also different spin–orbit coupling constants. So, the t_{2g} set of d orbitals can give rise to an orbital contribution. What happens when it contains a ligand component—which, in this case, has to be π? The arguments used to justify an orbital contribution from the metal orbitals apply equally to the ligand π orbitals—the appropriate ligand combinations can be interchanged by suitable rotations, although it is unlikely that they will overlap with each other in the same way. The real question, therefore, is whether there is any reason to expect the ligand contribution to the moment to be equivalent to that part of the metal contribution which it has replaced. There is no such reason and so we must allow for the difference somewhere within our model. The usual way is to multiply the orbital contribution, calculated for the free metal ion, by a parameter, choosing that value of the parameter which gives the best agreement with experiment. This parameter is usually denoted k and is called the orbital reduction factor because it generally turns out to have a value of less than unity.

9.9 An example

This Section gives an outline of an algebraic calculation of the magnetic properties of a complex ion. The treatment is repeated in more detail in Appendix 10. In real life, the calculations would be carried out by a computer and some of the approximations that will be made here would not be necessary. However, the language of the subject is so wedded to the algebraic treatment that it is essential to study it, at least superficially.

The problem that will be considered is very simple, at least in principle. It is that of an octahedral t_{2g}^1 complex. So, the theory is that of a strong field complex of Ti^{III}—strong field because the e_g orbitals are considered so high in energy that they can be ignored. It would be good at the end of our development to be able to say that the final equation gives an excellent account of the magnetic properties of Ti^{III} complexes. Unfortunately, it is not possible to be so enthusiastic—indeed, it is possible that the theory is fundamentally flawed, although it will only be possible to appreciate why this is so after the theory has itself been developed.

The t_{2g}^1 configuration gives rise to a $^2T_{2g}$ ground term—three orbital functions, each of which may be combined with either of two spin functions, giving a total of six functions to be considered. Eventually, we will arrive at a picture in which these six levels, although clustered together, are split apart. We will be concerned with the Boltzmann distribution over these separated levels and, in particular, the sensitivity of the distribution to the application of a magnetic field. In this situation, it would clearly be ridiculous to consider just the effects of the magnetic field and to ignore the larger effects of any distortion and of spin–orbit coupling. For simplicity, a distortion from octahedral which retains some orbital degeneracy will be considered. This means a distortion which retains either a fourfold or a threefold axis. Both are mathematically similar, only one additional low-symmetry field splitting parameter being involved in either case. Of the two it is simpler to consider the tetragonal case because we can then retain the coordinate axes of the octahedron. Such a tetragonal distortion leads to the orbital triplet (t_{2g}) splitting into an orbital singlet (b_{2g}) and an orbital doublet (e_g), the latter two labels being appropriate to D_{4h} symmetry. We will assume that the doublet is the lower in energy. Formally, we have two wavefunctions, a $^2B_{2g}$ term, above a set of four wavefunctions, a ^2E_g. Spin–orbit coupling splits the ^2E_g into two doublets. The final pattern is one with three distinct levels, each doubly degenerate, each a Kramer's doublet. These double degeneracies can only be split by a magnetic field.

All that remains is to remove the Kramer's degeneracies by the application of a magnetic field. Here, there is a clear divergence between the computer and algebraic solutions to the problem. In the algebraic approach, a distinction is made between the effect of a magnetic field on a particular level itself (the so-called first-order Zeeman effect) and its effect in mixing different levels together (the second-order Zeeman effect). The first-order Zeeman effect is proportional to the magnetic field strength, H, and the second order to the square of the magnetic field. Computer calculations show that this separation into terms proportional to H and to H^2 is only approximate, although for the strength of magnetic fields that are used in experimental work, the approximation is usually a good one.

The pattern of splittings that has just been described is shown in Fig. 9.2, where it has been assumed that the tetragonal field and the spin–orbit coupling produce effects of similar magnitudes. For clarity, in this figure, the effects of the magnetic field have been greatly exaggerated. In order to relate the energy level sequence shown in Fig. 9.2 to experiment, a value for the average magnetic moment per molecule χ_M is needed (see Section 9.2). Now, χ_M is

$$\chi_M = \frac{\displaystyle\sum_{\text{all molecules}} \text{(magnetic moment of a molecule)} \times \text{(number of molecules with this moment)}}{\text{(total number of molecules)}}$$

There is a Boltzmann distribution of molecules over the levels shown on the right-hand side of Fig. 9.2, so that the relative number of molecules occupying a level with energy E is $\exp(E/kT)$, a quantity which can be used in the expression above provided that we have explicit expressions for the energy levels—and Appendix 10 provides them. The Zeeman part of the calculation provides the value of the magnetic moment for the molecules in a particular level. Having thus obtained a complete expression for χ_M, this can be related to the quantity usually discussed, the Bohr magneton number μ_{eff} because, as was shown in Section 9.2 (eqns 9.4 and 9.5)

$$\mu_{\text{eff}}^2 = \frac{3kT\chi_M}{N_A\beta^2} \tag{9.6}$$

Rather than work with the rather unwieldy equations resulting from the expressions given in Appendix 10 we shall simplify. We ignore the tetragonal

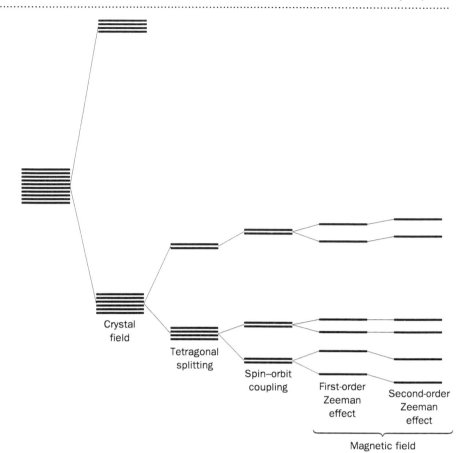

Fig. 9.2 Schematic energy level diagram illustrating the various splittings determining the ground state derived from a t_{2g}^1 configuration.

Crystal field

Tetragonal splitting

Spin–orbit coupling

First-order Zeeman effect

Second-order Zeeman effect

Magnetic field

distortion. With this step we obtain the Kotani model, in which just the effects of spin–orbit coupling and the magnetic field on a t_{2g}^1 configuration are considered. We thus obtain an important expression, derived in reasonable detail in Appendix 10:

$$\mu_{\text{eff}}^2 = \frac{\dfrac{3\zeta}{kT} - 8\exp\left(\dfrac{-3\zeta}{2kT}\right) + 8}{\dfrac{\zeta}{kT}\left[\exp\left(\dfrac{-3\zeta}{2kT}\right) + 2\right]}$$

or, by putting $\zeta/kT = x$ (a number) the equation simplifies to

$$\mu_{\text{eff}}^2 = \frac{(3x - 8)\exp\left(\dfrac{-3x}{2}\right) + 8}{x\left[\exp\left(\dfrac{-3x}{2}\right) + 2\right]}$$

another equation first derived by Kotani.

We can now plot the value of μ_{eff} (given by the positive square root of the above expression) against x, or, more conveniently, against $1/x$, to give a so-called Kotani plot. This has been done in Fig. 9.3. Also on this figure are indicated the values of ζ/kT for some d^1 ions at 300 K, ζ being given the

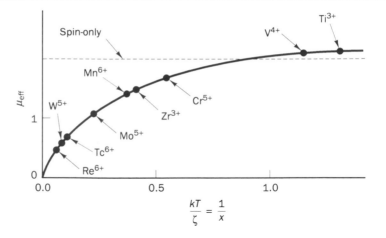

Fig. 9.3 A Kotani plot for the t_{2g}^1 configuration together with 300 K values for some t_{2g}^1 ions. The spin-only formula gives $\mu_{\text{eff}} = 1.73$; whereas first row transition metal ions roughly approximate to this value at room temperature, values for second and third row ions can be very different.

free-ion value. It is seen that μ_{eff} is almost independent of temperature at high enough temperatures (the larger ζ, the higher the temperature needed); room temperature measurements on first row transition metal ions give, essentially, their 'plateau' values. It is under these limiting conditions that simple formulae such as

$$\mu_{\text{eff}} = \sqrt{n(n+2)}$$

the spin-only formula (which will be derived in the next section), in which n is the number of unpaired electrons, become appropriate. For complexes of the second and third row transition series the spin-only formula is not applicable and magnetic measurements over a temperature range are absolutely essential, even if it is only the number of unpaired electrons which is to be determined. For measurements on such compounds at room temperature, the plateau has not been reached.

For electronic configurations other than the one discussed above, analogous calculations lead to relationships which are roughly similar to that shown in Fig. 9.3. μ_{eff} does not generally drop to zero at 0 K and, down to $\lambda/kT \approx 1.5$ it may increase slightly with decreasing temperature, decreasing as the temperature is lowered further. If the model is not simplified and k (the orbital reduction factor) and t (the factor describing the distortion to a tetragonal field), or related functions, are included in the final energy-level expressions, the temperature dependence of μ_{eff} is less than in the corresponding case in which they are omitted. As an illustration of this, in Fig. 9.4 is shown a comparison between the predicted and experimental results for $[VCl_6]^{2-}$, the cation being the pyridinium ion, $C_5H_5NH^+$. Simple Kotani theory is roughly followed if the spin–orbit coupling constant is reduced to 190 cm^{-1} from the free-ion value of 250 cm^{-1} (Fig. 9.4(a)). Much better agreement is obtained if distortion and covalency are allowed for using the full energy level pattern of Fig. 9.2. This better agreement is shown in Fig. 9.4(b). In that figure, the theoretical curve is that calculated for $k = 0.75$, $\zeta = 150 \text{ cm}^{-1}$ and $t = 150 \text{ cm}^{-1}$ in the expression given in Appendix 10. The negative sign on t implies that the orbital singlet lies lowest, not the orbital doublet—the distortion is the opposite to that assumed in Appendix 10 and, indeed, earlier in this chapter. The job of fitting a theoretical curve

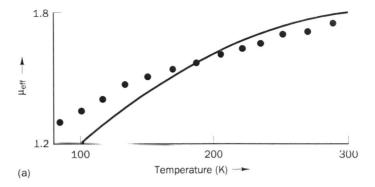

(a)

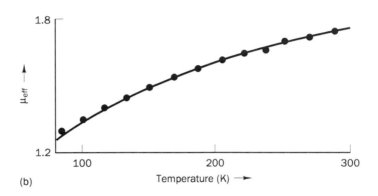

(b)

Fig. 9.4 (a) A simple Kotani plot for $[VCl_6]^{2-}$ ($\zeta = 190$ cm^{-1}). (b) A Kotani plot including best-fit tetragonal distortions and orbital reduction factors. In both cases experimental data are given as dots.

to the experimental results is best done by computer; the need for high experimental accuracy is evident.

This, then, is the algebraic model. What, if anything, is wrong with it? Look again at Fig. 9.4(b), where the best fit between the modified Kotani theory and the experimental data is shown. The lower temperature points seem to show a systematic divergence between experiment and theory. This is not surprising, for as either a qualitative consideration of the effect of decreasing temperature on the population of the levels in Fig. 9.2 or, equivalently, a study of Fig. 9.4(a) (the Kotani plot) shows, the most severe test of the theory is to be expected at the lowest temperatures, for here the average magnetic moment has its greatest temperature dependence. The low-temperature cut-off point in Fig. 9.4 is determined by the boiling point of liquid nitrogen. What if the temperature is taken down to that of liquid helium (3 K), an extension which is now normal. Inevitably, complications set in! These complications can be many-faceted. Small distortions away from true trigonal or tetragonal may become evident, as may long-range magnetic coupling between what would otherwise be regarded as isolated magnetic centres (some examples of this were given at the end of Section 9.1). Fortunately, at such low temperatures it is possible to measure the changes in thermal population of the magnetic-field split levels by another method, that of specific heat measurements. Such additional data are not without additional complications. In particular, the varying thermal population of the multitude of lattice vibrations has to be taken into account. Fortunately, at low temperatures this part of the specific heat of most solids has a T^3

dependence and may usually be allowed for with adequate accuracy. As this example and as the data available in the literature make clear, high temperature magnetic susceptibility data are of limited value—and some would say that 100 K is high—and that much more is revealed at low temperatures. An additional bonus is that at low temperatures it may be possible to use electron paramagnetic (spin) resonance spectroscopy (EPR, discussed in Section 12.6) to study transitions between levels such as those of Fig. 9.2 and so to have independent measures of them. A limited number of ions give room temperature EPR spectra which are well resolved but some which do not are found to give quite sharp lines at low temperatures.

It is here that our circle closes. Such measurements have been made on some Ti^{III} salts and have shown the presence of a major energy contribution which is not present in Fig. 9.2 and is also absent from the associated text and Appendix 10. This contribution arises from a phenomenon that was discussed in Section 8.5, the dynamic Jahn–Teller effect. Throughout our development of the t_{2g}^1 case, we assumed that the molecule under study was rigid. The dynamic Jahn–Teller effect introduces (some, but not all) metal–ligand vibrations into the picture. How important is all this? Well, it has been emphasized earlier that the magnetic properties of a complex are very sensitive to small distortions and therefore to vibrations. The lattice vibrations mentioned above and in Section 9.1 do not distort molecules—in them the molecules move as rigid bodies—and metal–ligand vibrations, also mentioned in Section 9.1, will normally be frozen out at very low temperatures. The dynamic Jahn–Teller effect is much less readily frozen out and so, indeed, it *is* a potential complication. When it is included in the calculation of the energy levels in the t_{2g}^1 case, patterns such as that in Fig. 9.5 are obtained. The difference from Fig. 9.2, although small, is significant. Just as a reasonably complete understanding of the electronic spectra of transition metal complexes requires the inclusion of vibronic (vibrational–electronic) effects so it seems likely that an understanding of magnetism will too. This aspect of the subject is still in its infancy, although it is clear that there are other aspects of the magnetism of coordination compounds in which vibronic effects are implicated (the dynamic Jahn–Teller effect is just one form of vibronic coupling).

The developments outlined in the above paragraph may well help to refute criticisms of that theory of magnetism which has been developed in the last few pages. When data from a large number of complexes were reviewed it was concluded that

1. the theory produces too many ligand field parameters which, in the event, often seem to have no obvious chemical relevance;

2. the models and parameters used for molecules with different geometries seem to bear little relationship with each other; they do not vary with geometry in a way that is obviously sensible;

3. molecules with very low symmetry could not be handled.

Since these criticisms were made the situation has improved, not only because the need to include the dynamic Jahn–Teller effect has been recognized, but also by theoretical developments. Rather than start with an octahedron and distort it, the present tendency is to, theoretically, build up

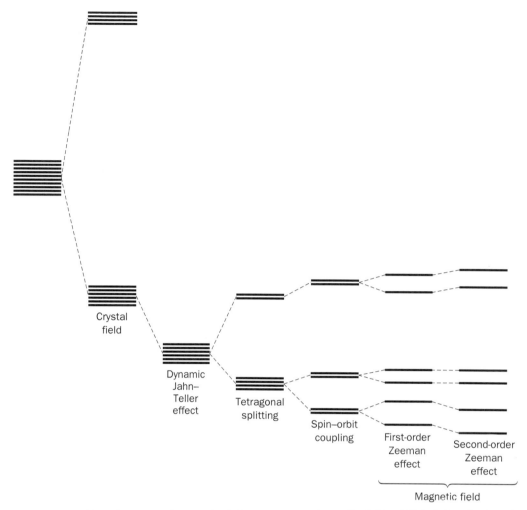

Fig. 9.5 A schematic energy level diagram corresponding to Fig. 9.2 but with inclusion of a dynamic Jahn–Teller effect.

a complex using geometry-sensitive but transferable parameters for each ligand. Surprisingly, an economy of parameters can result. This approach will not be developed here (although the Further Reading at the end of the chapter contains a reference to it) because it tends to be molecule-specific but the general aspects of it will be covered in the next chapter when the angular overlap model is outlined. The reference in the Further Reading at the end of Chapter 10 on modern developments of the angular overlap model provides an entry into the literature on the subject.

9.10 Spin-only equation

The previous section has shown that even if there is an orbital contribution to the magnetic moment it may be reduced by covalency and by distortions. What if we assume that there is no orbital contribution at all—that it is completely quenched? In this section this problem is considered and we will be led to the spin-only equation mentioned several times earlier in this chapter. The problem will be worked through in some detail to give the reader who has decided not to tackle Appendix 10 some idea of how the calculations are performed.

We are interested in the case in which there are n unpaired electrons, each with spin $\frac{1}{2}$, in an isolated ion which couple to give a resultant, denoted S, where $S = n/2$. The allowed components of S along the direction (z) of a magnetic field, when such a field is applied, are

$$S_z = S, \ (S - 1), \ (S - 2), \ \ldots, \ (S - (2S - 1)), \ (S - 2S)$$

That is, the allowed components run from S through to $-S$ and are $(2S + 1)$ in number. Their energies in a magnetic field will be proportional to the magnetic field strength and to the magnitude of the z component of S, the value of S_z. So, their energies will be of the form $-2S_z\beta H$, the negative sign indicating that the higher S_z values are stabilized and the 2 being the Landé g factor for the electron—the splittings in a magnetic field are twice as great as one would expect from the magnitude of the spin angular momentum. So, their energies are

$$-2S\beta H, \ -2(S - 1)\beta H, \ -2(S - 2)\beta H, \ \ldots, \ 2S\beta H$$

The average magnetization per molecule will be given by an expression of the form

$$\frac{\sum (\text{magnetic moment of a molecule}) \times (\text{number with this moment})}{(\text{total number of molecules})}$$

Assuming that in a macroscopic sample there is a Boltzmann distribution of ions over the various S_z levels, we have that the number of ions with moment $2S_z\beta H$ will be $\exp(2S_z\beta H/kT)$. Using this and eqn 9.6 together with eqn 9.1, we obtain

$$\mu_{\text{eff}}^2 = \frac{3kTN_A \sum\limits_{S_z=S}^{-S} 2S_z \times \beta \times \exp\left(\dfrac{2S_z\beta H}{kT}\right)}{N_A\beta^2 \sum\limits_{S_z=S}^{-S} \exp\left(\dfrac{2S_z\beta H}{kT}\right)}$$

Simplifying and expanding the exponentials using the equation $\exp(x) \approx 1 + x$,

$$\mu_{\text{eff}}^2 = \frac{6kT \sum\limits_{S_z=S}^{-S} S_z\left(1 + \dfrac{2S_z\beta H}{kT}\right)}{H\beta \sum\limits_{S_z=S}^{-S} \left(1 + \dfrac{2S_z\beta H}{kT}\right)}$$

This equation contains three standard summations:

$$\sum_{x=a}^{-a} x = 0; \quad \sum_{x=a}^{-a} x^3 = \frac{a}{3}(a + 1)(2a + 1); \quad \sum_{S_z=S}^{-S} 1 = (2S + 1)$$

so,

$$\mu_{\text{eff}}^2 = \frac{6kT}{H\beta} \times \frac{2\beta H}{kT} \times \frac{S}{3} \times (S + 1) = 4S(S + 1)$$

That is,

$$\mu_{eff} = 2\sqrt{S(S + 1)} = \sqrt{n(n + 2)}$$

As we have seen, this relationship holds quite well for the first row transition series. For the second and third row elements it gives totally misleading results because the spin–orbit coupling of these ions is much greater than kT—so that, usually, the plateau of the Kotani plot has not been reached.

9.11 Magnetically non-dilute compounds

When a simple Curie law plot of χ_M against T^{-1} (this pattern was discussed in Section 9.2) does not give a straight line it is often found that a modified form, the Curie–Weiss law, does:

$$\chi_M = \frac{C}{T + \theta}$$

Here, θ is a constant, the Weiss constant, which has the dimensions of temperature, and is measured in degrees. Again, despite its widespread applicability, it is difficult to give a quantum-mechanical interpretation of this relationship. In the limiting cases where this is possible, θ is associated with a breakdown of the Langevin assumption that $B = H$, and values of $\theta \neq 0$ are generally regarded as indicative of magnetic interaction between discrete molecules in condensed phases.

This brings us back to an assumption made at the beginning of this chapter, that we could restrict our discussion to magnetically dilute materials. Magnetically non-dilute materials fall into one of three main classes. These are ferromagnetic, antiferromagnetic and ferrimagnetic—terms that, strictly, apply to extended lattices. In addition we will discuss the phenomenon of ferromagnetic and antiferromagnetic coupling between individual ions, leading to an understanding of the consequences of local magnetic coupling.

In Fig. 9.6 are shown diagrams representing paramagnetism, ferromagnetism, antiferromagnetism and ferrimagnetism, both with and without a (very large!) magnetic field applied. In Fig. 9.7 is shown the temperature behaviour of the magnetism associated with each type. Of the four classes of magnetically non-dilute behaviour shown in Fig. 9.6, that of ferromagnetic behaviour is well known—everyday magnets are made of ferromagnetic materials. As with the other three classes, the individual magnetic centres must be considered together, not separately, the correct building block being the unit cell of the crystalline solid. In ferromagnetic materials, such as metallic iron, the electron spins of each of the atoms couple together to form a resultant unit cell magnetic moment. Iron has a cubic unit cell so let us suppose that the unit cell moment is perpendicular to one of the faces of the unit cell. There are three pairs of such faces and in a perfect crystal of non-magnetized iron there are domains—within each of which there is magnetic alignment between adjacent unit cells—but in the whole crystal there is a random distribution over all of the possible orientations. The process of magnetization brings these moments into alignment and a resultant permanent moment results. These features make ferromagnetic materials quite different from most of those which concern a chemist and

No magnetic field   In a magnetic field   Resultant

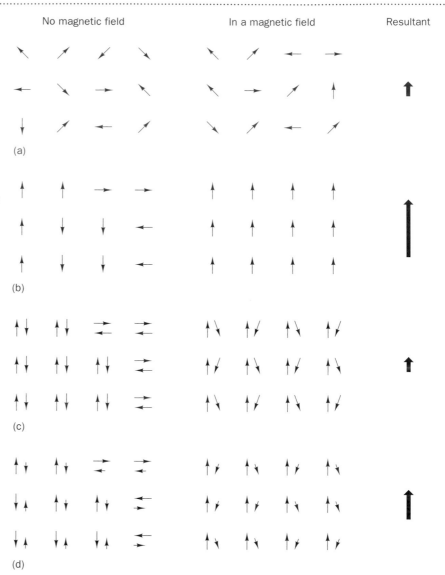

Fig. 9.6 Representations of the most important magnetic phenomena in solids. For simplicity of presentation, the diagrams are more classical than quantum mechanical. (a) Paramagnetism: magnetically dilute materials with zero coupling between adjacent spins. (b) Ferromagnetism: strong coupling leads to a parallel orientation of domains in a strong field. (c) Antiferromagnetism: two, opposed, arrangements which may lead to a domain structure (canted arrangements also occur). At low temperatures the right-hand member of the 'in a magnetic field' pairs settle down to oppose the left hand and the magnetization drops with decrease in temperature. (d) Ferrimagnetism: similar to antiferromagnetism except that the two sets of magnets are of different strength, leading to preservation of magnetism at low temperatures.

we shall not consider them further here, although we shall have something to say about molecular ferromagnetism.

The second—and most important case for the chemist because of its molecular counterpart, to which we shall turn shortly—is that of antiferromagnetism (although we will find that our discussion on this topic takes us to molecular ferromagnetism). In antiferromagnetic materials, transition metal ions are separated by (usually small) ligands, so that many transition metal oxides and halides show antiferromagnetic behaviour. In such compounds, adjacent metal ions couple with their spins antiparallel; there are always equal numbers with the two arrangements so that there is no resultant magnetization in the absence of a magnetic field. As an example of one complicating feature of antiferromagnetism consider the sets of four two-dimensional crystallographic unit cells, as determined by X-ray diffraction,

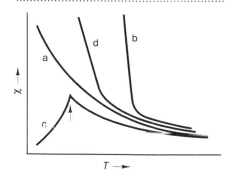

Fig. 9.7 Schematic temperature–susceptibility plots for various types of magnetic behaviour: (a) antiferromagnetism (the arrow indicates the maximum which has sometimes been called the Curie temperature, T_c, or the Néel temperature, T_N); (b) paramagnetism; (c) ferrimagnetism; (d) ferromagnetism. Note that this diagram does not extend to very low temperatures (see Question 9.5, for instance) or high magnetic fields (when saturation effects occur).

shown in Fig. 9.8. X-rays are blind to magnetism[5] and so the magnetic centres are represented by circles in the first diagram (ligands are omitted). We now admit the existence of antiferromagnetic coupling between some pairs of magnetic centres. In the other three sets of cells in this figure are given three possible antiferromagnetic arrangements of spin orientations within the block of four cells. It can be seen that these are all different; the bottom pair of arrangements contain only two *magnetic* unit cells. Compared with the 32 crystallographic point groups and 230 space groups of classical crystallography there are 90 crystallographic magnetic point groups and 1651 magnetic space groups. By using polarized neutrons (neutrons have an intrinsic spin and so behave like tiny bar magnets; in a beam of polarized neutrons these magnets are all essentially parallel), neutron diffraction data are capable of allocating an antiferromagnetic material to its correct magnetic space group. If this is not known, it is not possible to give a complete theoretical discussion of the magnetic properties of an antiferromagnetic material.

Antiferromagnetic materials show a maximum in a plot of χ_M against temperature, at the so-called Néel point—this is shown in Fig. 9.7. Below the Néel point the susceptibility is to some extent field-dependent. These properties provide an indication of the existence, or absence, of antiferromagnetism. Other tests include a comparison of the results of solution and solid-state measurements, where this is possible, and, where it is not, by the technique of the dilution of the magnetic ions in the lattice by their partial substitution with an isomorphous non-magnetic ion. At temperatures sufficiently above the Néel point, antiferromagnetic materials follow a Curie–Weiss law. Even if the Néel point is at too low a temperature to be measured conveniently, the observation that a material follows the Curie–Weiss law usually implies a residual antiferromagnetic coupling.

The third class is that of ferrimagnetic materials. Just as for materials

Fig. 9.8 Crystallographic and magnetic unit cells. X-ray diffraction would give for all of these systems the unit cell pattern of the top left-hand diagram. Polarized neutron diffraction would show doubled, magnetic, unit cells for two of them.

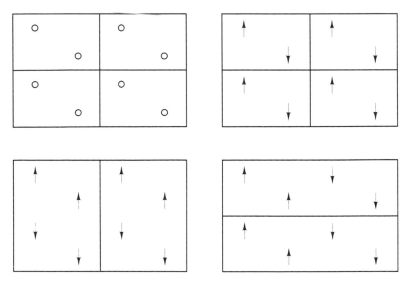

[5] This is a marginal overstatement. If the very intense X-ray beam from a synchrotron source is used, weak X-ray magnetic scattering—which is relativistic in origin—can be observed. The effect is rendered more observable by tuning the X-ray energy to coincide with an energy difference in the material under study, when a resonance enhancement occurs.

showing antiferromagnetic behaviour, ferrimagnetic materials have ions on two sets of lattice sites. These have opposed spin arrangements but as they do not cancel each other out there is a resultant permanent moment. The best known example is Fe_3O_4, the black mineral which used to be called magnetic oxide of iron, but which is a complicated case because one site contains (formally) Fe^{II} and Fe^{III} in equal amounts, whilst the other contains only Fe^{III}.

A topic which has gained considerable attention is that of molecular antiferromagnetism (because it potentially leads to the possibility of information storage at the molecular level). This occurs when two transition metal ions both have unpaired electrons and are bonded together, most usually by bridging ligands. In such a case it is common to find that the magnetic properties of the complex are not simply the sum of those of the two individual metal ions. There is a magnetic interaction between them. This situation has already been anticipated at two points in the text. At the end of Section 8.11 it was commented that such magnetic interactions can lead to spectral changes; at the end of Chapter 6 it was recognized that polarized neutron diffraction measurements indicate that the distribution of electrons in a ligand is affected by the unpaired electrons on the metal atom to which it is bonded. The following discussion could be based on the details of Fig. 6.31 but it is simpler to consider a case in which there is bonding between the metal orbital containing the unpaired electron and a ligand pure p orbital. This is shown in Fig. 9.9(a). Because we are interested in a ligand that bridges two transition metal atoms this is the situation pictured in Fig. 9.9(a). Start at the left-hand side of Fig. 9.9(a). The metal ion electron has spin up. Although the ligand p orbital formally contains two electrons, we have seen in Section 6.2.1 that it can nonetheless be involved in bonding with the metal orbital. Such bonding means a pairing between the metal unpaired electron and that of opposite spin in the p orbital. This leaves an election in the p orbital which is of the same spin as that in the left-hand metal d orbital. Pairing between this electron and the metal d electron in the orbital at the right-hand side of the diagram requires that this latter d electron has its spin in the *opposite* direction to that of the d electron in the orbital at the left-hand side of the diagram. That is, the two metal ions are *antiferromagnetically linked* by the ligand. In this example the two metal ions were implicitly taken to be identical. What if they are different? Suppose that one—the left-hand one—has its unpaired electron in an e_g orbital (we assume octahedral symmetry) and the other, the right-hand, has its unpaired electron in a t_{2g}. Starting with the left-hand d orbital, the argument follows that given above until we reach the right-hand d orbital, an orbital which is orthogonal to (that is, has a zero overlap integral with) the ligand p orbital. Unpaired electrons in orthogonal orbitals tend to align themselves with parallel spins (examples of this can be seen by looking back at Fig. 7.10)—this arrangement is exchange-stabilized. In Fig. 9.9(b), therefore, the electron in the right-hand d orbital is of the same spin as that in the left-hand. The two metal ions are *ferromagnetically* linked.

Key to the development of ferromagnetic coupling in the argument above was a step involving the orthogonality of two orbitals. The orthogonality need not be between metal and ligand orbitals. In Fig. 9.9(c) is shown an interaction pathway which includes two p orbitals on the same atom; these

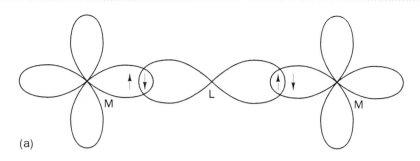

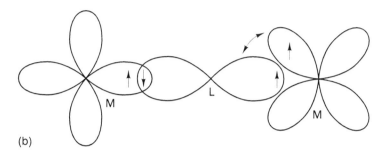

(a)

(b)

Fig. 9.9 (a) Ligand σ-orbital-mediated antiferromagnetic coupling; (b) and (c) show two ways in which orbital orthogonality can lead to ferromagnetic coupling.

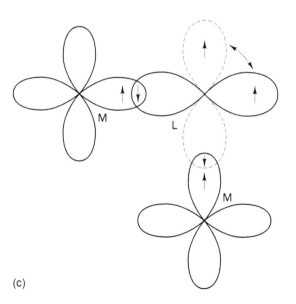

(c)

p orbitals, of course, are mutually orthogonal. Again, ferromagnetic coupling results; in Figs. 9.9(b) and 9.9(c) the point at which exchange stabilization leads to parallel spins is indicated by double-headed arrows. It would be sensible for the reader to stop reading at this point and to trace through the pathway of Fig. 9.9(c) and check that a ferromagnetic interaction, indeed, results.

Clearly, by careful choice of metal ions and bridging ligands (which have to be sufficiently rigid to provide rather fixed geometrical relationships between the interacting orbitals) an orthogonal step can be introduced so

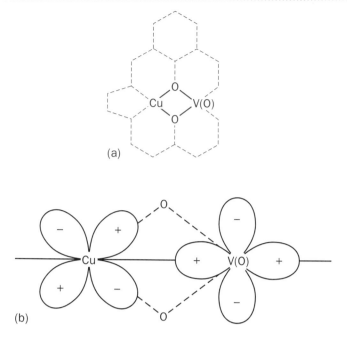

Fig. 9.10 (a) A Cu^{II}–VO complex showing ferromagnetic interaction between the copper and vanadyl ions. The ligand skeleton which holds the two ions together is shown dotted. (b) On the d^9 ion Cu^{2+} an unpaired electron is in an e$_g$ type of orbital; on the d^1 ion VO^{2+} an unpaired electron is in a t$_{2g}$ type of orbital. These orbitals are, respectively, antisymmetric and symmetric with respect to the vertical mirror plane shown as a solid line, and are therefore orthogonal.

that ferromagnetic coupling occurs between two metal atoms, leading them to have parallel spins. An example of such a ligand is shown in Fig. 9.10. Although with cunning choice of ligand and metal ion it is possible to obtain ferromagnetic complexes such as that shown in Fig. 9.10, in most cases the coupling between two linked metal atoms, each with unpaired electrons, will contain at least one antiferromagnetic coupling pathway—a pathway involving non-orthogonal orbitals—and, perhaps, a ferromagnetic coupling through another orbital sequence. In such cases, the antiferromagnetic coupling almost invariably dominates. Lastly, an important word, none the less important for having been left until the end. The discussion of the last few paragraphs hinged on the effects of exchange between electrons on different atoms. Irrespective of whether the outcome is ferromagnetic or antiferromagnetic coupling, the general mechanism involving ligand orbitals to mediate coupling between metal electrons is referred to as *superexchange*.

Finally, although strictly out of place in this chapter, we mention high-temperature superconductors. They are introduced at this point because, as in superexchange, superconductivity involves the properties of electrons on different atoms being correlated with each other. At the time of writing this book the detailed mechanism by which high-temperature superconductivity occurs is not known. The interested reader will find the problem explored in Appendix 11; they should be warned, however, that this appendix is likely to date rather rapidly!

9.12 Spin equilibria

In the final section of this chapter we briefly review some phenomena which, although involving no new concepts, nonetheless lead to some unexpected magnetic results. In Chapter 3 we met two forms of isomerism which have magnetic implications. Because the splitting of d orbitals is a function of coordination geometry, a change in geometry may well mean a change in the number of unpaired electrons. So, the existence of geometrical isomerism (discussed in Section 3.4.2) may well be magnetically evident (it will be spectrally apparent also). Many examples are provided by Ni^{II} complexes; this ion forms many square planar complexes—which are diamagnetic—which may isomerize to give tetrahedral species. The latter, being d^8, have two unpaired electrons; the reverse isomerization may also occur, of course. This pattern of behaviour is particularly common for Ni^{II} complexes of general formula $[NiP_2X_2]$, where X is a halogen and P is an organophosphine ligand. Typically, the square planar form is red or brown and the tetrahedral green, the Ni^{II} in the latter having a magnetic moment of ca. 3.5 (see Table 9.1).

The second form of isomerism which has magnetic implications is spin isomerism (see Section 3.4.12). This occurs because with the correct choice of ligand (and much effort has been spent on ligand design and variation) the value for Δ, for an octahedral complex can be made to be close to that appropriate to the change-over point from high to low spin complexes in d^4, d^5, d^6 and d^7 systems (see the Tanabe–Sugano diagrams in Appendix 7 for an idea of the magnitude of ligand field required). Most known examples come from the 3d series—Mn^{II}, Mn^{III}, Co^{II}, Co^{III} and, particularly, Fe^{II} and Fe^{III}. Because the two spin states in equilibrium involve different t_{2g}–e_g populations, the metal–ligand bond lengths would be expected to vary between the two spin isomers. X-ray measurements show that the M–L bond lengths tend to be roughly 0.2 Å longer in the high spin form because the e_g orbitals are weakly antibonding and have a higher occupancy in the high spin isomer. Not surprisingly, therefore, the spin equilibria are pressure sensitive, as well as sensitive to the choice of counterion, any solvent of crystallization as well as steric effects originating in the ligands. Many studies have been made of the lifetimes of each of the species in equilibrium. These lifetimes are about 10^{-7} s, but vary by at least one order of magnitude in either direction. Often, the magnetic susceptibility changes on warming a sample are not opposite of those on cooling, this hysteresis indicating some cooperative, perhaps domain, behaviour. Although, magnetically, both spin states are unexceptional, the study of the equilibria between them has been energetically pursued by many workers and has involved the use of almost all of the techniques described in this book (especially those described in Chapter 12), and more.

Further reading

The literature on this subject tends to divide between the old, which provides the best available treatment of the classical (including the classical quantum mechanical), and the recent, which provides an up-dating of the earlier. Older are:

- *Magnetism and Transition Metal Complexes* F. E. Mabbs and D. J. Machin, Chapman & Hall, London, 1973.

- 'The Magnetochemistry of Complex Compounds' B. N. Figgis and J. Lewis, Chapter 6 in *Modern Coordination Chemistry* J. Lewis and R. G. Wilkins (eds.), Interscience, New York, 1960.

- 'The Magnetic Properties of Transition Metal Complexes' B. N. Figgis and J. Lewis, *Prog. Inorg. Chem.* (1964) *6*, 37

- A useful review which also includes an extensive compilation of diamagnetic corrections is 'Magnetochemistry—Advances in Theory and Experimentation' C. J. O'Connor, *Prog. Inorg. Chem.* (1982) *26*, 203

- A still-current view of theory, experiment and problems is

• 'A Local View of Magnetochemistry' M. Gerloch, *Inorg. Prog. Chem.* (1979) *26*, 1.

Problems in the effects of magnetic coupling are reviewed in 'Magnetochemistry: a research proposal' R. L. Carlin, *Coord. Chem. Rev.* (1987) *79*, 215.

An excellent current book is *Molecular Magnetism* O. Kahn, VCH, Weinheim, 1993; a rather different account is 'Organic and Organometallic Molecular Magnetic Materials—Designer Magnets', a very long review by J. S. Miller and A. J. Epstein in *Angew. Chem., Int. Ed.* (1994) *33*, 385.

A detailed account of magnetic phases and phase transitions at low temperatures is 'Magnetic phase transitions at low temperatures' J. E. Rives, *Transition Met. Chem.* (1972) 7, 1. A more general account is 'Magnetic Symmetry' by W. Opechowski and R. Guccione, Chapter 3 in *Magnetism*, Volume IIA, G. T Rado and H. Suhl (eds.), Academic Press, New York, 1965.

More recent, and concentrating on systems of two transition metal ions, is 'Magnetism of the Heteropolymetallic Systems' O. Kahn, *Struct. Bonding* (1987) *68*, 89.

Spin equilibria are covered in

• 'Dynamics of Spin Equilibria in Metal Complexes' J. K. Beattie, *Adv. Inorg. Chem.* (1988) *32*, 1.
• 'Static and Dynamic Effects in Spin Equilibrium Systems' *Coord. Chem. Rev.* (1988) *86*, 245.

Questions

9.1. Show that the metal d orbital and the ligand σ orbitals that contain unpaired spin density in Fig. 6.31 are orthogonal to each other. (Hint: consider their symmetries.)

9.2. Show that the interactions shown in Fig. 9.9(c) lead to a ferromagnetic (parallel spin) coupling of the two metal ions.

9.3. If one step involving orthogonal interactions leads to a ferromagnetic interaction, then two such steps should lead to an antiferromagnetic. Check this conclusion by replacing one of the d orbitals shown in Fig. 9.9(c) by a t_{2g} d orbital and tracing through the coupling pathway.

9.4. Figure 9.8 does not exhaust the possible two-dimensional antiferromagnetic arrangements of two magnetic ions; three more are shown in Fig. 9.11. However, unit cells similar to those of Fig. 9.8 are not given. Insert them. To complete this task for the third diagram will require the recognition of a feature not discussed in the text.

9.5. In Fig. 9.9 and the associated discussion within the text, the ligand orbital involved in superexchange was a σ orbital. Ligand π orbitals may also be involved. Figure 9.12 is the ligand π orbital equivalent of Fig. 9.9(a). Give an account of Fig. 9.12 which parallels that given in the text for Fig. 9.9(a).

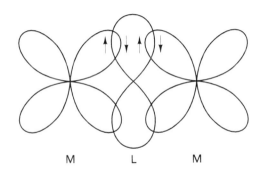

Fig. 9.12 Question 9.5.

9.6. There is a, perhaps, unexpected result contained in Fig. 9.4. At 0 K a system with a single unpaired electron is predicted to be non-magnetic. Give a qualitative explanation for this behaviour (remember—there are two sources of magnetism, orbital and spin).

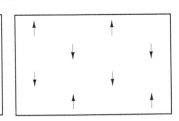

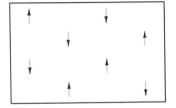

Fig. 9.11 Question 9.4.

10

Beyond ligand field theory

In this chapter an overview will be presented of many of the theoretical methods currently being used to obtain insights into the electronic structure of coordination compounds. The discussion is rather general but it is helpful to focus on one area. Some attention is therefore given to transition metal organometallic and related complexes, this class being selected because nowhere else in this book are they explicitly treated. The first part of the chapter therefore provides a background by outlining the simplest, mostly symmetry-based, treatment of their structure.

10.1 Bonding in transition metal organometallic complexes

The present-day qualitative approach to the bonding in transition metal organometallic complexes is based on a combination of two models. The first, the 18-electron rule, has already been met in Chapter 2 (see Table 2.8)—'ligands that bond to a transition metal in a low valence state normally do so in such a way that the metal atom is, formally, surrounded by 18-electrons'. Of course with a few exceptions, the classical complexes of the Werner type and which are the subject of most of this book do not obey this rule, so that it is clear that much hinges on the low valence state requirement. Equally, the fact that most transition elements form neutral bis-η^5-cyclopentadiene complexes, $M(C_5H_5)_2$, shows that even for organometallic complexes of transition metals it is only an approximation. Many examples of molecules with more, and many examples with fewer, than 18 valence-shell electrons are known. Those with more tend to be readily oxidized, those with fewer tend either to be sterically hindered or readily add further ligands.

The second model, which is widely applied to organometallic complexes of transition metal ions, is the requirement that the bonding interactions are symmetry-determined. A very common and important case is provided by the bonding of cyclic conjugated C_nR_n systems—of which C_5H_5 is perhaps the best known example. These are discussed in more detail in Appendix 13

Table 10.1 The C–C bond distance in C_2H_4 and its complexes. It is salutary to note that Zeise's salt, for which the Chatt–Duncanson bonding model was first suggested, shows the smallest effect. The last complex is one in which C_2H_4 acts as a bridge between two Zr and so is bonded to both

Molecule	C–C distance (Å)
C_2H_4	1.34
$[Pt(C_2H_4)Cl_3]^-$	1.35
$V(C_2H_4)(PMe_3)_2(C_5H_5)$	1.36
$Fe(C_2H_4)(PEt_3)_2$	1.38
$Ni(C_2H_4)(PPh_3)_2$	1.43
$Pt(C_2H_4)(PPh_3)_2$	1.43
$Ru(C_2H_4)(PMe_3)_4$	1.44
$Fe(C_2H_4)(CO)_4$	1.46
$Zr_2(C_2H_4)Cl_6(PEt_3)_4$	1.69

for the cases in which $n = 4$, 5 or 6, where two apparently rather different approaches to the problem are described. The approaches are actually extensions of the Chatt–Duncanson model for the bonding in Zeise's salt, described in Chapter 2—see Fig. 2.2 and the associated text. In general, the Chatt–Duncanson model is supported by the experimental data—see, for instance, Table 10.1. It therefore seems reasonable to extend it to other, related organic ligands. In this extension of the model, the lower nodality, occupied, π orbitals of the organic ligand donate electron density to the metal atom whilst the higher nodality, unoccupied, π orbitals accept electron density from the metal orbitals (which, because of the higher-nodality requirement, will be d orbitals). It would be expected that of particular importance in this bonding will be the ligand HOMO as electron donor and the ligand LUMO as electron acceptor. Because of the high symmetry of the isolated, planar, ligand, many of the π orbitals occur as degenerate pairs. When two such ligands are simultaneously coordinated then the normal pattern is that the ligand π orbitals again occur in pairs, the sum and difference of the corresponding π orbitals of the individual ligands (the sum and difference would normally have different symmetry labels). Overall, then, we have pairs of pairs. If one of the sum and difference pairs were more extensively involved in the bonding than the other, the molecular orbitals would be separated in the energy level diagram—although it is usually not difficult to recognize such pairings. Figure 10.1 shows such a qualitative energy level diagram for ferrocene, $Fe(C_5H_5)_2$. Care will have to be taken using this diagram because, to avoid unduly complicating it, not all of the energy-level tie lines have been included.[1] The pairs of pairs just discussed appear on the left-hand side of this figure. In Section 10.7 this diagram will be compared with the results of some detailed calculations. Figure 10.2 shows in more detail, the general bonding pattern for metal–C_n(ring) complexes, $n = 4$, 5 and 6, in which the matching of ligand and metal orbital nodal patterns is emphasized (this is a topic discussed in some detail in Appendix 13 and is summarized in Fig. 10.2). The fact that a ligand *can* bond symmetrically does not mean that it *does* bond symmetrically: Fig. 10.3 gives some of the ways that benzene, C_6H_6, has been found to bond to transition metals—it can be seen that it can function as a two-electron, a four-electron or as a six-electron donor; in such cases the electron count would normally be that which satisfies the 18-electron rule. In addition, many complexes are known in which the benzene molecule bridges two or more transition metal atoms; in these the benzene is seldom symmetrically bridged. Such asymmetric attachments of hydrocarbon ligands is even more probable for chain-conjugated hydrocarbons and related species, two examples of which are given in Fig. 10.4. However, as this figure shows, there is a persistence of a pattern of π orbitals which can be ordered according to the number of nodal planes, even when, as in the examples shown, the introduction of a hetero-atom (here, an O in place of a CH_2 group) produces considerable changes in the detailed forms of the ligand π orbitals.

---

[1] In the gas phase ferrocene has an eclipsed (D_{5h}) geometry; in the crystal it is staggered (D_{5d}). The energy difference between the two is small (ca. 4 kJ mol^{-1}, 0.8 kcal mol^{-1}). Although the symmetry labels change between the two, the various interactions involved in the bonding are little different. In this chapter the staggered configuration will be the one discussed because most of the theoretical and experimental work invoked assume this geometry.

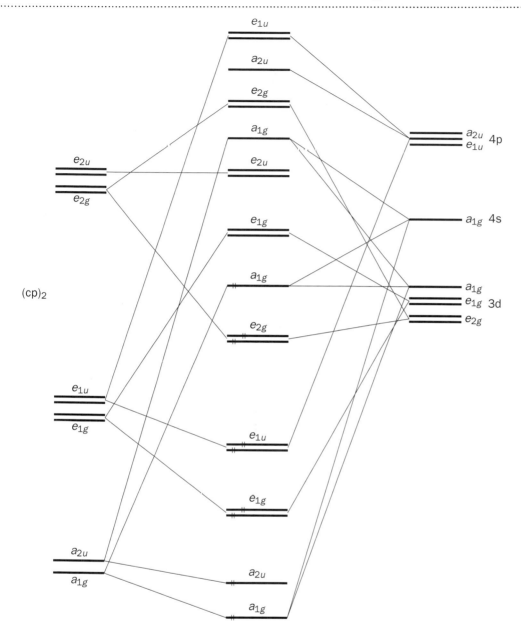

Fig. 10.1 Schematic molecular orbital energy level diagram for ferrocene. A more accurate diagram is given in Fig. 10.9. (Adapted and reproduced with permission from Shriver, Atkins and Langford, *Inorganic Chemistry*).

An example in which the 18-electron rule and the symmetry-determination of interaction patterns come together rather strongly is in the molecule hexacarbonylchromium, $Cr(CO)_6$. This is an octahedral complex, to which, in principle at least, all of the relevant arguments of Chapters 6 and 7 may be applied. That the 18-electron rule is satisfied is easily seen; a Cr^0 species has six valence-shell electrons—which will fill the t_{2g} set of d orbitals—and each CO ligand contributes two σ electrons from each carbon (formally, these electrons are σ-donated to the metal), giving a total of 18. It is perhaps not surprising that $Cr(CO)_6$ is colourless; CO, like CN^-, is a strong field ligand (we met this in Section 6.2.2) and so the $t_{2g} \rightarrow e_g$ transitions fall in

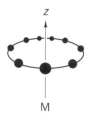

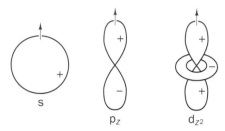

Metal orbitals with 0 nodes

Fig. 10.2 A classification of orbitals according to the number of their nodal planes which contain the z axis for metal and planar cyclic C_nR_n systems. The z axis is the axis of highest rotational symmetry (top).

(a) Different metal orbital sets differ in the nodality patterns they span: σ orbitals only have zero, π orbitals zero and one, δ orbitals, zero, one and two. *(continued)*

Metal orbitals with 1 node

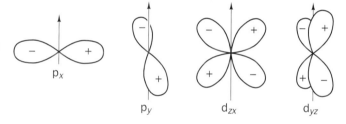

Metal orbitals with 2 nodes

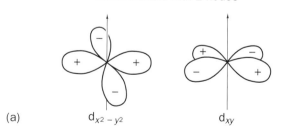

(a)

the near-ultraviolet region of the spectrum. This, then, is a molecule for which Figs. 6.11 and 6.16(c) must be brought together, and this is done in Fig. 10.5. It would be worthwhile for the reader to stop at this point and examine in some detail the connection between these three figures. One hidden problem relates to the simple picture of CO π bonding given in Section 2.2 and shown in Fig. 2.1. The simple picture has, effectively, two electron pairs back-donated from metal d orbitals into the π antibonding orbitals of the CO. The group theory associated with Fig. 6.16, in which it was shown that the 12 ligand π orbitals transform as $t_{1g} + t_{1u} + t_{2g} + t_{2u}$, is to be compared with the symmetries of the available metal orbitals; of these four sets, only two, $t_{1u} + t_{2g}$, find matching partners in the metal set. So, the simple picture

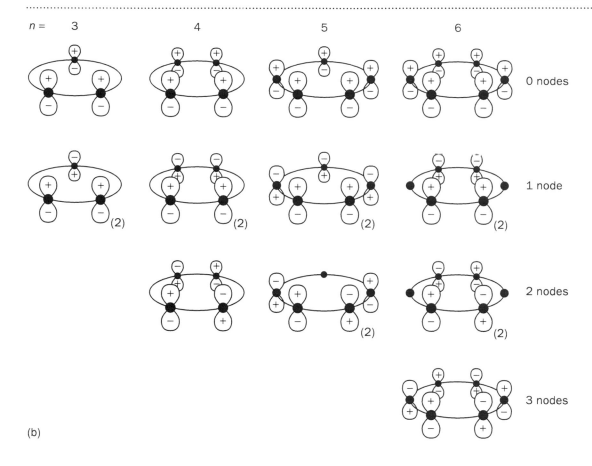

(b)

Fig. 10.2 (continued) (b) Different planar cyclic C_nR_n ring systems differ in their p_π orbital nodality patterns: $n = 2$, zero and one; $n = 3$, also zero and one. However, there are two different orbitals with one nodal plane containing the z axis (indicated by the (2) in the figure). The nodal plane shown is that parallel to the plane of the paper; the other orbital has its nodal plane perpendicular to the plane of the paper. The case $n = 4$ has zero, one (two of) and two nodal planes; the case $n = 5$ has zero, one (two of) and two (two of). The case $n = 6$ has zero, one (two of), two (two of) and three. The case $n = 7$ (not shown) has zero, one (two of), two (two of), three (two of) etc.

overestimates the extent of π back-bonding. In Section 10.7 the qualitative picture of bonding in $Cr(CO)_6$ will be compared with the results of detailed calculations.

10.2 Metal–fullerene complexes

In recent years some considerable excitement has been generated by the discovery of a series of carbon cage compounds, generated in low yields when an electric arc is struck between carbon rods in a low pressure of a gas such as argon. Although by no means the only species formed, there has been particular study of C_{60} and C_{70}, the major products of the preparation. Of these, C_{60} has an icosahedral structure and, because of the general resemblance to the geodesic domes designed by Buckminster Fuller, has come to be known as buckminsterfullerene; it is shown in Fig. 10.6. The whole family has come to be called fullerenes. In contrast to C_{70}, which has a D_{5h} structure and five structurally different types of carbon atom, all of the carbon atoms of C_{60} are structurally equivalent. Our discussion will therefore concern this molecule alone. The molecule is just over 7 Å in diameter, the central hole has a diameter of half of this. It is therefore large enough to accommodate an atom and such metal complexes have been prepared. They have been called endohedral complexes (as opposed to

Modes of six-electron donation

Modes of four-electron donation

Mode of two-electron donation

Fig. 10.3 The benzene molecule can function as a six-electron, a four-electron and a two-electron donor. It does not necessarily remain planar and may be involved in donation to more than one metal. The model of donation is usually clear from a crystal structure determination and, in all probability, also by application of the 18-electron rule. Examples of relevant molecules are, left to right:

six-electron donation—$[Mo(\eta^6\text{-}C_6H_6)(CO)_3]$, $[V_2(\eta^3,\eta^3\text{-}C_6H_6)(\eta^5\text{-}C_5H_5)_2H_2]$,
$\quad$ $[Os_3(\eta^2,\eta^2,\eta^2\text{-}C_6H_6)(CO)_9]$;
four-electron donation—$[Os(\eta^4\text{-}C_6H_6)\eta^6C_6H_6)]$, $[(Re(\eta^5\text{-}C_5Me_5)(CO)_2)_2(\eta^2,\eta^2\text{-}C_6H_6)]$,
$\quad$ $[Pd_2(\eta^2,\eta^2\text{-}C_6H_6)_2(AlCl_4)_2]$;
two-electron donation—$[(\eta^2\text{-}C_6H_6)Re(CO)_2(\eta^5\text{-}C_5Me_5)]$.

exohedral complexes in which a metal is attached externally to the carbon cage. The bonding in these latter is similar to that discussed above and in Chapter 2).

A symbol which has been suggested to denote endohedral complexes is @; so, La@C_{60}. The method of preparation is scarcely sophisticated; the carbon rods used in the preparation of C_{60} are impregnated with a salt of the metal and a small amount of the metal complex is formed in parallel with the preparation of C_{60}. The method is reminiscent of the use of similar impregnated carbon rods to produce luminous arcs in the early days of cinema projection—it may be that this was when the complexes were first produced! Endohedral complexes have been prepared for carbon cages C_{28} (U@C_{28}) and upwards; for the larger cages it has proved possible to incorporate up to three metal atoms endohedrally, as in Sc_3@C_{82}. However, it is found that such complexes are more difficult to prepare the more d electrons the endohedrally incorporated element possesses; most of the compounds that have been prepared are of the lanthanides, although barium and strontium compounds have also been reported. It is not difficult to see why this should be. The metal atom in such a complex must be stabilized by interactions between its orbitals and the orbitals of the carbon atoms which point towards the centre of the icosahedron. The latter will be of a σ type (viewed from the metal at the centre of the icosahedron) and, for C_{60}, 60 in

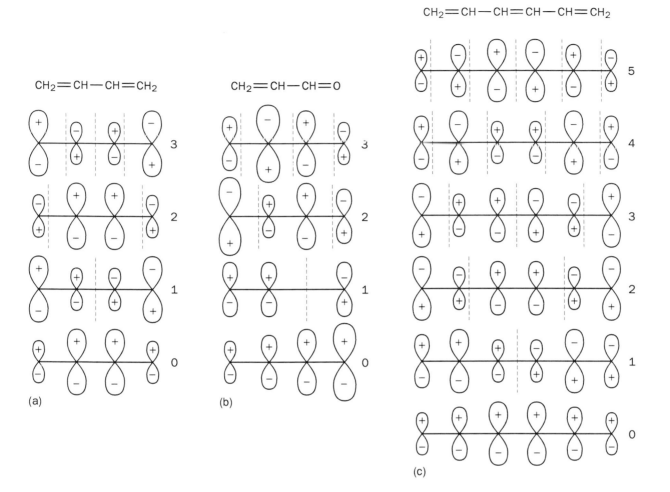

Fig. 10.4 p_π molecular orbitals of some linear conjugated molecules in the Hückel approximation (this is a crude, but adequate, model). Nodal planes are shown dotted and the total indicated. The more nodes, the higher the energy. (a) Butadiene (in the Hückel approximation the *cis* and *trans* species are the same). (b) Acrolein—this molecule may be regarded as butadiene in which a terminal CH$_2$ group has been replaced by O. The detailed consequences for the p_π molecular orbitals are considerable but the general pattern is unchanged. (c) Hexatriene—comparison of this pattern with (a) indicates the nature of the general pattern.

number. However, they are different when viewed from the perspective of the rings of carbon atoms that make up the C_{60} cage, each ring being approximately planar. From this point of view, it is a reasonable approximation to regard them as the counterparts of the π orbitals of the organic ring molecules discussed earlier in this chapter. They can be treated similarly and the MOs to which this gives rise ordered in energy sequence according to the number of inherent nodal planes that they possess. This has been done in Fig. 10.7.

Because of the high symmetry of the icosahedron, higher even than that of the octahedron, orbital degeneracies of four and five occur in icosahedral molecules and this will be seen in Fig. 10.7, along with their symmetry labels in the icosahedral group (although these labels add nothing to a discussion except at a rather detailed level). Paralleling the organic rings discussed earlier and in Appendix 13, the nodal patterns of the C_{60} cage orbitals have to match those of the orbitals of the endohedral atom. Fortunately, again because of the near-spherical symmetry of the icosahedron, this is rather simple (and can be shown to be so by looking at the detailed significance of the symmetry labels in Fig. 10.7). The C_{60} no-node combination interacts

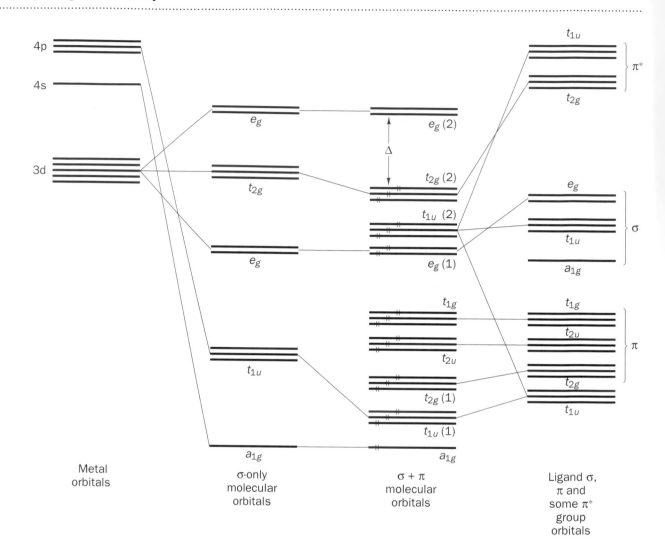

Fig. 10.5 A qualitative $\sigma + \pi$ bonding molecular orbital energy level diagram for $Cr(CO)_6$ based on Figs. 6.11 and 6.16(c) together with the associated discussion. This schematic diagram may be compared with Fig. 10.10 which is a more accurate molecular energy level diagram for $Cr(CO)_6$.

exclusively with the s orbital of the endohedral atom, the three C_{60} one-node combinations interact with the endohedral atom's p orbitals, the five C_{60} two-node combinations interact solely with the endohedral atom's d orbitals, the seven C_{60} three-node combinations (which split into two little-separated sets) interact with the endohedral atom's f orbitals, and so on. Now, just as in the neutral hydrocarbon rings discussed earlier, the total number of electrons to be accommodated in the C_{60} orbitals is equal to the number of atoms, 60. So, the bottom 30 orbitals are full, and this is indicated in Fig. 10.7. This means that the no-node, the one-node, the two-node and the three-node orbitals (and 14 others, too) are all full. When they interact with an endohedral metal atom the only interactions that can contribute a stabilization are those with *empty* endohedral atom orbitals. When an endohedral atom has occupied d orbitals in its valence shell then those d electrons are forced to occupy *antibonding* orbitals if interactions involving the d orbitals contribute significantly to the molecular stability. Similar arguments apply to the s and p orbitals, of course.

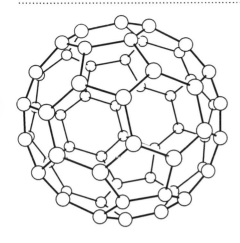

Fig. 10.6 The cage molecule C_{60}, buckminsterfullerene.

The experimental observation that the only atoms which form stable endohedral complexes are those with empty orbitals is at once explained. The content of Chapter 11 suggests that the f electrons of the lanthanides are expected to be too well tucked away inside the atom to be much involved in bonding and so, for this reason, the above discussion does not apply very strongly to the C_{60} three-node orbitals; Fig. 10.8 illustrates this argument. Although this seems reasonable, some recent rather detailed calculations on $Ce@C_{28}$ suggest that at least some f orbital covalency may exist, although not enough to alter the above general conclusions. Hints that f orbital covalency is not unreasonable will be found in Chapter 11, at the end of Section 11.2. Finally, it should be noted that there is a similarity between the fullerenes and the metal clusters which form the subject of Chapter 15. The methods which have been used to describe the bonding in these clusters are also applicable to the fullerenes although in the case of the spherical tensor method (Section 15.5—the method is to some extent anticipated by the above discusssion) the complication of d orbital involvement on the cage atoms is absent.

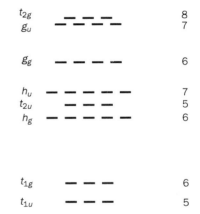

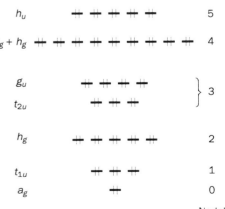

Fig. 10.7 The molecular orbital energy level pattern for those orbitals of C_{60} involved in bonding an endohedral atom.

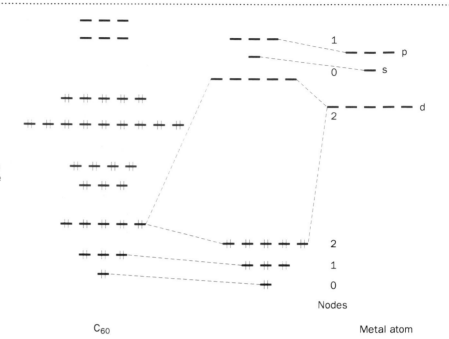

Fig. 10.8 Schematic diagram of the interaction between the orbitals shown in Fig. 10.7 and the atomic orbitals of an endohedral atom.

10.3 *Ab initio* and Xα methods

At the present time it is a routine matter to carry out *ab initio* calculations on small organic molecules. Such calculations attempt to solve the Schrödinger equation for the molecule to a high level of approximation. Essentially, one needs rather accurate mathematical descriptions of the orbitals involved— these are derived from work on atoms and approximated by a series of functions. This is followed by a proper mathematical treatment which transforms the Schrödinger equation into a form both adapted to the mathematical description of the orbitals used and to solution by computer. The commercial availability of programs which run on relatively small computers, coupled with molecular graphics programs to display the results, is changing organic chemistry. Such calculations can be expected to yield bond lengths and bond angles in good accord with experiment—bond lengths accurate to ca. 0.01 Å—but usually slightly short in the best calculations—and angles accurate to within 2–3°. What is not yet available is anything comparable for molecules containing transition metal atoms. The current situation is made evident when it is recognized that it is difficult theoretically to reproduce experimental data for the isolated nickel atom or the diatomic Cr_2 molecule—although this may be too pessimistic, for these two examples have particular problems, such as a very low-lying excited state which, because of the approximations made, could be calculated to be the ground state.

The problem for transition metals seems largely to arise from the difficulty of fully allowing for electron repulsion; the particular aspect of electron repulsion which is relevant is called *electron correlation* and it has been briefly mentioned several times in earlier chapters. The simplest method of appreciating the problem is to place the nuclei at the experimental positions in a molecule. All electrons except one are placed in the lowest orbitals, although one of these will have a hole because of the selected electron, which is ignored for the moment. The repulsive field generated by all of the included electrons is averaged and taken as that experienced by the selected electron, which modifies its orbital accordingly. The selected electron is then placed in this modified orbital. This procedure is repeated for all of the electrons, each being selected in turn, over and over again, until the input and output arrangements are essentially the same and a *self-consistent field* is obtained. A weakness of this model is that it effectively smears the density of the selected electron over the whole molecule; in reality, the other electrons will tend to stay away from where the selected electron *is*, not from a smeared-out distribution—the electron positions are *correlated* with one another. In calculations this problem can be dealt with, in some measure, by adding to the ground state wavefunction additional functions which change it so as to compensate for the error. The functions most evidently available are those involving what would formally be regarded as excited states and so different allocations of electrons to orbitals—that is, different electron configurations. In this way the interactions between these configurations is, has to be, incorporated. So, the process of improving the ground state wavefunction involves so-called configuration interaction. Other things being equal, the lower-lying the excited state, the more extensively it is likely to be involved in the ground state correction.

For transition metals there is a large number of low-lying excited states to be considered; many of these—and the ground state also—may not be spin singlets, a situation of particular difficulty (the theory is different). Further, the d electrons are confined to a small volume and it can be difficult to maintain a balance between these and the more diffuse ligand densities. Unless a proper balance is maintained, the more diffuse ligand orbitals may distort so as to compensate for the deficiencies in the less adaptable set. Add to this the fact that the correlation energies are likely to be comparable in magnitude to metal–ligand binding energies. Further, relativistic corrections, needed where spin–orbit coupling or heavy elements are involved, can alter bond lengths and total energies and you have a very difficult situation. Indeed, several examples are known in which apparent excellent agreement with experiment has disappeared when the calculation was improved, showing that the initial good agreement was, in fact, illusory. Lest this appear too pessimistic, it is to be emphasized that good calculations are beginning to appear, but they are far from routine. Ferrocene[2] provides an excellent example. After a calculation including configuration interaction with over a million excited state configurations and a correction for relativistic effects, the Fe–C bond length was still 0.07 Å longer than found experimentally, a discrepancy which should be compared both in magnitude and direction with that quoted above as resulting from far less sophisticated calculations on organic molecules. This error of 0.07 Å seems to be common—it has been found in the C–Cr bond length in di-η^6-benzenechromium and also in η^6-benzenetricarbonylchromium. The error probably reflects a common deficiency in the functions used to describe the orbitals of the metal atom, perhaps a need to include f orbitals. It is surely significant that it is only for

---

[2] See H. P. Lüthi, P. E. M. Siegbahn, J. Almlöf, K. Faegir and A. Heiberg, *Chem. Phys. Lett.* (1984) *111*, 1.

a main group cyclopentadienyl, $Mg(C_5H_5)_2$, that acceptable agreement with experiment has been obtained. In the case of ferrocene, the inclusion of extensive configuration interaction had the effect of showing the Fe–C bonding to be significantly more covalent than would otherwise have been concluded.

Although the situation is improving, the computer demands of *ab initio* calculations of the type just considered impose severe limitations. Alternative approaches which are less computer demanding but, hopefully, comparable in accuracy are attractive! One such method is very different in approach. The molecule is, pictorially and mathematically, divided up, a spherical shell being placed around each atom (the shell is usually called, pictorially, a 'muffin tin'—although 'cake tin' might be more appropriate in the UK). Each sphere is treated separately. What this means is that when, as before in the *ab initio* method, an electron is selected out, it is selected only from within the sphere, not some more distant part of the molecule. If the potential between the spheres is taken to be constant, it is not difficult to ensure that the resultant wavefunctions vary smoothly crossing the surface of the shell. Although this model sounds artificial, it is not, for within each sphere one is dealing, essentially, with an atom—and it is much easier to do calculations on isolated atoms than on the same atoms embedded within a large molecule. Of course, the electron density within any one sphere is dependent on that in the others, as we would expect. This so-called $X\alpha$ method has been widely used in calculations on transition metal complexes and is almost certainly the best method presently available for treating large molecules such as the metal clusters which will be discussed in Chapter 15 and the solids which will be the subject of Chapter 17. In the name $X\alpha$, the α is a number, which some treat as a variable and others as a fixed quantity, which has a value of between $\frac{2}{3}$ and 1. It occurs in the formalism by which the method treats the electron repulsion involving the selected electron. However, the results resulting from the use of the $X\alpha$ method have been somewhat erratic and contemporary work is turning increasingly towards more sophisticated, but related, approaches known as density functional methods. In these, the (local) density serves a role akin to that of the shells, the muffin tins, of the $X\alpha$ method. As this comment suggests, current research is aimed as much at producing reliable computational methods as producing results, so that it is scarcely possible to provide a survey of these results, although later in this chapter we shall look at ferrocene again.

10.4 Semiempirical methods

In the previous section, the *ab initio* method was presented as one that makes a serious attempt to obtain very close approximations to an exact solution of the Schrödinger equation. The best of such calculations are enormously demanding of the time of the fastest and most modern computers—many millions of integrals over three-dimensional space have to be calculated to high precision. If progress depended on such calculations being available for all molecules, progress would be slow indeed. Faced with a need for relatively quick, hopefully reasonably reliable, calculations, many have asked why reinvent the wheel?, or, in context, why try to calculate that which is already known? Specifically, rather than spend a lot of computer resources trying to include accurately, for example, all of the electron–electron repulsion and nucleus–electron attraction contributions to the atomic orbital energies involved in the calculation, why not simply take them from experiment? You may not be able to say much about how the energies arise, but at least you get the answer right! Of course, there are alternatives. One could set out to do the massive sums but simply not include those integrals for which there is good reason to believe that the answer would be very small. Or, a relatively small calculation could be undertaken but the results scaled to those of a more accurate calculation. Or, an approximate theory could be developed which could be applied to a wide range of experimental data but which contains parameters. Variations in the values of these parameters could then be systematically explored until a good fit with the data is obtained. One then asks whether the parameter values are reasonable and vary sensibly between related compounds. All of these various possibilities will not be explored in this chapter. Rather, two somewhat different approaches will be considered in outline. Others which the reader may well encounter in the literature and which make serious attempts to model the exact solutions are complete neglect of differential overlap (CNDO), mentioned in Table 10.2, intermediate neglect of differential overlap (INDO), and a method due to Fenske and Hall, mentioned in Table 10.3. They are excluded simply because a non-mathematical discussion is scarcely feasible for them; they differ in the ways they handle various integrals that arise in the detailed theory.

10.5 Extended Hückel method

The original Hückel method applied solely to unsaturated organic molecules, particularly hydrocarbons. It was concerned with the π electrons and the interactions between them which led to delocalized π molecular orbitals. For hydrocarbons just two quantities were involved, the energy of the isolated π electron and the energy of interaction between pairs of them. Of these, only the latter, usually denoted β, affected the results. It was a crude and simple model—but it worked! The results could be used to explain stabilities, spectra, reactivities and dipole moments; Fig. 10.4 is based on them. It is not surprising that chemists should seek to adapt the model to the more complicated molecules of inorganic chemistry, for the Hückel approach is perfectly applicable to molecules of low symmetry (it actually increases their symmetry in a subtle way—chemically distinct atoms are given the same parameters). The model which has resulted is called the extended Hückel method, although inevitably many variants exist (all, like their parent, given the opportunity also subtly increase symmetry).

In the most common species of the method encountered, each orbital to be included is given an energy, the value of which is taken from atomic spectral data. The energy of interaction between two orbitals is assumed to be related to their overlap integral (to evaluate this, of course, explicit algebraic forms for the orbitals must be assumed). An overlap integral is a number and an energy of interaction is needed. It is therefore necessary to multiply the overlap by an energy and many suggestions for this step have been made—the outcome of the calculation is dependent on the choice made. Simplest, and most widely used, is to multiply by the average of the energies taken as those of the interacting orbitals. The product (overlap integral ×

energy term) is taken to be proportional to the interaction energy. The two are related, then, by a proportionality constant, the practice being to use that value which gives best agreement with experiment (this constant has sometimes been denoted F, a notation which has become less common—perhaps because cynics have been heard to comment that F stands for fudge factor). This comment is unduly uncharitable not least because once the constant is given a value it is fixed, at least for a series of related compounds. Commonly, the value of 1.75 is used. It is important in the method that the proportionality constant is given the same value for each and every interaction within the molecule. In this way the method is made *rotationally invariant*. That is, the answer obtained does not depend on the direction taken for the molecular z axis nor does it change if, say, s and p orbitals are mixed to form sp hybrids and these used instead of separate s and p orbitals in the calculations.

The theory so far described suffers from one major defect. The maximum stabilization resulting from a particular interaction is a maximum when the corresponding overlap is a maximum. So, bond distances (and the total bonding) are determined by some sort of weighted average over all of the bonding interactions included. In reality, internuclear distances are determined by a balance between attractive and repulsive forces; so far the model lacks repulsive forces. There have been several suggestions for introducing them into the calculation. One of the most useful uses the same orbitals as used to obtain overlap integrals to calculate repulsion integrals. This model is called the atom-superposition electron-delocalization molecular orbital (ASED-MO) method, has given good results and is incorporated in some (but by no means all) of the computer packages available for extended Hückel calculations. Other modifications to the simple model write the proportionality constant as a product of factors which, for instance, introduce an explicit internuclear separation dependence to improve the predictions at larger internuclear separations or to reduce the extent to which low-lying, compact, filled orbitals are mixed with higher-lying diffuse orbitals (so-called counterintuitive orbital mixing, a problem akin to the need for balanced basis sets, mentioned above).

The value of extended Hückel theory is that it enables one to apply simple and acceptable ideas of energy differences and overlaps to low symmetry molecules and so is applicable to systems for which symmetry alone is of little help. In so doing it can both highlight major interactions and competitions between interactions, a concept which can then be separated from the model itself and discussed separately, and, perhaps, largely qualitatively. The danger of extended Hückel theory is that more meaning can easily be attached to the energy levels obtained than is warranted. That this is so can best be seen if some hidden assumptions of the model are highlighted. First, in carrying out the calculations, it is necessary to assume a molecular geometry. If the geometry were to be systematically changed, so as to locate the total energy minimum, this would probably be found to be well away from the assumed geometry—and quite possibly at a geometry which would be felt to be quite ridiculous (because of the absence of any repulsive terms, for instance). Unless specific steps are taken to remedy this situation, the calculations are made at some high point on a potential energy surface and, without some detailed justification, cannot be expected to give a reliable account of the energetic consequences of bond length changes. The remedy is to systematically change the input parameters until the input internuclear separations *are* minima; unfortunately, it is not always clear from published results whether this has been done. Strictly, the geometry should be one in which any small movement of any atom leads to a less stable molecule—and it is likely that few of the published calculations satisfy such a strict requirement. Second, in carrying out the calculations it is necessary to assume formal charges on the atoms in order to be able to make use of energies from atomic spectroscopy and to determine the precise form of the orbitals to be used to calculate overlap integrals. These charges are not unambiguous. For ferrocene, for example, do we take Fe^0 or Fe^{II}? Probably the former, because the latter would force us to work with $C_5H_5^-$, and C_5H_5 is easier. However, this way of determining the choice begs the Fe^0–Fe^{II} question, it does not solve it. Unfortunately, the results of an individual calculation are dependent on the choice made.

This highlights a third point. If, after completing the calculation, the formal charges on the atoms were calculated (and approximate simple ways are available for this) they would be found to differ from the starting charges. The new charges would probably be non-integers but this is no real problem since it is not difficult to interpolate reliably between free-ion energies and, perhaps, orbital parameters. We could then rerun the calculation and keep on repeating until the input and output atomic charges became essentially identical. It seems that such self-consistent charge and configuration (SCCC) calculations do not give better agreement with experiment than calculations which are not self-consistent. As a result, in most, if not all, current applications of the extended Hückel method there is no iteration; the calculations are stopped once the first output has been obtained (just as in the Hückel theory of organic chemistry). It follows that formal atomic charges calculated by this method are not self-consistent and that arguments based on such charges should only be accepted *cum grano salis*. The final charges depend on the initially chosen charges.

The extended Hückel method is of value because it addresses the problem of energy levels and energy level changes when symmetry has either been exhausted or symmetry does not exist. It is simple, so that it can be applied to molecular fragments and to crystalline solids alike, and has been applied extensively to both. It is neither computationally nor mathematically demanding; computer programs are readily available so that it can be—and is—used by inorganic chemists as an everyday tool. It is also very flexible, a feature which is at once both a weakness and an advantage. So, if the molecular orbital energy level sequence for a particular molecule were determined beyond all reasonable doubt, perhaps by a very detailed *ab initio* study, then there is little doubt that the results could, in large measure, be duplicated by an extended Hückel calculation. The advantage of the method is that it could then be applied to related molecules with considerable reliability. The virtues and weaknesses of the method are perhaps best summed up in the words of one of its leading practitioners: 'since all other methods are superior to it, it inculcates in its user a feeling of humility and forces him or her to think about why the calculations come out the way that they do'. However, as reference to any issue of a current journal dealing with inorganic chemistry will testify, the extended Hückel method is widely used both for molecules and solids. Its status is evident in the fact that when detailed *ab initio* or similar calculations are published it is normal to compare the results obtained with those given by the extended Hückel method. Later in this chapter there is an example which illustrates how, despite all its shortcomings, the method can provide insights which could scarcely have been obtained in its absence.

10.6   Angular overlap model

The angular overlap model was originally introduced as a way to extend the ligand field model to complexes of low symmetry without disproportionately increasing the number of unknowns, the number of parameters, in the problem. In its original form it did this by postulating that all bonding interactions are proportional to the square of the corresponding overlap integral (an overlap integral can be either positive or negative, the square can only be positive). As befits a development of the ligand field model, the angular overlap model usually concerns itself exclusively with the d orbitals on the transition metal. If the form of the atomic orbitals is known along with the molecular geometry—or reasonable approximations can be made—then the method contains only two unknowns, the proportionality constants relating overlap to interaction energy for σ and for π interactions. Commonly, these proportionality constants do not have to be evaluated explicitly. Often, the problem becomes trigonometric, dependent only on the angular component of the orbital wavefunctions, so that it is not necessary to evaluate overlap integrals either. A further advantage of the method is that, although normally applied to low-symmetry molecules, when applied to high-symmetry species it gives results equivalent to those coming from crystal or ligand field models. For instance, it gives the result that $\Delta_{oct} = -9/4\,\Delta_{tet}$, a relationship that we met in Chapter 7. So, it is a method with a minimum of parameters and has been found to give good agreement with experimental results, at least for transition metal complexes. Its development with time has been in the opposite direction to that of extended Hückel theory; its practitioners have sought to develop the model by its application in increasing detail to a limited range of molecules. So, it has been refined to the point to which it can predict with accuracy such diverse quantities as anisotropies in the magnetic properties and the intensities of electronic spectral bands. These developments have been accompanied by subtle changes in the model itself. For example, it has been found possible to incorporate within the, formally, d electron-only model, some of the effects of s and p orbital involvement in the bonding—and to do so in a reasonable way. One way in which this has been achieved is to distinguish between a free ligand and a ligand in a complex; to allow for a difference between the orbitals and orbital energies of the free and complexed ligands. Any attempt to deal with the ligands in any global sense has been abandoned by some workers; each ligand is treated individually (the most developed current model is a *cellular* angular overlap model). This progress has been made by a painstaking detailed study of individual molecules; the theory is only just reaching the point at which it can be generally applied without need for molecule-specific development work.

10.7 Three examples: ferrocene, hexacarbonylchromium and ethenetetracarbonyliron

10.7.1 Ferrocene

There have been numerous calculations reported on ferrocene, at a variety of levels of sophistication. Although there is agreement on the general bonding pattern in the molecule, this agreement is far from obvious. The problem is made evident when the results of *ab initio* calculations are compared with all of the others. This comparison is given in Table 10.2. Whereas all non-*ab initio* methods give the highest orbital as an a_{1g}, an orbital which largely consists of the metal $3d_{z^2}$, this same orbital is at the bottom of the *ab initio* list. The difference is deceptive; it arises from a different definition of what is meant by an orbital energy in the two cases. The origin of the difference is important because it has always to be remembered when *ab initio* and other calculations are compared.

When an electron is removed from a molecule—and such ionizations are implicit in sequences such as those of Table 10.2—then the remaining electrons must be expected to change their distributions and orbitals slightly. For the *ab initio* calculations in Table 10.2 this adjustment, the reorganization energy, has not been included because the readjustment has not taken place—the energy levels are those of the neutral ferrocene molecule. The other methods give sequences which answer the question 'what is the order of ease of ionization of the electrons?'—and so include the adjustments. The way in which this adjustment energy appears in the *ab initio* calculations is as follows. *Ab initio* calculations on the ferrocene cation, so that there is one electron less than for the neutral molecule, indicate that the ground state of the cation is one in which an $a_{1g}(3d)$ electron is removed from neutral

Table 10.2 Comparison of orbital energy level sequences resulting from *ab initio* and other calculations on ferrocene. The most readily ionized electrons are at the top (but with exceptions, see the text). Orbitals printed together are close in energy. The results of two different *ab initio* calculations are included

	Ab initio	Ab initio	Xα	Extended Hückel[a]	CNDO[a]
	$e_{1u}(\pi Cp)$	$e_{1u}(\pi Cp)$	a_{1g}	a_{1g}	a_{1g}
	$e_{1g}(\pi Cp)$	$e_{1g}(\pi Cp)$	e_{2g}	e_{2g}	e_{2g}
	$e_{2g}(3d)$	$e_{2g}(3d)$	e_{1u}	e_{1u}	e_{1g}
↑			e_{1g}	e_{1g}	
Energy	$e_{2u}(\sigma Cp)$		e_{2u}		e_{1u}
	$a_{2u}(\pi Cp)$	$a_{2u}(4p)$	e_{2g}		
	$e_{2g}(\sigma Cp)$		a_{2u}		e_{2g}
	$a_{1g}(3d)$	$a_{1g}(\pi Cp)$			
		$a_{1g}(3d)$			
				a_{2u}	
				a_{1g}	

[a] The σ C_5H_5 were not included in the calculation.

ferrocene. That is, it comes from the orbital which is at the bottom of the *ab initio* list in Table 10.2. The reorganization energy compensates for the fact that the $a_{1g}(3d)$ orbital in ferrocene is relatively low in energy. The reason that e_{2g} is the second orbital in the non-*ab initio* list, but not in the *ab initio* list is similar. There is a basic and important reason for all of this. Both of the more readily ionized electrons, a_{1g} and e_{2g}, are largely metal in character. In the neutral molecule they are confined to a small volume of space and it is not surprising that the molecular electron density distribution should change significantly when one of these electrons is removed. It is the consequential change in energies that result in the apparently anomalous position of the a_{1g} and e_{2g} orbitals in the neutral molecule energy level sequence. In contrast, orbitals which are higher lying in the ferrocene *ab initio* energy level sequence are largely ligand in nature. This means that their electron density is spread over at least 10 atoms and so is relatively diffuse. As a result, the change in the molecular electron density distribution following the removal of such an electron is relatively small and so the rearrangement energy is small also. This is not just some subtle difference between *ab initio* and other calculations. If, as well may be the case, we are interested in the bonding in the ferrocene molecule, then the energy level pattern given by the *ab initio* method is that which is appropriate—it more accurately shows the stabilizations resulting from the bonding interactions. If, on the other hand, we are more interested in the chemical reactions of ferrocene—that is, in situations in which electron density is displaced, then the orbital energy level diagrams of the other methods become more relevant. Because of its relative ease of ionization, a_{1g} and e_{2g} behave as if they are the HOMOs, even if, strictly, they are not. This, of course, is a situation which is very similar to that met at the end of Section 6.2, where for classical coordination complexes it was found that the (incompletely filled) d orbitals are not, in fact, the highest lying occupied orbitals.

At the present time it does not seem to be unambiguously determined whether the lowest ionization potential of ferrocene corresponds to the loss of an electron from the a_{1g} orbital or whether it comes from the e_{2g}. As Table 10.2 shows, most theoretical work favours a_{1g}—and there are electronic spectroscopic arguments in favour of this assignment—but the *ab initio* calculations point to the alternative, e_{2g} assignment and find support in both EPR and photoelectron spectroscopic data (the latter will be given and discussed in Section 12.7). It is likely that the *ab initio* assignment is correct.

The results of an *ab initio* calculation on ferrocene are also shown in Fig. 10.9. It is perhaps more useful than the same data in Table 10.2 in addressing the question of whether these calculations support the simple picture of the bonding in ferrocene given in Section 10.1. The answer, comfortingly, is yes. This conclusion is perhaps most easily seen by looking at the empty, antibonding orbitals in Fig. 10.9. As we have seen, occupied π orbitals in a hypothetical $(C_5H_5)_2$ unit which could act as electron donors to the iron atom are of $A_{1g} + A_{2u} + E_{1u} + E_{1g}$ symmetries. If these interactions do indeed contribute to the metal–ligand bonding then we would expect this to be signalled by destabilized, antibonding, orbitals of these same symmetries. As Fig. 10.9 shows, such orbitals do, indeed, exist. Looking at antibonding orbitals in this way is a simple, approximate, way of avoiding the problems posed by the plethora of C–C and C–H bonding orbitals in the bonding set.

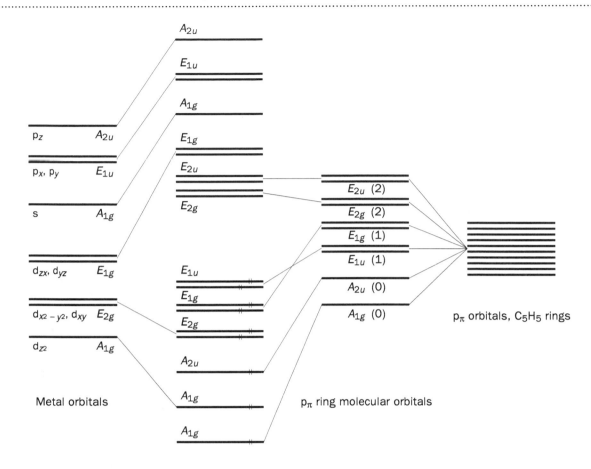

Fe (cp)$_2$ molecular orbitals

Fig. 10.9 Schematic *ab initio* molecular orbital energy level diagram for ferrocene (data from T. E. Taylor and M. B. Hall, *Chem. Phys. Lett.* (1985) *114*, 338). This figure differs from Table 10.2 in that the table, but not this figure, includes C$_5$H$_5$ ring orbitals. This diagram does not demonstrate the fact that because of configuration interaction the average occupancy of the doubly occupied bonding orbitals is slightly below 2 (ca. 1.98) and that of the empty antibonding slightly above 0 (ca. 0.02).

Rather similar arguments hold for the potentially π acceptor orbitals on the hypothetical (C$_5$H$_5$)$_2$ unit, $E_{2g} + E_{2u}$. There is no metal orbital of E_{2u} symmetry but there is one of E_{2g}. Again, there is an empty E_{2g} orbital, corresponding to an occupied one, just as required by the simple picture of Section 10.1.

10.7.2 Hexacarbonylchromium

Section 10.1 contains a schematic molecular orbital energy level diagram for Cr(CO)$_6$, Fig. 10.5, which was derived from a model in which it was regarded as a complex in which extensive π back-bonding occurs from metal t_{2g} to $\pi^* t_{2g}$. There is abundant evidence that this back-bonding is a real phenomenon. Theoretically, all detailed calculations concur on its reality; experimentally, evidence for its presence is revealed by a detailed analysis of the electron distribution about the Cr–C–O axis, obtained from accurate, low-temperature X-ray diffraction measurements. Perhaps the simplest evidence is provided by the changes in the average ν(CO) stretching frequencies (cm^{-1}) along the isoelectronic series

$Ti(CO)_6^{2-}$	$V(CO)_6^-$	$Cr(CO)_6$	$Mn(CO)_6^+$
1780	1897	2017	2113

The argument is that the different frequencies reflect different electron populations in the CO π^*, antibonding, orbitals, electron density which originates in donation from the metal atoms to the CO groups. The greater the antibonding electron population, the lower the frequency. If, as seems entirely reasonable, the formal charges on, for instance, Mn^+ and V^-, respectively, inhibit and enhance the metal → ligand π electron donation then an entirely self-consistent picture is obtained. The picture makes the assumption that the σ electron density does not share the sensitivity to formal metal charge shown by the π, but there is ample evidence in support of this assumption.

Table 10.3 summarizes the results of some of the available calculations on $Cr(CO)_6$ and Fig. 10.10 is a molecular orbital energy level diagram based on the *ab initio* data in Table 10.3. All of the calculations in Table 10.3 agree about the HOMO—it is a t_{2g} molecular orbital which is extensively involved in the π bonding that was discussed at the end of the previous paragraph. An electron in one of these t_{2g} orbitals is approximately one-third located in ligand π^* orbitals. Apart from these t_{2g} orbitals, all of the other orbitals listed are largely located on the ligands. Because they are distributed over six ligands they are relatively diffuse and Koopmans' theorem holds (this theorem will be discussed in Section 12.7; it equates an ionization energy with an orbital energy). It follows that there are no complications of the sort encountered with ferrocene, where some of the *ab initio* results could not immediately be compared with the others. However, this is not to say that electron correlation is unimportant in hexacarbonylchromium. For the *ab initio* calculations it has the effect of decreasing the electron density in the metal t_{2g} orbitals and to decrease it in the e_g, a step that serves to bring the *ab initio* and Xα results closer together.

Table 10.3 Orbital energy level sequences resulting from various calculations on $Cr(CO)_6$. Orbitals printed close together are close in energy

	Ab initio	Xα	Extended Hückel	ca. Fenske–Hall
	t_{2g}	t_{2g}	t_{2g}	t_{2g}
	t_{1u}	t_{1u}	e_g	t_{1u}
	e_g	t_{1g}	t_{1u}	t_{1g}
		t_{2u}		t_{2u}
Energy ↑	t_{1g}		a_{1g}	
	t_{2u}	t_{2g}		t_{1u}
			t_{1g}	
	t_{2g}	t_{1u}	t_{2u}	a_{1g}
	t_{1u}			
		e_g		e_g
	a_{1g}		t_{1u}	
		a_{1g}		t_{2g}
			t_{2g}	

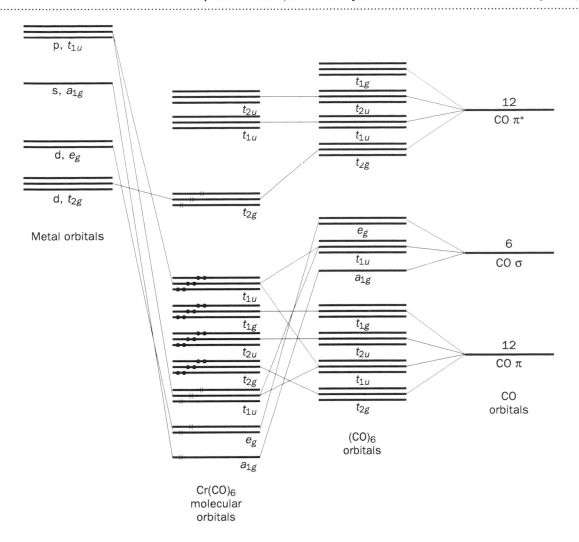

Metal orbitals

$Cr(CO)_6$ molecular orbitals

$(CO)_6$ orbitals

CO orbitals

Fig. 10.10 Schematic Xα molecular orbital energy level diagram for $Cr(CO)_6$ (data from R. Arratia-Perez and C. Y. Yang, *J. Chem. Phys.* (1985) 83, 4005). Electrons which would be counted in an application of the 18-electron rule are shown as lines, those that would be excluded are shown as dots.

As was discussed in Chapters 6 and 7, the occupied ligand orbitals of an octahedral complex can be classified as either σ or π. The latter span the irreducible representations $T_{1g} + T_{1u} + T_{2g} + T_{2u}$. Of these, the T_{1g} and T_{2u} are non-bonding because there are no orbitals of these symmetries on the metal. It is therefore not surprising to see in Table 10.3 that they are found to have almost identical energies—any differences are due to ligand–ligand interactions. Three of the entries in Table 10.3 show a cluster of orbitals of $T_{1g} + T_{1u} + T_{2g} + T_{2u}$ symmetries which it is tempting to equate with the entire π set. This identification would not be entirely valid for the T_{1u} orbitals. There are two sets of orbitals of this symmetry and both are mixtures of ligand σ and ligand π. The ligand σ contribution is the greater for the higher lying, not for the lower. For the lower, the π contribution dominates. Both of the approximate methods place an orbital of T_{2g} symmetry as the lowest listed; the actual wavefunctions show a significant metal t_{2g} component. The *ab initio* and Xα calculations do not give this result, and we must conclude that it is an artefact introduced by the approximations of the less rigorous methods.

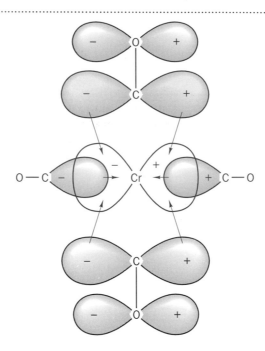

Fig. 10.11 Simultaneous donation from CO σ and π occupied orbitals into a chromium 4p orbital in $Cr(CO)_6$. Only orbitals of T_{1u} symmetry are involved in this simultaneous donation but it occurs for both of the bonding T_{1u} orbital sets of Fig. 10.10.

The ligand σ orbitals span the $A_{1g} + E_g + T_{1u}$ irreducible representations. Both the *ab initio* and Xα calculations agree that the a_{1g} is the lowest but differ in their placing of the e_g orbitals. The origin of this disagreement is not clear, although it is associated with different metal orbital contributions (about 20% for the *ab initio* and 33% for the Xα); it would probably be expected that the *ab initio* result is the more reliable. If the *ab initio* results are taken as a criterion, the approximate methods do not fare well, although as has been pointed out, some inherent flexibility may well enable them to duplicate the *ab initio* results more closely, if this were to be taken as the goal. Details apart, however, all the methods agree in that the qualitative picture presented in Section 10.1 is supported. However, there is one unexpected result that merits mention. In Section 7.2.2, we recognized the possibility of interaction between a set of σ bonding and a set of π bonding t_{1u} orbitals but did not explore the possible outcomes in detail. As we have indicated above, the t_{1u} molecular orbitals actually contain contributions from ligand σ, ligand π and metal p orbitals. The detailed calculations reveal an unexpected picture, one in which there is simultaneous electron donation from adjacent CO groups, the donation from one CO group being σ whilst that from its neighbours is π. This interaction is shown in Fig. 10.11; it occurs for all of the bonding t_{1u} orbitals of Table 10.3 and Fig. 10.10.

10.7.3 Ethenetetracarbonyliron

This compound, $Fe(CO)_4C_2H_4$, is prepared by the reaction between a high pressure of C_2H_4 and $Fe(CO)_5$ and has the structure shown in Fig. 10.12. It is a yellow oil at room temperature and is relatively unstable, decomposing into C_2H_4 and $Fe_3(CO)_{12}$, the latter presumably being formed from three $Fe(CO)_4$ units. This section will be concerned with the answers to two questions, to which we will find that the insights provided by extended

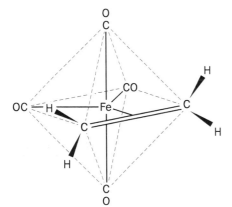

Fig. 10.12 The molecule $Fe(C_2H_4)(CO)_4$. The arrangement of atoms around the Fe is approximately octahedral (the edges of the octahedron are shown dotted).

Hückel theory provide answers:

- Why is this compound stable at all?
- Why is it not more stable?

For simplicity, it will be assumed that decomposition occurs by $Fe-C_2H_4$ bond fission, an assumption which means that we can restrict our discussion to the interaction between $Fe(CO)_4$ and C_2H_4 units. As Fig. 10.13(a) shows, the $Fe(CO)_4$ fragment is rather like part of an octahedral molecule and we could well be tempted (incorrectly) to assume that the metal has two additional orbitals available to complete an octahedral bonding pattern. If this were so, then, using the molecular C_{2v} point group, it is a simple matter to show that these two orbitals give rise to combinations of $A_1 + B_1$ symmetries, combinations that are shown in Fig. 10.13(b). It would have been logical to have concluded that the A_1 combination is the more stable because it is the no-node combination, whereas the B_1 has a planar node. However, we would have been wrong in a way that is readily revealed by the extended Hückel calculations—the A_1 combination is Fe–CO antibonding,

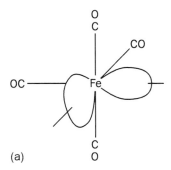

Fig. 10.13 (a) The $Fe(CO)_4$ unit regarded as a fragment of an octahedron (drawn from the same viewpoint as Fig. 10.12). The Fe orbitals required to complete the octahedron are shown. (b) The A_1 and B_1 combinations of the octahedron-completing Fe orbitals. The symmetry elements of the C_{2v} point group are shown inset. The orbital combinations are drawn from a viewpoint lying in the σ'_v mirror plane.

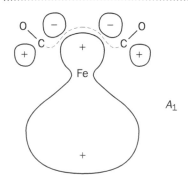

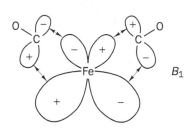

Fig. 10.14 The A_1 and B_1 combinations of Fig. 10.13(b) as given by the extended Hückel method, viewed along the linear OC–Fe–CO axes of Fig. 10.13(b). The nodal plane indicating Fe–C antibonding in the A_1 orbital is dotted. The Fe–C overlaps responsible for the bonding interactions in the B_1 orbital are shown as double-headed arrows.

whereas the B_1 combination is Fe–CO bonding. The extended Hückel calculations reveal this fact and thus that the orbital ordering is the opposite of that given by the simple model; in fact, the A_1 orbital is above the B_1. These orbitals are shown schematically in Fig. 10.14.

The picture is completed by noting that when it spans the two empty octahedral sites of the Fe(CO)$_4$ unit, the ethene π bonding orbital has A_1 symmetry and the π antibonding orbital B_1 symmetry. The partial molecular orbital energy level diagram shown in Fig. 10.15 is thus obtained, and, with it the existence of the molecule explained—there are appropriate bonding orbitals occupied. But why is it not more stable, our second question? Most probably because the bonding within the Fe(CO)$_2$ unit shown in Fig. 10.14 is weakened by the addition of C$_2$H$_4$. The occupied B_1 orbital of the fragment can only become involved in C$_2$H$_4$ bonding at the expense of the Fe–(CO)$_2$ bonding. Put another way, electron density in the A_1 fragment orbital means population of an Fe–(CO)$_2$ antibonding orbital. In retrospect, it can be seen that the existence of Fe(CO)$_5$, a trigonal bipyramidal molecule, provides an argument in favour of the 'A_1 high energy, B_1 low energy' pattern. This molecule differs from Fe(CO)$_4$C$_2$H$_4$ by a CO group taking the place of the C$_2$H$_4$. Following the usual model of CO bonding to a metal, σ donation from the carbon of the CO group to the metal requires an empty A_1 orbital on the iron. Similarly, back bonding into an empty π antibonding orbital of the CO requires a filled B_1 orbital on the iron. This A_1 high, B_1 low, pattern is just that revealed by the extended Hückel calculations.

Before concluding our discussion of the Fe(CO)$_4$ unit it is worthy of comment that it is often thought of as a member of an *isolobal* series such as that given in Fig. 10.16. Members of an isolobal series are always shown as related to each other by the symbol $\longrightarrow$. Membership of such a series is based on experimental evidence as much as theoretical, but members of an isolobal series usually have similar orbital patterns, both in terms of orientation and energy. Experimentally, they form similar compounds—so, each member of the middle series shown in Fig. 10.16 forms an H$_2$ compound. However, of the members of this series only Fe(CO)$_4$ has the A_1 high, B_1 low energy level pattern. Theoretically, this species is the only member of

Fig. 10.15 Schematic molecular orbital diagram showing the interaction between the π orbitals of C$_2$H$_4$ and the Fe(CO)$_4$ fragment orbitals.

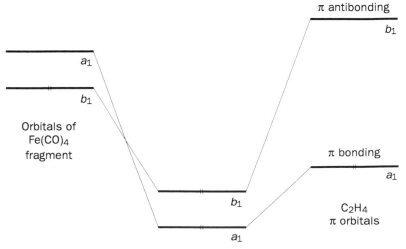

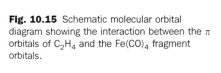

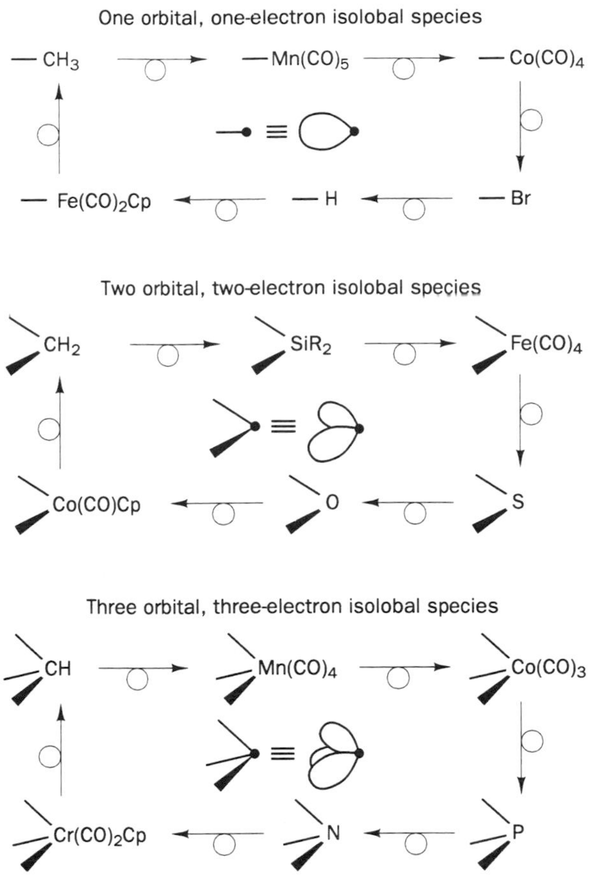

Fig. 10.16 Examples of isolobal series.

the series for which the two electrons accommodated in the pair of orbitals shown would normally be regarded as paired. This might be thought of as a weakness in the isolobal concept, but, in fact, the value of the concept lies in its flexibility. It has come to be widely used in inorganic chemistry.

10.8 Final comments

There is every indication that there will be many changes in the area of this chapter over the next few years. The increasing availability of ever-increasing computer power will surely mean that higher-level calculations become more routine. If this is true, it could well be that the present time (1994) is close to the time at which the extended Hückel method is at its zenith. Current results coming from methods as diverse as the cellular angular overlap and the density functional models are impressive. Then there are developments on the horizon which presage change. The first is the wider incorporation of relativistic phenomena. They are not trivial; the spin–orbit coupling which has featured so often in Chapters 8 and 9 is relativistic in origin. It has recently been shown that the reason that $Bi(C_5H_5)_5$, a molecule with C_{4v} symmetry, is an intense violet colour is because of relativistic effects. In C_{4v} symmetry, but not D_{3h}, relativistic effects cause the LUMO to drop in energy, moving an allowed transition into the visible (in the former symmetry, but not the latter, the bismuth 6s orbital can participate in the LUMO; for a discussion on the relevance of this see the footnote at the beginning of Section 11.3). One must anticipate similar, if often less dramatic, phenomena for many compounds of the heavier elements. An example of one of the more dramatic phenomena is the prediction that relativistic effects may well lead to the molecule HgF_4 having some stability (to date, it has not been prepared). Finally, there is the development of spin-coupled valence bond theory. This book, like most of its generation, has concentrated on molecular orbital theory. Valence bond theory was mentioned at the beginning of Section 6.2.1 but then forgotten. Subsequently, problems were found with the molecular orbital theory model arising from electron repulsion. Such problems are much less severe for valence bond theory, but it has problems of its own. These are in large measure overcome in spin-coupled valence bond theory and so this method offers the prospect of a quite new approach to the electronic structure of many of the molecules of this chapter. This development is for the future because, at the moment, the method is only being applied to diatomic transition metal species.

Further reading

This Chapter has deliberately avoided the mathematical complexities associated with the methods it describes. In practice, for most readers who wish to explore the topics at greater depth, the problem will rapidly translate into one of working intelligently with a suitable computer program. A helpful book, with seven chapters related to the topics of this chapter (and part of Chapter 13) is *Computational Chemistry Using the PC* by D. W. Rogers, VCH, Weinheim, 1990.

General accounts of the chemical bonding concepts discussed in this chapter are to be found in the following two books, although only the second covers isolobality.

- J. N. Murrell, S. F. A. Kettle and J. M. Tedder *The Chemical Bond*, Wiley, Chichester, 1985.
- T. A. Albright, J. K. Burdett and M-H Wangko *Orbital Interactions in Chemistry*, Wiley, New York, 1985.

An excellent source of *ab initio* results relevant to the present chapter is 'Ab initio Calculations of Transition-metal Organometallics: Structure and Molecular Properties', A. Veillard, *Chem. Rev.* (1991) *91*, 743. The paper contains not a single equation and so is very easy for the general reader to read; unfortunately, it contains not a single diagram either.

A very readable account of the bonding in ferrocene is given by M. M. Rohmer and A. Veillard in *Chem. Phys.* (1975) *11*, 349. Also of interest is 'The Transition Metal–Carbon Bond', by E. R. Davidson, K. L. Kunze, F. B. C. Machado and S. J. Chakravorty, in *Acc. Chem. Res.* (1993) *26*, 628. This contains an account of the bonding in $Cr(CO)_6$ which is both a bit more advanced than and different from that in the text (whilst being readable and consistent with that in the text).

A paper that uses the general approach developed in this chapter but which extends both it and its applicability is by T.

Ziegler, V. Tshinke, L. Fan and A. D. Becke, *J. Amer. Chem. Soc.* (1989) *111*, 9177. The work reported used density functional methods, which represent a development and improvement on Xα calculations. The Xα method is readably described in the first few pages of 'The Self-Consistent Field for Molecules and Solids', which is volume 4 of the series *Quantum Theory of Molecules and Solids*, J. C. Slater, McGraw-Hill, New York, 1974. Perhaps even more readable (the only, two, equations are in the first paragraph) is 'Electronic Structure Calculations using the Xα Method', D. A. Case, *Ann. Rev. Phys. Chem.* (1982) *33*, 151. An overview of the density functional method can be gained by a browse through *Density Functional Methods in Chemistry*, J. K. Labanowski and J. W. Andzelm (eds.) Springer-Verlag, New York, 1991. In general easier to read (in the second half, at least) is 'Approximate Density Functional Theory as a Practical Tool in Molecular Energetics and Dynamics', by T. Ziegler, *Chem. Rev.* (1991) *91*, 651. A paper which includes not only *ab initio* but also density functional results is 'Selected Topics in *ab initio* Computational Chemistry in Both Very Small and Very Large Chemical Systems', by E. Clementi, G. Corongiu, D. Bahattacharya, B. Feuston, D. Frye, A. Preiskorn, A. Rizzo and W. Xue in *Chem. Rev.* (1991) *91*, 679. It includes a detailed discussion of buckminsterfullerene and also a topic not included in the present chapter—the use of theoretical methods to calculate the way that molecular systems evolve with time.

The first paper to describe the extended Hückel method in its present form is by R. Hoffmann, *J. Chem. Phys.* (1963) *39*, 1397, but this covers its application to organic molecules. A closely related, very readable blow by blow account for inorganic molecules is by H. D. Bedon, S. M. Horner and S. Y. Tyree *Inorg. Chem.* (1964) *3*, 647.

The angular overlap model in its simplest form is described readably and in detail in Appendix 3 of *Inorganic Electronic*

Spectroscopy, A. B. P. Lever, Elsevier, Amsterdam, 1984. A more recent overview is provided by 'The Angular Overlap Model as a Unified Bonding Model for Main Group and Transition Metal Compounds' by D. E. Richardson, *J. Chem. Educ.* (1993) *70*, 372. More developed forms of the model are the subject of a review by M. Gerloch and R. G. Wooley in *Prog. Inorg. Chem.* (1984) *31*, 371. Most recent developments are covered by C. A. Brown, M. J. Duer, M. Gerloch and R. F. McMeeking in *Mol. Phys.* (1988) *64*, 825 and M. J. Duer,

S. J. Essex, M. Gerloch and K. M. Jupp *Mol. Phys.* (1993) *79*, 1147, an article which is useful for the non-mathematically minded because towards the end it contains a review of the physical significance of the parameters contained within the developed angular overlap model.

Finally, the spin-coupled valence bond theory is described in 'Applications of Spin-Coupled Valence Bond Theory', D. L. Cooper, J. Gerratt and M. Raimondi, *Chem. Rev.* (1991) *91*, 929.

Questions

10.1 The ring system used at the top of Fig. 10.2 to define the z direction is a planar eight-membered ring such as cyclooctatetraene, C_8H_8. Extend Fig. 10.2(b) to include both it and the seven-membered ring mentioned in the caption.

10.2 It has been suggested that there is a similarity between the bonding of C_2H_4 to CH_2 in cyclopropane, C_3H_6, and the Chatt–Duncanson model of the bonding of the C_2H_4 to Pt in Zeise's salt. Critically assess this suggestion.

10.3 On a relatively small modern computer it is now possible to carry out approximate *ab initio* calculations on quite large organic molecules. However, with none of the programs available for this, is it possible to include a transition metal atom? Outline the reasons for this and explain how approximate methods attempt to circumvent the problem.

10.4 Crystal and ligand field theories predict that the d^0 ion $[TiH_6]^{2-}$ will be octahedral. Extended Hückel, however, predicts a C_{2v}, bicapped tetrahedron, structure. Initially, *ab initio* methods predicted an octahedral structure but with the inclusion of configuration interaction a trigonal prismatic structure is indicated (see *Inorg. Chem.* (1989) *28*, 2893). Using this discordance of results as a basis, suggest those situations in which each method may be expected to make reasonably reliable geometry predictions (if ever!).

10.5 In the last example of Table 10.1 an ethene molecule bridges two zirconium atoms. Assume that the Zr_2C_2 unit is planar with D_{2h} symmetry and that the ethene hydrogens lie in a plane perpendicular to the Zr_2C_2. Extend the Chatt–Duncanson model to cover this case and thus suggest why the C–C bond length is the longest in Table 10.1.

10.6 In the text it was shown that the simple picture of π back-bonding in metal–Cr bonding given by pictures such as Fig. 2.1 overestimates the extent of this bonding in $Cr(CO)_6$. Show that it similarly overestimates it in the tetrahedral molecule $Ni(CO)_4$. For this, the character table of the T_d point group will be needed.

T_d	E	$8C_3$	$3C_2$	$6S_4$	$6\sigma_d$
A_1	1	1	1	1	1
A_2	1	1	1	-1	-1
E	2	-1	2	0	0
T_1	3	0	-1	1	-1
T_2	3	0	-1	-1	1

f electron systems: the lanthanides and actinides

11.1 Introduction

In recent years there has been an increasing study of compounds of the lanthanides and, to a lesser extent, of the actinides. These two groups of elements have varying numbers of electrons in their f orbitals (4f for the lanthanides and 5f for the actinides), thus inviting a comparison with the transition metal elements, discussed in Chapters 6–9, with their varying number of d electrons. However, as we shall see, such a comparison is not particularly helpful, a situation which has contributed to an attitude commonly encountered—that f electron systems are difficult to understand, that the theory is difficult. It is hoped that it will be possible to demonstrate in this chapter that this is not the case. Indeed, it is hoped to convince the reader that a study of f electron systems is not only of value in its own right but that such a study helps in the understanding of d electron systems. It does this by its concern with phenomena which also exist in d electron systems but which are currently largely ignored when discussing them.

What are the difficulties with f electron systems? First, they involve f orbitals and these are unfamiliar. Fortunately, they have already been met in this book, in Section 7.3, where it was found useful to assess their relative energies, the f orbital splittings, in an octahedral crystal field. Secondly, spin–orbit coupling is important. Again, this phenomenon has already been met (in Sections 8.4 and 9.5) but because we shall have need of a deeper understanding, pictures of spin–orbit functions will be introduced. Thirdly, crystal and ligand field effects are small, something which poses problems when one is more familiar with molecules in which they are large. Finally, coordination number and geometry are much more varied (and more uncertain) than for d electron systems. Fortunately, the last two problems tend to cancel each other out. When crystal field effects are small, knowledge of the detailed ligand arrangement becomes less important.

Although the discussion in this chapter will be largely concerned with theoretical matters, it is helpful to first take a brief look at the chemistry of the lanthanides and actinides. The solution and solid state chemistry of the lanthanides is dominated by the trivalent state. Other valence states exist— cerium(IV) is widely used as an oxidizing agent, for instance—but although their chemistry has been much studied, the knowledge of their spectroscopic and magnetic properties is still relatively primitive. It seems that these properties resemble those of the corresponding trivalent ions. Just as the spectroscopic and magnetic properties of iron(III) resemble those of manganese(II), because both are d^5 (although high spin–low spin complications occur), so one would expect europium(II) to resemble gadolinium(III) because both are f^7 (and without spin complications). Our discussion will therefore be confined to the trivalent lanthanides and actinides. This simplification is much more justified for the lanthanides than the actinides. All the lanthanides, but not all the actinides, have stable trivalent species. Although this valence state is stable for the simple actinides towards the end of the series, americium(III) onwards, for the early members it is usually strongly reducing and difficult to characterize. Thorium(III) probably does not exist except in the gas phase. Nonetheless, we shall confine our discussion of the actinides to the trivalent state, for then we can treat them and the lanthanides together.

Apart from oxidation–reduction behaviour, the chemistry of the earlier lanthanides resembles that of calcium(II)—the oxides absorb water to give hydroxides and they absorb carbon dioxide to give carbonates. The hydroxides tend to be slightly soluble in water giving alkaline solutions, the carbonates tend to be insoluble and so on. The chemistries of the later lanthanides tend to be more like that of aluminium, although the hydroxides are not amphoteric. These similarities with the properties of more common ions is not just an *aide-mémoire*; it can be exploited. Thus, calcium(II) ions are biologically very important—their movement is involved in nerve action, for example. Unfortunately, the study of such calcium ions is very difficult because calcium(II) lacks any convenient spectroscopic property—a spectroscopic 'handle'—by which it can be studied. Because of their similar chemistries, it is possible to replace the calcium(II) with an ion such as terbium(III), which does have convenient handles, as we shall see. Study of the terbium(III) then gives the information we could not get directly for calcium(II). Such replacements are a popular trick in bioinorganic chemistry (see Chapter 16).

The smooth change in chemical properties of the trivalent ions of the lanthanides across the series is associated with the so-called *lanthanide contraction*—the size of the trivalent ions becomes progressively smaller with increasing atomic number. Although the ionic radii often quoted in support of this statement are not free from objection,[1] the phenomenon seems real enough. Because the 4f orbitals are highly nodal and all of the nodes pass through the nucleus, the probability of finding an f electron close to the nucleus is low. The 5d and 6s orbitals—which are formally empty orbitals— may well accept (paired) electrons donated by ligands and in so doing help to determine the ionic radius of the lanthanide. But electrons in these orbitals have a much higher probability of being very close to the nucleus than do

---

[1] See footnote on page 303.

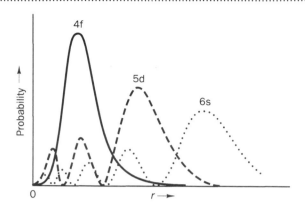

Fig. 11.1 Radial probability functions for hydrogen-like orbitals. Although the 4f (solid) has its maximum closer to the nucleus than 5d (dashed) and 6s (dotted), close to the nucleus the density of the latter two both exceed that of 4f.

the 4f (see Fig. 11.1). Moving across the lanthanide series, the addition of a 4f electron to balance each increase in positive charge on the nucleus is not sufficient to prevent electrons in the 5d and 6s (and, for that matter, the 5s and 5p) orbitals feeling the increased nuclear charge.[2] The effect on the 5p, 5d and all other outer electrons is a progressive orbital contraction and, with it, a decrease in the ionic radius. Although the data are incomplete, it is clear that the actinides exhibit an actinide contraction which closely parallels the lanthanide contraction.

A study of the electric spark discharge spectra of the elements (and, so, of the electronic energy levels) shows that the ground states of most atoms of the lanthanide elements have an outer-shell electronic configuration of the form . . . $4f^n 6s^2$, although some have . . . $4f^{n-1}5d^1 6s^2$. In the trivalent ions, also seen in the arc spectra, the two s electrons and one other are lost; the other one being an outer d electron if one were present in the atom. The electronic ground states of the trivalent lanthanides therefore have filled 5s and 5p shells, empty 5d and 6s and a number of 4f electrons which varies from none in lanthanum(III) through to 14 in lutecium(III). These configurations are detailed in Table 11.1. Following the lanthanides, the trivalent actinide ions contain a variable number of 5f electrons, no 6d and no 7s. The actual electronic configurations of the ground states of the atoms and trivalent ions are given in Table 11.2.

11.2 Shapes of f orbitals

Because of their importance to us, the object of the present section is to give the reader more familiarity with the shapes of f orbitals than that provided in Section 7.3. In that section the f orbitals were used to indicate the way that F terms (arising from d^n configurations, $n = 2$, 3, 7 or 8) split in an octahedral ligand field. The labels that were used to describe these f orbitals were abbreviated, just as the label d_{z^2} is an abbreviation for $d_{2z^2 - x^2 - y^2}$. The complete labels for the f orbitals accurately describe the lobes of the orbitals, their relative phases and their positions.[3] Drawings of the cubic set of f orbitals are repeated in Fig. 11.2, together with their abbreviated and

---

[2] Some textbooks treat the lanthanide contraction as involving only 4f electrons. This cannot be correct because these electrons do not determine the ionic radius—they are too well tucked away inside the ion.

[3] Here, as usual, we consider only the angular part of the complete orbital.

Table 11.1 The electronic configurations of the neutral atoms and trivalent lanthanide ions. Core electrons have been omitted

Atomic number	Name	Symbol	Isolated atom electron configuration	M^{3+} f electron configuration
57	Lanthanum	La	$5d^16s^2$	$4f^0$
58	Cerium	Ce	$4f^15d^16s^2$	$4f^1$
59	Praseodymium	Pr	$4f^36s^2$	$4f^2$
60	Neodymium	Nd	$4f^46s^2$	$4f^3$
61	Promethium	Pm	$4f^56s^2$	$4f^4$
62	Samarium	Sm	$4f^66s^2$	$4f^5$
63	Europium	Eu	$4f^76s^2$	$4f^6$
64	Gadolinium	Gd	$4f^75d^16s^2$	$4f^7$
65	Terbium	Tb	$4f^96s^2$	$4f^8$
66	Dysprosium	Dy	$4f^{10}6s^2$	$4f^9$
67	Holmium	Ho	$4f^{11}6s^2$	$4f^{10}$
68	Erbium	Er	$4f^{12}6s^2$	$4f^{11}$
69	Thulium	Tb	$4f^{13}6s^2$	$4f^{12}$
70	Ytterbium	Yb	$4f^{14}6s^2$	$4f^{13}$
71	Lutecium	Lu	$4f^{14}5d^16s^2$	$4f^{14}$

Note: Promethium, effectively, does not occur in nature. It is a fission product of uranium and may be made, for example, by neutron bombardment of neodymium to give an isotope with a half-life of just under 4 years.

Table 11.2 The electron configuration of the neutral atoms and trivalent actinide ions. Core electrons have been omitted

Atomic number	Name	Symbol	Isolated atom electron configuration	M^{3+} f electron configuration
89	Actinium	Ac	$6d^17s^2$	$5f^0$
90	Thorium	Th	$6d^27s^2$	$5f^1$
91	Protactinium	Pa	$5f^26d^17s^2$	$5f^2$
92	Uranium	U	$5f^36d^17s^2$	$5f^3$
93	Neptunium	Np	$5f^46d^17s^2$	$5f^4$
94	Plutonium	Pu	$5f^67s^2$	$5f^5$
95	Americium	Am	$5f^77s^2$	$5f^6$
96	Curium	Cm	$5f^76d^17s^2$	$5f^7$
97	Berkelium	Bk	$5f^86d^17s^2$ or $5f^97s^2$	$5f^8$
98	Californium	Cf	$5f^{10}7s^2$	$5f^9$
99	Einsteinium	Es	$5f^{11}7s^2$	$5f^{10}$
100	Fermium	Fm	$5f^{12}7s^2$	$5f^{11}$
101	Mendelevium	Md	$5f^{13}7s^2$	$5f^{12}$
102	Nobelium	No	$5f^{14}7s^2$	$5f^{13}$
103	Lawrencium	Lr	$5f^{14}6d^17s^2$	$5f^{14}$

complete labels. The reader may have noted the use of the phrase cubic set in the preceding sentence. This is because the f orbitals of Fig. 11.2 are only appropriate for cubic molecules (in practice, this means molecules with O_h or T_d symmetries or related point groups—molecules with symmetries such that the x, y and z axes are symmetry-related). For non-cubic geometries one uses a different set of f orbitals. Actually, the situation is not all that unfamiliar. Consider a d_{z^2} orbital in an octahedron. It is so normal to choose the z axis to coincide with a fourfold axis that we seldom consider any

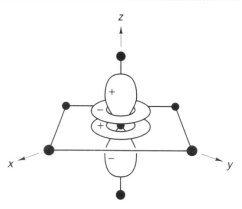

$$f_{z3} = f_{z(2z^2 - 3x^2 - 3y^2)}; \text{ similar are } f_{x3} = f_{x(2x^2 - 3y^2 - 3z^2)} \text{ and } f_{y3} = f_{y(2y^2 - 3z^2 - 3x^2)}$$

Fig. 11.2 The cubic set of f orbitals together with their shortened and detailed Cartesian angular forms.

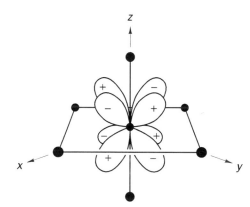

$$f_{z(x^2 - y^2)}; \; f_{x(y^2 - z^2)} \text{ and } f_{y(z^2 - x^2)} \text{ are similar}$$

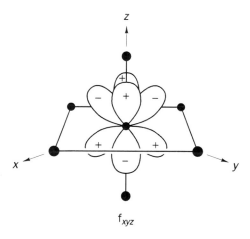

$$f_{xyz}$$

alternative. However, in real life an octahedral complex is often slightly distorted and this distortion is frequently a distortion along a threefold axis (this case was discussed in Section 7.10). The threefold axis is then the axis of highest symmetry and so becomes the z axis and the d_{z^2} orbital has to be oriented along it. This d_{z^2} orbital cannot be the same as that which is oriented along a fourfold axis in the undistorted octahedron, although they look the same. In fact, the d_{z^2} (threefold axis) is a mixture of the t_{2g} set, d_{xy}, d_{yz} and d_{zx} (fourfold axis). For d_{z^2} the existence of two choices of axes does not mean a change in shape; for the f orbitals a change in axis set *does* mean a change in shape.

There is another way of looking at this. In O_h symmetry the f orbitals transform as $A_{2u} + T_{1u} + T_{2u}$. Reduce the symmetry by a distortion (compression or elongation, for example) along a fourfold axis so that the symmetry is now D_{4h}. The f orbitals now transform as $(B_{1u}) + (A_{2u} + E_u) + (B_{2u} + E_u)$, where the brackets correspond, in order, to the symmetries in O_h given above. It can be seen that two different sets of f orbitals have E_u symmetry. Because they have the same symmetry they can mix and it proves convenient to let them do this and to work with combinations of them. As a result, some differently shaped f orbitals arise.[4] The members of this so-called general set of f orbitals are shown in Fig. 11.3, where both their abbreviated and complete labels are given.

The lanthanide and actinide trivalent ions commonly occur with high coordination numbers, typically 7, 8, or 9 but up to 12, in low-symmetry geometries. Simple theoretical models, such as the extended Hückel and angular overlap models (Sections 10.5 and 10.6) have therefore been applied to them since symmetry alone does not give much insight. In such calculations the general set of f orbitals would be used. It is perhaps appropriate to comment that if the bonding in lanthanide and actinide complexes were entirely ionic then it would be nonsense to apply such models to them—they depend on the existence of a (covalent) overlap between the metal and ligand orbitals. The fact that the f electrons in these compounds have properties which are very close to those of the isolated M^{3+} ions should not lead one to conclude that the same is true of all other electrons.

11.3 Electronic structure of the lanthanide and actinide ions

Although the idea can be over-emphasized,[5] there is little doubt that f electrons are tucked away well inside the atoms of the lanthanides and

---

[4] The argument used here is correct for all but one of the geometries of interest. It does not hold for the group $D_{\infty h}$ (which is appropriate for the important ion UO_2^{2+}), where the f orbitals split into $A_{2u} + E_{1u} + E_{2u} + E_{3u}$, and no symmetry species appears twice.

[5] A note of caution is appropriate because we have ignored relativistic effects. These effects most directly concern 1s electrons because s orbitals do not contain any node at the nucleus. The 1s electrons are distinguished because they may be very close to the nucleus and only avoid capture because of their high speed, which for the heavier elements approaches the speed of light. This, relativistically, increases the mass of the 1s electrons which in turn means a smaller orbital (in the Schrödinger model of H-like atoms the most probable distance of a 1s electron away from the nucleus is inversely proportional to the electron mass; the essentials carry over into more complicated atoms). So, the 1s electrons screen the nucleus a bit more effectively than expected on a non-relativistic model. Seeing a smaller positive charge on the nucleus, f and d electrons occupy orbitals which are both larger and have lower ionization potentials than we might have expected, thus increasing their availability for bonding. In the context of the present chapter this is particularly important in the context of the f orbitals. Although the primary relativistic effect concerns the 1s electrons, all s orbitals have to remain orthogonal to each other. So, if the 1s contracts, so too must all other s orbitals, to maintain orthogonality. There therefore is an effect on outer s electrons also, an effect which seems particularly pronounced for the elements Pt, Au and Hg.

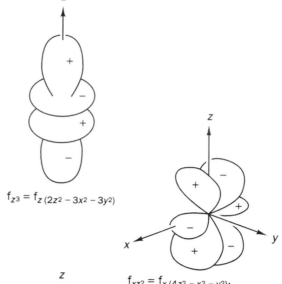

$$f_{z3} = f_{z\,(2z2\,-\,3x2\,-\,3y2)}$$

$$f_{xz2} = f_{x\,(4z2\,-\,x2\,-\,y2)},$$
$$f_{yz2} = f_{y\,(4z2\,-\,x2\,-\,y2)} \text{ is similar}$$

Fig. 11.3 The general set of f orbitals together with their shortened and detailed Cartesian angular forms.

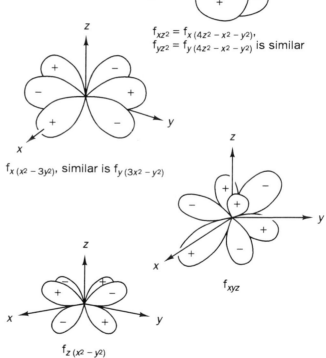

$$f_{x\,(x2\,-\,3y2)}, \text{ similar is } f_{y\,(3x2\,-\,y2)}$$

$$f_{xyz}$$

$$f_{z\,(x2\,-\,y2)}$$

actinides. This means that they are rather insensitive to their molecular environment. So, the crystal field splittings of f orbitals are about one-hundredth of those of d orbitals. That is, in the language of crystal field theory, all lanthanide and actinide complexes are weak field, high spin. Similarly, the f–f electronic spectra of the complexes are very similar to those of the free ions as seen in arc spectra. The fact that the spectra of, nearly, isolated ions can be seen in crystalline materials has led to extensive studies aimed at their understanding. We will return to these spectra later in this chapter; it is the purpose of the present section to begin to assemble the background which will make such a discussion possible.

Clearly, our discussion of the f electrons of lanthanides and actinides must be appropriate to the free ion with, initially, total neglect of crystal fields. If we insisted, we could work with the f orbitals of the previous section—and they have the clear advantage of representing static charge densities. However, we will prefer to use rotating charge densities, for they are then associated with an angular momentum and it turns out that this angular momentum is quantized. The connection between static and rotating[6] charge densities and with angular momentum is simple and connected with the number of planar nodes of the static orbitals seen down the z axis. (In the case of the free ion it is helpful to think of it as being subjected to a weak electric field which thus enables a z axis to be defined; in a molecule it is even easier to think of such electric fields defining a z axis.) Figure 11.4 indicates the answer. It shows that one pair of general f orbitals, $f_{x(x^2-3y^2)}$ and $f_{y(3x^2-y^2)}$, have three planar nodes containing the z axis. Static charge distributions are obtained from a pair of rotating charges, members of the pair being equivalent except that they are rotating in opposite senses. The two rotating charge densities are combined in phase and out of phase to give the two static distributions. So, the two static f orbitals of Fig. 11.4 are combinations of rotating distributions, rotating distributions which have angular momentum quantum numbers of 3 and -3. As Fig. 11.4 shows, the number 3 reappears in the two static orbitals as the number of planar nodes containing the z axis. Similarly, the static orbitals f_{xyz} and $f_{z(x^2-y^2)}$, each with two nodal panes containing the z axis, are combinations of those rotating orbitals with angular momentum quantum numbers 2 and -2. The orbitals f_{xy^2} and f_{yz^2}, with one relevant nodal plane, are derived from the rotating orbitals with angular momentum quantum numbers 1 and -1. Finally, the static orbital f_{z^3} is also the function with angular momentum 0; as it has 0 angular momentum it is a static function already.

We can now move towards finding the electronic ground states of all the trivalent lanthanides and actinides. We already know that these are weak field, high spin, species, with the maximum number of unpaired f electrons possible. This means that as far as possible the f electrons occupy different orbitals. It is at this point that the use of rotating charge densities, orbitals classified according to their angular momentum, becomes convenient. An example shows how this works. Consider the f^2 configuration. Electron repulsion means that the electrons will stay as far apart as possible and this means that they will occupy the maximum number of different orbitals. So, we allocate each electron to the orbital with the highest angular momentum available. For the f^2 configuration this means that the first electron is allocated to the orbital with angular momentum 3 and the second to the orbital with angular momentum 2. The total angular momentum is the sum, $3 + 2 = 5$. It is this simplicity of addition which makes rotating charge densities so much simpler to work with. The relationship between the total angular momentum and the term of the ground state is the usual one:

Total angular momentum	0	1	2	3	4	5	6	7	8
Term symbol	S	P	D	F	G	H	I	J	K

[6] Note that we use the terms static and rotating charge densities (or orbitals) in preference to the more conventional names real and complex because they are felt to be easier to understand by most readers.

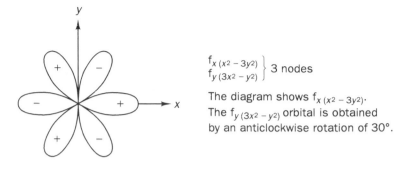

$\left.\begin{array}{l} f_{x(x^2 - 3y^2)} \\ f_{y(3x^2 - y^2)} \end{array}\right\}$ 3 nodes

The diagram shows $f_{x(x^2 - 3y^2)}$.
The $f_{y(3x^2 - y^2)}$ orbital is obtained
by an anticlockwise rotation of 30°.

Fig. 11.4 The nodal patterns of the general set of f orbitals, viewed down the z axis (cf. Fig. 10.2(a)).

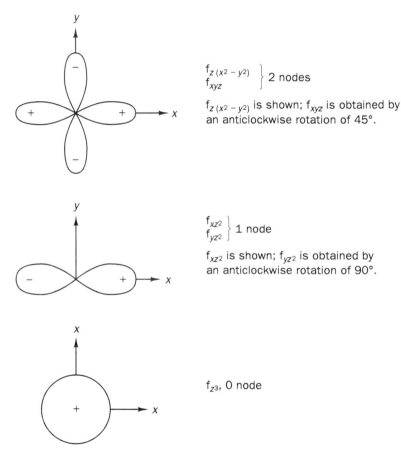

$\left.\begin{array}{l} f_{z(x^2 - y^2)} \\ f_{xyz} \end{array}\right\}$ 2 nodes

$f_{z(x^2 - y^2)}$ is shown; f_{xyz} is obtained by an anticlockwise rotation of 45°.

$\left.\begin{array}{l} f_{xz^2} \\ f_{yz^2} \end{array}\right\}$ 1 node

f_{xz^2} is shown; f_{yz^2} is obtained by an anticlockwise rotation of 90°.

f_{z^3}, 0 node

It follows that the ground state term of an f^2 ion is H. Because we have two unpaired electrons it is ^3H. Moving to the f^3 configuration, the extra electron is allocated to an orbital with angular momentum 1, giving a total angular momentum of 6 and a ^4I term. Continuing this pattern gives the data in Table 11.3, which also shows the details of each calculation.

Had the f electrons in complexes of the lanthanides and actinides experienced crystal fields comparable to those in complexes containing d electrons, we would next have to include these fields, just as was done in Chapters 7 and 8. But these fields are small—much more important

Table 11.3 Trivalent lanthanide ion ground state characteristics. For all of these ions the f electrons are the ones that determine the (many electron) term of the ground state of the ion

Ion	f electron configuration	Angular momentum of relevant orbitals	Total orbital angular momentum	Term associated with the ground state
La^{III}	f^0	none	0	^1S
Ce^{III}	f^1	3	3	^2F
Pr^{III}	f^2	3, 2	5	^3H
Nd^{III}	f^3	3, 2, 1	6	^4I
Pm^{III}	f^4	3, 2, 1, 0	6	^5I
Sm^{III}	f^5	3, 2, 1, 0, −1	5	^6H
Eu^{III}	f^6	3, 2, 1, 0, −1, −2	3	^7F
Gd^{III}	f^7	3, 2, 1, 0, −1, −2, −3	0	^8S
Tb^{III}	f^8	3^a	3	^7F
Dy^{III}	f^9	3, 2	5	^6H
Ho^{III}	f^{10}	3, 2, 1	6	^5I
Er^{III}	f^{11}	3, 2, 1, 0	6	^4I
Tm^{III}	f^{12}	3, 2, 1, 0, −1	5	^3H
Yb^{III}	f^{13}	3, 2, 1, 0, −1, −2	3	^2F
Lu^{III}	f^{14}	3, 2, 1, 0, −1, −2, −3	0	^1S

^{a} From Tb^{III} onwards, for simplicity, the half-filled shell that is also present is not detailed.

are the effects of spin–orbit coupling and it is these that we now consider. As will be seen, the effects of spin–orbit coupling are most simply covered using rotating orbitals, although later an attempt will be made to give their static equivalents.

11.4 Spin–orbit coupling

The phenomenon of spin–orbit coupling was first met in Section 8.4, where we were interested in its spectroscopic consequences and, again, in Section 9.5 when it was relevant to the magnetic properties of transition metal complexes because its effects are much greater than those of the magnetic field. In this latter section it was found convenient to regard spin–orbit coupling as resulting from the coupling of orbital, solenoid-like, magnets and spin, bar-like, magnets. There is a fairly evident connection between rotating orbitals, solenoid-type magnets and angular momenta. Now, the phrase 'the electron has an intrinsic spin' is actually equivalent to the statement that 'the electron has an intrinsic angular momentum'. Angular momentum is a characteristic of both orbital and spin magnets. So, equivalent to the statement that 'the orbital and spin magnets couple together in the phenomenon of spin–orbit coupling' is the statement that 'the coupling of spin and orbital angular momenta gives a resultant angular momentum'. This is a more useful statement because it means that we can treat spin–orbit coupling with no more difficulty than we had in compiling Table 11.3. Let us look at an example, that of a single f electron. An individual electron can have spin, that is angular momentum, of either $+\frac{1}{2}$ or $-\frac{1}{2}$. In Table 11.3 the f^1 case seems almost trivial; a single f electron, one electron with spin $\pm\frac{1}{2}$ in an orbital set which has a maximum angular momentum of 3, gives rise to a ^2F term. Bringing these angular momenta together, there are only two possible resultants, $3 \pm \frac{1}{2}$, i.e. $\frac{7}{2}$ and $\frac{5}{2}$. So, spin–orbit coupling splits the ^2F ground state term of cerium(III), an f^1 ion, into two levels, $^2F_{7/2}$ 'doublet F seven halves' and $^2F_{5/2}$ 'doublet F five halves'. The subscripts are the two possible values of j, where j denotes a sum of spin and orbital angular momenta, $j = l + s$. It is perhaps helpful to think of this splitting resulting from the interaction of two magnets (which, for simplicity can be thought of as bar magnets), side by side. The lowest energy arrangement is that in which the north pole of one magnet is next to the south pole of the other. So, here, the $^2F_{5/2}$ is the more stable level. It is a general pattern for atoms that an angular momentum of m implies a $(2m + 1)$ degeneracy. For instance, spin functions with $s = \frac{1}{2}$ are doubly degenerate; s functions with angular momentum of 0 are singly degenerate; p functions with an angular momentum of 1 are threefold degenerate and so on. So, here the $j = \frac{7}{2}$ functions are eightfold degenerate and the $j = \frac{5}{2}$ are sixfold, a total of 14 functions. This is just the number implied by the parent symbol, ^2F; a 2×7 fold degeneracy. The reader may reasonably complain that we have treated the two angular momenta in different fashion in the above development. We have taken the *maximum* orbital angular momentum and added to it, in turn, each *component* of the spin angular momentum. It is by no means self-evident that this procedure is valid; unfortunately it would break the continuity of the discussion too much to present a justification for it here but one is given in Appendix 12. Using this procedure we arrive at the spin–orbit split levels given in Table 11.4.

The final question to be answered for each configuration is which of the spin–orbit levels becomes the ground state? An answer has already been given for the f^1 case, where it was concluded that the smallest value of j becomes the ground state (an analogy was drawn with a pair of magnets being most stable when the north pole of one is next to the south of the other). This result is general for the first half of the lanthanide series, the *lowest* value of j is the most stable. Just as was found for d electron systems, so too for f electron systems; for more than half-filled shells it is simplest to work in terms of holes rather than electrons. Now, if in an orbital there is a single electron with spin up, then if we describe this situation using the hole formalism then we have to talk of a hole with spin down (this is the spin of the electron which is absent). It follows that the spin–orbit state which is most stable when talking about electrons will be the least stable

Table 11.4 Spin–orbit levels arising from f electron ground state Russell–Saunders terms

f electron configurations	Term	Spin–orbit levels
f^1, f^{13}	^2F	$^2F_{5/2}$, $^2F_{7/2}$
f^2, f^{12}	^3H	$^3H_4$, $^3H_5$, $^3H_6$
f^3, f^{11}	^4I	$^4I_{9/2}$, $^4I_{11/2}$, $^4I_{13/2}$
f^4, f^{10}	^5I	$^5I_4$, $^5I_5$, $^5I_6$, $^5I_7$, $^5I_8$
f^5, f^9	^6H	$^6H_{5/2}$, $^6H_{7/2}$, $^6H_{9/2}$, $^6H_{11/2}$, $^6H_{13/2}$, $^6H_{15/2}$
f^6, f^8	^7F	$^7F_0$, $^7F_1$, $^7F_2$, $^7F_3$, $^7F_4$, $^7F_5$, $^7F_6$
f^7	^8S	$^8S_{7/2}$

Table 11.5 Ground and low-lying electronic levels of the lanthanides and actinides

f electron configuration	Relevant ions	Ground state	Low-lying excited levels
f^0	La^{III}, Ac^{III}	$^1S_0$	–
f^1	Ce^{III}, Th^{III}	$^2F_{5/2}$	$^2F_{7/2}$
f^2	Pr^{III}, Pa^{III}	$^3H_4$	$^3H_5$, $^3H_6$
f^3	Nd^{III}, U^{III}	$^4I_{9/2}$	$^4I_{11/2}$, $^4I_{13/2}$, $^4I_{15/2}$
f^4	Pm^{III}, Np^{III}	$^5I_4$	$^5I_5$, $^5I_0$, $^5I_7$, $^5I_0$
f^5	Sm^{III}, Pu^{III}	$^6H_{5/2}$	$^6H_{7/2}$, $^6H_{9/2}$, $^6H_{11/2}$, $^6H_{13/2}$, $^6H_{15/2}$
f^6	Eu^{III}, Am^{III}	$^7F_0$	$^7F_1$, $^7F_2$, $^7F_3$, $^7F_4$, $^7F_5$, $^7F_6$
f^7	Gd^{III}, Cm^{III}	$^8S_{7/2}$	–
f^8	Tb^{III}, Bk^{III}	$^7F_6$	$^7F_5$, $^7F_4$, $^7F_3$, $^7F_2$, $^7F_1$, $^7F_0$
f^9	Dy^{III}, Cf^{III}	$^6H_{15/2}$	$^6H_{13/2}$, $^6H_{11/2}$, $^6H_{9/2}$, $^6H_{7/2}$, $^6H_{5/2}$
f^{10}	Ho^{III}, Es^{III}	$^5I_8$	$^5I_7$, $^5I_6$, $^5I_5$, $^5I_4$
f^{11}	Er^{III}, Fm^{III}	$^4I_{15/2}$	$^4I_{13/2}$, $^4I_{11/2}$, $^4I_{9/2}$
f^{12}	Tm^{III}, Md^{III}	$^3H_6$	$^3H_5$, $^3H_4$
f^{13}	Yb^{III}, No^{III}	$^2F_{7/2}$	$^2F_{5/2}$
f^{14}	Lu^{III}, Lr^{III}	$^1S_0$	–

Note the symmetry in this table (compare the first level listed for an f^n ion with the last listed for the f^{14-n}). This symmetry is detailed in the text.

when talking about holes. So, for the second half of the lanthanide series it is the spin–orbit state with the *highest j* value which becomes the ground state. This is not the easiest of arguments to follow and so an example is given in some detail in Appendix 14. For the f^7, the half-filled shell case, where the orbital angular momentum contributions sum to zero (although the spins most certainly do not) there is no spin–orbit splitting.

We are now in a position not only to detail the ground states of all the lanthanide and actinide trivalent ions but also to give some of the low-lying excited states when these result from spin–orbit splitting. Excitation to these low-lying levels corresponds to energies in the infrared or near infrared regions of the spectrum for the lanthanides, the splittings resulting from spin–orbit coupling being of the order of 1000 cm^{-1}. These states are detailed in Table 11.5. This Table has an underlying symmetry, perhaps most readily revealed if the ^8S, f^7, entry is moved sightly to the right and then regarded as an approximate centre of symmetry.

11.5 Spin–orbit coupling in pictures[7]

As has been pointed out, for the lanthanides and actinides spin–orbit coupling has a greater effect on f electron energy levels than does the crystal field generated by the surrounding ligands in their complexes. In the previous section it has been seen that it is not difficult to introduce spin–orbit coupling mathematically if rotating orbitals are used. The question inevitably arises of whether it is possible to give any sort of static orbital picture of the

---

[7] This section is an attempt to make understandable the subject of so-called *double groups*. These are important for an understanding of systems with an odd number of electrons, not only for f electron systems but for d also.

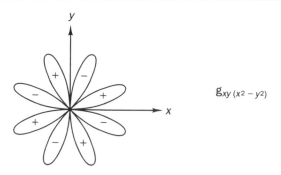

$g_{xy\,(x^2-y^2)}$

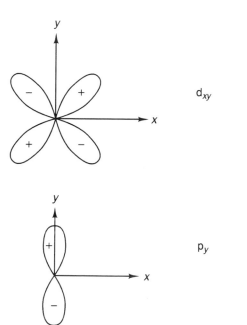

d_{xy}

p_y

Fig. 11.5 View down the z axis of orbitals with 4, 2 and 1 nodes containing the z axis.

phenomenon. One would expect the answer to be in the affirmative. After all, the (static) p and d orbitals with which the chemist is so familiar actually first appear as rotating orbitals, when the Schrödinger equation for the hydrogen atom is solved. In this section is described the static orbital pictures of spin–orbit coupled functions. Unfortunately, the pictures contain a feature which makes them less immediately useful than those of p, d and f orbitals.

We start with an unexpected step. We introduce the orbital $g_{xy(x^2-y^2)}$. The angular part of this orbital is shown in Fig. 11.5 along with d_{xy} and p_y. This series is chosen because the g orbital is a static orbital corresponding to angular momentum 4 (it has four planar nodes containing the z axis), the d orbital corresponds to angular momentum 2 and the p orbital to angular momentum 1. Each corresponds to one-half of the angular momentum of the orbital which precedes it. Figure 11.5 has been used to compile Table 11.6, a Table in which we have included the next member in the geometric series 4, 2, 1, $\frac{1}{2}$. We can learn something about the static orbital picture of the angular momentum $=\frac{1}{2}$ case by extrapolation from the pattern revealed by the other three. The reader should now study Table 11.6 and complete it by filling in the missing entries, by replacing the question marks by the appropriate number and angle.

Table 11.6

Property	Function			
	$g_{xy(x^2-y^2)}$	d_{xy}	p_y	Spin $\frac{1}{2}$
$\begin{vmatrix} \text{angular} \\ \text{momentum} \end{vmatrix}$	4 $\xrightarrow{\div 2}$	2 $\xrightarrow{\div 2}$	1 $\xrightarrow{\div 2}$	$\frac{1}{2}$
Number of angular nodes (Fig. 11.5)	4 $\xrightarrow{\div 2}$	2 $\xrightarrow{\div 2}$	1 $\longrightarrow$	?
Smallest angle between nodes (Fig. 11.5)	$45°$ $\xrightarrow{\times 2}$	$90°$ $\xrightarrow{\times 2}$	$180°$ $\longrightarrow$	?

The number, of course, is one-half and the angle three-hundred and sixty degrees. This seems nonsense. What meaning can be given to half a node and what does it mean to have three-hundred and sixty degrees between complete nodes? Actually, both are entirely sensible. The essential step is to change the identification of the identity operation (the E at the head of character tables, the 'leave alone' operation). This operation can be equated with a rotation of $360°$ (although this equation is not often explicitly stated). Instead, we now equate it with a rotation of $720°$. It is easy enough to make this change for the functions shown in Fig. 11.5. This is done in the top row of Fig. 11.6 which shows the angles $0°$ through to $360°$, compressed into the (actual) region $0°$ to $180°$. The diagrams are completed in the middle row, where the angular pattern from $360°$ through to $720°$ simply duplicates that for $0°$ to $360°$. In this row the static counterpart of the angular momentum $\frac{1}{2}$ case has been included, which can be seen to comply with the requirements imposed on it by the completed Table 11.6. It is evident that these diagrams are becoming rather congested and so at the bottom of Fig. 11.6 are given simplified pictures, drawn in a pattern that will be followed for the remainder of this section.

Although a picture of an angular momentum $= \frac{1}{2}$ function (more accurately, a $|\frac{1}{2}|$ function) has been obtained, it still seems somewhat artificial. It can be given more physical reality using a Möbius strip. The top of Fig. 11.7 shows a long strip of paper, creased so that it can easily be bent (away from the viewer) along its central axis (indicated by the arrows). Lobes are drawn with phases as shown, the lobes being terminated by nodes at the edge of the paper. Suppose that the back of the strip of paper is coated with a contact adhesive, so that as soon as the strip is folded back it sticks to itself. Join the ends of the strip together but in doing so twist the strip so that the back comes to the front. The result is shown at the bottom of Fig. 11.7. Starting at any point on this Möbius strip it is necessary to go round twice, akin to rotating by $720°$, before regaining the starting point. The nodal pattern encountered in traversing the Möbius strip is just that shown in Fig. 11.6 for the angular momentum $= \frac{1}{2}$ function. Having pictured a static angular momentum $\frac{1}{2}$ function it is not difficult to extend the approach to other half integer functions and this is done in Fig. 11.8, which also shows the corresponding unfolded, unstuck, Möbius strips. A few points remain to complete this section. Just as in all other cases except the angular momentum $= 0$ case, all functions appear in pairs (only one member of each

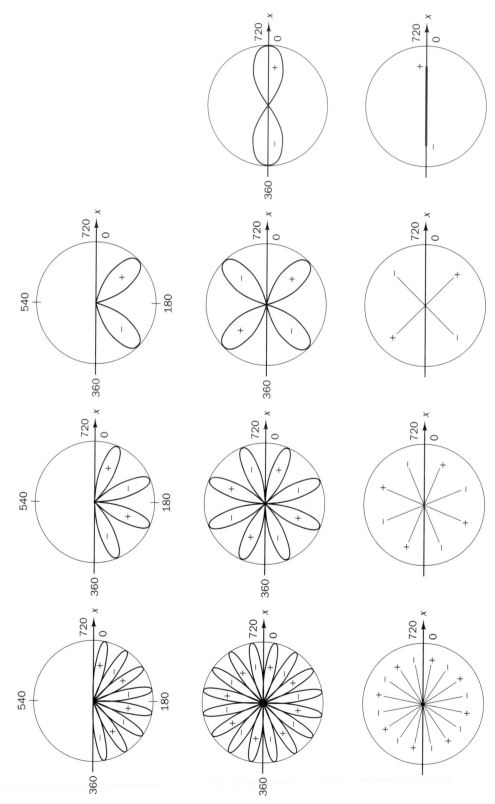

Fig. 11.6 The identity operation = 720° representation of the orbitals of Fig. 11.5, together with the spin $= \frac{1}{2}$ function.

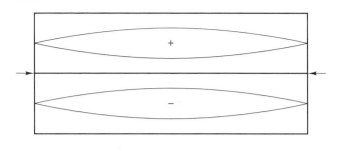

Fig. 11.7 The Möbius strip representation of a spin $= \frac{1}{2}$ function. There is a node on each side of the strip at the point indicated by an arrow (this is the point of contact of the two ends of the top diagram). In the bottom diagram only the signs of the function are given. There is no attempt to indicate amplitude. Note that two different Möbius strip representations exist—the twist can be made in either of two senses.

Stick, twist ↓ and join

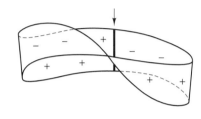

Angular momentum $= \left| \frac{3}{2} \right|$

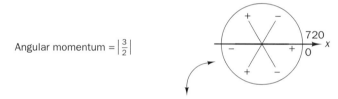

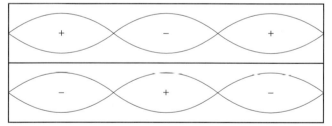

Fig. 11.8 The Möbius strip constructions for angular momentum $= \frac{3}{2}$ and $\frac{5}{2}$ functions.

Angular momentum $= \left| \frac{5}{2} \right|$

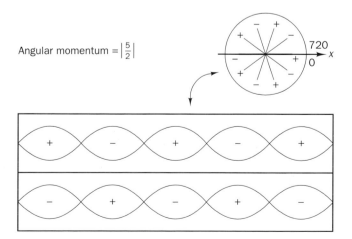

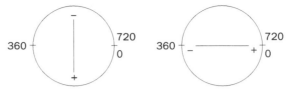

The two static functions corresponding to angular momentum $= \pm \frac{1}{2}$

Fig. 11.9 The pairs of static (real) functions corresponding to angular momentum $= \pm \frac{1}{2}$ and $\pm \frac{3}{2}$.

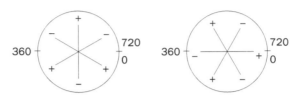

The two static functions corresponding to angular momentum $= \pm \frac{3}{2}$

pair is shown in Fig. 11.5). So, too, for the angular momentum $= \frac{1}{2}, \frac{3}{2}, \ldots$, and so on cases. Figure 11.9 shows both members of the $\frac{1}{2}$ and $\frac{3}{2}$ pairs. Finally, this section has been entirely concerned with half-integer values of the angular momentum. Spin–orbit functions may also have integer values of angular momentum; for these the pictures look the same as for the familiar orbitals with the same angular momentum, although their meaning, of course, is somewhat different.

11.6 Excited states of f electron systems

When d electron systems were considered in Chapter 7 it was found that a detailed study of the d^1 and d^2 configurations could easily be extended to cover all the low-lying excited states of the same spin multiplicity as the ground state for all d^n configurations. Even if spin–orbit coupling could be ignored, the f electron case is more complicated. As Table 11.5 shows, not only f^1 and f^2 but also f^3 configurations would have to be included before all the possible ground terms F, H and I, had been covered. But spin–orbit coupling cannot be ignored and this means that we cannot even talk about spin allowed and spin forbidden transitions and thus restrict the discussion to terms with the same spin multiplicity. Faced with this complexity, the f^2 problem and its solution will be outlined. This will enable us to see how to tackle the other configurations—the extension of our arguments to include them is more tedious than difficult. Despite what has just been said, we shall at first appear to ignore spin–orbit coupling. However, we shall also ignore any restriction on spin multiplicity (in Chapter 7 the discussion was restricted to terms with the same spin multiplicity as the ground state) and this will enable us, later on, to mix the different spin multiplicities by the mechanism of spin–orbit coupling.

In Section 11.3 a ^3H ground term for the f^2 configuration was obtained by feeding electrons with spins parallel into rotating orbitals characterized by angular momenta of 3 and 2. Had we been prepared to pair these electrons they could both have been placed in the orbital with angular momentum 3

Table 11.7 Terms arising from the f^2 configuration (left-hand column) and states consequently resulting from spin–orbit coupling

Russell–Saunders term	Spin–orbit states
^1S	$^1S_0$
^3P	$^1P_0$ $^1P_1$ $^1P_2$
^1D	$^1D_2$
^3F	$^3F_2$ $^3F_3$ $^3F_4$
^1G	$^1G_4$
^3H	$^3H_4$ $^3H_5$ $^3H_6$
^1I	$^1I_6$

to give a ^1I term. This ^1I term and the ^3H are the start of a neat pattern based, ultimately, on the pattern that the angular momenta of the two electrons can either add or subtract,[8] but never go negative. The greatest orbital angular momentum they can have as a pair is 6 and the smallest 0, seven values in all. Write down, in order, all the term symbols corresponding to all values of the orbital angular momenta from 0 to 6. We obtain

$$S + P + D + F + G + H + I$$

Now add the only two possible spin labels, 1 and 3, alternately, in such a way as to include ^3H and ^1I. We obtain:

$$^1S + ^3P + ^1D + ^3F + ^1G + ^3H + ^1I$$

This is a complete list of the terms that arise from the f^2 configuration. A check on the correctness of the result can be obtained by counting wavefunctions. For the f^2 configuration the first electron can be inserted in any one of 14 ways (we have seven orbitals, the spin can be either up or down); the second electron can be fed in in any one of 13 ways (it cannot be in the same orbital as the first and also have the same spin). Because the electrons are indistinguishable we have counted every possibility twice—so, ↑↓ and ↓↑ have been counted as different. It follows that the total number of wavefunctions is $14 \times 13/2 = 91$. This number, 91, is the same as the number of functions contained in the terms that we generated above:

$$^1S + ^3P + ^1D + ^3F + ^1G + ^3H + ^1I$$
$$1 + 9 + 5 + 21 + 9 + 33 + 13 = 91$$

It is now time to introduce spin–orbit coupling and this is done using the procedure that has already been described in Section 11.4 and Appendix 12. We simply add each component of the spin angular momentum to the orbital angular momentum. The result is given in Table 11.7, where the final levels are stacked according to their total angular momenta, according to their j values, this being given by the subscripts. Now comes an important point;

[8] The reader may reasonably object—what about cases such as -3 combined with -3, this gives us -6? In fact such cases are included in the discussion in the text. So, for example, the I term, from the ^1I, comprises 13 ($2L + 1$; here $L = 6$) different functions, differing from each other because the z component of the orbital angular momentum spans the 13 values from $+6$ through to -6. We do not have freedom to use the -6 function a second time; the angular momenta which are reflected in term symbols such as $S, P, D, F, \ldots$, are never negative.

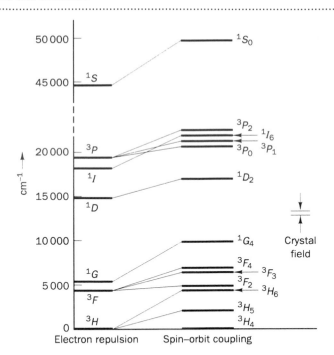

Fig. 11.10 The electronic energy level diagram for Pr^{III} (4f^2). A typical crystal field splitting is shown at the right; its effects are much smaller than those of spin–orbit coupling. As with Tanabe–Sugano diagrams, the horizontal axis is taken as the ground state ($^3H_4$). This enables the effects of interactions between levels with the same total angular momentum to be more clearly seen. Note, for instance, the relative upward displacements of levels with the j quantum number 4 ($^3H_4$, $^3F_4$, $^1G_4$).

levels which are in the same column can interact together because of spin–orbit coupling; they have the same values of j. In terms of the pictorial representation of the previous section, the spin orbitals that interact have the same number of nodal planes containing the z axis; they have the same symmetry. As is evident from Table 11.7, some levels stand in isolation and so remain unchanged by spin–orbit coupling ($^1P_1$ for instance); for others, mixing is between functions with the same spin ($^1S_0$ and $^1P_0$). However, the fact that the ground state, $^3H_4$, mixes with $^1G_4$ ensures that all of the f → f electronic transitions of an f^2 ion such as praseodymium(III) are of mixed spin character (a mixture of spin triplet and singlet), emphasizing the fact that because of spin–orbit coupling it is not useful to attempt to talk of spin on its own.

The above discussion has been relatively superficial but has covered all of the important points. In Fig. 11.10 is given an actual energy level diagram, which is not relatively superficial, and which shows how the primary separation between levels is due to electron repulsion, a topic which was considered in some detail in Chapter 8, in particular. Note the way that spin–orbit coupling changes the levels. So, although the $^3H_4$ is the most stable level originating in the ^3H, $^3F_4$ is the least stable of those coming from ^3F; the splitting between $^3F_4$ and $^1G_4$ is much greater than that between ^3F and ^1G. Several other similar patterns can be discerned in Fig. 11.10.

This completes our discussion of the excited states of f electron systems. For completeness, however, Table 11.8 gives a more complete list of the Russell–Saunders terms arising from all f^n configurations. It does not include the effects of spin–orbit coupling, which have to be worked out for each case individually. The reader may find it a helpful exercise to select a configuration from Table 11.8 and use it to construct a table similar to Table 11.7.

Table 11.8 Terms arising from f^n configurations

f electron configuration	Number of functions listed	Total number of functions	Terms arising
f^1, f^{13}	14	14	^2F
f^2, f^{12}	91	91	$^3H + {}^3F + {}^1G + {}^1D + {}^3P + {}^1I + {}^1S$
f^3, f^{11}	294	364	$^4I + {}^4F + {}^4S + {}^2H + {}^2G + {}^2K + {}^4G + {}^2D + {}^2P + {}^2I + {}^2J + {}^4D$
f^4, f^{10}	394	1001	$^5I + {}^5F + {}^5S + {}^3K + {}^5G + {}^3H + {}^3G + {}^3J + {}^3D + {}^3P + {}^3I + {}^5D$
f^5, f^9	238	2002	$^6H + {}^6F + {}^4H + {}^4D + {}^6P + {}^4I + {}^4S$
f^6, f^8	49	3003	^7F
f^7	478	3432	$^8S + {}^6P + {}^6I + {}^4S + {}^4D + {}^4H + {}^4I + {}^4K + {}^6D + {}^6G + {}^6F + {}^6H$

For f^n, $3 < n < 11$, only a selection of terms is given, listed in approximate energy sequence (lowest first). A limited number is given because mixing of the type shown in Fig. 8.16 occurs and particularly complicates the more highly excited states of those configurations for which the total number of functions is large.

11.7 Electronic spectra of f electron systems

As a class, the complexes of the lanthanides are both beautifully and delicately coloured. There is a symmetry in the colours of the ions in aqueous solution, a symmetry which seems to be accidental. These colours are given in Table 11.9, listed in a way which shows the colour parallel between f^n and f^{14-n} ions. Also shown are the colours of some organo-lanthanide compounds, which do not show anything like the same pattern, colours which support the idea that the colour pattern of the aqua ions is accidental.

The actual absorption bands that occur in the spectra of the lanthanides and actinides and which are associated with electronic transitions have been divided into three types.

1. **f → f transitions** These transitions are localized entirely within the f shell and so, like d → d transitions, are formally forbidden. However, like d → d, they actually occur and give rise to a large number of weak, sharp bands from the infrared through to the visible region. The absorption band patterns obtained from species in solution are closely related to the emission spectrum of the corresponding ion, obtained from the arc spectra of the elements.

2. **nf → (n + 1)d transitions** Here, the n and $(n + 1)$ are principal quantum numbers so these are allowed bands in which a 4f (lanthanides) or 5f (actinides) electron is promoted to a 5d and 6d orbital, respectively. They give rise to quite intense and broad bands, lowered by ca. $15\,000\,\text{cm}^{-1}$ in solution compared to the gaseous ion.

3. **Ligand → metal f (electron-transfer bands)** These are usually intense, broad bands which lie in the ultraviolet region. The charge transfer means that in the excited state there is one more f electron than in the ground state.

Of these three types of transition, it is the first which has received, by far, the greatest study, even though they are the weakest bands. To put this in perspective, Fig. 11.11 shows the electronic spectra of aqueous Ce^{III}, $4f^1$, and Pr^{III}, $4f^2$. Only in the latter are f → f transitions evident; this is because for the Ce^{III} the only transition is between the two spin–orbit components of

Table 11.9 Typical colours of lanthanide complexes

Configuration	Ion	Colour	Configuration	Ion	Colour
Aqueous ions					
f^1	Ce^{III}	colourless	f^{13}	Yb^{III}	colourless
f^2	Pr^{III}	yellow–green	f^{12}	Tm^{III}	light green
f^3	Nd^{III}	red–violet	f^{11}	Er^{III}	lilac
f^4	Pm^{III}	pink	f^{10}	Ho^{III}	yellow
f^5	Sm^{III}	yellow	f^9	Dy^{III}	yellow–green
f^6	Eu^{III}	pale pink	f^8	Tb^{III}	pale pink
f^7	Gd^{III}	colourless			
Lanthanide tricyclopentadiene complexes[a]					
f^1	Ce^{III}	orange	f^{13}	Yb^{III}	dark green
f^2	Pr^{III}	greenish	f^{12}	Tm^{III}	yellow–green
f^3	Nd^{III}	blue	f^{11}	Er^{III}	pink
f^4	Pm^{III}	yellow–orange	f^{10}	Ho^{III}	yellow
f^5	Sm^{III}	orange	f^9	Dy^{III}	yellow
f^6	Eu^{III}	brown	f^8	Tb^{III}	colourless
f^7	Gd^{III}	yellowish			

[a] The crystal structures of these complexes show three cyclopentadiene rings η^5-coordinated to each lanthanide, one of these rings additionally being bonded in η^1 or η^2 fashion to an adjacent lanthanide.

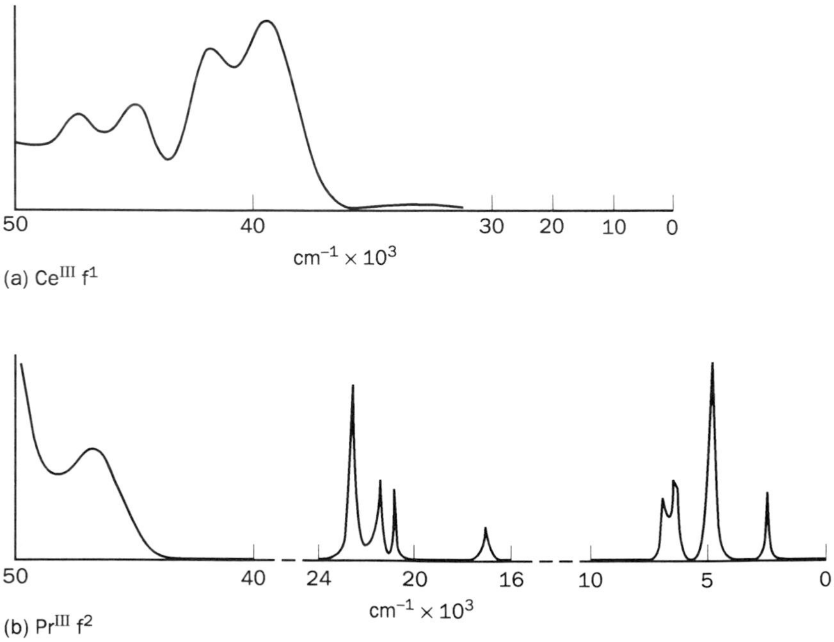

Fig. 11.11 (a) The electronic spectrum of Ce^{III}, f^1 is dominated by charge-transfer bands. The f → f spectrum is expected to be very simple, consisting of a single transition between the two spin–orbit levels given in Table 11.4. This transition will be both weak and in the infrared—towards the right-hand side of Fig. 11.11(a)—where it is lost somewhere in the forest of vibrational bands that occur in this spectral region. (b) The electronic spectrum of Pr^{III}, an f^2 ion, is more characteristic of lanthanides, showing both charge transfer and f → f transitions (sharp). The involvement of f–f electron repulsion moves some of these transitions into the visible region of the spectrum. The relevant energy level diagram is given in Fig. 11.10.

(a) Ce^{III} f^1

(b) Pr^{III} f^2

the ^2F term. It falls in the infrared region, where it is difficult to distinguish it from the forest of vibrational bands; not all of the electronic transitions of an ion may be available for study. As a simplification our discussion will be restricted to the lanthanides. For the actinides the spin–orbit coupling is so great that the spin–orbit components of different Russell–Saunders terms overlap (only the ^1I overlaps with the ^3P in Fig. 11.10; a corresponding diagram for an actinide would be much more complicated). Secondly, crystal fields are about twice as great for the actinides as for the lanthanides so that it is less acceptable to ignore them in the way that we shall do.

When a lanthanide ion is at a centre of symmetry, in the hydrated perchlorates where the lanthanide is octahedrally surrounded by water molecules, for instance, the $f \rightarrow f$ transitions are an order of magnitude weaker than in low-symmetry complexes, indicating that one intensity-generating mechanism is the mixing of f with other orbitals (presumably d, because this would lead to an allowed $f \rightarrow d$ component in the transition) by the low-symmetry crystal field, a mechanism analogous to that discussed for low-symmetry transition metal complexes in Chapter 8 (where d–p mixing was invoked). Just as for transition metal complexes, too, it seems that the mechanism by which the transitions in centrosymmetric complexes gain intensity is through the dynamic distortions caused by molecular vibrations (for transition metal complexes this mechanism was discussed in Section 8.7). Support for this explanation comes from experiments such as those in which a low concentration of a lanthanide ion is doped into crystalline Cs_2NaYCl_6. The yttrium is an f^0 ion and octahedrally surrounded by Cl^- ions. Particularly with low-temperature samples, it is possible to observe mixed electronic + vibrational (vibronic) transitions for the doped lanthanide ion, consistent with a vibrational mechanism for generating the intensity.

The point in our discussion has now been reached at which the parallel with $d \rightarrow d$ transitions is little help; new mechanisms for generating the band intensities have to be introduced. The reason for introducing such new mechanisms is that those given above prove inadequate when really tested. For instance, detailed evidence that a vibrational mechanism is involved in making $f \rightarrow f$ transitions weakly allowed in centrosymmetric complexes has just been presented. But what of the band origin which is seen under high-resolution conditions? It has no vibrational component and so no vibrational explanation can be given for its appearance. The explanation which is usually given for it is that it is *magnetic dipole allowed*, an explanation which itself calls for an explanation! In the Maxwell model, a monochromatic polarized beam of light consists of two vectors, an electric vector and a magnetic vector. These are mutually perpendicular to each other and both are perpendicular to the direction of propagation of the light wave. Normally, attention is confined to the electric vector because calculations indicate that the interaction between the magnetic vector and a molecule is much weaker than that involving the electric vector.[9] Magnetic vectors, and magnetic fields, cause circular displacements of charge (think of the path followed by a charged particle in a conventional mass spectrometer). Such rotations of charge, like all rotations, are symmetric with respect to inversion in a centre of symmetry. This is in contrast to an electric vector, and the linear charge displacements to which it gives rise; they are anti-symmetric with respect to inversion in a centre of symmetry (Fig. 11.12). So, magnetic-dipole-allowed transitions are g in nature whereas electric-dipole-allowed transitions are u. Now, $f \rightarrow f$ transitions are $u \rightarrow u$ and $u \times u = g$. So, although $f \rightarrow f$ transitions are electric-dipole-forbidden they are magnetic-

---

[9] Not that it is zero—all magnetic resonance measurements rely on the fact that the interaction between an oscillating magnetic field and a molecular system is non-zero.

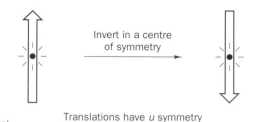

Fig. 11.12 Inversion in a centre of symmetry changes a translation into a translation in the opposite direction (a), whereas the sense of a rotation is left unchanged (b).

(a) Translations have *u* symmetry

Rotations have *g* symmetry (the rotation is in the same sense after the inversion)

(b)

dipole-allowed. This is why it is believed that a magnetic-dipole mechanism is responsible for the appearance of the band origin, with no vibrational component, in high resolution studies of the spectra of lanthanide ions doped in host lattices. Note that a similar explanation could apply to the corresponding d → d transitions and has, indeed, been suggested.

The fact that a beam of light travels more slowly through a totally non-absorbing block of glass more slowly than through a vacuum shows that light interacts with matter, whether or not absorption takes place. That such an interaction should take place is not surprising—as has just been pointed out, light waves behave, in part, like oscillating electric dipoles. They therefore cause electrons to move, to oscillate in sympathy with the electric vector of the light wave. A fascinating f → f band intensity mechanism has been proposed based on this phenomenon and is currently attracting much interest. It has been suggested that the light wave used in absorption studies polarizes the ligands and that the lanthanide ion feels the dipoles created. That is, it is suggested that the ligands provide a non-centric field at the lanthanide ion not by virtue of any vibrational movement of their nuclei but because of the instantaneous polarization of their electron density caused by the light wave itself.

11.8 Crystal fields and f → f intensities

As has been made clear, crystal field splittings are not an important aspect of f electron systems. In solution and the solid the intrinsic half-widths of the bands observed in the f → f spectra are typically 50 cm^{-1} and the crystal field splittings seem much the same, so that there is little information on crystal fields to be gained from f → f spectra. Nonetheless, there is clear evidence that crystal field effects do play a role, albeit not a dominant role—and not a well-understood one either.

The first evidence of crystal field effects is the recognition that lanthanide and actinide ions in solution and the solid state show a red shift of their spectral bands. This has been interpreted in terms of a nephelauxetic effect

Table 11.10 Some f → f hypersensitive transitions. In this table it has been recognized that the Eu^{III} $^7F_1$ state is low lying and is thermally populated at room temperature so that the hypersensitive transition could involve it as the ground state

Ion	Ground state term	Excited state levels of hypersensitive transitions
La^{III}	$^1S_0$	none
Ce^{III}	$^2F_{5/2}$	none
Pr^{III}	$^3H_4$	$^3H_5$, $^3F_2$
Nd^{III}	$^4I_{9/2}$	$^4G_{5/2}$, $^4G_{7/2}$, $^2G_{7/2}$, $^2K_{13/2}$
Pm^{III}	$^5I_4$	$^5G_2$, $^5G_3$
Sm^{III}	$^6H_{5/2}$	$^6F_{1/2}$, $^6F_{3/2}$, $^4H_{7/2}$
Eu^{III}	$^7F_0$($^7F_1$)	$^7F_2$
Gd^{III}	$^8S_{7/2}$	none
Tb^{III}	$^7F_6$	$^7F_5$
Dy^{III}	$^6H_{15/2}$	$^6H_{13/2}$, $^6H_{11/2}$, $^6F_{11/2}$
Ho^{III}	$^5I_8$	$^5G_6$, $^3H_6$
Er^{III}	$^4I_{15/2}$	$^2H_{11/2}$, $^4G_{11/2}$
Tm^{III}	$^3H_6$	$^3H_5$, $^3H_4$, $^3F_4$
Yb^{III}	$^2F_{7/2}$	none

(a cloud expanding effect; the occurrence of this phenomenon for transition metal ions was outlined at the end of Section 8.3), and this, in turn, has been explained by invoking metal–ligand covalent bonding. Whether the explanation is valid is unclear—one author has commented that 'the nephelauxetic parameters resulting seem not to have much meaning'—but the existence of the phenomenon is real enough and its explanation clearly has to involve the interaction of the f electron system with its environment. Secondly, no less clear and no better explained, are the so-called *hypersensitive transitions*. These are f → f transitions which have environment-sensitive intensities. Their intensities may vary by over three orders of magnitude with change of environment, with change of ligand for instance. The effect of the ligand environment is made evident by a band intensity change rather than the band splitting which experience with d → d systems would lead us to expect. A list of some hypersensitive transitions is given in Table 11.10; as this table shows, there is no evident pattern to them. Not surprisingly, the problem posed by the fact that some, but not all, f → f transitions are hypersensitive has been the subject of much study. Whilst the answer is still not clear, a common feature seems to be that the hypersensitive transitions are all quadrupole-allowed. A quadrupole is not a particularly familiar animal for most chemists, although the increasing use of quadrupole mass spectrometers is changing the situation. There are several ways of picturing a quadrupolar arrangement of charges of which the simplest is to think of it as a side-by-side arrangement of two opposite dipoles; two dipoles head-to-tail. So, in seeking an explanation for hypersensitive transitions it is natural to look for a mechanism involving two dipoles. Several such mechanisms have been suggested. The first is one which builds on the 'polarization induced by the electric vector of the light wave' model which was introduced at the end of the previous section. In this explanation, this induced polarization provides one dipole and the electric vector of the light wave itself provides the second.

Of course, the only light wave which is relevant here is that which has an energy corresponding to the energy of the hypersensitive transition observed. On this model, the sensitivity of hypersensitive transitions to different environments arises from the different polarizabilities of the different ligand environments.

A second mechanism which has been suggested to explain hypersensitive transitions applies only to low-symmetry molecules, and so, low-symmetry crystal fields. The mechanism postulates that one of the dipoles is inherent in the ligand arrangement; the overall ligand environment is so distorted that it is dipolar. The second dipole required for a quadrupole again comes from the incident light wave. On this model, different hypersensitivities arise because of different intrinsic dipoles in the ligand arrangements. This second model can be modified to cover the case of high-symmetry arrangements, arrangements which are centrosymmetric and so cannot be intrinsically dipolar, for example. This modification supposes that the high-symmetry molecule is distorted by a vibration, such as those vibrations used to explain the intensities of d → d transitions in Section 8.7. These, for an octahedral complex, are vibrations which destroy the centre of symmetry. Examples, again for an octahedral complex, are vibrations of T_{1u} or T_{2u} symmetries. The vibrationally distorted molecule is then non-centric, at least whilst the distortion persists (although only for a T_{1u} vibration would it have a—transient—dipole). This, then, is the source of one dipole; again, the second is that originating in the light wave itself. The hypersensitivity differences would, on this model, originate in the different vibrational properties of the different species.

We have then three different explanations for the phenomena of hypersensitive transitions. In general, they are not necessarily mutually exclusive although the relative importance of each would surely vary from complex to complex (one could not invoke the second for a centrosymmetric species, for instance). Unfortunately, although hypersensitive transitions show that a lanthanide ion is sensitive to its surroundings, until more is known about the mechanism of hypersensitivity it is not possible to use the phenomenon to learn more about this sensitivity. One final point, also puzzling: the actinide ions, despite the greater importance of ligand fields for these ions, do not seem to exhibit hypersensitive transitions. Even so, some of their transitions show small, but undoubtedly real, band intensity changes with change in ligand environment.

11.9   f → d and charge-transfer transitions

In Section 11.7 were listed two categories of electronic transitions which occur in the lanthanides and actinides in addition to the f → f transitions discussed in the preceding section. It is convenient to treat these two classes, f → d and charge-transfer transitions, together since, spectrally, they occur together, although generally at lower energies in the actinides than in the lanthanides. Much of the study of these bands has taken the form of seeking criteria which enable the two different sorts of transition to be distinguished. One such distinction is that in the lanthanides, the 4f → 5d transitions tend to be narrower than the electron-transfer bands (their half-widths are ca. 1000 and 2000 cm^{-1}, respectively). Although either type of band may be the lower lying, in the lanthanides the charge-transfer tends to occur at the lower energy when addition of an electron to an f-shell leads to a half or full electron shell. The f^6 ion europium(III) and the f^{13} ion ytterbium(III) are species for which this pattern holds. The converse pattern is also general. That is, when an f → d transition (a transition which means that the number of f electrons is reduced by one) leads to an empty or half-filled shell then these f → d transitions are the lower. The f^1 ion cerium(II) and the f^8 terbium(III) provide examples of this pattern. It seems that the f^2 ion praseodymium(III) is another example, although this could not have been predicted. When the lanthanide or actinide is one for which more than one valence state may be studied, then it is relatively easy to distinguish between electron-transfer and f → d bands. As the valence state increases, f → d bands move to higher energy with increase in valence state whereas the—ligand to metal—electron-transfer move to lower. Of course, the spectra of two differently charged ions will be far from identical but, nonetheless, bands which appear similar in the two spectra show these relationships. Another distinction, within a given valence state, is based on the fact that, as one would expect, the d orbitals are much more sensitive to the ligand environment than are the f, so that the energies of the f → d bands are dependent on coordination number whilst the charge-transfer are not. Conversely, for a fixed coordination number, the energies of the f → d bands are less sensitive to change in ligand than are the corresponding charge-transfer bands.

All of these criteria more-or-less follow simple common sense ideas about the characteristics of the two types of transition. Together, they enable distinctions to be drawn in most cases. Although the detailed study of the bands is in its infancy, it is interesting to note that f → d transition energies seem to follow a spectrochemical series, just as do d → d in the transition metal ions. So, the following 5f → 6d transition energies, all × 10^3 cm^{-1}, have been reported for complexes of U^{III} and other similar ions:

Ligand:	I$^-$	<	Br$^-$	<	Cl$^-$	<	SO$_4^{2-}$	<	H$_2$O	<	F$^-$
ν(f → d):	13		ca. 17		ca. 19		ca. 22		25		ca. 25

11.10 Lanthanide luminescence

It has long been known that many lanthanide ions fluoresce under ultraviolet light, the fluorescence coming from f → f transitions; some ions which do not normally fluoresce at room temperature do so when they are cooled. This fluorescence property has led to lanthanide ions being incorporated in the phosphor of domestic fluorescent tubes and in the screens of colour televisions.[10] When an ion is in an electronically excited state there is a competition between deactivation by radiative and non-radiative processes. For an ion to be a good emitter, any non-radiative process must be a poor second in the competition. If studies are carried out using aqueous solutions, it is found that the lanthanide ions at the centre of the lanthanide series are

---

[10] These screens contain a multitude of tiny clusters of carefully placed patterns of red, green and blue light-emitting phosphors. A metallic mask immediately behind the screen has matching tiny holes or slots which allow only electrons from the green cathode to hit the green light-emitting phosphor in each cluster, the electrons from the blue cathode to hit the blue phosphor and so on.

the most efficient emitters; not surprisingly, their observed radiative lifetimes are found to be anything up to a factor of 10^3 greater than those of the ions at either end of the series. Much more surprisingly, the energies of the excited states of the ions in the centre of the series are higher than those for the end members by a factor of two or three—as has been seen from Table 11.4. With the exception of gadolinium(III), the ions at the centre of the series have more f electron excited states and so the energy levels are more widely spread. One might have expected that the more energy there is to lose, the more readily it would be lost—but, clearly, this is not the case. It seems clear that a key part in this phenomenon is played by the solvent (water). So, if D_2O is used as the solvent in place of H_2O, the observed excited state lifetimes are all increased by an order of magnitude. The explanation seems clear; an important non-radiative deactivation process is one in which the energy in the electronic excited state is transferred to an overtone of the $v(O–H)$ stretching vibration of the solvent water. A lower overtone level is needed for the deactivation of the end members of the lanthanide series than for the central members. Because the frequency of the $v(O–D)$ vibration is only about three-quarters of that of the $v(O–H)$, a much higher overtone is needed in D_2O than H_2O. Remembering that the transitions involved are f → f, it is evident that this deactivation mechanism indicates something that we have met several times before, that f → f transitions are not totally insulated from their environment.

Amongst the most efficient of current commercial phosphors are those based on the red emission of europium(III), the green emission of terbium(III) and the blue emission of europium(III). These ions are at the centre of the lanthanide series so that it seems that the phenomena found in aqueous solution extend to the solid state. In addition, gadolinium(III) may well be a component of a phosphor but this is because of its excited state lifetime, not because of any emission. The actual matrix into which the lanthanide ions are incorporated in a phosphor are oxide or glass ceramic-forming oxyanions—borates, silicates, aluminates and tungstates. The choice is important because one step in the emission process is transfer of the absorbed ultraviolet radiation (absorbed by a carefully chosen impurity species, commercially called a sensitizer) through the crystalline lattice to the emitter. An ordered lattice facilitates this energy transfer but may simply serve to allow transfer to an (inadvertent) impurity ion which provides a mechanism for deactivation without emission. The degree of order of the host lattice has to be optimized, as too does the concentration of the chosen emitter. Incorporation of an ion such as gadolinium(III) at quite a high concentration (up to 20% of total cations) in the lattice is found to facilitate long-range energy transfer. The current understanding of the processes involved may best be illustrated by an example. Consider a borate lattice containing a high concentration of gadolinium, GdB_3O_6. A typical sensitizer is bismuth(III) and europium(III) will be chosen as emitter. Current notation would write this composition as GdB_3O_6:BiEu. The sequence of events seems to be:

Ultraviolet
light   Gd(III) → energy transfer   Eu(III) → **emission**
 ↓   ↑   ↓   ↑
  Bi(III) → excitation   Gd(III) → energy transfer

11.11 Magnetism of lanthanide and actinide ions

In Chapter 9 a review was given of the approach adopted in the calculation of the magnetic properties of transition metal ions and the same general philosophy applies to the lanthanides and actinides. First, one has to identify the ground state and then to make allowance for all those interactions in which it is involved and which, energetically, are larger than the effect of the magnetic field. Then, this corrected ground state is subjected to the magnetic field perturbation, the effect of which is usually to split it into a family of sublevels. These split levels are thermally populated with a Boltzmann distribution.

As has been seen in Section 11.4, for the lanthanides the most important factor determining the ground state energy level is the value of J ($=(L + S)$) (here, J is the sum of the individual one-electron j values, $j = (l + s)$). Indeed, in practice it is found that a J-only model, with no allowance at all for minor things such as crystal fields, gives a good agreement between experiment and theory. What this means is not that crystal fields are smaller than the effects of the magnetic field, it has been seen above that they are not, but that their inclusion does not significantly modify the magnetic properties of the ground state or its relationship with other thermally populated states. For the actinides the J-only model does not give such good agreement with experiment, in keeping with the fact that crystal field effects are larger and so do more than produce the simple additive correction found for the lanthanides.

It will be recalled that in Section 9.1 it was shown that the spin-only model for molecular paramagnetism gave rise to the equation

$$\mu_{eff} = \sqrt{4S(S + 1)}$$

It will therefore not be surprising that, essentially by replacing S by J in the derivation, the J-only model gives rise to the equation

$$\mu_{eff} \propto \sqrt{J(J + 1)}$$

The equality sign in the spin-only equation has been replaced by a proportionality because, whereas there can be no ambiguity about how a given value of S arises—one simply has to count the number of unpaired electrons and divide by two—as has been seen, a given J value can arise from a variety of different spin and orbital components. Spin magnets and orbital magnets are not immediately interchangeable and so the constant of proportionality in the J-only equation, denoted g, has to reflect the particular mix involved. The equation for g is

$$g = \frac{3}{2} + \frac{S(S + 1) - L(L + 1)}{2J(J + 1)}$$

The quantity corresponding to g in the spin-only formula is the factor 2 (which appears as 4 when placed inside the square root). In Table 11.11 is given a comparison of observed and calculated μ_{eff} values for the lanthanides. As has already been commented, the general agreement is very good. Two additional points should be made. First, a small contribution from temperature independent paramagnetism (TIP, the mixing of excited states into

Table 11.11 μ_{eff} values for lanthanide ions

Ion	g	μ_{eff} (calculated)	Approximate μ_{eff} (observed)
La^{III}	1	0.0	0.0
Ce^{III}	6/7	2.54	2.5
Pr^{III}	4/5	3.58	3.5
Nd^{III}	8/11	3.62	3.6
Pm^{III}	3/5	2.68	–
Sm^{III}	2/7	0.84	1.5
Eu^{III}	1	0.0	3.4
Gd^{III}	2	7.94	8.0
Tb^{III}	3/2	9.72	9.3
Dy^{III}	4/3	10.63	10.6
Ho^{III}	5/4	10.60	10.4
Er^{III}	6/5	9.59	9.5
Tm^{III}	7/6	7.57	7.4
Yb^{III}	8/7	4.54	4.5
Lu^{III}	1	0.0	0.0

the ground state by the magnetic field, discussed in Section 9.6) is expected and has, in fact, been included in the calculated values in Table 11.11. Secondly, the ions europium(III) and samarium(III) do not give good agreement. The reason for this is known. These two ions possess very low-lying excited states which are so low in energy that they are thermally populated. The *J* values for these low-lying states differ from those of the ground states and so the *J*-only equation above has to be modified to take account of this. When this correction is made, good agreement with experiment is obtained.

It is evident from Table 11.11 that several of the lanthanides have rather high values of μ_{eff}. Physically, this means that when salts of these ions are placed in a strong magnetic field, they become slightly warm. The system is stabilized, the salts are attracted into the magnetic field, and the energy of stabilization has to go somewhere; it appears as heat. At room temperature this effect is small but at low temperatures it becomes rather significant. This has led to the use of gadolinium salts, especially $Gd_2(SO_4)_3 \cdot 8H_2O$, and salts of dysprosium for obtaining low temperatures. The salts are placed in a container, cooled with liquid helium and a strong magnetic field applied. As has just been seen, heat is evolved. When the salts have cooled down to liquid helium temperature again, the magnetic field is switched off. Cooling—*adiabatic demagnetization*—occurs as the salts lose their magnetic orientation. There is an alternation of natural abundances of the lanthanides, starting with lanthanum (low),[11] cerium (high), so that both gadolinium and dysprosium are highs, explaining their selection for this application—gadolinium is not otherwise, from Table 11.11, the most obvious choice. Clearly, the total cooling effect depends on having a relatively large amount of the lanthanide salt available.

---

[11] Low and high here are relative to adjacent elements only. So, lanthanum is four or five times more abundant than either gadolinium or dysprosium.

11.12   f orbital involvement in bonding

The question of the extent to which f orbitals are involved in the bonding of complexes of the lanthanides and actinides is one that has long interested chemists. The final answer has yet to be given. On the one hand there is the experimental evidence provided by anions such as $[UF_8]^{3-}$, mentioned in Section 3.2.6, which have the fluoride ligands arranged at the corners of a cube, which is not the arrangement sterically favoured. The geometry becomes understandable if f orbitals are involved in the bonding because in an O_h molecule the f_{xyz} orbital can participate. This it cannot do in the most probable alternative geometry, the D_{4d} antiprism, a point which is explored in more detail in a footnote towards the end of Section 3.2.6. This apparently unambiguous evidence is in contrast to the results of theoretical calculations. The fact that spin-orbit coupling is of key importance for the lanthanides and, even more so, for the actinides, is a clear indication that it is important to include relativistic effects in an accurate calculation. And this is in what is already a difficult problem, because of the large number of electrons involved. Not surprisingly, a large number of different methods have been investigated, ranging from relativistic corrections to the extended Hückel model of Chapter 10 through to similarly modified *ab initio*, Xα and density functional methods. Although much of the work has, in reality, been as much concerned with the exploration of methods of tackling such a difficult problem, some of the results have shown impressive agreement with such measurables as bond lengths and spectroscopic transitions. Unfortunately, as was pointed out in Chapter 10, it is far from unusual for such agreement to disappear when a calculation is improved and this could be the case here. Nonetheless, some reasonably well accepted generalities seem to have emerged. First, f orbital involvement is greater for the actinides than for the lanthanides. This conclusion seems clear, despite the greater difficulty inherent in calculations on the actinides. Secondly, whilst most calculations on the actinides *do* indicate at least a limited f orbital involvement, what is most unclear is whether this involvement has any chemical consequences—whether it contributes significantly to bond energies, for instance. The

Table 11.12 A comparison of the results of relativistic (rel) and non-relativistic (non-rel) calculations on the uranium atom

Orbital	Non-rel	Rel	Rel/Non-rel
Energies (eV)			
7s	4.54	5.51	1.21
6d	7.25	5.69	0.78
5f	17.26	9.01	0.52
Radii (A)			
7s	2.67	2.30	0.86
6d	1.52	1.71	1.12
5f	0.67	0.76	1.13

In this example the inclusion of relativistic effects does not change the sequence of energy levels. In the case of thorium, in contrast, it seems that the 6d and 5f energies interchange when relativistic effects are included.

evidence of $[UF_8]^{3-}$ is that it does—and it is to be remembered that, as the footnote at the beginning of Section 10.3 indicated, the consequence of the inclusion of relativistic effects on f orbitals is that of making them more available for chemical bonding. This is illustrated in Table 11.12, which shows the results obtained on some rather good calculations on the uranium atom. The calculations were carried out twice, with and without allowance for relativistic effects. The profound consequences of these effects is evident, the 7s orbitals becoming smaller and electrons in them more difficult to ionize. On the other hand, the 6d and the 5f become larger and electrons occupying them much easier to ionize. Even so, the radius of the 5f electrons remains small, although changes in the effective nuclear charge of the uranium, perhaps caused by chemical bonding, could change it yet again. It is not surprising that opinions about the importance of the f orbital contribution to that bonding remain mixed.

Further reading

A good, if at some points rather dated, survey is *Complexes of the Rare Earths*, S. P. Sinha, Pergamon, Oxford, 1966. An excellent recent account of the chemistry of the lanthanides and actinides is to be found in *Lanthanides and Actinides*, S. Cotton, Macmillan, London, 1991. An excellent series of brief review articles is to be found in *Radiochim. Acta* (1993) *61*. Three examples are 'Overview of the Actinide and Lanthanide (the f) Elements' by G. T. Seaborg (p 115), 'Systematics of Lanthanide Coordination' by E. N. Rizkalla (p 118) and, for those who are interested in learning how to study the chemistry of a couple of dozen atoms, 'Atom-at-a-Time Chemistry' by D. C. Hoffman (p 123).

The shapes of f orbitals is discussed in readable fashion in two adjacent papers: H. G. Friedman, G. R. Choppin and D. G. Feuerbacher, *J. Chem. Educ.* (1964) *41*, 354, and C. Becker, *J. Chem. Educ.* (1964) *41*, 358. See also C. A. L. Becker, *J. Chem. Educ.* (1979) *56*, 511, and O. Kikuchi and K. Suzuki, *J. Chem. Educ.* (1985) *62*, 206.

For spectra, a useful compilation is *Spectra and Energy Levels of Rare Earth Ions in Crystals*, G. H. Dieeke, Interscience, New York, 1968. Particularly valuable is an article 'Optical Properties of Actinide and Lanthanide Ions' by J. P. Hessler

and W. T. Carnall in the American Chemical Society Symposium Series, 1980, Vol 131, 349. The interpretation goes well beyond the content of the present chapter but the article contains energy level diagrams and other very useful and accessible data.

Many of the topics touched on in this chapter are treated in more detail and depth in some issues of *Structure and Bonding*, notably 13, 22, 25, 30 and 59/60(actinides). A reference that contains more than its title indicates is 'Lanthanide Ion Luminescence in Coordination Chemistry and Biochemistry', W. D. Horrocks and M. Albin, *Prog. Inorg. Chem.* (1984) *31*, 1.

Two articles that provide an overview of the current situation on f orbital involvement in bonding in the actinides are: 'Cyclopentadienyl-Actinide Complexes: Bonding and Electronic Structure' by B. E. Bursten and R. J. Strittmatter, *Angew. Chem. Int. Ed.* (1991) *30*, 1069 (in its last couple of pages this provides a brief overview of all the theoretical methods being employed) and 'The Electronic Structure of Actinide-Containing Molecules: A Challenge to Applied Quantum Chemistry' by M. Pepper and B. E. Bursten, *Chem. Rev.* (1991) *91*, 719. A reference given at the end of Chapter 6 is also relevant, although some of the relevant articles are more readable than others— *The Challenge of d and f Electrons*, D. R. Salahub and M. C. Zerner (eds.) American Chemical Society Symposium Series, 1989, Vol. 394.

Questions

11.1 Select any non-trivial row from Table 11.5 and show that the degeneracy implicit in the simple Russell–Saunders term (e.g. $^6H = 6 \times 11 = 66$) is equal to the sum of degeneracies, $\sum_j (2j + 1)$, of the corresponding spin–orbit levels. (Hint: the ^2F case has been worked through in Section 11.4.)

11.2 The f^1 ion Ce^{III} is colourless in aqueous solution, whereas the d^1 ion Ti^{III} is purple. Outline the reason for this difference.

11.3 The lanthanides and actinides form complexes with planar cyclic C_nR_n organic species. Use Fig. 11.4 together with Fig. 10.3 to detail how the bonding in such complexes might be expected to differ from those of the corresponding transition metal complexes.

11.4 Table 11.9 shows that both Ce^{III}, f^1, and Yb^{III}, f^{13}, are colourless in aqueous solution. However, the spin–orbit splitting of these two ions are ca. 650 cm^{-1} and ca. 3000 cm^{-1}, respectively. Explain why these observations are mutually compatible.

12

Other methods of studying coordination compounds

12.1 Introduction

In preceding chapters the theories underlying some of the methods which have been used to characterize coordination compounds have been discussed in some detail. The present chapter provides a less detailed survey of other methods, stretching from indicators of complex formation through to some which provide insights into physical structure, others into electronic structure, yet others into the forces between atoms and molecules.

If one is solely concerned with the question of whether or not a complex is formed in a particular system, relatively crude methods often suffice—although, historically, such methods have played an important part in the development of the subject. The evolution of heat, the crystallization of a product, a change in chemical properties—the failure to undergo a characteristic reaction, for instance—are all simple but useful indicators. More sensitive methods often involve the study of some physical property of the system as a function of its composition and some of these were outlined in Chapter 5. Other examples of methods used for non-transition metal complexes are the measurement of colligative properties such as vapour pressure, boiling point or freezing point. Less commonly, measurements of quantities such as viscosity and electrical conductivity have been employed. Another electrical measurement which has been much used is that of pH change. Most ligands may be protonated at their coordination site so that, for example, corresponding to the NH_3 ligand there is the protonated species, NH_4^+, the ammonium ion. This means that, if in the same solution there is free ligand L, an acceptor species A (which could well be a metal ion), and acidic protons—which for simplicity will be written as H^+, there will be two important equilibria involving L:

$$H^+ + L \leftrightarrows LH^+$$

and

$$A + L \leftrightarrows [AL]$$

If the second equilibrium exists, that is, if a complex is formed, addition of more A will cause the first equilibrium to be displaced to the left and the pH of the solution will decrease. Measurement of pH not only allows the formation of complexes to be detected but may also be used to determine the stability constants of the species formed (see Chapter 5). As will be seen, half-cell potentials and polarographic measurements may be used both to detect complex formation and to determine stability constants but, before turning to these a variety of techniques will be outlined which not only indicate the formation of coordination complexes but also provide information on their structure.

12.2   Vibrational spectroscopy

For most workers, vibrational spectroscopy means infrared and Raman,[1] although the availability of a third method should not be overlooked. This is *inelastic neutron scattering*; in the simplest form, a beam of monochromatic neutrons is bounced off a solid sample. The energies of the rebounding neutrons are measured, the differences from the energy of the incident neutrons give vibrational frequencies. The measurements are of relatively low resolution and require national or international facilities but have the advantage that they may reveal bands which are both infrared- and Raman-forbidden; they are particularly good at revealing modes involving the motion of hydrogen atoms. However, the improvements provided by modern Fourier transform infrared spectrometers have made weak bands much more accessible; the extension of these methods to Raman spectroscopy has made this technique both easier to use and applicable to more samples (laser-induced fluorescence, which can readily swamp the weak Raman scattering, has been largely overcome by the use of infrared lasers, which also enable an infrared spectrometer to be used to measure Raman spectra).

The infrared and Raman frequencies of isolated molecules are determined by the arrangement of the atoms in space and by the forces between the atoms. When a ligand coordinates, all of these change. So, changes are to be expected both in the vibrational spectral features associated with the free ligand and in those of the system to which it becomes attached. An example is provided by the thiocyanate anion, SCN^-. In simple ionic thiocyanates, such as KCNS, there is a $\nu(C-N)$ stretching vibration at ca. 2060 cm^{-1}, a $\nu(C-S)$ stretch at ca. 746 cm^{-1} and a $\delta(NCS)$ bend at ca. 480 cm^{-1}. In the anion $[Co(NCS)_4]^{2-}$, where the SCN^- is N-bonded, the $\nu(C-N)$ mode rises to ca. 2070 cm^{-1}, the $\nu(C-S)$ increases dramatically to ca. 815 cm^{-1}

---

[1] Infrared-allowed transitions are dipole-allowed. That is, the (electric) dipole of the incident radiation couples in resonance with a vibration which changes the (electric) dipole of the molecule. So, an infrared-allowed vibration behaves like a dipole and that means like a coordinate axis. In contrast, Raman involves two light waves, one in, the other out. So, Raman selection rules reflect this difference and a vibration which behaves like a product of *two* dipoles, like a product of coordinate axes, are Raman-allowed. Clearly, infrared and Raman have different selection rules. In the statements above, the word behaves means 'has the same symmetry properties as'.

but the bend only drops by ca. 5 cm^{-1}. In contrast, in the $[Hg(SCN)^4]^{2-}$ anion, where the SCN$^-$ anion is S-bonded, the v(C–N) mode rises to ca. 2100 cm^{-1}, the v(C–S) drops to ca. 710 cm^{-1} whilst the bend drops to ca. 450 cm^{-1}. It is obvious that these different patterns can be used to determine the way that a SCN$^-$ ligand is coordinated. Even more obvious would be to look for the appearance of v(M–N) or v(M–S) features, but this is not always easy if there are other ligands present with metal–ligand stretching modes in the same spectral regions. This explains why there is a particular interest in ligand-specific modes falling in otherwise relatively clear spectral regions.

In favourable cases vibrational spectra can be used to determine the ligand geometry around a central metal atom.[2] Best known is the fact that a centre of symmetry means that bands active in the infrared are not Raman active and vice versa, although this is just one of many symmetry–spectra relationships. Such relationships are not infallible. So, a band may exist but be too weak to be observed separately from other bands in the same spectral region. Alternatively, vibrational couplings may be small. So, if two vibrators are uncoupled it is irrelevant that they happen to be related by a centre of symmetry—a single band will appear, coincident in infrared and Raman. Because of this element of uncertainty, vibrational methods of structure determination are sometimes referred to as sporting methods—they are not infallible—although for compounds for which crystals, and thus X-ray crystallography, cannot readily be made available, they may be the only methods which it is feasible to use.

As a typical example of the application of vibrational methods, consider square planar complexes of Pt^{II} of general formula PtL$_2$X$_2$, where X is a halogen and L is a polyatomic ligand. A study of the spectral activity of the Pt–X stretching modes usually shows whether a particular complex is *cis* or *trans*. If it is *cis* then the molecular geometry is approximately C_{2v} and the v(Pt–X) vibrations are of $A_1 + B_1$ symmetries so that both modes are both infrared and Raman active. It follows that two v(Pt–X) bands are expected in each spectrum, the bands in one spectrum coincident with those in the other, to within experimental error. The *trans* isomer has approximate C_{2h} symmetry and the v(Pt–X) vibrations have $A_g + A_u$ symmetries. Of these, the former is Raman active and the latter is infrared. One band is predicted in each spectrum, at a different frequency in each. The difference in the spectral predictions between the two isomers is so great that it should be possible to use the method even if the spectra are incomplete or less than ideal. An application of these results is illustrated in Fig. 12.1, which shows the Raman spectra of *cis* and *trans* [PtCl$_2$(NH$_3$)$_2$] in the v(Pt–Cl) and v(Pt–N) stretching regions. The presence of two peaks in the v(Pt–N) region is sufficient to establish the lower spectrum as that of the *cis* isomer, without aid of the infrared spectrum, even though the v(Pt–Cl) only shows one (the expert might well argue that the greater breadth of the v(Pt–Cl) peak in the

---

[2] This has been exploited in transition metal carbonyl chemistry. Simple metal carbonyl derivatives often have quite high symmetries and, equally important, a variety of possible geometric arrangements. The vibrational coupling of v(C–O) groups is so great—and the resulting bands so well separated—that simple group theoretical methods may often be used to distinguish between them. For instance, an M(CO)$_3$ unit could have D_{3h}, C_{3v} or C_s symmetries. These could be distinguished because they give rise to one, two and three infrared bands in the v(C–O) region, respectively.

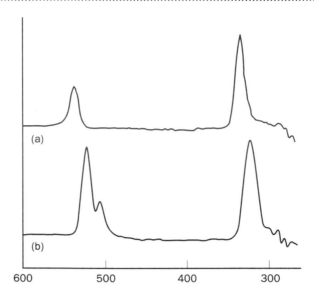

Fig. 12.1 The Raman spectra of (a) *trans*-[PtCl$_2$(NH$_3$)$_2$] and (b) *cis*-[PtCl$_2$(NH$_3$)$_2$]. The v(Pt–N) mode is at ca. 500 cm^{-1} and the v(Pt–Cl) at ca. 300 cm^{-1} (adapted and reproduced with permission from M. J. Almond, C. A. Yates, R. H. Orrin and D. A. Rice. *Spectrochim. Acta A* (1990) *46A*, 177).

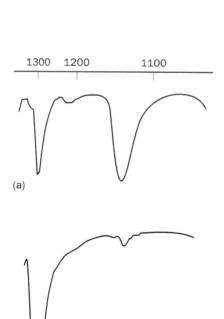

Fig. 12.2 The infrared spectra of (a) [Co(NH$_3$)$_5$–ONO$_2$]$^{2+}$ and (b) [Co(NH$_3$)$_5$–NO$_2$]$^{2+}$. The v(N–O) peak in the latter at ca. 1300 cm^{-1} overlaps with a peak common to both spectra.

cis isomer is indicative of a band splitting which is not resolved in the spectrum). Another origin of band splitting which may conveniently be mentioned at this point is that which occurs in a ligand mode when there is a decrease in ligand symmetry on coordination. For instance, the infrared active v(S–O) stretches of the isolated (tetrahedral, T_d) sulfate anion appear as a single band (the mode is of T_2 symmetry), which splits, loses degeneracy, on coordination.

Two other examples of the application of the general approach are shown in Figs. 12.2 and 12.3. Figure 12.2 provides a further illustration of the way that vibrational spectra can indicate a mode of coordination. So, in complexes of the NO$_2^-$ anion, a v(N–O) mode appears at ca. 1300 cm^{-1} when the ligand is N-bonded and at ca. 1150 cm^{-1} when it is O-bonded. This is a much-studied and much-exploited pattern which is illustrated for two cobalt(III) complexes in Fig. 12.2. This particular example was chosen because an experiment in many undergraduate inorganic chemistry courses is a study of the way that the two species in Fig. 12.2 isomerize, using infrared spectroscopy to follow the reaction.[3]

Figure 12.3 shows two Raman spectra of the anion [Fe(CN)$_6$]$^{4-}$. Although a relationship between them exists, they are surprisingly different. The first thing to note is that some of the bands appear above 2080 cm^{-1}, the position of the v(C–N) band in KCN; bands do not invariably drop in frequency when a ligand coordinates, although this is the common pattern. The differences between the two spectra in Fig. 12.3 show that vibrational interactions can occur between different complex molecules in the solid state. In the present example, between different [Fe(CN)$_6$]$^{4-}$ anions. It is an unfortunate fact of life that such splittings can easily be confused with splittings resulting from molecular geometry. The general way that solid-state splittings originate is shown in Fig. 12.4. This figure shows four single

[3] Visible spectroscopy and NMR can also be used; the reaction in solution shows base catalysis and provides a useful illustration of this phenomenon. A word of caution, however, the *back* reaction is photocatalysed so that working in bright sunlight can produce unexpected results!

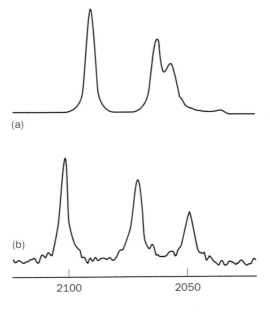

(a)

(b)

2100   2050

Fig. 12.3 The Raman spectra of crystalline samples of (a) $K_4[Fe(CN)_6]$ and (b) $Na_4[Fe(CN)_6]$. This figure illustrates several of the points made in the text. In the $[Fe(CN)_6]$ anion the $\nu(C{\equiv}N)$ vibrations are coupled. The higher frequency bands in both spectra are the A_{1g} (totally symmetric, 'breathing' mode) and the lower E_g, the components of which are split apart. This latter splitting—and all the frequency differences between the two spectra—show the importance of solid-state effects, in which they originate. (Adapted and reproduced with permission from T. N. Day, P. J. Hendra, A. J. Rest and A. J. Rowlands, *Spectrochim. Acta A* (1991) *47A*, 1251).

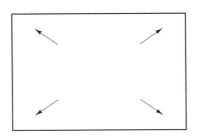

Fig. 12.4 A unit cell of a crystal in which the unit cell contains four molecules. Corresponding vibrations of each molecule (top) couple together to give four unit cell vibrations (lower). The coupling between the modes of the individual vibrations is represented by the dotted lines drawn between them. Of the four unit cell vibrators the upper two are Raman active and the lower two infrared active. The diagrams show two-dimensional unit cells but the principle carries over unchanged into three-dimensional unit cells.

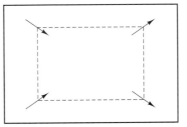
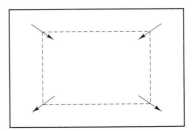
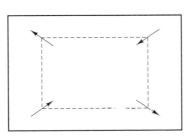
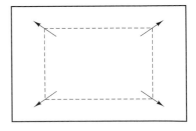

vibrators in a unit cell coupling together to give four coupled vibrations; if the coupling is large enough then the four coupled vibrations have different frequencies and, selection rules willing, appear as separate bands. Fortunately, the vibrational analysis of a crystalline solid is no more difficult than that of an isolated molecule—one treats the contents of the unit cell as a giant molecule—although the number of cases in which such spectra have been used to make predictions about crystal structure is rather small! The theory of these solid state effects is covered more fully in Chapter 17 (Section 17.6).

It is interesting that it is possible to use vibrational spectroscopy as a probe of more than molecular symmetry; in favourable cases the detailed molecular structure can be explored. Most of this work has been done on the infrared spectra of transition metal carbonyls—their $v(C–O)$ stretching modes occur in a spectral region (ca. 2000 cm^{-1}) almost free from other features. Further, the individual CO vibrators couple together strongly giving well-separated bands which are associated with intense infrared absorptions. Finally, the dipole moment change associated with each CO stretch seems essentially colinear with the bond axis. Suppose we have an $M(CO)_2$ unit, each CO group being linear, as is usually the case to quite a good approximation, and at an angle θ to each other, as shown in Fig. 12.5. The in-phase vibration of the two CO groups will have a resultant dipole moment that bisects θ; its magnitude will be $2d\cos(\theta/2)$, where d is the bond dipole moment change. The dipole associated with the out-of-phase combination will be perpendicular to that of the in-phase and will have a resultant dipole moment change of $2d\sin(\theta/2)$. Infrared band intensities are proportional to the square of the corresponding dipole moment changes (dipole moment changes can be either positive or negative but band intensities can only be positive). If the intensity of the band associated with the in-phase vibration

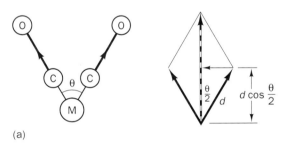

(a)

Fig. 12.5 (a) The in-phase vibrations of two CO groups in an $M(CO)_2$ unit subtend a resultant vector (shown dotted) of $2d\cos(\theta/2)$. (b) The out-of-phase vibrations of two CO groups in an $M(CO)_2$ unit subtend a resultant vector (shown dotted) of $2d\sin(\theta/2)$. In both cases the vibrations are shown and, to the right, the corresponding vector sums.

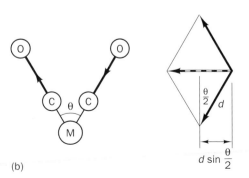

(b)

is I_i and that associated with the out-of-phase is I_o then it follows that

$$\frac{I_o}{I_i} = \frac{\left(2d \sin \frac{\theta}{2}\right)^2}{\left(2d \cos \frac{\theta}{2}\right)^2} = \tan^2 \frac{\theta}{2}$$

So, measurement of relative band intensities leads to a C–M–C bond angle, θ. This procedure can be generalized to other metal carbonyl systems; for simple systems, rather good agreement with crystallographically determined angles is obtained. Unfortunately, few ligands other than CO prove amenable to such an analysis.

12.3   Resonance Raman spectroscopy

In normal Raman spectroscopy a sample is placed in a (monochromatic) laser beam and the very weak scattered light of lower frequency is studied. In such a study the colour of the laser light is usually chosen to be away from any absorption band of the sample because such a choice reduces the risk that the focused laser beam will destroy the sample by heating it. In the resonance Raman effect the laser beam colour is deliberately chosen to coincide with an absorption band—an electronic transition—of the sample. Whilst this may lead to the destruction of the sample, for favourable cases it leads to Raman scattering which is much stronger than normal. This, in turn, means that the laser power can be reduced, improving the chances of sample survival. The spectra obtained from compounds showing such a resonance Raman effect are both simpler and more complicated than normal Raman spectra. They are simpler because, often, only totally symmetric vibrational modes are seen. The reason for this is that if the electronic

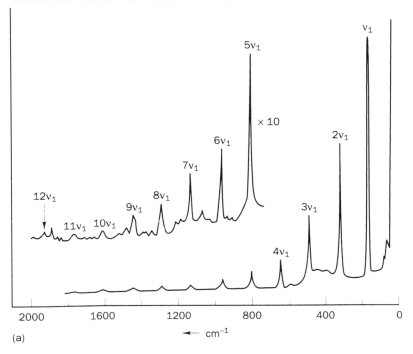

Fig. 12.6 Resonance Raman spectra. (a) Crystalline TiI_4; v_1 is the $v(Ti–I)$ totally symmetric breathing mode, reproduced with permission from R. J. H. Clark and P. D. Mitchell, *J. Amer. Chem. Soc.* (1973) 95, 8300.

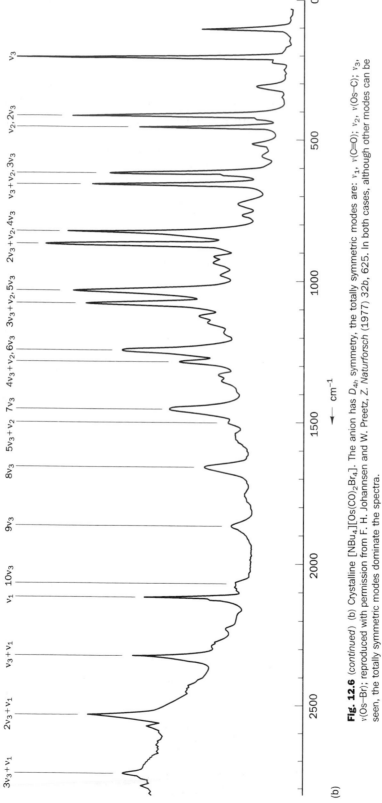

Fig. 12.6 (*continued*) (b) Crystalline [NBu$_4$][Os(CO)$_2$Br$_4$]. The anion has D_{4h} symmetry, the totally symmetric modes are: ν_1, ν(C≡O); ν_2, ν(Os–C); ν_3, ν(Os–Br); reproduced with permission from F. H. Johannsen and W. Preetz, *Z. Naturforsch* (1977) 32b, 625. In both cases, although other modes can be seen, the totally symmetric modes dominate the spectra.

(b)

excitation involved is one in which the number of bonding electron changes then the excited molecule will find itself, be produced, in a geometry well away from its equilibrium. Typically, if the electronic transition which is the source of the absorption band is located on a metal atom, then a breathing mode of the surrounding ligands is the only vibrational mode excited—totally symmetric modes are the only ones that produce volume changes of the type needed to reach the equilibrium geometry. The spectra are more complicated because it is not just the fundamental frequency of the breathing mode which is excited but also a host of overtones—frequently up to the 10th but cases are known in which overtones approaching the 20th are observed. Two examples of resonance Raman spectra are shown in Fig. 12.6. In the literature there are examples of spectra labelled as 'resonance Raman spectra' which simply show generally enhanced intensities rather than the specific characteristics highlighted in Fig. 12.6, perhaps because the electronic excitation involved is one in which no great change in equilibrium geometry occurs. It may be that, de facto, a more relaxed definition of the name 'resonance Raman' is emerging but in this section the restricted definition is used.

Resonance Raman spectra are important for two main reasons. First, they enable the ready determination of potential energy curves (one associated with each of the modes excited) to unusually high levels. Secondly, they can be used to target small parts of a large molecule. So, if in a bioinorganic molecule containing an Fe^{III} ion the laser can be tuned to an absorption band associated with this iron atom (perhaps a $d \rightarrow d$ band, Chapter 8), then just vibrations originating in the atoms coordinated to the Fe^{III} are to be expected. This means that it is possible to probe these atoms—just what is their valence state and in what geometry they are coordinated? Such data can be most valuable when dealing with a complicated molecule. Indeed, it is sometimes possible to probe more than one site in a biomolecule by judicious choice of exciting laser frequencies and obtaining a different spectrum for each frequency. Another measurement possible by resonance Raman but not by conventional Raman is the measurement of the vibrational spectra of electronically excited states. Typically, a high-power pulse laser of a frequency corresponding to an electronic transition of the target molecule is used to excite a large proportion of the molecules into the excited state. This pulse is immediately followed by another which excites the resonance Raman spectrum. So, the anion $[Re_2Cl_8]^{2-}$ has a δ bond between the two metal atoms (this will be dealt with more fully in Chapter 15). Excitation to an excited state in which an electron is promoted to the δ (antibonding) orbital leads to the Re–Re stretching vibration dropping from 274 to 262 cm^{-1} and the Re–Cl increasing from 359 to 365 cm^{-1}. The changes are small, indicating that the δ bonding is not strong, but they are real and in accord with what would be expected.

12.4 Spectroscopic methods unique to optically active molecules

If a weight at the end of a piece of string is set in motion so that it traverses a circular path, this motion may be regarded as composed of two mutually

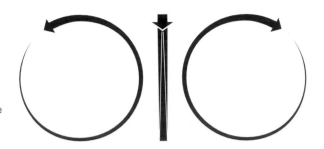

Fig. 12.7 A simple harmonic motion (centre) may be regarded as a sum of two circular motions. The motion represented takes place in the plane of the paper. The thickness of the lines represents time, rather like an oscilloscope trace on a tube with a persistent phosphor.

perpendicular simple harmonic motions combined out of phase with each other by 90°. In a similar way, a single back-and-forth harmonic vibration of the weight may be regarded as an in-phase combination of two circular motions in opposite directions. An attempt to show these alternatives in pictorial fashion is given in Fig. 12.7. Thus, there is always a choice. A circular motion may be described as a combination of two simple harmonic motions and, similarly, a simple harmonic motion may be expressed in terms of two circular motions. This choice is analogous to those met elsewhere: between real and complex atomic and molecular orbitals—we preferred to call them static and rotating in Chapter 11; between linearly and circularly polarized light (and this will be our concern in this section) and a similar choice has to be made in areas remote from chemistry—in some aspects of electrical circuit theory, for instance. Which of the alternative descriptions is adopted is usually determined by that which leads to the simpler mathematics, even if it involves some initial conceptual difficulties. So, whilst it is true that optically active compounds rotate the plane of linearly polarized light, it turns out that it is more convenient to discuss the phenomenon in detail in terms of circularly polarized light. Figure 12.8 shows how linearly polarized light may be regarded as made up of two circularly polarized components.

Consider what happens when each of the circularly polarized components is propagated along the threefold axis of the optically active ion $[Co(en)_3]^{3+}$. From Fig. 12.9 it is evident that one of the circularly polarized components tends to pass along the backbone of the ethylenediamine ligands whilst the other tends to cut the backbone perpendicularly. This description is inaccurate in that the wavelength of visible light (and it is in this that we shall be interested) is of the order of a thousand times greater than the molecular dimensions. Nevertheless, the vital point remains—the two

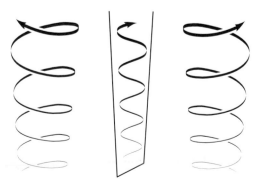

Fig. 12.8 Linearly polarized light (centre) regarded as a sum of two circularly polarized components. Figure 12.7 may be regarded as a view of Fig. 12.8 along the axis of propagation, the thickness of the lines then indicating distance.

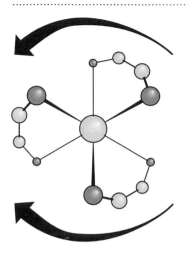

Fig. 12.9 Two circularly polarized beams of light passing through a molecule of $[Co(en)_3]^{3+}$, the threefold axis of the latter being coincident with the axis of propagation of the light. One circularly polarized beam (the top one) tends to cut the backbone of an ethylenediamine molecule, the other to pass along it.

circularly polarized components will encounter slightly different electron density distributions in passing through the $[Co(en)_3]^{3+}$ ion. As is well known, the interaction between a light wave and electron density leads to a reduction in the velocity of propagation of the light wave—refractive indices are greater than unity—and the greater the electron density (roughly), the higher the refractive index and the slower the velocity of propagation. This means that one circularly polarized component must be expected to pass more slowly through the $[Co(en)_3]^{3+}$ ion than the other; the faster might be that which tends to cut the ligand backbone. Although this last statement is, in fact, not generally true, it correctly suggests that polarized light may be used to determine the chirality of an optically active species. Because one circularly polarized component travels faster than the other, one of the helical 'springs' shown in Fig. 12.10 will be more extended than the other. This means that after traversing the molecule the two components will be slightly out-of-phase. Combining them, as in Fig. 12.10, leads to a rotation in the plane of polarization compared with the incident light, as observed. If a graph is plotted of the angle of rotation against wavelength, the angle of rotation being suitably corrected for path length and concentration of the optically active species, a curve of optical rotatory dispersion, an ORD spectrum, is obtained. At wavelengths at which the species is transparent, where there is no electronic transition, the ORD spectrum is unexciting, varying little with wavelength. It is only when absorption occurs that things begin to happen!

It is well known that the absorption of light by a molecule is frequently anisotropic—and this anisotropic absorption occurs simply because molecules are frequently anisotropic. Thus, a light wave incident on a molecule in one direction may excite a transition but, light of the same wavelength, incident

Fig. 12.10 The passage of a beam of linearly polarized light through an optically active molecule. Note the rotation of the plane of polarization and correlate this with the different behaviour of the two circularly polarized components within the molecule.

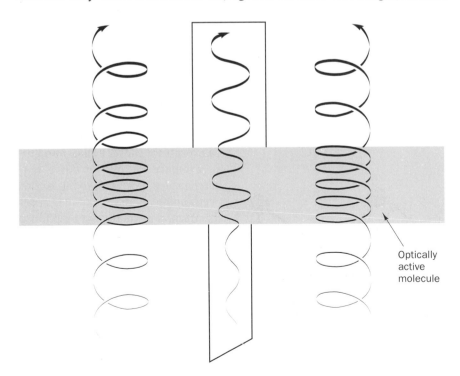

Optically
active
molecule

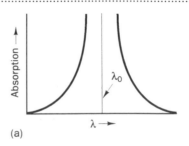

(a)

Fig. 12.11 Idealized diagrams of (a) an electronic absorption band of an optically active species and (b, c) the corresponding ORD (solid line) and CD (broken line) plots for the two different hands of the species.

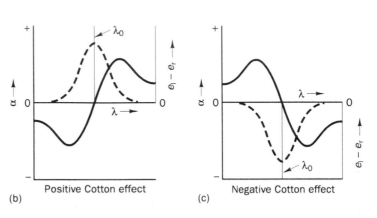

(b) Positive Cotton effect

(c) Negative Cotton effect

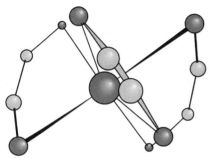

Fig. 12.12 The $[Co(en)_3]^{3+}$ ion viewed perpendicular to the threefold axis (cf. Fig. 12.9).

in another direction, may not be absorbed. In a region of absorption of light by an optically active species, plane-polarized light must be expected to be attenuated, but what of its circularly polarized components; are they equally attenuated? Since these follow slightly different paths through a molecule, as seen earlier for $[Co(en)_3]^{3+}$, it would seem reasonable that they should be absorbed to different extents. This prediction is confirmed by measurements made with circularly polarized light—it is possible to pass right (clockwise, viewed in the direction of propagation) and left (anticlockwise) circularly polarized light alternately through a solution and to compare their relative absorptions. The difference in extinction coefficients between left and right circularly polarized light, $e_l - e_r$ is small but is measurable. It is called the circular dichroism, CD. Together, ORD and CD are collectively called the *Cotton effect*, after the Frenchman who did the fundamental work. A CD spectrum shows a maximum at the position of maximum absorption of an ordinary absorption curve (so that if a CD spectrum shows several maxima it indicates that the ordinary absorption curve consists of several overlapping bands). The ORD spectrum behaves as the derivative of the CD spectrum, passing through a point of inflection at the absorption maximum. These relationships are illustrated in Fig. 12.11; as this figure shows, the curves for one species are the negative of those of its enantiomorph. A complication arises, however. Although the $[Co(en)_3]^{3+}$ cation is shown behaving like a left-handed screw in Fig. 12.9, the same molecule, when viewed in a perpendicular direction, behaves as a right-handed screw (Fig. 12.12)! This duality is not unique to this example but occurs for all optically active molecules.

When a molecule absorbs light and undergoes an electronic excitation this occurs through a displacement of electron density within the molecule.

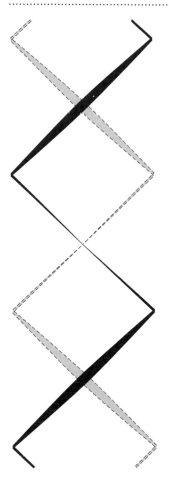

Fig. 12.13 When the grey and black helices intersect they are at right angles to each other. The two helices, one corresponding to a helical molecule and the other to a transition occurring perpendicular to the molecular helix, are of opposite hand.

It turns out that the absolute configuration of a molecule can be related to the Cotton effect observations only if the direction of charge displacement is known. For an optically active molecule this displacement occurs along a helical path[4] and, as has just been seen, within a given molecule there are always both right- and left-handed helical paths available. This is made yet more complicated when we recognize that if the displacement is perpendicular to, say, a left-handed helical backbone, then the displacement itself has the character of a right-handed helix (Fig. 12.13)! Yet only if we know something about the nature of the electronic transition can we hope to deduce a molecular configuration from Cotton effect measurements. Evidently, it is advisable to study a species with as few electronic transitions as possible, if one has choice, for this reduces the probability of error. It is for this reason that much more detailed work has been done with complexes like $[Co(en)_3]^{3+}$ than with others such as $[Co(en)_2Cl_2]^+$. The former has the higher symmetry and, therefore, fewer absorption bands. If the same d $\rightarrow$ d transition can be observed in a series of closely related compounds then, remembering that in the crystal field model, at least, the ligands generate an electrostatic crystal field but are not otherwise involved, if the absolute configuration of one of the members can be established, so too can that of each of the others. In practice, it is usually possible to identify transitions by band intensity criteria and, fortunately, the absolute configuration of $(+)$-$[Co(en)_3]^{3+}$ and of several other key species have been determined by X-ray methods. An alternative would be to be able to have a detailed and reasonably accurate quantum mechanical treatment of the phenomenon, for these would also provide the structure–spectra relationship. Such, *ab initio*, calculations are beginning to appear but have yet to reach the point at which they fulfil the above objective.

Finally, it is to be emphasized that the sign of an ORD curve at any arbitrary point—like the sodium D-line wavelength—is composed of a superposition of tail absorptions from all the ORD bands in the spectrum. The sign of the specific rotation at such a point is therefore an uncertain guide to molecular structure, although, historically, many have sought to use them to find reliable specific rotation–structure relationships.

12.5   Nuclear spectroscopies

In this section three methods will be outlined which have been used to study the properties of nuclei in coordination compounds. None of the techniques is of universal application and they all suffer from the disadvantage that the connection between the spectra obtained and the molecular bonding is seldom simple so that in this application they are generally best used to compare two compounds rather than discuss either in isolation. However, the spectra can be outstandingly useful in obtaining details of molecular geometry and reactions.

---

[4] The requirement for optical activity is that electronic charge displacement occurs along a helical path, for then optical isomers will be associated with bands of opposite helical charge displacement. Group theoretically, this means that translation along and rotation about an axis transform isomorphically—as the same irreducible representation (and, if the irreducible representation is degenerate, as the same component).

12.5.1 Nuclear magnetic resonance (NMR)

The use of NMR in kinetic studies will be covered in Section 14.5 and in particular the way that it is used to study intramolecular gymnastics (fluxionality). At this point, it has to be emphasized that spectra obtained at room temperature may be misleadingly simple because of dynamic exchange processes in which two (or more) nuclei interconvert their chemical environments so rapidly that they give an averaged spectrum.

NMR has developed to the point at which it is the second most important technique in chemistry for determining molecular structure (the most important being X-ray diffraction). As a technique, it is not applicable to all elements, although it is increasingly rare to find a compound in which at least one element cannot be studied. When there are two or more such elements, the range of experiments available, and so the information available, increases considerably. In Table 12.1 are listed the elements which, with a suitable spectrometer—or array of spectrometers—can be studied. Just as infrared spectroscopy has been revolutionized by Fourier Transform techniques, so too has NMR. In practice, a pulse of high-frequency (in the MHz range) radio waves is applied to the sample, which is contained in a magnetic field. The way that the nuclei respond after the pulse is completed is monitored. In more subtle experiments, other pulses are applied after the first. The time interval between the pulses, the length of the pulses, the frequency content of the pulses, the power level of the pulses; all can be varied, giving rise to a plethora of different measurements. In addition, it is possible to irradiate at more than one frequency. So, for instance, by using high power at the correct radio frequency it is possible to excite one set of nuclei so that the relevant ground and excited state populations become equal. Because the observation of an NMR spectrum depends on the

Table 12.1 Some of the nuclei which have been studied by NMR spectroscopy; spin $= \frac{1}{2}$ unless otherwise stated

Nucleus (spin)	Natural abundance (%)	Relative sensitivity[a]
^{1}H	99.9	1.0
^{2}H(1)	0.02	1.5×10^{-6}
^{11}B($\frac{3}{2}$)	80.4	0.13
^{13}C	1.1	1.8×10^{-4}
^{14}N(1)	99.6	1.0×10^{-3}
^{15}N	0.4	3.9×10^{-6}
^{12}O($\frac{5}{2}$)	0.04	1.1×10^{-5}
^{19}F	100.0	0.83
^{29}Si	4.7	3.7×10^{-4}
^{31}P	100.0	0.066
^{33}S($\frac{3}{2}$)	0.76	1.7×10^{-5}
^{35}Cl($\frac{3}{2}$)	75.5	3.6×10^{-3}
^{37}Cl($\frac{3}{2}$)	24.5	6.6×10^{-4}
^{81}Br($\frac{3}{2}$)	49.5	4.9×10^{-2}
^{119}Sn	8.6	4.4×10^{-3}
^{127}I($\frac{5}{2}$)	100.0	9.3×10^{-2}
^{195}Pt	33.8	3.4×10^{-3}

[a] Relative sensitivity includes a number of factors of which percentage abundance is one.

difference between such populations, high powers lead to spectral changes, usually simplifications.

Although measurements such as the variety indicated above are available, they require a considerable amount of time and expertise—in order to arrive at the optimum experimental pattern for a given measurement a fair amount of trial and error may be involved. Not surprisingly, the vast majority of NMR measurements remain of the relatively simple variety. Fortunately, too, much of the sophisticated work is aimed at taking complicated spectra and making them simple. In this section, therefore, the simpler aspects of the subject will be briefly reviewed. In the more general case, where such simplifications cannot be achieved by use of experimental ingenuity, use of computer programs to match observed and calculated spectra is normal. It is usually possible to be confident that such a match is unique and so the relevant molecular quantities—number of nuclei of a particular type, chemical shifts and coupling constants—are uniquely determined. Finally, mention should be made of the fact that it is becoming increasingly possible to study solids by NMR. The method depends on spinning the sample inclined at the correct angle to the applied magnetic field. In favourable cases, narrow lines are obtained—normally solids give broad lines—but in some cases it is not physically possible to spin fast enough to achieve liquid-like spectra. The method works well for amorphous solids and so can generally be expected to address questions such as whether the structure in the solid is the same as in solution. The following paragraphs provide a brief indication of the ways in which NMR finds application in coordination chemistry by a study of selected examples.

Complexes of WF_6, $[WF_6L]$ where L is a ligand, exist and in which the tungsten atom might be seven coordinate. A low-resolution ^{19}F spectrum shows three lines of relative intensity 1:1:4, which does not appear to be consistent with a seven-coordinate structure. The most likely structure is $[WF_5L]^+ F^-$, in which the tungsten atom is octahedrally coordinated by six ligands, five of which are fluorines. The four coplanar fluorines give rise to the largest peak and the axial fluorine to one of the others. The fluoride anion gives the final peak. The fine structure of the peaks confirms this assignment as does the conductivity of the compounds in liquid sulfur dioxide (Section 12.7) and, finally, X-ray crystallography. This example, as presented, depends on an interpretation of the number of observed peaks. In practice, empirical relationships between structure, chemical shifts and coupling constants would also be used in such structure determinations.

It is not usually possible to observe the NMR spectra of paramagnetic species because the lines are extremely broad. In particular cases, however, the spectra can be observed but cover a frequency range perhaps 50 times as great as that encountered with similar diamagnetic compounds. As was seen in Chapter 6, unpaired electrons which, from crystal field theory, would be expected to be localized on a transition metal, are in fact delocalized onto the ligands (main group elements rarely form paramagnetic coordination compounds). If, in particular, the ligands are conjugated systems, the unpaired electron density will also be delocalized over the entire ligand. The additional shifts observed in the NMR spectra of paramagnetic compounds are directly proportional to the unpaired electron densities on the nuclei giving rise to the various peaks. Unpaired electron densities may be either

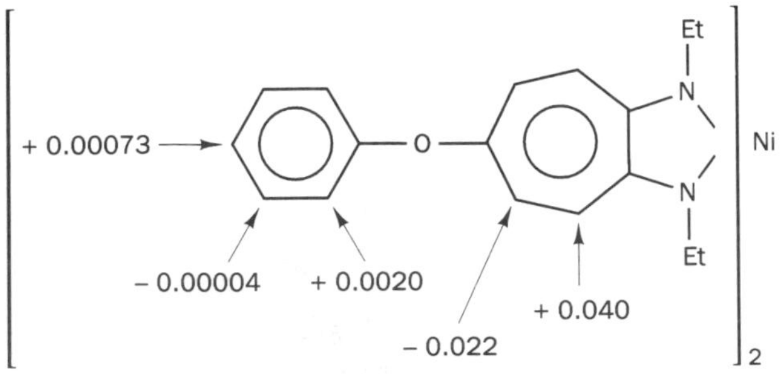

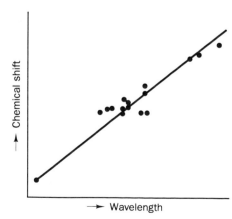

Fig. 12.14 Nickel(II) aminotroponeimineato complex.

positive or negative depending on whether the unpaired electron density is of α or β spin (spin densities usually alternate in sign along a conjugated system in a manner similar to the superexchange mechanism discussed at the end of Chapter 9; at the end of Chapter 6 an alternation of spin densities was also encountered). So, in a nickel(II) aminotroponeimineato complex the unpaired spin densities shown in Fig. 12.14 have been calculated from the observed spectra. In this compound the shifts of the proton resonances were studied. The unpaired spin densities given in Fig. 12.14, however, are those at the corresponding carbon atoms. This is because there is a direct proportionality between these two quantities. In principle, the spin density at the carbons could be measured directly, by use of ^{13}C NMR. However, this does not seem to have been done.

The opening-out of spectra due to paramagnetism is exploited by the use of so-called *shift reagents*. These are paramagnetic compounds, soluble in organic solvents, which have a through-space effect on the species in solution. By careful choice of shift reagent, chemical shift differences in the solute molecules are enormously increased, thus simplifying spectra, without any undue broadening of the peaks. The detailed mechanism of the effect is not fully known but it is likely that a small proportion of the solute molecules form a transient, weak, complex with the shift reagent, the effect perhaps affecting all solute molecules by a process akin to those described in Section 9.11. For ^{1}H spectra, a lanthanide complex of the ligand $Me_3CCOCHCOCMe_3$ with an ion such as Eu^{III} or Pr^{III} is frequently used.

Earlier in this section the use of empirical correlations involving chemical shifts were mentioned as an aid to structure determination. There has long been known a particularly good illustration of the origin of one contribution to such correlations in coordination compounds. This example concerns ^{59}Co resonances, for which it has been predicted that the chemical shift shown by octahedral cobalt(III) complexes should be inversely proportional to the energy separation between the $^1A_{1g}$ ground state and the lowest excited $^1T_{1g}$ term, a separation which may, of course, be measured from the d $\rightarrow$ d spectra. In Fig. 12.15 is given a plot of chemical shift against the wavelength of this transition (the wavelength is proportional to the inverse of the energy separation) for some octahedral cobalt(III) complexes. The amount of the (small) mixing of the $^1T_{1g}$ term into the $^1A_{1g}$ term by the magnetic field is inversely dependent on their energy separation, thus explaining the observation. This same mixing was met at the end of Section 9.3, where it was seen that it gives rise to temperature independent paramagnetism (TIP).

Fig. 12.15 The ^{59}Co chemical shifts shown by cobalt(III) complexes plotted against the wavelength of the corresponding $^1T_{1g} \leftarrow {}^1A_{1g}$ electronic transition.

Chemical shift →

→ Wavelength

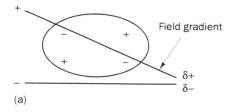

(a)

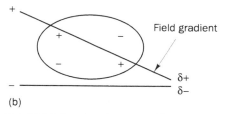

(b)

Fig. 12.16 A nucleus (represented by an ellipse) with a quadrupole moment (represented by the charges within the nucleus) in (a) a more stable and (b) a less stable orientation within an applied electrical field gradient.

12.5.2 Nuclear quadrupole resonance (NQR)

Atoms which have nuclear spins greater than $\frac{1}{2}$ behave as if the distribution of charge within the nucleus is non-spherical. The nucleus does not behave as a dipole because the nuclear charge distribution remains centrosymmetric. It does, however, possess an electrical quadrupole moment (CO_2, which is centrosymmetric and linear, is an example of a molecule which has an electric quadrupole moment). In an applied non-uniform electrostatic field, the non-uniformly charged nucleus can take up at least two orientations, one of which is more stable than the others (the number of orientations depends on the magnitude of the nuclear quadrupole moment). A diagram which gives some idea of the origin of this splitting is shown in Fig. 12.16. It is possible to excite the nucleus from a lower to an upper state by application of suitable radio-frequency radiation. This is a classical description of the phenomenon, but the essentials are carried over into a quantum mechanical treatment. In practice, the non-uniform electrostatic field is generated by the charge distribution around, but very close to, the nucleus. A more detailed discussion shows that any variation in field gradient is almost entirely due to unequal occupancy of the p orbitals of the atom—so that if there is to be a field gradient, the p orbitals cannot be symmetry-related. Clearly, this is a phenomenon which is only applicable to atoms in a low-symmetry environment (but, note carefully, this does not automatically mean a low-symmetry complex). The most-studied of the nuclei which exhibit quadrupole resonance spectra are ^{35}Cl and ^{37}Cl. The method is inherently insensitive and the high concentrations of these isotopes in samples such as solid $K_2[PtCl_6]$ is a great advantage. Only solids can be studied, anyhow, because the molecular tumbling in a liquid or gas averages the effect to zero. The sensitivity of the method has increased in recent years by the advent of pulse techniques (such as those used in NMR) and also, in suitable cases, by not looking at the NQR nucleus itself but, rather, at one which is energetically coupled to it and which is NMR-active.

Other, currently more specialist but of potential wide applicability, methods include the optical detection of quadrupole resonances—a sample is laser-excited to an electronically excited state, the return to the ground state is by phosphorescence; the intensity of the phosphorescence is sensitive to whether or not concurrent microwave radiation matches an energy separation in some quadrupole-split intermediate state. Yet another method depends on correlations between successive β or γ emissions from excited quadrupolar nuclei (where the excitation can be achieved by suitable nuclear bombardment). These do not exhaust the list of current developments—they have been chosen to illustrate the wide front on which new techniques are emerging. It is likely that because of these developments the future will see a wider use of NQR spectroscopy. It is also likely that the interpretation of the data will become more sophisticated. Traditionally, the experimental data have been interpreted to give the percentage ionic character of a bond. This is because, for example, in the Cl^- ion all of the p orbitals are equally occupied whilst in Cl_2 the σ bond, if composed of p orbitals only, corresponds to one electron in the p_σ orbital of each chlorine atom, and so Cl^- and Cl_2 differ in their resonant frequencies. Interpolation allows a value for the ionic character of a Cl–M bond to be determined from the chlorine resonance

Table 12.2 Ionic character of the M–Cl bond in $[MCl_6]^{2-}$. Although these ions have O_h symmetry, the symmetry at the chloride is, at most, C_{4v} (adjacent cations may reduce this symmetry)

Species	Ionic character of the M–Cl bond (%)
$[PtCl_6]^{2-}$	44
$[PdCl_6]^{2-}$	43
$[IrCl_6]^{2-}$	47
$[OsCl_6]^{2-}$	46
$[ReCl_6]^{2-}$	45
$[WCl_6]^{2-}$	43
$[SnCl_6]^{2-}$	66
$[TeCl_6]^{2-}$	68
$[SeCl_6]^{2-}$	56

frequencies in Cl^- and Cl_2. Some correction may be applied to allow for the fact that a pure chlorine p orbital may not be involved in the M–Cl bond. In this way, Table 12.2 was compiled. When there are two non-equivalent NQR nuclei in the unit cell of a solid these give rise to separate resonances which may be resolvable. In this way NQR spectroscopy gives structural information. Both bromine, ^{79}Br and ^{81}Br, and iodine, ^{127}I, but not fluorine, give NQR spectra, as too may ^{14}N, ^{55}Mn, ^{59}Co, ^{63}Cu, ^{65}Cu, ^{75}As, ^{121}Sb, ^{123}Sb, ^{201}Hg, and ^{209}Bi.

12.5.3 Mössbauer spectroscopy

Just as there are ground and excited states of atoms and molecules (electronic, vibrational and the like), so too there exist both ground and excited states of nuclei—they have just been mentioned as sometimes providing a method of measuring NQR spectra. In decaying from an excited state a nucleus may emit light, just as an atom or molecule may. In the case of nuclei, this light is of very short wavelength, it is γ radiation. If this γ radiation falls on another, identical, nucleus it may be absorbed, leaving the second nucleus in an excited state. Nuclei of any one element which are in different chemical environments will have slightly different energy levels, but the environment-induced changes are so small that it is possible to compensate for them with a Doppler shift of the γ radiation, achieved by moving the emitting nucleus either towards or away from the absorber. In Mössbauer spectroscopy the absorption of γ rays by the sample is recorded as a function of the velocity of the source. Solid samples are used; the source may be moved by attaching it to the diaphragm of a loudspeaker driven by a suitable signal generator. The effect has been observed for relatively few nuclei at the concentrations at which they occur in most coordination compounds, of which ^{57}Fe and ^{119}Sn have been the most widely studied. A more complete list is given in Table 12.3. The difference in absorption velocity and that of a suitable reference standard is called the isomer (or chemical) shift; it is denoted δ and is usually expressed in units of mm s^{-1} or cm s^{-1}. The chemical environment affects the nuclear energy levels through those electrons which are in orbitals which allow them to make contact with the nucleus. This means that only s electrons can directly affect isomer shifts since for all other orbitals the nucleus is contained in a nodal plane (electrons in p, d, or f orbitals can only influence isomer shifts through their incomplete shielding of the nucleus, leading to a change in effective nuclear charge which is felt by the s electrons). Examples of isomer shifts, relative to an iron foil standard, of ^{57}Fe in some iron compounds are given in Table 12.4. As this table illustrates, the isomer shift decreases with increase of negative charge and increases with increase of coordination number. Both of these generalizations find application in the observation that the isomer shift of $Fe(CO)_5$ is greater than that of $Fe(CO)_4^{2-}$. If the same nucleus has two different chemical environments in a compound, these will normally give rise to separate resonances. In this way it has been shown that the iron atoms in insoluble Berlin blue, $Fe_4[Fe(CN)_6]_3$, which is formed by the reaction of Fe^{III} salts and $[Fe(CN)_6]^{4-}$ ions in aqueous solution, are not all equivalent but retain their distinct oxidation states. Indeed, a particular use of Mössbauer data has been to indicate the valence state of an atom—empirical parameters

Table 12.3 Some nuclei which have been studied by Mössbauer spectroscopy

Isotope	Natural abundance (%)
^{57}Fe	2.2
^{99}Ru	12.0
^{119}Sn	8.6
^{121}Sb	57.3
^{125}Te	7.0
^{127}I	100.0
^{129}Xe	26.2
^{197}Au	100.0

Table 12.4 Isomer shifts for some iron compounds[a]

Compound	Isomer shift (mm s^{-1})
High spin Fe^{III}	ca. 0.3–0.5
FeF$_3$ (O_h)	0.49
FeCl$_3$ (O_h)	0.46
[FeF$_4$]$^-$ (T_d)	0.30
Low spin Fe^{III}	
[Fe(CN)$_6$]$^{3-}$ (O_h)	−0.12
High spin Fe^{II}	ca. 0.9–1.5
FeF$_2$ (O_h)	1.48
FeCl$_2$ (O_h)	1.16
FeBr$_2$ (O_h)	1.12
[FeCl$_2$(H$_2$O)$_4$] ('O_h')	1.36
[FeCl$_4$]$^{2-}$ (T_d)	0.90
Low spin Fe^{II}	
[Fe(CN)$_6$]$^{4-}$ (O_h)	−0.04

[a] Isomer shifts show a slight temperature dependence which has been ignored in this compilation. Data from N. N. Greenwood and T. C. Gibb *Mössbauer Spectroscopy* Chapman & Hall, London, 1971.

are available which compensate for the effects of change of substituent and coordination number. So, isomer shift data have been used to conclude that the π bonding ability of ligands decreases in the order

$$NO^+ > CO > CN^- > SO_3^{2-} > PPh_3 > NO_2^- > NH_3$$

A complication arises in that if the resonant nucleus experiences a non-zero electrostatic field gradient, a quadrupole splitting of the resonance may be observed (nuclear quadrupoles were discussed in the previous section). In an octahedral environment, such as that in [Fe(CN)$_6$]$^{4-}$, there is no field gradient (the chlorine atoms considered in [PtCl$_6$]$^{2-}$ in the previous section were not in an octahedral environment although in an octahedral complex), but in [Fe(CN)$_6$]$^{3-}$ the extra hole in a t_{2g} orbital has the effect of introducing a very temperature-dependent quadrupole splitting, which presumably occurs through vibrational distortions of the octahedron. For [Fe(CN)$_5$NO]$^{2-}$, where there is a built-in asymmetry, the quadrupole splitting is almost temperature-independent. The [Fe(CN)$_6$]$^{4-}$ and [Fe(CN)$_5$NO]$^{2-}$ ions are compared in Fig. 12.17. There is evidence that quadrupole splittings can be influenced by relatively distant ions in the crystal lattice—change a counterion, for instance, and there is a small change in quadrupole splitting.

An excellent example of the use of Mössbauer spectroscopy in structure determination is provided by Fe$_3$(CO)$_{12}$. Although the dark green crystals of this compound are easy to prepare—dissolve Fe(CO)$_5$ in aqueous alkali to give the Fe(CO)$_4^{2-}$ anion and oxidize this with solid MnO$_2$ to give Fe$_3$(CO)$_{12}$—its structure was uncertain for over 30 years.[5] Although eight structures, all incorrect, had been proposed for Fe$_3$(CO)$_{12}$, it was the ninth, suggested on the basis of its Mössbauer spectrum, which eventually proved to be correct. The spectrum is reproduced in Fig. 12.18 (peaks are downwards). The most evident thing is that it corresponds, approximately, to three peaks of equal intensity. It would be wrong to conclude that each corresponds to a different type of iron atom. If the iron atoms were all different then they could not all be in high symmetry environments and so quadrupole splittings would be expected on most of the peaks. Given that the chemical nature of the three iron atoms is so similar—and so similar isomer shifts are to be expected—the only reasonable interpretation of the spectrum is that the outer two lines are the quadrupole-split components arising from a peak of intensity two, centred at about the same position as the central peak (which itself has but a small quadrupole splitting and is of intensity one). So, it seems that there are two equivalent, low-symmetry, iron atoms and one of high symmetry. The correct structure, since confirmed by X-ray studies—which were made difficult by disorder in the crystal—is shown in Fig. 12.19. The high-symmetry iron atom is on the right (it has but two types of bond, one to terminal CO ligands, the other to Fe atoms). The low-symmetry pair are on the left. The X-ray crystallographic work showed the bridging CO ligands to be asymmetric, off-centre. This means that the low-symmetry iron atoms are each involved in four different bonds, two to bridging CO ligands—one long, one short—one to the terminal CO ligands and one to

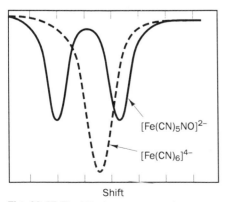

Fig. 12.17 The Mössbauer spectra of [Fe(CN)$_5$NO]$^{2-}$ (d^5, ligand assymetry) and [Fe(CN)$_6$]$^{4-}$ (d^6, symmetric ligand field).

[5] The story makes fascinating reading—see 'The Tortuous Trail Towards the Truth' by R. Desiderato and G. R. Dobson, *J. Chem. Educ.* (1982) 59, 752.

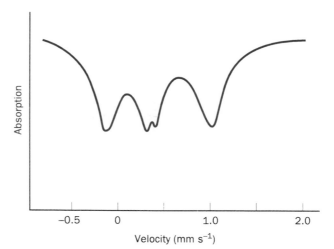

Fig. 12.18 The Mössbauer spectrum of $Fe_3(CO)_{12}$.

the unique iron atom. A more recent study of the Mössbauer spectrum over a temperature range has provided explanations for the asymmetries in the peak-intensity patterns of Fig. 12.18 which are entirely in accord with the accepted structure.

Finally, mention should be made of the fact that additional information can be obtained from measurements made with samples in a magnetic field, a magnetic field that can even be self-generated because of an inherent magnetism within the sample—a point of particular relevance to Fe^{III}, d^5 complexes. In Fig. 12.20 is shown the quadrupolar splitting that occurs when a ^{57}Fe nucleus is in a magnetic field, together with an energy level diagram that both explains the observed spectrum and also allows the relevant selection rules to be deduced.

12.6 Electron paramagnetic (spin) resonance spectroscopy (EPR, ESR)

As was emphasized in Chapter 9, the energy level splittings caused by placing a paramagnetic ion in a magnetic field are small, and in order to give a complete account of them it is necessary to include all of the effects which give rise to comparable or larger splittings. This is particularly important in electron spin resonance spectroscopy (or, perhaps a wider and therefore better, name, electron paramagnetic resonance spectroscopy[6]). In this, transitions are induced and observed between pairs of levels, split by a magnetic field.

Suppose that we have a single crystal containing paramagnetic ions, held in a fixed orientation in the lattice. At any orientation of the crystal within an applied magnetic field there will, in general, be a splitting of the energy levels of the paramagnetic ions. For the moment, let us select just one of these ions. If we now rotate the crystal, keeping the magnetic field constant, then the magnetic splittings of the ground state will vary with the orientation of the crystal, unless both the ion and crystal have perfectly octahedral or

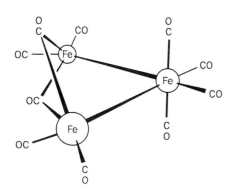

Fig. 12.19 The molecular structure of $Fe_3(CO)_{12}$.

---

[6] The name electron spin resonance is that which is the more appropriate to the particular discussion which follows.

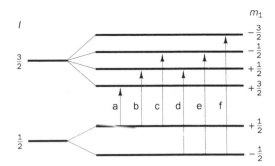

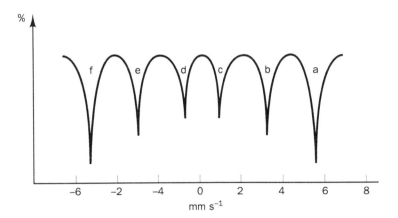

Fig. 12.20 The Mössbauer spectrum of ^{57}Fe in a magnetic field (reproduced courtesy of Prof. K. Burger).

tetrahedral symmetry, a highly improbable situation. Consequently, the frequency at which resonance will occur varies with crystal orientation. This splitting may be thought of as represented by an ellipsoid, with the paramagnetic ion at its centre. As the ion is rotated, so the ellipsoid rotates with it, but at all points faithfully represents the relative splittings of the energy levels between which transitions are being observed. The ellipsoid is described, mathematically, as a tensor and is referred to as a *g tensor*, the splitting of the energy levels at any particular orientation being of the form $g\beta H$, β being the Bohr magneton and H the applied magnetic field. It is the job of the theoretician to calculate and interpret the *g* tensor in terms of a suitable model. In a real crystal, the ions or molecules seldom have their corresponding axes parallel to each other. In such a real crystal there will be a family of molecules within a unit cell and a corresponding family of orientations of molecular *g* tensors. The result will be distinct sets of lines that will smoothly move and interchange as the crystal is rotated.

But this is too simple. As has been commented before in this book, it is helpful to interpret electron spin as the electron behaving like a tiny bar magnet. The transitions that we have been talking about are, then, associated with turning the bar magnet end-over-end in the applied magnetic field. But this is to neglect the presence of other bar magnets, the presence of nuclei each with their own nuclear spin. These nuclear magnets must also adopt a particular orientation within the magnetic field and the variety of possible orientations will be felt by the electron magnet. The result is that the election spin transitions have an additional structure, the so-called *hyperfine structure*

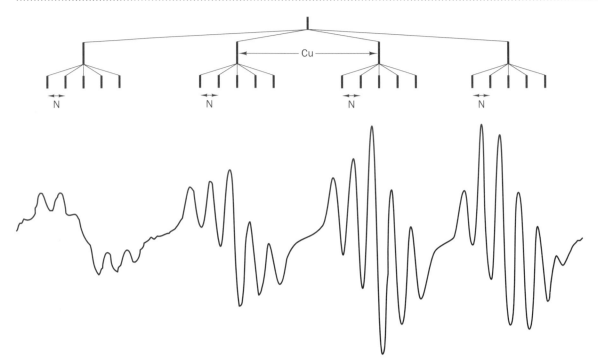

Fig. 12.21 The EPR spectrum of 5-chlorosalicylaldoxinecopper(II). In this complex there are two N atoms, each with spin 1, coordinated to the copper, so that the resultant N spin can be 2, 1, 0, −1 or −2. It follows that a five-line peak structure is expected. The Cu spin of $\frac{3}{2}$ gives rise to a four-line structure $(\frac{3}{2}, \frac{1}{2}, -\frac{1}{2}, -\frac{3}{2})$ and so the spectrum is interpreted as indicated. Clearly, the unpaired electron interacts with both copper and nitrogen nuclei. This spectrum is presented in derivative form, the normal presentation for EPR spectra. In this presentation the fine structure is made most evident. For comparison, Fig. 12.22 is presented in integrated form (reproduced courtesy of Prof. K. Burger).

originating in interactions between the unpaired electron(s) and nuclei with non-zero spin. So, copper has two naturally occurring isotopes, each of spin $\frac{3}{2}$. For each, we expect $(2S + 1) = 4$ nuclear levels. The EPR spectra of copper(II), d^9, complexes therefore show four peaks rather than one. Actually, the two copper isotopes do not give precisely superimposed lines; rather, two inter-related sets of four lines each, with intensities determined by the relative isotopic abundances, 70:30. The (single) four-line pattern will be seen at the very top of Fig. 12.21.

The interpretation of an electron paramagnetic resonance spectrum may involve all of the parameters introduced in our discussion of magnetic susceptibilities in Chapter 9 and Appendix 10—so that EPR measurements can give information on these parameters. Such measurements have the advantage that they give the individual components of the g tensor, whereas magnetic susceptibility measurements normally lead only to an average g value. In suitable cases, individual g values can also be obtained from EPR measurements on samples in solution. There are problems, however. EPR spectra can usually only be observed for ions which have spin degenerate ground states in the absence of a magnetic field—and that means only for molecules with an odd number of electrons. The spectra observed consist of transitions between members of this so-called *Kramer's doublet* (there is more discussion of Kramer's doublets in Section 9.10 and in Appendix 10). Further, there are often peak-broadening phenomena which can only be overcome by working at very low temperatures, perhaps as low as that of liquid helium. These peak-broadening phenomena are particularly severe when the ground state has another orbital level not far above it in energy (not far means comparable with kT). Thus, it is very difficult to observe the spectra of octahedral complexes of titanium(III), d^1, for which the ground state is derived from $^2T_{2g}$ by application of spin-orbit coupling and low-symmetry

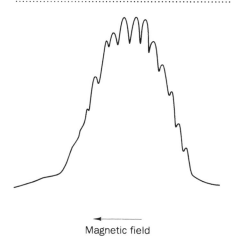

Fig. 12.22 The EPR spectra of a crystal of $Na_2[PtCl_6] \cdot 6H_2O$ containing a small amount of the isostructural $Na_2[IrCl_6] \cdot 6H_2O$ as an impurity. This spectrum is presented in integrated form. More common is for the spectrum to be differentiated and the first derivative plotted, as in Fig. 12.21.

Magnetic field

crystal field perturbations. Here, the ground state is usually such that it has two orbital (four spin–orbital) terms not far removed from it because of their common $^2T_{2g}$ parentage. Manganese(II), d^5, however, with its high spin $^6A_{1g}$ ground state, gives room-temperature spectra. Another broadening mechanism may occur if the paramagnetic species is at all concentrated. For this reason, single crystal data are usually obtained on a crystal of a diamagnetic compound, isomorphous with that under study, in which a very low concentration of the paramagnetic species is incorporated as an impurity.

It was mentioned earlier that EPR spectra may show fine structure due to interaction between the unpaired electron(s) and nuclei with non-zero spin. These nuclei may be those of the paramagnetic ion (for example, ^{55}Mn, with spin $\frac{5}{2}$ and 100% natural abundance) or those of a ligand. EPR spectra of copper(II) complexes with nitrogen ligands show evidence for interactions of this sort and have been extensively studied—an example is shown in Fig. 12.21, where the nitrogen fine structure is additional to the quartet pattern mentioned above and is indicated by N. The best known example, however, is that of the EPR spectrum of the $[IrCl_6]^{2-}$ ion, involving a strong-field complex of iridium(IV), with a d^5 configuration, incorporated as an impurity in a crystal of $Na_2[PtCl_6] \cdot 6H_2O$. The fine structure observed, shown in Fig. 12.22, can be interpreted in terms of interaction of the unpaired electron with the ^{35}Cl and ^{37}Cl nuclei of the ligands. Such observations provide excellent evidence for covalency in transition metal complexes, emphasizing the superiority of the ligand field model over the crystal field approach—it appears that the unpaired electron density in $[IrCl_6]^{2-}$ is about 5% on each ligand. As for several other of the techniques described above, EPR is currently undergoing a rapid development which can only increase its future impact. Perhaps most important has been the introduction of pulse (and therefore Fourier transform) techniques, which in turn has led to the same sort of variety of pulse sequences that has transformed NMR. In addition, the method is no longer confined to the microwave region but has moved into the far infrared.

12.7 Photoelectron spectroscopy (PES)

The consequences of the interaction of a beam of light with a molecule depend on the energy of the light beam. If it is low it can only lead to the molecule rotating or vibrating. At higher energies, electronic excitations occur, these becoming more violent with increase in energy of the light beam. Ultimately, in the vacuum ultraviolet, the interaction leads to the ejection of electrons. If a monochromatic ultraviolet source is used, the energy of the incident photons is known. As measurements have to be made under high-vacuum conditions—this *is* the vacuum ultraviolet—the energy of the ejected photoelectrons can be measured by passing them through a magnetic or electrostatic field and measuring their change in path direction. The difference in energy between that of the incident photons and of the emergent electrons is the energy required to ionize the electrons from the molecule. Such measurements are the basis for photoelectron spectroscopy, a technique which has provided great insights into molecular structure, particularly for simple molecules.

It would be ideal to use a vacuum ultraviolet laser as the source of monochromatic light but, for all practical purposes, such lasers do not exist.[7] Instead, a helium discharge is used. One would expect a multitude of lines, and indeed they do occur, but with careful control of conditions it is found possible to have over 99% of the emission in the He[I] 584 Å (21.22 eV) line. Under different conditions, it is possible, alternatively, to obtain a high intensity in the He[II] 304 Å (40.81 eV) line. The use of helium gas as a discharge medium in a ultrahigh vacuum setup is not too much of a problem—the light from the discharge is admitted into the spectrometer through a length of capillary tube. Under high vacuum conditions, atoms and molecules have mean free paths of several centimetres so that a length of 0.5 mm diameter capillary represents a real obstacle. More of a problem is removal of the gaseous sample under study after it has passed the narrow shaft of ultraviolet light. Cunning design and good vacuum pumps provide the solution!

A PES spectrum consists of many lines. Ignoring fine structure for the moment, each line in the spectrum may be associated with photoionization of an electron of different energy. So, if there are four occupied orbitals with energies above 21.2 eV but six above 40.8 eV, then the He[I] PES will normally show four lines and the He[II] six. Note the use of the word orbitals in the preceding sentence. The equation of an ionization energy with an orbital energy is usually, but not always, valid. It is usually referred to as *Koopmans' theorem*, although Koopmans' approximation would be more accurate. Koopmans' theorem is of greatest validity when an electron is photoionized from a very diffuse, delocalized, molecular orbital. The adjustment of the remaining electron density to take account of the loss of an electron is minimal. In contrast, when an electron is photoionized from a very compact, localized orbital the local readjustment of electron density may well have considerable energetic consequences. In such cases it is no use to compare the numerical results of PES with those from a very accurate *ab initio* calculation on the molecule. Rather, the PES data have to be compared with the differences in total energy between calculations on M and on M^+ in the appropriate state. As has been seen in Section 10.7—and will be seen again, below—this point is of particular importance in molecules such as ferrocene.

So far, fine structure has been neglected. In fact, fine structure is of key importance in PES. Suppose that the photoionization observed is from a bonding orbital. Then the ionized molecule will be more weakly bonded than the parent and will be produced in what, for it, is a compressed structure. That is, not only is it produced vibrationally excited—and so there will be vibrational fine structure in the PES spectrum—but the vibrational frequencies will be lower than their counterparts in the parent. Conversely, photoionization from an antibonding orbital gives an ionized molecule with higher vibrational frequencies than the parent. Photoionization from a non-bonding orbital gives rise to small or zero vibrational frequency changes.

---

[7] The nearest that one can get in practice is to use monochromatic ultraviolet light from a synchrotron source. Such sources have the advantage of being tunable—so that the wavelength of the radiation can be chosen to be the optimum for the measurement—but the disadvantage that they are only available at a relatively small number of national or international facilities.

Careful analysis of such vibrational frequencies can even indicate the symmetry of the orbital from which the electron was ionized.

Other causes of fine structure in the spectrum are spin–orbit coupling and the Jahn–Teller effect operating in the ionized molecule. These complications will not be further discussed at this point because they are adequately covered in Sections 8.4, 11.5 and 8.5, respectively. Two other general points should be noted: first, that PES is of limited resolution—at best, about 20 meV, 160 cm^{-1}, peak half widths (at this resolution a good peak gives about 100 counts per second for a typical organometallic). This means that the effect of the so-called fine structure is often to give broad bands and thus band overlap. Secondly, band intensity changes occur going from He^{I} to He^{II} which make more difficult attempts to relate band intensity to the degeneracy of the orbital set from which an electron is photoejected (although the intensity changes may be related to changes in cross-section; so the metal 3d cross-section increases and the carbon 2p decreases from He^{I} to He^{II}).

In Fig. 12.23 are shown the He^{II} photoelectron spectra of $Cr(CO)_6$ (a) and $W(CO)_6$ (b). Their general similarity is obvious. The interpretation of these spectra is aided by the data given in Chapter 10 in Table 10.4. At

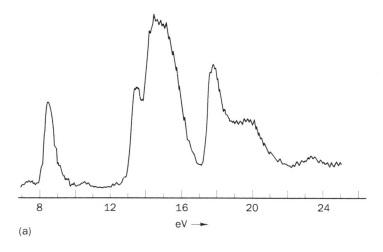

(a)

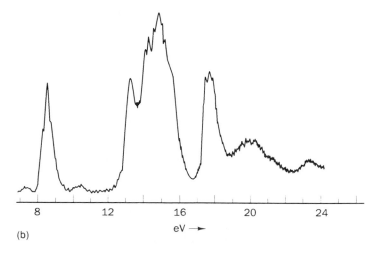

(b)

Fig. 12.23 The He^{II} PES of (a) $Cr(CO)_6$ and (b) $W(CO)_6$. Adapted and reproduced with permission from B. R. Higginson, D. P. Lloyd, P. Burroughs, D. M. Gibson and A. F. Orchard, *J. Chem. Soc., Faraday Trans. 2* (1973) 69, 1659.

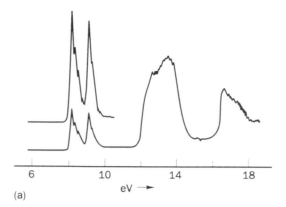

(a)

Fig. 12.24 The He^I PES of (a) Mg(C$_5$H$_5$)$_2$ and (b) Fe(C$_5$H$_5$)$_2$. Adapted and reproduced with permission from S. Evans, M. C. H. Green, B. Jewitt, A. F. Orchard and C. F. Pygall, *J. Chem. Soc. Faraday Trans. 2* (1972) 68, 1847.

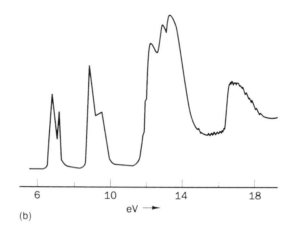

(b)

about 8.5 eV in the PES, ionization from the metal t_{2g} d orbitals is seen. For W(CO)$_6$ this band has a shoulder, clearly resolved in the He^I PES. This splitting is attributed to spin–orbit splitting in the t_{2g}^5 orbitals of [W(CO)$_5$]$^+$. Between 13.3 and 14.4 eV lie the CO π bonding combinations t_{1u}, t_{1g} and t_{2u}. The t_{2g} member of the π bonding set appears within the span of the CO σ bonding combinations e_g, t_{1u} and a_{1g} at lower energies (they vary in going from Cr(CO)$_6$ to W(CO)$_6$ but are centred at ca. 15.6 eV). In the region 17–22 eV lie the oxygen σ lone-pair combinations a_{1g}, t_{1u} and e_g (these are not included in Table 10.4). It is to be noted that although the oxygen σ (outward-pointing) orbitals are physically well separated in the molecule, the involvement of, and mixing with, what are more inner orbitals—of the same symmetry—is sufficient to cause large splittings.

As a final example consider the case of ferrocene, Fe(C$_5$H$_5$)$_2$, for which the He^I PES is given in Fig. 12.24 (b) together with that of Mg(C$_5$H$_5$)$_2$ (a). Although there are marked similarities between the spectra there are also clear differences. A pair of bands at ca. 7 eV in ferrocene have no counterpart in magnesocene and are to be associated with ionization from the d orbitals of the transition metal. The pair of bands between 8 and 10 eV are closer together in ferrocene whilst for this molecule the broad band between 12 and 14 eV is more structured. The pair of bands at ca. 7 eV have been assigned to e_{2g}(3d) and an a_{1g}(3d)—yet in Section 10.7, and in particular Table 10.3, we saw that in *ab initio* calculations on ferrocene these are rather *low*-lying orbitals and not those that would be expected to be the most readily ionized. We have a breakdown of Koopmans' theorem. The reason for this pattern—in terms of the high degree of localization of the electrons in these orbitals and so the large readjustment energy of the remaining electrons— has been discussed in Section 10.4. Returning to the photoelectron spectrum of ferrocene, the 8–10 eV bands have been assigned to $e_{1u}(\pi\text{-Cp})$ and $e_{1g}(\pi\text{-Cp})$ whilst the 12–14 eV region is assigned, qualitatively, to ionization from ligand σ and π orbitals. It is illuminating to compare the experimental patterns with the alternative sequences given in Table 10.3.[8]

One final comment: in this section we have been concerned with vacuum photoelectron spectroscopy. A similar X-ray photoelectron spectroscopy exists in which the ejected electrons come from inner electron shells. The ejected electron energy is sensitive to the chemical environment from which it originates and so gives information on this.

12.8 Evidence for covalency in transition metal complexes

This is a convenient point at which to bring together the various pieces of evidence indicating covalency in transition metal complexes although it must be remembered that not all are applicable to every species. The pieces of evidence are listed below.

1. Ligand hyperfine splittings in EPR spectra, discussed in Section 12.6.

2. The percentage of ionic character in metal–chlorine bonds as measured by NQR spectroscopy, discussed in Section 12.5.2.

3. Unpaired electron spin densities on ligand atoms as measured by NMR spectroscopy, discussed in Section 12.5.1.

4. The superexchange mechanism of antiferromagnetic coupling which enables the electrons on one atom to see the spin of those on another and which operates through their mutual overlap with a ligand atom, discussed in Section 9.11.

5. The orbital reduction factor k, which is introduced to allow for the effects of covalency on the magnetic properties of paramagnetic species, mentioned in Chapter 9 and discussed in some detail in Appendix 10.

6. The reduction in spin–orbit coupling constant in a complex compared with the free ion value, discussed in Section 8.4.

7. The reduction in electron repulsion energies (the nephelauxetic effect) in a complex compared with their values for a free ion, covered in Section 8.3.

8. Accurate electron density maps obtained from X-ray diffraction data, which show that there is no point of zero electron density on metal–ligand axes. It is therefore not possible to regard metal and ligand electrons as in any way separate.

[8] For a more detailed discussion see G. Cooper, J. C. Green and M. P. Payne, *Mol. Phys.* (1988) *63*, 1031.

9. Polarized neutron diffraction measurements, which show electron spin densities on the ligands, albeit not always of the spin expected, discussed in Section 7.6. These data not only show the need for a model which includes covalency but also indicate the limitations of ligand field and other one-electron models.

10. The use of shift reagents in NMR, outlined in Section 12.5.1, which show that unpaired electron density in both transition metal and lanthanide complexes can be felt by molecules outside the coordination sphere.

11. The most precise theoretical calculations, described in Chapter 10, supported by the results of photoelectron spectroscopy—see Section 12.7—unambiguously support the general conclusions of ligand field theory. However, a general tendency on the part of inorganic chemists to qualitatively over-estimate the involvement of d orbitals in bonding, particularly for main group elements[9] but also for transition metals, is to be noted.

12. Band intensities: some spectral bands involving electronic transitions have intensities other than those which might be expected in terms of a purely ionic model. This evidence tends to be more speculative than the others listed above and so it has not been discussed earlier in the text. There are four intensity anomalies which may be taken to indicate covalency.

    (a) d → d electronic transitions which are more intense when close to charge-transfer bands. This is indicative of a mixing between the two classes of transition which, in turn, implies covalency within the M–L bonds; see Section 8.10.

    (b) Some metal–ligand stretching vibrations have much lower infrared intensities than would be expected for vibrating, non-overlapping, ions; see Section 12.2.2.

    (c) The optical activity introduced into formally d → d electronic transitions by an optically-active ligand may be much greater than expected; see Section 12.3.

    (d) In Chapter 11 the mathematical treatment of the origin of f → f band intensities in lanthanides and actinides was not discussed. The basic theory, the so-called Judd–Ofelt model, contains a number of parameters. Of these, at least two are generally regarded as associated with covalency.

12.9 Molar conductivities

Electrochemical methods have been part of coordination chemistry from its earliest days. In recent times there has been a revival of interest in such techniques, so we start with a brief overview of the traditional as a prelude

---

[9] As an example the case of SF_6 may be cited; the bonds are rather polar, a feature which explains its properties without the need to invoke significant d orbital participation on the sulfur. See J. Cioslowski and S. T. Nixon, *Inorg. Chem.* (1993) *32*, 3209.

to considering the more modern in the next section. In water, equivalent conductivities—the contribution of an ion to the equivalent conductance of a salt at infinite dilution—are about 60 ohm^{-1} at 20 °C for each species (large inorganic ions give lower values whilst H$^+$ and OH$^-$, for which chain conduction mechanisms are available, give higher values). As a first approximation, then, one would expect the contribution to the molar conductivity of an ion I^{r+} to be about 60r ohm^{-1}. For ions M^{m+} and X^{n-} in the salt M$_n$N$_m$, the contributions will be 60m ohm^{-1} (from M^{m+}) and 60n ohm^{-1} (from X^{n-}). Multiplying by the number of ions of each type and adding leads to the conclusion that a salt M$_n$N$_m$ will have a molar conductivity of about 120 ohm^{-1} at 20 °C. So, a compound analysing as CoCl$_3$5NH$_3$ is found to have a molar conductivity of 261 ohm^{-1} and it is concluded that it is [Co(NH$_3$)$_5$Cl]Cl$_2$ with $nm = 2$. Similarly, CoCl$_3$5NH$_3\cdot$H$_2$O has a molar conductance of 390 ohm^{-1} and so is [Co(NH$_3$)$_5$(H$_2$O)]Cl$_3$, with $nm = 3$. Water is a coordinating solvent and, particularly for labile complexes, is best avoided for conductivity measurements. Nitrobenzene and nitromethane are commonly used alternatives, although solubility problems can arise. The usual procedure is to measure the conductivity at a concentration of ca. 10^{-4} molar and, assuming a molecular weight, to compare the value with that obtained for similar complexes of known ion type (1:1, 1:2, etc.). A rather better method is to make measurements over a range of concentrations and to plot λ_o and $\lambda_e(c)$, where $\lambda_e(c)$ is the equivalent conductivity at infinite dilution, against $\sqrt{c}$. Again, this plot is compared with those obtained for similar species in the same solvent. The equivalent weight of the complex is needed for this method and this may be obtained by choosing a suitable counterion for the complex species. The potassium ion is a suitable cation and the perchlorate has been widely used as a suitable anion, although, following the discussion of Section 4.2.2, the triflate anion CF$_3$SO$_3^-$ may well be preferred. None of these species is involved in strong coordination and so, assuming that one can ignore their complex-forming ability for the compound under study, the equivalent weight may be obtained from its empirical formula.

12.10 Cyclic voltammetry

As is well known, oxidation and/or reduction of ionic species in solution can occur on electrolysis. This is the basis of electroplating and, in reverse, the mechanism of action of many batteries. Controlled potential electrolysis as a preparative method was met in Section 4.2.3. Cyclic voltammetry is a method of studying such oxidation–reduction processes in detail. It is a method that has gained much in popularity in recent years since among other things it enables the measurement of thermodynamic redox potentials. The word cyclic in the name refers to the fact that if in the measurement A → B is an electrolytic oxidation, in the measurement it is immediately afterwards followed by B → A in an electrolytic reduction. This might sound rather like a pointless exercise but it, in fact, enables the fate of B to be probed immediately after it is formed. If it is found that there is less B to be reduced than there was B formed, experiments can perhaps be designed to investigate its fate. The method also provides information on the reversibility of electron gain/loss processes, their kinetics and on the strategies that might be successful in the bulk preparation of species for which there is cyclic voltammetric evidence.

At its simplest, cyclic voltammetry (CV) consists of applying a saw-tooth voltage across two electrodes (often of platinum but carbon or mercury are also used), usually in an aqueous solution containing a supporting electrolyte as well as the ion under study. The supporting electrolyte is involved in electrical current transfer across the bulk of the cell but is not involved in the electrode reactions. In order to ensure that this is so, the voltage range of the saw-tooth is limited. To avoid complications caused by the electrolytic reduction of dissolved oxygen, the solutions are degassed; when non-aqueous solvents are used great care must be taken to remove traces of water. The concentration of the ion under study is usually less than 1% of that of the supporting electrolyte. This means that as the electroreduction (say) of A → B progresses, the layer of the solution adjacent to the electrode becomes depleted of A and so the electroreduction, ultimately, becomes diffusion-controlled. If the potential across the cell is now reversed, as it will be if the applied saw-tooth spans both positive and negative potentials, then, ultimately, the electrooxidation B → A will occur. If B has not been reduced in concentration by some reaction, a similar pattern will be followed—a maximum current followed by a drop to a diffusion-controlled limit. The sequence is illustrated in Fig. 12.25.

Although a sequence A → B; B → A has just been followed, it is clear that if some B has reacted en route then the final condition is not the same as the starting. That is, successive voltammagrams will not be identical. A further reason for a difference would be if the products of the side-reactions of B were themselves electroactive. An important variable in studies aimed at investigating such phenomena is the slope of the saw-tooth waveform. Rates of up to ca. $100 \, \text{V s}^{-1}$ are common—which means that a complete scan takes only about $10^{-2} \, \text{s}$—although they can be very much slower. The method is a very useful one for the qualitative, rather than quantitative, characterization of compounds. It has found common application in bio-inorganic chemistry, where species often show a family of oxidation–reduction steps. If one were trying to synthesize a model complex analogous to a naturally occurring bioinorganic, one question that it would be natural to ask is, does it have a similar CV behaviour?

Two final comments: first, although the above discussion has been about a two-electrode process, the real-life experiment involves at least three, the third being a reference electrode (usually calomel or silver/silver chloride). If the supporting electrolyte contributes significantly to the current—as may well happen if, for instance, the pH is well away from 7, then a four-electrode cell can be arranged in such a way that the contribution from the supporting electrolyte is cancelled. Finally, as hinted above, not all electrode reactions are completely or immediately reversible. Such irreversibility is indicated by the separation between the cathodic maximum current potential (E_c) and the anodic maximum current potential (E_a) being separated by a potential which is greater than expected (which is ca. $0.059/n$, where n is the number of electrons involved in the electrode reaction).

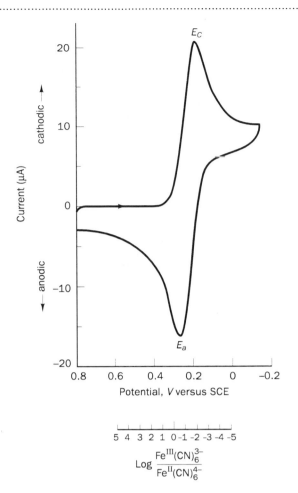

Fig. 12.25 A cyclic voltammagram of 6 mM $K_3[Fe(CN)_6]$ in 1 M KNO_3. Adapted and reproduced with permission from P. T. Kissinger and W. R. Heineman, *J. Chem. Educ.* (1983) *60*, 702.

12.11 X-ray crystallography

Undoubtedly the most important method of structure determination at the present time is that of X-ray crystallography. It is a large subject, of which only a few points can be highlighted in the present section. In section 11.7 the classical model of the way that a light beam interacts with matter was detailed; in particular, the light beam is slowed down. This happens because the light beam induces a dipole within each atom or molecule, which then re-emits. The same happens with X-rays but with the important difference that the re-emissions occur from atoms separated by distances comparable to or greater than the X-ray wavelength. The result is that both constructive and destructive interference can occur between the re-emitted beams. For crystalline samples, destructive interference is the norm, constructive only occurring at very special angles between the incoming (monochromatic) X-rays and the crystal. These angles are determined by the details of the regular, repeated, geometrical arrangement of the atoms in the crystal and also by the X-ray wavelength. The intensities of the emitted, diffracted, beams depends on the position of each atom in the regular array and, most important, on the chemical identity of the atoms, for this determines the strength with which each atom re-emits (basically, the more electrons the stronger the re-emission). So, measurement of the angles of the diffracted beams gives details of the regular array, and thus to some extent structure, and measurement of their intensities give the information of greatest chemical interest, the identities of the atoms and of their positions relative to other atoms. Clearly, measurement of both angles and of intensities is important. In most laboratories the process is now highly automated. A tiny single crystal is selected and mounted on a fibre of lithium borate glass (low atomic weights, low X-ray scattering). Some preliminary diffraction measurements enable the computer to orient the crystal and to orchestrate the collection of data. After some data manipulation, the intervention of the researcher is generally needed to choose the most probably correct, but approximate, crystal structure, which is then refined by the computer and the full data set deconvoluted. It sounds too easy! Whilst, in favourable cases, little knowledge is needed to solve a crystal structure, learning how to control the computer can be the major hurdle—there are traps for the unwary. First, the theory assumes that the crystal is uniformly bathed in X-rays. No scattering, no diffraction, no absorption. This is why the crystal has to be tiny, perhaps almost too small to see. This points to a major problem but one that is not always explicitly addressed, that of whether the crystal studied is representative of the sample from which it is selected. Next, heavy atoms dominate the scattering process, so that in structures containing them it is difficult to accurately place light atoms nearby and sometimes even to identify these light atoms. In many structures, hydrogen atoms are not detected (neutron diffraction, best on *deutero* materials, solves this problem). Again, at room temperature the atoms in a crystal will be, thermally, vibrationally excited. This motion 'blurs out' the structure, reducing precision. However, for some samples, cooling—which would normally give better precision—shatters the crystal (this was a problem for ferrocene). Particularly if rather rounded molecules are involved, disorder can occur in a structure, a disorder which has to be modelled in some way before the crystal structure can successfully be refined. Rounded molecules can often be accommodated in several different ways in the structure, all arrangements being of comparable energy. In such a case the model would be one in which all of the possible arrangements are superimposed, each weighted according to its occurrence (the weightings would be varied to obtain the best fit). A different type of disorder occurs when solvent molecules occupy some, but not all, equivalent sites. Finally, at several points in this chapter, as well as elsewhere in the book, mention has been made of spectroscopic measurement on crystals. Apart from the far from trivial fact that these often demand crystals which are very much larger than those for X-ray structure determinations (so that it is not a trivial problem to check that the large crystal has the same structure as that on which X-ray work was carried out), it is important to recognize that all spectral interpretations demand the use of a *primitive* unit cell. This is because almost all spectral measurements use wavelengths which are large compared to typical interatomic separations. Molecules or ions related by pure translations experience the same electric vector originating in the incident light wave, for example. To deal with this, the pure translations must be correctly chosen—and the correct choice is that set which interrelates primitive unit cells. If, as may well happen, crystallographers find it convenient to work with a centred unit cell (body or face-centred), the spectroscopist—and theoretician—has to work with the corresponding primitive unit cell (there always is one). This particular point will be met again in Chapter 17.

12.12 Conclusion

In this chapter some of the more important methods of studying coordination compounds have been reviewed. The detailed interpretation of the experimental results is often rather difficult, and, for the phenomena of optical activity and NQR, for example, the theory of the method has only been worked out incompletely. Discussion has therefore been confined to the qualitative level, it being considered important that the student should have a pictorial idea of the phenomena considered. In this way he or she should both have been made aware of which technique is likely to be of use in tackling a particular problem and also of some of the difficulties associated with its application. Inevitably, it has been necessary to be selective and this has been done on the basis either of techniques which the student may well meet in the laboratory or of techniques which are of particular importance. Finally, Chapter 16 will provide examples of the application of some of the techniques described in the present chapter, as well as a few more which have proved to be of particular value in the study of bioinorganic molecules.

Further reading

Two books which cover the majority of the methods described in this Chapter are *Structural Methods in Inorganic Chemistry* by E. A. V. Ebsworth, D. W. H. Rankin and S. Cradock, Blackwell, Oxford, 1987 and *NMR, NQR, EPR and Mössbauer Spectroscopy in Inorganic Chemistry*, R. V. Parish. Ellis Horwood, Hemel Hempstead, 1990. An older book, but one that covers a wider subject area, with some good chapters is *Physical Methods in Advanced Inorganic Chemistry* H. A. O. Hill and P. Day, eds., Interscience, London, 1968. Readable insights into a limited number of particular areas can be found in *Spectroscopy of Inorganic-Based Materials* (Advances in Spectroscopy Vol 14). R. J. H. Clark and R. E. Hester, eds. John Wiley, Chichester, 1987.

Other, complementary, useful references are:

- 'A Primer on Fourier Transform NMR' by R. S. Macomber in *J. Chem. Educ.* (1985) *62*, 213.

- *Fourier Transform Raman Spectroscopy. Instrumentation and Chemical Applications*, P. Hendra, C. Jones and G. Warnes, Ellis Horwood, Chichester, 1991. A general introduction, not confined to coordination compounds.

- 'Polarographic Behaviour of Coordination Compounds' by A. A. Vlcek, *Prog. Inorg. Chem* (1963) *5*, 211.

- 'Circular Dichroism of Transition Metal Complexes' by R. D. Peacock and B. Stewart, *Coord. Chem. Rev* (1982) *46*, 129.

- 'An Introduction to Cyclic Voltammetry' by G. A. Mabbott, *J. Chem. Educ.* 60 (1983) 697.

- 'Cyclic Voltammetry–Electrochemical Spectroscopy' by J. Heinze, *Angew. Chem. Int. Ed.* (1984) *23*, 831.

- 'Cyclic Voltammetry' by P. T. Kissinger and W. R. Heineman, *J. Chem. Educ.* (1983) *60*, 702.

- 'Molecular Electrochemistry' by R. G. Compton and A. R. Hillman, *Chemistry in Britain* (1986) 1088.

- 'Nuclear Quadrupole Interactions in Solids' by J. A. S. Smith, *Chem. Soc. Rev.* (1986) *15*, 225.

- 'Modern Techniques in Electron Paramagnetic Resonance Spectroscopy by J. H. Freed, *J. Chem. Soc., Faraday Trans.* (1990) *86*, 3173.

- *Electron Paramagnetic Resonance of d Transition Metal Compounds* (Studies in Inorganic Chemistry, Vol 16). F. E. Mabbs and D. Collison. Elsevier, Amsterdam, 1992. A large reference book that at first sight appears mathematical. In fact, there are some easy-to-read chapters; the derivations are presented relatively simply and certainly completely.

- 'Pulsed Electron Spin Resonance Spectroscopy: Basic Principles, Techniques and Examples of Application' by A. Schweiger, *Angew. Chem., Int. Ed.* (1991) *30*, 265.

- 'Some Aspects of the Electron Paramagnetic Resonance Spectroscopy of d-Transition Metal Compounds' by F. E.

- *NMR, NQR, EPR and Mössbauer Spectroscopy in Inorganic Chemistry* by R.V. Parish, Ellis Horwood, Chichester, 1990. A book with a minimum of theory which provides an excellent follow-up for the material in the present chapter and a bridge to more theoretical treatments.
- 'NMR and the Periodic Table' R. K. Harris and B. E. Mann (eds.), Academic Press, New York, 1978; provides an older, but excellent and reasonably readable, specialist treatment.

A very readable review of modern aspects and applications of NMR is to be found in the July 1993 (page 589 on) issue of *Chemistry in Britain*.

There are two key books on the applications of vibrational spectroscopy: *Metal–Ligand and Related Vibrations* by D. M. Adams, Edward Arnold, London, 1984; and *Infrared and Raman Spectra of Inorganic and Coordination Compounds*, by K. Nakamoto, J. Wiley, New York, 1986.

A useful source is Volume 1 of *Comprehensive Coordination Chemistry* G. Wilkinson, R. D. Gillard and J. A. McCleverty (eds.), Pergamon Press, Oxford, 1987, Chapter 8.1 'Electrochemistry and Coordination Chemistry' by C. J. Pickett.

Questions

12.1 Figure 12.26 shows the infrared spectra of *cis* and *trans* $[Pd(NH_3)_2Cl_2]$ in the $\nu(Pd–X)$ region. Which spectrum corresponds with which isomer? (Figure adapted from R. Layton, D. W. Sink and J. R. Durig, *J. Inorg. Nucl. Chem* (1966) *28*, 1965).

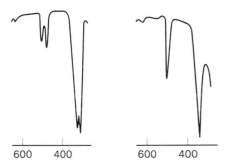

Fig. 12.26 Infrared spectra, Question 12.1.

12.2 Based on a study of Fig. 12.3 and using a group theoretical approach, suggest, qualitatively, what band patterns might be observed in the infrared spectra of crystalline $K_4[Fe(CN)_6]$ and $Na_4[Fe(CN)_6]$.

12.3 In the footnote on page 271 it was claimed that infrared spectroscopy could distinguish three different $M(CO)_3$ geometries. Substantiate this claim by working out the predicted allowed bands for each geometry (it will be necessary to use group theory). Could a similar claim be made for Raman spectroscopy?

12.4 It was found that a complex containing an $Fe(CO)_2$ unit, when in solution, gives an infrared spectrum with two bands of equal intensity (to within experimental error) in the $\nu(C–O)$ region. What is the C–Fe–C bond angle?

12.5 There are reports in the literature—including a (very early) crystal structure—of a compound $[Pt(CH_3)_4]_4$, containing a tetrahedron of platinum atoms with a CH_3 spanning each face. More recent work suggests that the compound is $[Pt(CH_3)_4OH]_4$, with faces centred by OH groups. Which of the methods described in the present chapter might be expected to distinguish between the two formulations?

12.6 You have been given the task of equipping a general inorganic research laboratory with a limited but ill-defined budget. Place the instrumentation associated with the techniques described in this chapter in order of priority on your shopping list. For each instrument write a short paragraph explaining the reasons for its position on your list.

13

Thermodynamic and related aspects of ligand fields

13.1 Introduction

One of the earliest applications of crystal field theory was its use to explain irregularities in thermodynamic and related properties of a series of transition metal complexes as the transition metal was varied. These applications are usually dealt with in one of the early chapters of a book such as this; their consideration has been deferred in order to be able to draw on the background and additional insights provided by Chapters 6, 7 and 11.

Although the notation that will be used implies the use of the crystal field model, the discussion is not limited to this, as the title of the chapter indicates. Rather, the concern will be with energy level separations, conveniently but not always accurately, thought of as resulting from orbital energy level differences—hence the convenience of using the language of simple crystal field theory, even if we do not really mean it! Thermodynamic aspects of coordination chemistry have already been encountered in this book—they represent the major theme of Chapter 5. The present chapter is distinguished from Chapter 5 because here we deal with the fine detail; the earlier chapter was more concerned with gross effects.

13.2 Ionic radii

The crystals of simple salts such as NaCl and ZnS have lattices which are often regarded as ionic. That this is a plausible approximation is indicated by, amongst other things, the fact that it is possible to assign ionic radii and to use the approximate additivity of these radii to predict interatomic distances.[1] Crystal field theory similarly assumes that the forces between a

---

[1] It is perhaps not sufficiently well recognized that although the ionic radius approach is reasonably self-consistent, the radii discussed may not correspond to physical reality. For example, in the KCl crystal the minimum of electron density along the K–Cl axis is ca. 1.45 Å

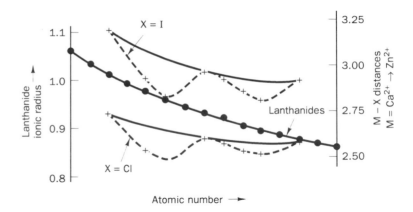

Fig. 13.1 The lanthanide contraction and metal–halogen distances for divalent first-row transition and related metal ions. The thin solid lines indicate the probable values for the M–X distances corrected for crystal field effects.

transition metal cation and the surrounding ligands are purely electrostatic. The theory suggests, therefore, that it should be both valid and possible to obtain values for the ionic radii of transition metal ions from crystallographic data. How might these radii be expected to vary from one ion to the next? If crystal field effects are neglected, the essential difference between adjacent members of a series of ions such as Ti^{3+}, V^{3+}, Cr^{3+} and so on, is that each successive member has an extra positive charge on its nucleus and an extra, compensating, d electron. As a first approximation, therefore, no change in ionic radii along the series might be expected. Recognizing the incomplete screening of the additional positive charge by the additional electron, however, a small decrease in ionic radius seems more probable. These qualitative ideas find support in the gradual decrease in ionic radius exhibited by successive trivalent ions in the lanthanide series—the so-called lanthanide contraction, although here, of course, it is f electrons which are involved. This particular ionic radius decrease is shown in Fig. 13.1 together with the metal–ligand separation in halides of the first transition series (in these the metal is octahedrally surrounded by halide ions). By plotting metal–ligand separations an assumption about the ionic radius of the halide ion is avoided. Values for copper(II) compounds have been omitted from this diagram. These complexes are usually highly distorted, so that Cu–X distances both greater and smaller than the values predicted by interpolation of the data for adjacent ions have been reported. The smooth curves which were anticipated do not appear in Fig. 13.1. There is a simple explanation for this which becomes evident when crystal field effects are included. In the absence of a crystal field all of the metal d orbitals are degenerate and so, effectively, equally occupied. Application of an octahedral crystal field removes the degeneracy, so that the t_{2g} orbitals are preferentially filled. This means that the d electrons are preferentially placed in orbitals which do not screen the ligands from the increased (attractive) nuclear charge as we go from one metal to the next. When an electron is placed in an e_g orbital, however, it has an enhanced screening effect, because these orbitals are concentrated along the metal–ligand axes. For a d^5 (high spin) configuration all of the d

---

from the potassium and 1.70 Å from the chlorine, compared with the usually quoted ionic radii of 1.33 Å and 1.81 Å respectively. However, the (presumably) more accurate radii, such as that of 1.70 for Cl^-, are not constant but vary from compound to compound and so the additivity is lost.

orbitals are equally occupied and the screening is the same as it would have been in the absence of a crystal field. It follows that the additional screening of an electron in an e_g orbital compensates for the deficiencies in the screening of one and a half electrons in the t_{2g} orbitals. The deviation of ionic radii from that given by the simple picture is therefore expected to be in the order:

$$t_{2g}^6 > t_{2g}^5 > t_{2g}^6 e_g > t_{2g}^4 > t_{2g}^3 \approx t_{2g}^6 e_g^2 > t_{2g}^2 > t_{2g}^6 e_g^3$$

$$\approx t_{2g}^3 e_g > t_{2g} \approx t_{2g}^4 e_g^2 > t_{2g}^3 e_g^2 \approx t_{2g}^6 e_g^4 \approx 0$$

Remembering that all of the examples given in Fig. 13.1 are high spin, it will be seen that this series is followed quite well. The only exception appears to be d^7 (high spin) cobalt(II) for $X = I^-$. However, it has already been recognized in Section 7.5 that cobalt(II) has the configuration $t_{2g}^{24/5} e_g^{11/5}$ in the weak field limit—and I^- is a weak field ligand. If the weak-field limit values are taken, then this configuration for cobalt(II), along with d^2 which has a configuration $t_{2g}^{9/5} e_g^{1/5}$, should be placed with the configurations $t_{2g}^6 e_g^3$ and $t_{2g}^3 e_g$ in the above series. In practice, high spin cobalt(II) complexes have configurations intermediate between $t_{2g}^{24/5} e_g^{11/5}$ and $t_{2g}^5 e_g^2$ and so occupy variable positions in the series. A consequence of this general discussion is that if the distribution of electrons between t_{2g} and e_g orbitals changes, then there will be a synchronous change in ionic radius. So, the ionic radii of transition metal ions in electronically excited states are expected to differ from their ground state values (to be greater, because the e_g population increases for spin-allowed d–d transitions). Conversely, when the population of the t_{2g} orbitals increases at the expense of the e_g a decrease in ionic radius must be expected. It will be seen in Section 16.2 that just such a change seems to be involved as a trigger in a mechanism which causes iron(II) ions some 30 Å or more apart in hemoglobin to be sensitive to whether or not their partners in the molecule are coordinated to O_2.

13.3 Heats of ligation

Crystal field theory was introduced in Chapter 7 by considering a free, gaseous, ion and then placing it in an octahedral crystal field. The heat released in this process is known as the heat of ligation (the ligands are also assumed to originate as free, gaseous, molecules). Values of the heat of ligation may be obtained from experimental data on reactions in solution by the use of suitable energy cycles. The heat of ligation is conveniently broken down into several parts. Thus, within a purely electrostatic model, the most important term is the stabilization resulting from the attraction between the ligand and the transition metal ion. A destabilization results from the electrostatic repulsion between the ligands themselves—the effect of this term is to cause a decrease in the increments of energy liberated as additional ligands are added to the central atom. Because the energy released when two point charges of opposite sign are brought together from infinity depends inversely on their final separation, it would be expected that the major, stabilizing, term would be modulated by the changes in ionic radius discussed in the previous section. However, there is another term which changes from one transition metal to another, as the following hypothetical sequence shows. Form an octahedral complex ion from its infinitely separated

Table 13.1 Crystal field stabilization energies for weak field and intermediate field octahedral complexes. Where alternative configurations are given, the fractional values are the weak-field limit and the integer values are the strong-field limit (between them they give the intermediate field range)

d^n configuration	Crystal field configuration	Crystal field stabilization energy
d^0		0
d^1	t_{2g}^1	$-2/5\,\Delta$
d^2	t_{2g}^2	$-4/5\,\Delta$
	$t_{2g}^{9/5}e_g^{1/5}$	$-3/5\,\Delta$
d^3	t_{2g}^3	$-6/5\,\Delta$
d^4	$t_{2g}^3e_g^1$	$-3/5\,\Delta$
d^5	$t_{2g}^3e_g^2$	0
d^6	$t_{2g}^4e_g^2$	$-2/5\,\Delta$
d^7	$t_{2g}^5e_g^2$	$-4/5\,\Delta$
	$t_{2g}^{24/5}e_g^{11/5}$	$-3/5\,\Delta$
d^8	$t_{2g}^6e_g^2$	$-6/5\,\Delta$
d^9	$t_{2g}^6e_g^3$	$-3/5\,\Delta$
d^{10}	$t_{2g}^6e_g^4$	0

components, so that the final metal–ligand distance is that observed in the complex. Energy will be liberated, most of which will be lost from the molecule. Suppose that sufficient is retained for the split d orbitals all to be equally populated, so that the electron distribution is identical to that in the free ion. At this state in the process the energy lost will differ from one transition metal ion to another only by virtue of

- their different effective nuclear charges;
- the different metal–ligand distances.

Of these, the first would be expected to vary smoothly along the series. Now, let the d electrons assume their ground-state configurations. An additional increment of energy will be liberated if, in the ground state, the d orbitals are not equally occupied. This energy, the so-called *crystal field stabilization energy*, will vary with the metal ion. It is a simple matter to calculate crystal field stabilization energies for they depend only on the electron configuration and the crystal field splitting Δ. An electron in the t_{2g} set contributes a stabilization, of $-2/5\,\Delta$, one confined to the e_g set contributes a destabilization of $3/5\,\Delta$ (see Fig. 7.9). Crystal field stabilization energies for weak field octahedral complexes are listed in Table 13.1. A similar table can be constructed for strong field complexes but an additional destabilizing term, allowing for the pairing energy of two d electrons, would have to be added. The presence of this term makes the interpretation of the experimental data more difficult and so our discussion is confined to weak field complexes. In Fig. 13.2 are shown the variation of heats of ligation of transition metal ions with water as ligand. The data are most complete for this ligand but the available data for other ligands indicate a similar behaviour. The deviations from a curve drawn through the data points for calcium(II), manganese(II) and zinc(II) are in good agreement with the

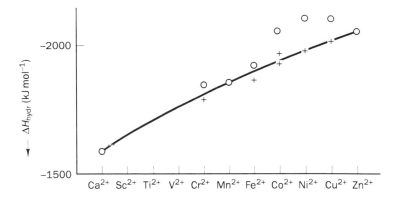

Fig. 13.2 Heats of hydration of first row transition metal ions. Experimental values (which are subject to varying errors) are shown by circles. When corrected for the contribution made by crystal field stabilization energy (crosses) they conform to a smooth behaviour (solid line).

crystal field stabilization energies given in Table 13.1. Indeed, if one corrects for the crystal field stabilization energies, using this table and the spectroscopic values of Δ, then the resulting points fall on an almost straight line. For cobalt(II), d^7, two points are shown, corresponding to stabilizations of $-3/5\,\Delta$ and $-4/5\,\Delta$, for the reason discussed above. The curve passes between these limits. This relationship between experimental and corrected values of the heat of ligation has led to the suggestion that it may be used as a method of obtaining approximate values of Δ. However, our discussion suggests that the agreement that is found with the spectroscopic values of Δ is somewhat fortuitous, for the ionic radius effect must be superimposed on the crystal field stabilizations in Fig. 13.2. Further, several assumptions have been made in the above discussion. In particular, one fact has been overlooked— that when the randomized arrangement of d electrons changes to the ground state arrangement there is an increase in the effective charge seen on the cation by the ligands—the number of e_g electrons screening the nucleus decreases. Accordingly, there is a change in the electrostatic energy. This emphasizes the interrelationship between effects which have been discussed separately. However, since the contributions to the total energy arising from the variation in metal–ligand distances and the crystal field stabilizations act in the same direction—compare the inequalities given in Section 13.1 with the relative magnitude of stabilization energies seen in Table 13.1—the essential point is that the seemingly rather erratic nature of the experimental data in Fig. 13.1 can be understood.

13.4 Lattice energies

In the previous section an isolated complex ion was discussed. Ionic lattices, in which the structure consists of interconnected octahedra, tetrahedra or other geometrical arrangements of ligands, when regarded as coordination compounds, may be similarly treated. Here, the lattice energy replaces the heat of ligation as the experimental quantity under discussion. A complication arises when the members of a series of compounds which one wishes to study are not isomorphous; they crystallize with different crystal structures, possibly due to the irregularities in ionic radii that were discussed in the previous section. A detailed consideration of this problem for the first row

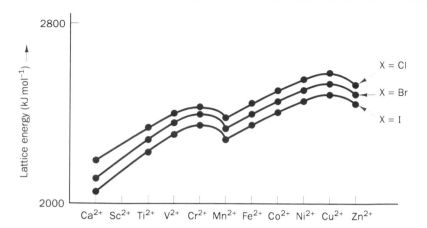

Fig. 13.3 Lattice energies of the dihalides, MX$_2$, of the first row transition series and related elements.

dihalides indicates that here, at least, the differences in lattice energies resulting from these complications are small and may be neglected as a first approximation. In this way the data shown in Fig. 13.3 were compiled. The now-familiar pattern is repeated; its interpretation closely follows that of the previous section.

13.5 Site preference energies

It has been seen that the crystal field stabilization energy is a relatively small component of the total energy involved in the formation of a complex in the gas phase—compare the corrections in Fig. 13.2 with the total energy involved, as indicated by the scale of the ordinate. In a more realistic situation, when, say, the hexaaquanickel(II) cation is converted into tris-ethylenediaminenickel(II), there will be a change in crystal field stabilization energy because water and ethylenediamine exert rather different crystal fields (the complexes are green and purple, respectively). However, the major change is due to the difference in electrostatic attraction between Ni^{2+} and a water and an ethylenediamine molecule (there are also entropy and heat terms which favour the formation of the ethylenediamine complex and which were discussed in Section 5.5). Unless it is certain that these changes are zero, or at least very small, any argument based on crystal field stabilization energies alone must be regarded with suspicion. If they lead to predictions which agree with experiment, this may be because the changes of crystal field stabilization energy parallel the major energy changes rather than because the crystal field stabilization energy is the determining factor. However, as has been seen, this reservation is less applicable when one compares a series of related ions for which all energy factors other than crystal field stabilization energies would be expected to vary smoothly along the series. In this case, the deviations of individual members of the series from the pattern may largely be determined by crystal field stabilization energies.

An example of this use of crystal field stabilization energies is the following. Just as a table of stabilization energies for weak field octahedral complexes was obtained above, so one can be obtained for tetrahedral

Table 13.2 Crystal field stabilization energies for tetrahedral complexes and comparison with high spin octahedral complexes. As indicated in the text, all quantities involving Δ are moduli, absolute values without regard to sign (this means that we do not need explicitly to take account of the fact that Δ_{tet} and Δ_{oct} are of opposite sign, although it is convenient to include negative signs in this table to correctly indicate the fact that it is stabilization energies which are listed)

d electron configuration	Tetrahedral field configuration	Crystal field stabilization energies		(Octahedral) site preference energies
		Tetrahedral field	Octahedral field	
d^0		0	0	0
d^1	e^1	$3/5\,\Delta_{tet} = -4/15\,\Delta_{oct}$	$-6/15\,\Delta_{oct}$	$2/15\,\Delta_{oct}$
d^2	e^2	$-6/5\,\Delta_{tet} = -8/15\,\Delta_{oct}$	$-12/15\,\Delta_{oct}$	$4/15\,\Delta_{oct}$
d^3	$e^2t_2^1$	$-4/5\,\Delta_{tet} = -16/15\,\Delta_{oct}$	$-54/45\,\Delta_{oct}$	$38/45\,\Delta_{oct}$
d^4	$e^2t_2^2$	$-2/5\,\Delta_{tet} = -8/45\,\Delta_{oct}$	$-27/45\,\Delta_{oct}$	$19/45\,\Delta_{oct}$
d^5	$e^2t_2^3$	0	0	0
d^6	$e^3t_2^3$	$-3/5\,\Delta_{tet} = -4/15\,\Delta_{oct}$	$-6/15\,\Delta_{oct}$	$2/15\,\Delta_{oct}$
d^7	$e^4t_2^3$	$-6/5\,\Delta_{tet} = -8/15\,\Delta_{oct}$	$-12/15\,\Delta_{oct}$	$4/15\,\Delta_{oct}$
d^8	$e^4t_2^4$	$-4/5\,\Delta_{tet} = -16/45\,\Delta_{oct}$	$-54/45\,\Delta_{oct}$	$38/45\,\Delta_{oct}$
d^9	$e^4t_2^5$	$-2/5\,\Delta_{tet} = -8/45\,\Delta_{oct}$	$-27/45\,\Delta_{oct}$	$19/45\,\Delta_{oct}$
d^{10}	$e^4t_2^6$	0	0	0

complexes. A simplified version is given in Table 13.2 in terms of Δ_{tet} and, remembering that

$$|\Delta_{tet}| \approx 4/9\,|\Delta_{oct}|$$

also in terms of $|\Delta_{oct}|$. The corresponding values for weak field complexes are also included. It can be seen that the octahedral stabilization energies are invariably greater than the corresponding values for tetrahedral complexes, unless both are zero. The difference—the (octahedral) site preference site energy—is given in the final column of Table 13.2. For d^3 and d^8 ions it is particularly large. This has led to the suggestion that it is the cause of the relative rarity of tetrahedral complexes of chromium(III), d^3, and nickel(II), d^8. Since this suggestion was first made, many tetrahedral complexes of nickel(II) have been prepared (the simplest to make is perhaps $[NiCl_4]^{2-}$ which is formed when $NiCl_2$ is added to molten NaCl) and examples of tetrahedral chromium(III) complexes have also been claimed, so that the evidence on which the original suggestion was based has weakened with time.

To put the problem into perspective, consider the complexes $[Cr(H_2O)_6]^{3+}$ and $[Cr(H_2O)_4]^{3+}$. The heat of hydration for the former is approximately $4400\ kJ\ mol^{-1}$ and its crystal field stabilization energy $250\ kJ\ mol^{-1}$, calculated using the spectroscopic value of Δ ($17\,400\ cm^{-1} \equiv 208\ kJ\ mol^{-1}$). Assuming that $4400/6\ kJ\ mol^{-1}$ is the energy of a Cr–H_2O bond, the difference in heats of hydration of the two species is $1467\ kJ\ mol^{-1}$. To this must be added the crystal field stabilization contribution of $38/45 \times 208 = 176\ kJ\ mol^{-1}$, to give a total heat of hydration of $[Cr(H_2O)_4]^{3+}$ to $[Cr(H_2O)_6]^{3+}$ of $1643\ kJ\ mol^{-1}$. Of this, about 11% is the contribution from the crystal field stabilization term. This is a crude, order-of-magnitude calculation, but it indicates that, whilst the crystal field stabilization term is not negligible, it is by no means the dominant factor in determining the relative stabilities of the two species (note that the entropy term, a term which is also of importance, has been neglected in this calculation). However, such data may be used to explain why, for example, cobalt(II) forms tetrahedral complexes more readily than does nickel(II)—the octahedral site preference energy is much smaller for cobalt(II) than nickel(II).

Another use of site preference energies is their application to the spinels. Spinel itself is a mineral of composition $MgAl_2O_4$ and may be considered the progenitor of a class of mixed oxides of composition $M^{2+}M_2^{3+}O_4^{2-}$ which are often referred to as the spinels. The oxide anions are arranged in a cubic close-packed arrangement. As in all close-packed arrangements, there are two sorts of hole between the close-packed atoms. These are tetrahedral and octahedral holes, so-called because they are at the centres of the contact between four and six anions, sets of contacts which have these geometrical arrangements. In the close-packed oxygens there are twice as many tetrahedral holes as there are octahedral holes, and there is one of the latter for each anion. The holes are large enough to be filled by cations. In MgO, for example, each octahedral hole in a cubic close-packed arrangement of oxide anions is filled by a magnesium cation to give a sodium chloride-like structure. One might reasonably expect a similar arrangement in the spinels, with the divalent cations occupying octahedral holes, for these are larger than the tetrahedral holes. In fact, the structures are not as simple as this. There are two main classes of spinels. In normal spinels, the divalent cations are in tetrahedral holes—the opposite of that which we expected. Because filling all of the tetrahedral holes would lead to a structure of general formula M_2O and the divalent ions in a spinel are only $M^{II}O_4$, the divalent cations occupy one-eighth of the available tetrahedral holes. In normal spinels the trivalent cations occupy one-half of the available octahedral holes. The second class of spinels are the so-called inverted spinels. In the inverted spinels the divalent cations are where we expected them, in octahedral holes, having changed places with one half of the trivalent cations. It is found that spinels with $M^{3+} = Cr^{3+}$ are normal and most of those with $M^{2+} = Ni^{2+}$ are inverted. Similarly, Fe_3O_4 ($= Fe^{2+}Fe_2^{3+}O_4^{2-}$) is inverted, but Mn_3O_4 and Cr_3O_4 are normal. The problem is to decide what determines the structure that is adopted.

First, we must consider the large cation–anion attraction term. This is more difficult than for an isolated complex ion, for this term, together with the cation–cation and anion–anion repulsion terms, must be summed over the whole structure to give the Madelung constants for normal and inverted spinels. Such calculations have been made and indicate that the two spinel structures have very similar energies. It is therefore reasonable that site preference energies should play an important part in determining which structure is adopted. As Table 13.2 demonstrates, it is at once understandable that both Cr^{3+} (d^3) and Ni^{2+} (d^8) should be located in octahedral holes, and their spinels be normal and inverted, respectively, for d^3 and d^8 ions are those that have the largest octahedral site stabilization energies.

The essential difference between normal and inverted spinels is the interchange of M^{2+} and half of the M^{3+} between octahedral and tetrahedral sites. If site preference energies are the only important factor, the spinels will be inverted if the octahedral site preference energy of M^{2+} is greater than that of M^{3+}. Consider spinels such as Fe_3O_4, where M^{2+} and M^{3+} are of the same element, differing only in charge. As a (crude) approximation, set

Table 13.3 Site preference energies, used to predict spinel type. In this table it has been assumed that $\Delta(M^{3+}) \sim 3/2\,\Delta(M^{2+})$

M^{2+}		M^{3+}			
High spin configuration	Octahedral site preference energies	High spin configuration	Octahedral site preference energies	Difference in site preference energies	Predicted spinel type
d^1, d^6	$2/15\,\Delta(M^{2+})$	d^0, d^5	0	$2/15\,\Delta(M^{2+})$	inverted
d^2, d^7	$4/15\,\Delta(M^{2+})$	d^1, d^6	$2/15\,\Delta(M^{3+}) \approx 1/5\,\Delta(M^{2+})$	$1/15\,\Delta(M^{2+})$	inverted
d^3, d^8	$38/45\,\Delta(M^{2+})$	d^2, d^7	$4/15\,\Delta(M^{3+}) \approx 2/5\,\Delta(M^{2+})$	$4/9\,\Delta(M^{2+})$	inverted
d^4, d^9	$19/45\,\Delta(M^{2+})$	d^3, d^8	$38/45\,\Delta(M^{3+}) \approx 19/15\,\Delta(M^{2+})$	$-38/45\,\Delta(M^{2+})$	normal
d^5, d^0	0	d^4, d^9	$19/45\,\Delta(M^{3+}) \approx 19/30\,\Delta(M^{2+})$	$-19/30\,\Delta(M^{2+})$	normal

$\Delta_{oct}(M^{3+}) = 3/2\,\Delta_{oct}(M^{2+})$. Using Table 13.2, the data in Table 13.3 are derived, in which the type of spinel lattice adopted is also predicted. The prediction is that Mn_3O_4 (Mn^{2+} is d^5) is normal and that Fe_3O_4 (Fe^{2+} is d^6) is inverted, both as found. Co_3O_4 (Co^{2+} is d^7) is normal but is predicted to be inverted. However, the energy difference between the two forms for d^2 and d^7 is only $1/15\,\Delta_{oct}(M^{2+})$, so that the prediction is hardly to be regarded as reliable, particularly when the variability of t_{2g} and e_g occupancy for d^7 ion is recalled. For Co_3O_4 and other cases which do not follow the simple predictions, more detailed analyses have led to agreement with experiment. It will be noted that high spin configurations have been used throughout this discussion. The presence of low spin configurations would introduce the complication of pairing energies.

13.6 Stability constants

The dependence of the heat terms on the crystal field stabilization energies discussed in the previous sections will presumably be reflected in the corresponding free-energy values. Now, free-energy differences are related to equilibrium constants by equations of the form

$$\Delta G = -Rt\ln k$$

and so *differences* in crystal field stabilization energies might well be reflected in equilibrium constants. Figure 13.4 shows the variation of K_1 for $[M(en)_3]^{2+}$ across the first transition series. Here, it is the differences between the crystal field stabilization energies of $[M(en)_3]^{2+}$ and $[M(H_2O)_6]^{2+}$ which are relevant. The data seem to show a clear effect, except that the Cu^{2+} value is greater than expected. However, as is well known, copper(II) complexes are seldom octahedral—the static Jahn–Teller effect (Section 8.5) is usually invoked to explain the considerable distortions observed—and so the fact that copper(II) is out-of-line is not too surprising. It also was omitted from the discussion in Section 13.2. It is to be noted in passing that Fig. 13.4 provides a partial illustration of the Irving–Williams series (Section 5.4)—the generalization that stability constants vary in the order

$$Ba^{2+} < Sr^{2+} < Ca^{2+} < Mg^{2+} < Mn^{2+} < Fe^{2+} < Co^{2+} < Ni^{2+} < Cu^{2+} > Zn^{2+}$$

an order that is relatively insensitive to the ligand involved.

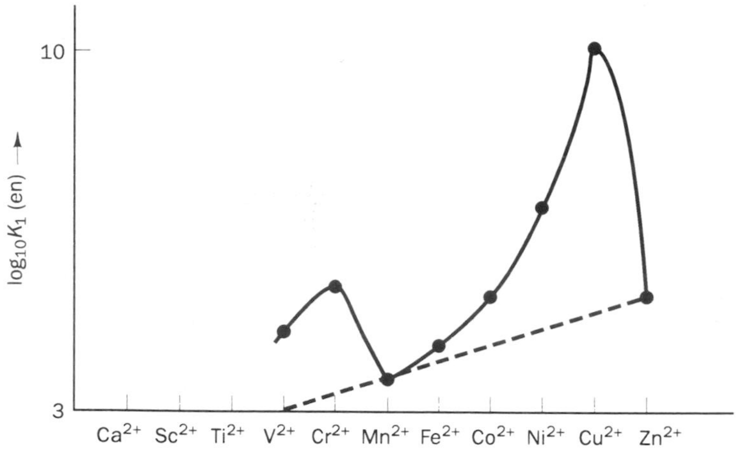

Fig. 13.4 Variation of $\log_{10} K_1$ across the first transition series for the reaction

$$[M(H_2O)_6]^{2+} + en \rightarrow [M(H_2O)_4(en)]^{2+} + 2H_2O$$

13.7 Lanthanides

The recognition that crystal field stabilization energies play a role in the chemistry of the first row transition elements has prompted a search for similar effects elsewhere in the periodic table. In particular, the question has been asked 'do similar effects occur for the lanthanides'? It was recognized in Chapter 11 that crystal field effects are small for the lanthanides and so, too, therefore must be any consequent stabilization. In Chapter 7 it was seen that the seven f orbitals split into $a_{2u} + t_{1u} + t_{2u}$ sets in an octahedral ligand field. If we make the—significant—assumption that this is the relevant symmetry for lanthanide complexes, then we need the relative energies of these three sets. We actually have them. In Section 7.5 we found that in the octahedral weak field limit—surely the limit applicable to the lanthanides—the ^3F term arising from the d^2 configuration splits into $^3A_{2g}$ $(6/5\,\Delta)$, $^3T_{2g}$ $(1/5\,\Delta)$ and $^3T_{1g}$ $(-3/5\,\Delta)$ components. In that so much of ligand field theory is symmetry-determined, it is not surprising to learn that the energies of these components are directly proportional to those of the split f orbitals in an octahedral crystal field. The differences arise from the fact that we are dealing with seven f orbitals, not five d. This means that g suffixes have to be replaced by u and that Δ has to be replaced by $-\Delta$. This latter point is most readily seen by considering, the f_{xyz} orbital. This orbital, of a_{2u} symmetry, is the most stable f orbital in an octahedral crystal field because it points away from all the ligands. In contrast, for the d^2 configuration it was shown in Chapter 7 that the $^3A_{2u}$ is the least stable ^3F component. So, the splitting of the f orbitals in an octahedral ligand field is as given in Fig. 13.5. It follows that the lanthanide crystal field stabilization energies will be

$$\left[-\tfrac{6}{7}n(a_{2u}) - \tfrac{1}{7}n(t_{2u}) + \tfrac{3}{7}n(t_{1u})\right]\Delta$$

where $n(a_{2u})$ is the number of electrons in the a_{2u} orbital, and so on. Feeding electrons into the orbitals of Fig. 13.5 in a high-spin manner, using the above equation together with a typical value of Δ, leads to the stabilizations of Fig. 13.6. Also in this figure are plotted data for some complexes of the lanthanides. The 2,2'-bipyridine-dicarboxylate anion shown in Fig. 13.7 forms complexes with the trivalent lanthanides in which three of the anions

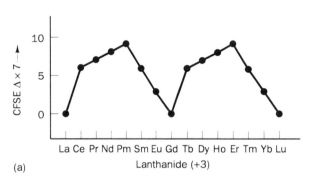

Fig. 13.5 The relative energies of f orbitals in an octahedral crystal field.

t_{1u} $\frac{3}{7}\Delta$

Free-atom f orbitals

t_{2u} $-\frac{1}{7}\Delta$

a_{2u} $-\frac{6}{7}\Delta$

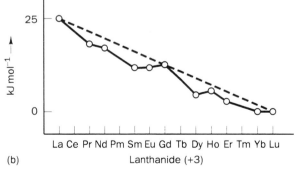

Fig. 13.6 (a) Octahedral crystal field stabilization energies of the tripositive lanthanide ions. (b) Enthalpy data related to the formation of (octahedral) complexes of the ligand shown in Fig. 13.7 (see also the caption to Fig. 13.2).

coordinate to give species such as $[Gd(bpydicarb)_3]^{3+}$, in which there is, essentially, octahedral coordination.

The argument just developed is weak in that in the calculation of crystal field stabilization energies individual f electrons have been allocated to individual f orbitals. It was seen in Chapter 7 that this allocation could not always be made for d electrons and d orbitals, and this fact was used in the discussion in Section 13.2. In Chapter 11 it was seen how much less it is a

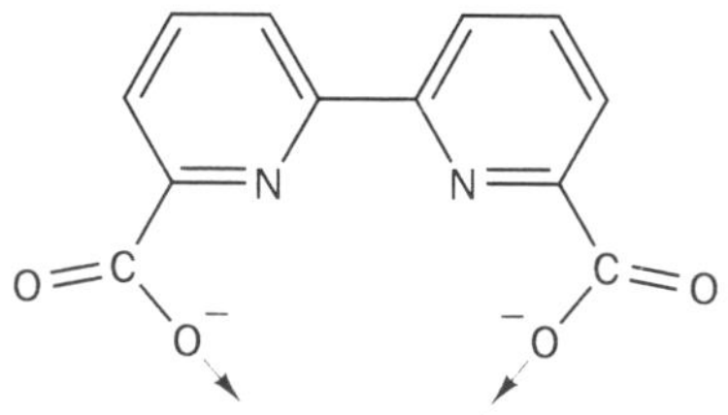

Fig. 13.7 The 2,2′-bipyridine-6,6′-dicarboxylate anion.

valid step for f electrons, where the effects of electron repulsion and spin–orbit coupling are much more important than those of the crystal field. Nonetheless, it seems from Fig. 13.6, and certainly has been argued, that some very small crystal field stabilization occurs for the lanthanide bipyridine-dicarboxylates. Given the magnitude of any such stabilization, it is not surprising that no crystal field modulation of the lanthanide ionic radii was observed in Fig. 13.1.

13.8 Molecular mechanics

So far in this chapter the concern has been to attempt to understand and predict relatively small structural or related differences. Although it was found that crystal field stabilization energies are often of vital importance it was recognized that this can only be so when the much larger contributions to the total energy are relatively constant across the variations under discussion. In this situation, it is inevitable that attempts should be made to assess these larger contributions and their variations. In principle this is an area to which quantum mechanics would be expected to make a major contribution but, as has become evident in earlier chapters, the application of quantum methods without major simplifications is a task to be carried out molecule by molecule and is not likely to be able to address the problems raised in this chapter for many years. On the other hand, as was asked in Chapter 10 'why re-invent the wheel'? Specifically, the questions of immediate concern are those about the way atoms and molecules interact with each other, how geometries are determined and so on. The problems of core electrons, often an area in quantum mechanics in which approximations are made, seem scarcely of relevance. After all, most chemists feel that they understand why there is a barrier to free rotation about the carbon–carbon bond in ethane, and do so with no need of detailed quantum mechanics. It is this philosophy which lies behind the subject of molecular mechanics. Is it possible to find relatively simple expressions for each of the forces acting between the atoms of a molecule and, by including all that are believed to be relevant, to balance them against one another and so make structural predictions? For instance, most of vibrational spectroscopy is built around the tenet that bond stretching is a simple-harmonic process. Of course, this does not cover very large bond length changes, but there one can turn to a Morse function or something similar—more complicated perhaps, but not much. Similarly, there are simple expressions for such terms as van der Waals interactions, repulsive forces between non-bonded atoms, bond angle changes, rotational barriers (such as that in ethane) and so on. It is not difficult to see that a computer program can be devised in which the energetic consequences of a molecule twisting, turning and stretching against itself are explored. When a reasonably stable arrangement has been found, it can then itself be varied until the optimum arrangement is obtained. Subsequently, alternative minima can be explored until there is confidence that all feasible minima have been investigated and the best then selected. Such methods are well developed in organic chemistry, where the parameters that arise have been fine-tuned. As a result, it is now possible to apply molecular mechanics methods to calculate most organic structures and obtain bond lengths that are within about 0.01 Å, and bond angles that are within about 1°, of those observed. It is not surprising that considerable efforts have been made to extend these calculations to inorganic chemistry, particularly at a time when, as became evident in Chapter 2, there is heightened interest in large organic ligands and ligands with interesting steric properties, both of which are properties which can already be treated rather well. Applied to inorganic species, molecular mechanics is not yet able to yield the accuracy found in organic chemistry. Not that it is without success—an error in a crystal structure determination was discovered when molecular mechanics predicted a different ligand conformation. Indeed, progress is encouraging but the transition metals exhibit a variability of geometry not matched in organic molecules and the prediction of this is a major problem. Nonetheless, in well-defined areas such as the subject matter of Appendix 1, predictions are usually in good accord with experiment. Similarly, when the outcome is dominated by the properties of the organic ligand, good results are achieved. A common problem, however is the occurrence of different predicted molecular geometries with energy separations which are within the errors inherent in the calculations. One criticism of many of the models adopted is that they make no specific allowance for those factors which, earlier in this chapter, were highlighted as holding the key to the particular geometry adopted. Attempts to circumvent such problems are beginning to be made, for example by the union of molecular mechanics and angular overlap calculations (see Section 10.6).

13.9 Conclusions

In this chapter some rather crude approximations based on crystal/ligand field theory were used to make a variety of predictions or correlations. Although no one piece of evidence is really definitive, taken together they provide support for the general crystal field approach. However, it must be emphasized that arguments based on crystal field stabilization energies must always be used with great care and with due regard for other energetic and entropic factors involved in the processes considered. Such factors are at the heart of the application of molecular mechanics calculations to inorganic systems. These calculations have minimal computer demands compared with those of their detailed quantum mechanical counterparts and represent a rapidly growing field in which significant advances are to be expected.

Further reading

An excellent source of much thermochemical data is S. J. Ashcroft and C. T. Mortimer, *Thermochemistry of transition metal complexes*, Academic Press, London and New York 1979, although, today, computer databases are better if access can be gained to one.

The classic reference for the material in all but the last two sections of this chapter is P. George and D. S. McClure in *Prog. Inorg. Chem.* (1959) *1*, 381.

The approach to crystal field stabilization energies adopted in this chapter is one that bypasses the problem of pairing energies. It is arguable that it is more correct to include them; however, to do so considerably complicates the discussion. See S. S. Parmar, *J. Chem. Educ.* (1981) *58*, 1035.

Molecular mechanics is covered in: 'Molecular Mechanics Calculations as a Tool in Coordination Chemistry' R. D. Hancock, *Prog. Inorg. Chem.* (1989) *37*, 187; 'Methods for molecular mechanics modelling of coordination compounds' B. P. Hay, *Coord. Chem. Rev.* (1993) *126*, 177; and a review with a wide variety of examples 'The relation between ligand structures, coordination stereochemistry, and electronic and thermodynamic properties' by P. Comba, *Coord. Chem. Rev.* (1993) *123*, 1.

Questions

13.1 Explain carefully what is meant by 'crystal field stabilization energy' and compare this term with the alternative, 'ligand field stabilization energy' used by some textbooks.

13.2 Write an essay on the structure of Fe_3O_4 (Section 9.11 should also be consulted).

13.3 Give a detailed explanation of the lattice energy data shown in Fig. 13.3.

13.4 Give an account of the factors influencing the stereochemistry of complexes of (a) main group and (b) transition group elements.

14

Reaction kinetics of coordination compounds

14.1 Introduction

In the majority of chapters of this book the concern has been with an understanding of the properties of individual molecules, properties which are regarded as essentially time-independent. However, no less important are the chemical reactions of these molecules and, here, changes as a function of time are of the essence. This chapter is devoted to a review of our present understanding of some of the reaction types which are characteristic of coordination compounds. It is as well to recognize the complexity of the problem. Suppose we are interested in a reaction such as the aquation of an ion such as $[Co(NH_3)_5Cl]^{2+}$, a much studied system:

$$H_2O + [Co(NH_3)_5Cl]^{2+} \rightarrow [Co(NH_3)_5H_2O]^{3+} + Cl^-$$

Because $[Co(NH_3)_5Cl]^{2+}$ salts are soluble in few other solvents, the study will probably be carried out with water as a solvent, as well as reactant. In this case, each $[Co(NH_3)_5Cl]^{2+}$ ion in solution will undergo a constant battering from water molecules incident from all directions. No doubt, the coordinated ammonia molecules will move in response, although perhaps restrained somewhat by hydrogen bonding to the solvent. For the reaction to occur, a Co–Cl bond has to be broken and this could be because vibrations of the coordinated ammonia ligands temporarily expose the Co–Cl bonding electrons to attack by a water molecule. Equally, it could be that a transient, strong hydrogen bond is formed between the Cl ligand and a water molecule, so facilitating breaking of the Co–Cl bond. Again, Co–Cl bond breaking could be dependent on much of the vibrational energy within the molecule localizing itself briefly in the Co–Cl bond stretching mode. Alternatively, some combination of all three factors could be involved. It seems likely that a multitude of slightly different reaction pathways exist and, since we study a large number of molecules simultaneously, all we can measure is some sort of average. Hopefully, there will be only one, clearly defined, lowest energy reaction route and the average will be dominated by this and its minor deviants. However, the three contributions listed above (and others could be added) will presumably have different temperature characteristics and so the (average) reaction route will also be slightly temperature dependent. Fortunately, this complication can usually be ignored —it seems to be of lesser importance than experimental error. Attention therefore focuses on a single reaction pathway and, in particular, the different potential energy profiles associated with alternative pathways. The task of the worker in the area is to use the experimental data, and, in particular, rate laws, to deduce the most probable reaction pathway. It is not an easy task. As will be seen, the existence of pre-equilibria (which mean that the real reactant species is present in much lower concentration than expected), the involvement of the solvent (which, because its concentration scarcely changes, will not be evident from the rate law), the ability to work over limited temperature and concentration ranges, all impose problems. Fortunately, more techniques are becoming available; thus, the ability to study reaction rates in solution as a function of the external pressure applied to the solution offers information as to whether the reaction pathway involves compression or extension of the reactant species and, thus, insights into its molecularity.

Only in one area can theory be said to have led experiment. This is in one aspect of oxidation–reduction, electron-transfer, reactions, a topic to which we shall return. First, however, we shall give an overview of the very important topic of reaction rates. Perhaps the most evident thing about them is their enormous range. They (or, perhaps easier to think of, the time to half-completion of a reaction) span a range of at least 10^{15}. Clearly, a theory is to be regarded as acceptable if it succeeds in correctly predicting— or interpreting—an order of magnitude, exact agreement is too much to hope for. The range of values emphasizes one thing. For a reaction to proceed it must be thermodynamically feasible. The range of rate and $t_{1/2}$ values can arise from variations in mechanism—the availability of a facile mechanism or the absence of one (thermodynamically, CCl_4 should react violently with water)—but a given mechanism can be associated with very different rates also.[1] Ultimately, the hope is that the mechanism can be related to the electronic properties of the molecule(s) under study and relative rates thus explained. As will be seen in this chapter, considerable progress has been made, but much remains to be done.

In Section 2.1 the distinction between inert and labile complexes was encountered. Several attempts have been made to formalize this distinction, of which the most popular seems to be Taube's definition: 'if no delay is noted in the substitution reaction under ordinary conditions (i.e. room temperature, ca. 0.1 M solutions) the system will be described as labile'. However, for most chemists inert complexes are effectively those for which their reactions may be studied by classical techniques, such as monitoring the change in intensity of a visible or ultraviolet spectral peak with time. Such reactions are half complete in about one minute or longer at 25 °C for

---

[1] So, the rates of substitution of Ni^{II} and Co^{III} are very different (fast and slow, respectively) but the mechanisms involved are probably rather similar.

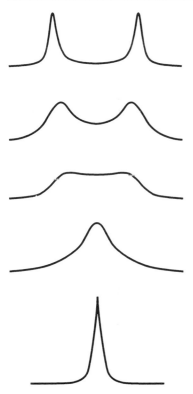

Fig. 14.1 Dynamic exchange in NMR. The top spectrum is the low-temperature, no exchange, limit. The bottom is the high-temperature, rapid exchange, limit. The central spectrum was recorded at ca. 200 °C. Those immediately below and above were recorded at ca. 190 °C and 210 °C, respectively. This particular example arises from two proton environments within a molecule becoming rotationally equivalent with increase in temperature. For the case described in the text the two low-temperature peaks would be of different areas and the high temperature at the area-weighted average position.

concentration of ca. 0.1 M. In the past, labile complexes were not so readily studied but current techniques have provided a wealth of data on them. Reactions of inert complexes are usually studied by mixing solutions of the reactants and monitoring either the appearance of a product species or disappearance of a reactant. It is difficult to mix two solutions completely in less than a millisecond (and only then in small volumes and in some sort of flow system) and this limits the extent to which technical ingenuity may be used to apply classical techniques to fast reactions. Some further extension, enabling the study of some not-too-labile complexes, is possible by working at low temperatures with very dilute solutions, but for very fast reactions quite different methods must be used. The simplest has already been mentioned (Section 5.3): if a stepwise equilibrium constant is known, together with the rate of either the forward or backward reaction, then enough data are available to enable the unknown rate to be determined. Other methods study an equilibrium as the following two examples illustrate.

Suppose that a diamagnetic complex containing coordinated trimethylamine is dissolved in trimethylamine and the proton magnetic resonance spectrum of the solution studied. If the coordinated trimethylamine exchanges with the solvent slowly, then two resonances will be observed (provided that complicating features are absent), one due to coordinated and the other due to free trimethylamine. If the exchange is rapid, only one resonance will be observed, at some sort of average position. In favourable cases, if the temperature of the sample is varied, at one temperature slow, and at another, fast exchange will be observed (when it will be found that the position of the single resonance is a concentration, i.e. peak-area-weighted, average of those of the two) as will be the intermediate region, in which broad peaks occur (Fig. 14.1). From such measurements the rate of exchange at the temperature at which intermediate behaviour is observed may be obtained. Strictly speaking, the terms slow and rapid in this example are relative to the NMR timescale and this depends on the separation between the peaks due to coordinated and solvent trimethylamine. If the complex is paramagnetic, additional line-broadening and shifts occur. These may be analysed to give rate data but the results tend to be somewhat ambiguous, since it is the slower of two processes which is measured. These are the exchange processes of interest and that of a paramagnetically induced change in the nuclear spin state of the protons (for proton magnetic resonance) of the coordinated ligand. Measurements with a different but similar ligand may establish which process is the slower, but in such cases one often has to be content with only being able to put a limit on the rate of the exchange process.

A most important approach to the study of the fast reactions of labile complexes involves relaxation phenomena. The position of dynamic equilibrium in a system depends not only on reactant concentrations but also on such quantities as temperature, pressure and an even electric field gradient (if present). If one of these is suddenly changed the position of chemical equilibrium will also change slightly and, with rapid response and sensitive instruments, this change may be detected. The speed with which the new equilibrium is reached depends on reaction rates which may thus be measured. In this way Eigen and his co-workers have determined unimolecular rate constants greater than $10^9 \, s^{-1}$, which is approaching the limit for diffusion-controlled reactions.

14.2 Electron-transfer reactions

The simplest class of reaction of coordination compounds which has been studied is that of electron-transfer reactions. Suppose that a solution of potassium ferrocyanide (hexacyanoferrate(II)) is mixed with one of potassium ferricyanide (hexacyanoferrate(III)), then if an $[Fe(CN)_6]^{4-}$ anion loses an electron and an $[Fe(CN)_6]^{3+}$ anion gains one, a chemical reaction has occurred, although there is no change in the composition of the mixture. If one of the atoms in just one of the complex ions is labelled in some way—with ^{14}C for example—then the reaction may be studied by seeing how quickly it appears in the other complex ion. In this particular case the reaction is fast, the second-order rate constant being ca. $10^3\,M^{-1}\,s^{-1}$ at $25\,°C$. This rate is much greater than those of reactions involving ligand exchange of either species, so a simple electron-transfer mechanism is indicated. Because there is no net chemical reaction there is no heat change associated with the electron transfer. The Fe–C bond length in $[Fe(CN)_6]^{3-}$ is very slightly shorter than that in $[Fe(CN)_6]^{4-}$, so, if an electron were to be transferred between the anions in their ground-state equilibrium configurations, by the Franck–Condon principle, the product $[Fe(CN)_6]^{3-}$, would be expanded and the $[Fe(CN)_6]^{4-}$ compressed relative to these configurations. That is, the products would be of higher energy than the reactants, contradicting the requirement of zero heat change. It follows that an electron-exchange reaction will only occur between precisely matched molecules which are vibrationally excited. It should be noted, however, that transfer would not occur between $[Fe(CN)_6]^{4-}$ and $[Fe(CN)_6]^{3-}$ anions, which are, respectively, compressed and expanded into each other's equilibrium geometry because this would mean that two vibrationally excited molecules become two vibrational ground-state molecules. It is only totally symmetric vibrations—breathing modes—which lead to the synchronous contraction or expansion of all symmetry-related bonds in a molecule and these are the modes which have to be excited to achieve the required matching (Fig. 14.2).

It is not essential that two matched anions are in contact at the instant of electron transfer. The question which then arises is just how far apart can two ions be and yet participate in an electron-transfer reaction? The answer to this question is particularly important in bioinorganic chemistry (Chapter 16), for there two potential participants may be unable to approach each other very closely because of the constraints imposed by the large molecule(s) within which they are found. For most systems the answer to the above question seems to be less than 10 Å but it is becoming clear that much

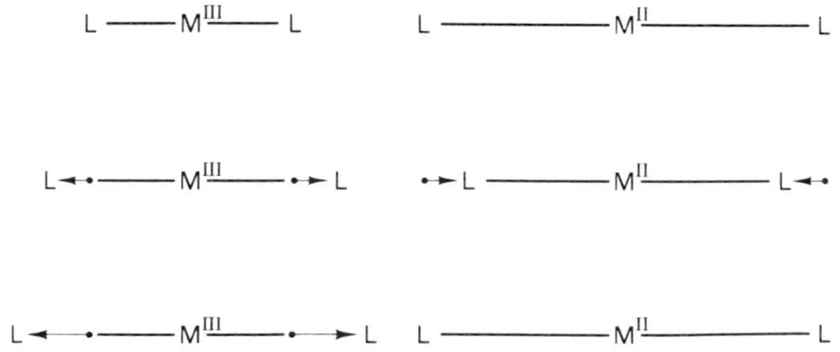

Fig. 14.2 For simplicity, this figure shows two centrosymmetric ML$_2$ molecules. The M^{III}–L (top left) is drawn much shorter than the M^{II}– (top right). By appropriate totally symmetric (breathing) M–L extensions or contractions the two molecules become of identical size (centre). The totally symmetric mode can involve just one molecule but then the amplitude is greater (bottom). Evidently, a continuous range of matching bond lengths exists.

larger separations are possible. At the time of writing the record seems to be ca. 40 Å. Such long-distance electron-exchange reactions are often referred to as occurring by an *outer-sphere* mechanism, that is, as not involving the immediate coordination sphere of either metal. However, it is reasonable to expect that electron transfer will occur most readily when the two reacting species are relatively close together. That this is so is indicated by the observation that outer-sphere electron-transfer reactions are more rapid for complexes containing ligands such as *o*-phenanthroline and the cyanide anion than for corresponding complexes with ligands such as H_2O or NH_3. That is, a ligand over which a metal electron may be extensively delocalized (cf. Section 12.8) significantly reduces the magnitude of the barrier to electron transfer (one may draw an analogy with a current flowing through a piece of resistance wire: replacing part of the resistance wire by a piece of copper wire increases the current).

In the example discussed above, the electron configuration of the iron atom in the $[Fe(CN)_6]^{4-}$ anion is t_{2g}^6 and that in the $[Fe(CN)_6]^{3-}$ anion is t_{2g}^5. Removal of an iron electron from $[Fe(CN)_6]^{4-}$ leaves a t_{2g}^5 configuration and addition of one to $[Fe(CN)_6]^{3-}$ gives a t_{2g}^6 configuration. It is not always as simple as this. Consider an electron-transfer reaction between a molecule of a cobalt(III) complex and one of cobalt(II), both octahedrally coordinated. Cobalt(III) complexes are usually low spin and cobalt(II) are high spin, so the electron configuration of the cobalt ions will be cobalt(III) t_{2g}^6, and cobalt(II) $t_{2g}^5 e_g^2$. After transfer of an electron from cobalt(II) to cobalt(III) these configurations will presumably become $t_{2g}^6 e_g$ and $t_{2g}^5 e_g$ (cobalt(II) and cobalt(III), respectively). However, these are not the ground-state configurations of the ions. That is, after the electron-transfer reaction both complexes will be electronically excited (this excess of energy will rapidly be lost either by radiation or, more probably, it will be converted into thermal energy). Because this electronic energy contributes to the activation energy of the process, the rate of electron-transfer reactions between cobalt(II) and cobalt(III) complexes is much slower than that between the $[Fe(CN)_6]^{3-}$ and $[Fe(CN)_6]^{4-}$ anions.

There is yet another reason for this. The original cobalt(III) ion in its normal octahedral ground state would be $^1A_{1g}$. It becomes an octahedral cobalt(II) ion with a $t_{2g}^6 e_g$ configuration. It therefore has a ^2E_g term (Table 7.5), but the ground state of octahedral cobalt(II) is $^4T_{1g}$ (derived from the $t_{2g}^5 e_g^2$ configuration). That is, in passing from the as-formed cobalt(II) to ground-state cobalt(II) the spin multiplicity must change. The only available mechanism is spin–orbit coupling, operating in a way similar to that shown in Fig. 8.16. Reference back to the text associated with that figure will show that such spin-multiplicity changes require special, and rare, conditions if they are to be facile. It is not surprising that an exchange such as

$$[Co(NH_3)_6]^{3+} + [Co(NH_3)_6]^{2+} \rightarrow [Co(NH_3)_6]^{2+} + [Co(NH_3)_6]^{3+}$$

should be extremely slow, with time for half-reaction measured in hours. The reader may find it helpful to reconsider the use of charcoal in the preparation of cobalt(III) complexes (Section 4.2.2) in the light of the above discussion.

At first sight, an attempt to calculate rate constants for electron transfers of the sort we have been discussing would seem an impossible task. Charged species, separated by varying amounts of solvent, a solvent which will react to the sudden change of charge on each of the two ions—and a proper description of this solvent polarization change proves to be a key component to a successful theory—the need for there to be a matching of vibrational energy levels of the two complex ions, all have to be considered. In fact, the problem is approximately soluble. This is, in part, because so many factors are involved and, so, fairly crude approximations for each will work reasonably well. Getting the overall picture approximately correct is more important than individual accuracy. Further, more than one set of approximations lead to the same general result. It is clear that a potential energy surface has to be modelled because vibrations are important. One model, then, describes the electron transfer as resulting from a coupling of two potential energy surfaces. If there is no coupling then there is no electron transfer. The problem of a mathematical description of the process thus becomes one of a description of a coupling between two potential energy surfaces. Such a theory contains many parameters but, fortunately, these largely cancel out when a series of closely related reactions is studied, so that much of the literature is concerned with reaction series. Inevitably, the resulting theory is somewhat mathematical but it seems that the Marcus–Hush theory (the name coming from the two workers who arrived at essentially the same result following somewhat different routes) correctly predicts to within an order of magnitude electron-transfer rates of the type so far considered. This is not at all bad when the rates studied vary by 10^{14}. But there is more. The type of electron-transfer reaction which has so far been covered is that in which two species differ by one electron—$[Fe(CN)_6]^{3-}$ and $[Fe(CN)_6]^{4-}$ and the pair $[Co(NH_3)_6]^{3+}$ and $[Co(NH_3)_6]^{2+}$, for example. In such reactions the concentration of neither species changes with time (so that some care has to be taken in defining the rate of reaction; in practice the first-order rate for oxidized → reduced, or its reverse, is commonly reported). However, if the reaction is between two such pairs and the oxidized form of one pair is reacted with the reduced of another, then concentrations *will* change with time, until equilibrium is reached. It turns out that Marcus–Hush theory can make quite good predictions about the rate of such so-called *cross-reactions*. Consider the two reactions, each having rate constants k_{aa} M^{-1} s^{-1} ($a = 1$ or 2) as defined above:

$$[Ru(NH_3)_6]^{3+} + [Ru(NH_3)_6]^{2+} \xrightarrow{k_{11}} [Ru(NH_3)_6]^{2+} + [Ru(NH_3)_6]^{3+}$$

and

$$[V(H_2O)_6]^{3+} + [V(H_2O)_6]^{2+} \xrightarrow{k_{22}} [V(H_2O)_6]^{2+} + [V(H_2O)_6]^{3+}$$

Then, one of the cross-reactions is

$$[Ru(NH_3)_6]^{3+} + [V(H_2O)_6]^{2+} \xrightarrow{k_{12}} [Ru(NH_3)_6]^{2+} + [V(H_2O)_6]^{3+}$$

The equilibrium constant for this last reaction, obtained from emf measurements, is K_{12}. For reactants and products of the same size and charge type the simplest form of the Marcus cross-relationship is

$$k_{12} \approx (k_{11}k_{22}K_{12})^{1/2}$$

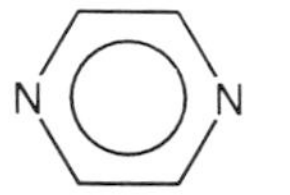

Fig. 14.3 The ligand pyrazine.

(If the size and charge requirements are not met, a further factor, f_{12}, appears within the bracket on the right hand side.)[2] For the above reactions, $k_{11} = 4 \times 10^3$, $k_{22} = 1.2 \times 10^{-2}$ and $K_{12} = 1.07 \times 10^6$. It follows that the calculated value of k_{12} is 7.2×10^3. This is to be compared with the experimental value of 4.2×10^3. In general, the equation gives values correct to within one or two orders of magnitude. When a self-exchange or cross-reaction does not obey the Marcus–Hush predictions, at least approximately, it is a reasonably reliable indication that some complicating feature is involved—a stabilization caused by hydrogen bonding between ligands in a cross-reaction, for instance.

So far, the electron-transfer reactions which have been discussed are those in which only the formal valence states of the metal ions involved changes, reactions which can occur when there is no intimate contact between the two reactants (although this absence of contact is a point which it may be difficult to prove). There is another class of electron-transfer reaction, in which intimate contact between the two reacting molecules leads to reaction by a bridge or *inner-sphere* mechanism. A class of compound which has been much studied in this connection is one based on a species first studied by Creutz and Taube, $[(NH_3)_5Ru(pyz)Ru(NH_3)_5]^{5+}$, in which the two reacting centres (one is Ru^{II} and one is Ru^{III} are linked by a bridging ligand, in this case pyrazine (Fig. 14.3) although variation in choice of bridging ligand is common. Inner-sphere mechanisms often have indicators which reveal their presence. Reactions of the type

$$\text{chromium(II)} + \text{cobalt(III)} \rightarrow \text{chromium(III)} + \text{cobalt(II)}$$

have been extensively studied, again, notably by Taube and his co-workers. Cobalt(III) and chromium(III) form inert complexes whilst the corresponding divalent ions give labile complexes. This means that if a ligand is transferred from cobalt(III) to chromium(III) in the reaction it will be possible to show that this transfer has occurred.

Consider the reaction:

$$[Co(NH_3)_5Cl]^{2+} + [Cr(H_2O)_6]^{2+} \rightarrow [Co(NH_3)_5(H_2O)]^{2+} + [Cr(H_2O)_5Cl]^{2+}$$

The reaction is carried out in water and the final cobalt(II) product is actually $[Co(H_2O)_6]^{2+}$, but this is immaterial to our discussion. It was found that if the solution contained labelled chloride ion ($^{36}Cl^-$) none of the activity appeared in the $[Cr(H_2O)_5Cl]^{2+}$ product; the reaction does not involve the free chloride ions present in the solution. This observation indicates that there must be an intimate contact between the reacting species, a Cl^- ion being transferred from cobalt(III) to chromium(II) and an electron migrating in the opposite direction. It therefore seems likely that a species something similar to that shown in Fig. 14.4 must be involved. Similar transfer of the ligand X from $[Co(NH_3)_5X]^{n+}$ to chromium(II) occurs for X = Cl^-, Br^-, N_3^-, acetate, SO_4^{2-} and PO_4^{3-}. That there is a transient intermediate something like that in Fig. 14.4 is supported by the observation that for X = NCS^- (the complex having a Co–N bond) the initial product is

[2] f_{12} is a function of k_{11}, k_{22}, K_{12} and a collision frequency which itself is usually taken as a constant. In practice, therefore, the presence of f_{12} causes no difficulty but it is simplest to omit it in the present discussion.

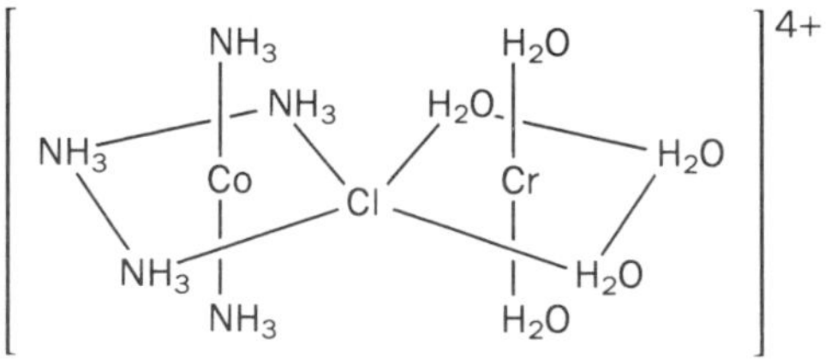

Fig. 14.4 The essentials of the intermediate species formed during the transfer of Cl⁻ ion from cobalt(III) to chromium(II).

$[Cr(NH_3)_5SCN]^{2+}$, with a Cr–S bond, which subsequently rearranges to $[Cr(NH_3)_5NCS]^{2+}$ (with a Cr–N bond).

Further supporting evidence for the suggested mechanism comes from the observation of fairly stable intermediate species in some inner sphere electron-transfer reactions. For example, reaction between vanadium(II) and vanadyl (VO^{2+}) complexes, to form vanadium(III) species, rapidly give a brown intermediate, believed to contain a V–O–V bridge which, relatively slowly, gives the final product. In some cases there is evidence that there are two or even three bridges in the intermediate (for example, in some inner-sphere electron-transfer reactions between chromium(II) and chromium(III) complexes containing the azide anion as ligand). An example of the ingenious methods that have been used to probe such reactions is provided by the used of an optically active inert species, such as a resolved isomer of $[Co(en)(ox)_2]^-$, reacting with a labile species which consists of two enantiomers, such as $[Co(en)_3]^{2+}$; and determining to what extent the inert product $[Co(en)_3]^{3+}$ is optically active—an activity which must have been acquired along with the electron transfer.

In the majority of cases there is a transfer of a (bridging) ligand concomitant with electron transfer. Examples are known, however, where although there is strong evidence for a bridge intermediate in the electron-transfer process, no ligand transfer occurs. Although not developed for it nor obviously adapted to it, there is evidence that the Marcus–Hush theory may be applicable to inner-sphere electron transfer. Perhaps this serves to emphasize the fact that a continuum of electron transfer processes exists, although it is convenient to divide them into two main categories. The same point is evident in another class of compounds which has been studied, those in which both metal atoms participating in an electron transfer reaction are contained within a single complex ion, being linked by a bridging ligand. Undoubtedly the most important series of compounds which has been studied are species derived from the Creutz–Taube ion, mentioned above, by choosing bridging ligands related to, but different from, the pyrazine of the parent species. Another example is provided by the molecule shown in Fig. 14.5 which contains one iron(II) and one iron(III). Such complexes can be made by partial oxidation or reduction of a symmetrical species. They can be studied by a variety of methods, most notably by electronic spectroscopy where an additional metal-to-metal charge transfer band is observed, the frequency and intensity of which turn out to be related to the

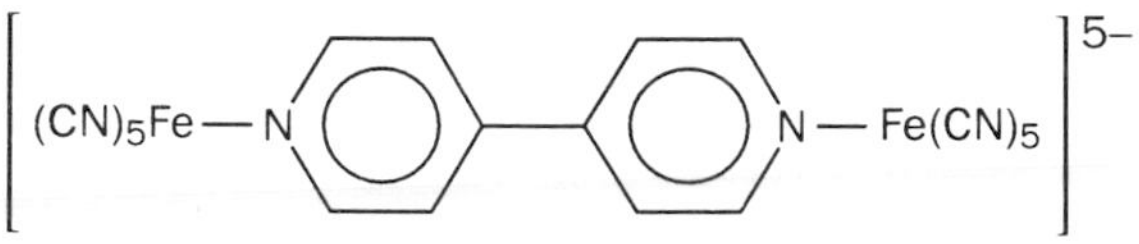

Fig. 14.5 Complex anion containing both iron(II) and iron(III).

rate constant of electron transfer. Clearly, the Robins–Day classification (Section 8.11) is relevant here. When both metal ions are not identical it is sometimes possible to prepare a molecule with one metal ion in the oxidized form and then measure the slow transfer of an electron from what initially is the reduced metal to that which initially is the oxidized.

14.3 Mechanisms of ligand substitution reactions: general considerations

In this section reactions of coordination compounds in which one ligand is replaced by another will be discussed, i.e. ligand substitution reactions. It is convenient to start by a consideration of a general potential energy profile, a topic which, whilst relevant to the electron transfer reactions of the previous section, has been deferred until the present because of its importance to it. In the introduction it was seen that it is convenient to make the approximation that all molecules undergoing a particular reaction follow identical energy pathways. As shown pictorially in Fig. 14.6, in the simplest case a potential energy barrier separates reactants from products, along the pathway of atomic movements that leads the reactants to become products (the so-called *reaction coordinate*). The rate of reaction is determined in part by the height of this barrier (another factor is the accessibility of the reaction coordinate—this is where energy-matching requirements, such as those met in the last section, come in). The equilibrium position is in part determined by the relative heights of the reactant and product potential energy minima. The 'in part' in the last sentence arises because the equilibrium position is determined by the relative free energies, and so there is also an entropic contribution. It is often found, both in organic and in inorganic chemistry, that there is a proportionality between the free energy 'hill' and the free energy difference between products and reactants. The existence of such proportionalities gives rise to the study of linear free-energy relationships.

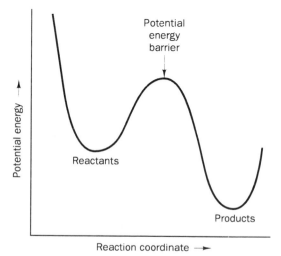

Fig. 14.6 A simple potential energy diagram relating reactants and products. The reaction coordinate can involve quite a complicated set of atomic motions, of which minor variants are admissible. It follows that this diagram is just one of a large family, of which all the acceptable variants are members.

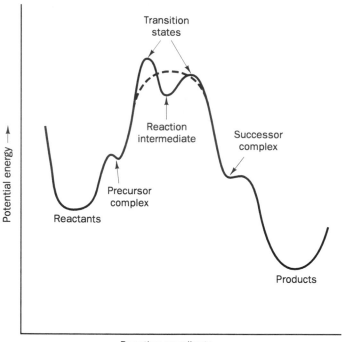

Fig. 14.7 A more realistic potential energy diagram than that shown in Fig. 14.6. An additional complication may be that the reaction coordinate may well not be the smooth, continuous, motion implied by this diagram. So, for example, the motions that lead from reactants to reaction intermediate may be rather different from those involved in the passage from reaction intermediate to products.

A little consideration soon shows that Fig. 14.6 is likely to be somewhat idealized. It implies that the path from products to reactants is smooth and uneventful. There are no sticky spots. But consider the reaction between a cationic complex, $[Ni(H_2O)_6]^{2+}$ for example, and an anion, Cl^- say, to give the complex ion $[Ni(H_2O)_5Cl]^+$. Simple electrostatic considerations suggest that the ion pair $[Ni(H_2O)_6]^{2+} \cdot Cl^-$ may well have some stability, particularly in solvents with a relatively low dielectric constant. The reactants are 'sticky'; the ion pair is likely to exist until atomic positions and momenta are either such as to allow the reaction to proceed or the ion pair to dissociate. Such an intermediate, formed between the reactants but in advance of reaction between them, is called a *precursor complex*. Similarly, when the reaction is one which involves loss of an anionic species by the complex, a so-called *successor complex* may be an intermediate on the way to the final product. This pattern is shown in Fig. 14.7 which also includes the possibility of a reaction intermediate of some stability. Fig. 14.7 shows a situation which is complicated and would therefore have a complicated rate law. Most systems studied either are, or are assumed to be, rather simpler.

Despite its complexity, Fig. 14.7 begs one important question. Consider either the reaction intermediate in that figure or, when no such intermediate is formed, the corresponding transition state (shown dotted in Fig. 14.7). Compared to the coordination number of the metal in the reactant complex, at this point in the potential energy surface the coordination number of the metal could have increased by one, decreased by one or stayed the same. The rate-determining step in the reaction mechanism involves association, dissociation or an interchange, respectively, and these labels, A, D and I, are used to describe the reaction type. The job of classifying a particular reaction is an experimental one. While it is easy to provide experimental criteria which enable such a distinction they refer to ideal cases and real life examples can be more complicated. Thus, there are cases where it seems that A and D mechanisms operate in parallel.

The distinction between an I mechanism and either A or D is that in the latter there should be proof of a reaction intermediate. In a D mechanism there is no dependence of the reaction rate on the incoming group. Similarly, in an A mechanism there is no dependence of the rate on the leaving group. These criteria require the ability to study a closely related series of reactions which follow similar reaction pathways. Of course, one cannot know in advance that all members of the series will behave as expected and so if one or two behave differently, a problem arises that can only be resolved by appeal to other data. Fortunately, the most readily available additional data refer to individual reactions and to the properties of the transition state. These are the entropy of activation, $\Delta S^{\neq}$, obtained from the intercept of a $\log k$ against $1/T$ plot (T = absolute temperature), and the volume of activation, $\Delta V^{\neq}$, obtained from the slope of a $\log k$ against P plot ($\partial \ln k/\partial P = -\Delta V^{\neq}/RT$, where P = pressure). Of these, $\Delta V^{\neq}$, although the more difficult to obtain, is perhaps the more reliable. Negative values of $\Delta V^{\neq}$—the rate increases with increasing pressure—implies an A mechanism, positive values a D mechanism. In the case of solvent exchange, for a limiting A mechanism the $\Delta V^{\neq}$ value equals the partial molar volume of the solvent and for a limiting D, it is negative. Smaller, positive, values indicate an I_d (d = dissociation) mechanism of solvent exchange; smaller and negative values an I_a (a = association) mechanism.

Type D mechanisms imply an increase in the effective number of particles in the system, an increase in the number of possible arrangements, and so $\Delta S^{\neq}$ is positive. Conversely, negative $\Delta S^{\neq}$ values imply A mechanisms. Type I mechanisms are associated with very small magnitudes of $\Delta S^{\neq}$. However, the view just adopted is somewhat simplistic in that there are many contributions to the $\Delta S^{\neq}$—translational, rotational, vibrational—of the reactants, together with a contribution from solvent ordering, and it is their sum which is determined. For this reason, and because a long graphical extrapolation is needed to obtain them, $\Delta S^{\neq}$ values are perhaps less useful than $\Delta V^{\neq}$. However, they are easier to obtain.

This far from exhausts the list of ways of gaining insight into reaction mechanism by kinetic studies—changes in the nature of the solvent and solution composition are obvious additional variables. Less evident are techniques such as carrying out measurements in the intense magnetic field available from a superconducting coil. Such experiments have revealed that a reaction intermediate can have quite different magnetic properties from either reactants or products. As a result, a magnetic field, which interacts with any unpaired electron present, can affect a reaction rate. This is particularly relevant to electron-transfer reactions—for at least one reactant and one product species must be paramagnetic.

It is perhaps not surprising that $\Delta S^{\neq}$ (and $\Delta V^{\neq}$ where available) values quite often indicate that there are I mechanisms with a rate dependence on either entering or leaving ligands. That is, I mechanisms which have some A character and others which have some D. The I class is therefore usually subdivided into I_a and I_d, labels we have already met. Although a constant research interest is where a particular reaction falls in the sequence:

$$A \rightarrow I_a \rightarrow I_d \rightarrow D$$

and we shall certainly need to refer to the topic,[3] it will not be a major concern in this chapter. Rather, we shall look at two particular, well researched, areas in some detail. The first, the reactions of square-planar platinum(II) complexes, provides an excellent example of a set of reactions which are essentially A-type. The second, the reactions of octahedral cobalt(III) complexes, provide examples of reactions which are in the I_d to D region. Two words of caution. First, it should not be thought that in all respects they are representative of their type—that all square planar complexes behave like platinum(II) and octahedral like cobalt(III). For instance, gold(III) forms square planar complexes but can also be reduced to gold(I) by the iodide anion, for example. Not surprisingly, this opens up reaction pathways and some of the reactions of gold(III) have no parallel in the chemistry of platinum(II). This apart, there seems to be a similarity between the kinetic patterns of square planar complexes of Pt^{II}, Pd^{II} and Au^{III}—although rates are much greater for the last two. The same cannot be said of square planar complexes of nickel(II), because nickel(II) occurs in a wide range of coordination numbers and geometries, features which set it apart. In general, however, the only safe viewpoint seems to be that 'where reactions of complexes of two different elements are superficially— stoichiometrically— similar on paper this is no guarantee of similar kinetic patterns'. This is also seen in the second word of caution. Most reactions have been studied with water as solvent. As will be seen, molecules of the solvent are seldom disinterested spectators of a reaction—they commonly get involved, unless they are both nonpolar and without features, such as carbon–carbon double bonds or donor oxygen atoms, which can function as ligands. Changing solvents in a way which reduces solvent participation commonly reveals features of the kinetics concealed when less innocent solvents are used. So, much of that which follows is a simplification. We shall often ignore a back-reaction, for example. Yet in the right situation such a reaction may become of importance. Quite a few workers find a real challenge in engineering molecule and solvent in such a way as to create the 'right situation', not only for this but also to force a D mechanism in platinum(II) chemistry, for example. However, the resulting exotic species are difficult to fit comfortably into a relatively brief account such as this.

14.4 Substitution reactions of square planar complexes

As has just been said above, there seems no doubt that the substitution reactions of square planar complexes, particularly of platinum(II) (the most-studied set of complexes) are A, associative, in type, the substitution being by a nucleophile. However, even for platinum(II), in rare and unusual

---

[3] For instance, as will be seen later in the text, there is general agreement that substitution reactions of pentaammine–Co^{III} complexes follow mechanisms which are mostly I_d with a leaning towards D; however, there is not the same agreement over the corresponding reactions of, say $[Cr(NH_3)_5H_2O]^{3+}$ for which I_a, borderline I_a/I_d and I_d have all been proposed. The most recent evidence favours the latter.

systems, electrophilic A-type substitution and even dissociation-controlled pathways can occur. Dominance of A-type nucleophilic substitution is what one would expect, for in a square planar complex there is no large steric constraint opposing bonding of the incoming ligand whilst the outgoing ligand is still attached. In practice, the rate law most generally observed for substitution reactions of platinum(II) complexes for the reaction:

$$[PtL_3X] + Y \rightarrow [PtL_3Y] + X$$

is

$$-\frac{d[PtL_3X]}{dt} = k_1[PtL_3X] + k_2[PtL_3X][Y]$$

Here, for simplicity, the ligands not involved in the substitution have been represented by L_3; however, they need not all be identical. Although in a particular study only one term of the rate law may be important, changing the reaction conditions somewhat can lead to the other becoming evident. The use of polar solvents leads to dominance of the k_1 term, of apolar solvents to the k_2. Conversely, when Y is a strong nucleophile the k_2 term is favoured; weak nucleophiles favour the k_1. As the ligands L_3 are made more and more bulky, so both k_1 and k_2 are reduced in magnitude. Similarly, when Y is made more bulky both k_1 and k_2 are again reduced. Although, from the rate expression alone, one might guess that the k_1 term represented a D mechanism, the evidence just presented strongly suggests that the k_1 and k_2 mechanisms are very similar. The detailed data are consistent with very similar mechanisms indeed; that in k_2 Y, and in k_1 solvent, are involved in the rate-determining A step.

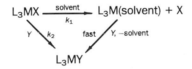

In this reaction sequence, no back-reactions are included. With varying degrees of ingenuity they can be made to be important (for example, by carrying out the reaction in the presence of a large excess of X). The kinetics, accordingly, become more complicated.

An incompletely understood aspect of these reactions is that they proceed with a remarkably high retention of configuration. This is surprising. An A mechanism implies a five-coordinate intermediate and so a, basically, square pyramidal or trigonal bipyramidal geometry is likely. The energy difference between the two geometries in most five-coordinate systems is found to be small and the systems fluxional.[4] If this were the case here, retention of configuration would be rare. However, as has been seen in Section 4.2.7, the retention is so reliable that it is given a name and offers the possibility of planned inorganic syntheses. As Section 4.2.7 illustrates, not only is configuration retained but, given a choice, the ligand displaced is determined not so much by its own nature as that of the ligand *trans* to it—the

---

[4] Fluxional = stereochemically non-rigid; see Section 14.8.

so-called *trans effect*. Because the *trans* effect is kinetic—that is, it is a generalization of the relative rates of the various alternative substitution reactions—explanations of the effect which do not take account of the A nature of the processes involved must be incomplete. As seen briefly in Section 4.2.7, the current popular theology of the *trans* effect distinguishes two contributions. The first is a static, ground state, effect, the *trans influence*. It is seen, for example, as metal–ligand bond lengths being sensitive to the *trans* ligand. The simplest explanation is to note that, if there is competition for electron density along a common (L–Pt–L′) axis, what one ligand gains, the other loses. This model offers an explanation for the large *trans* effect of the H^- ligand, for which the second contribution, from π-bonding, is scarcely likely to be relevant. The π bonding model suggests that a major contribution to the *trans* effect comes from preferential stabilization of a particular trigonal bipyramidal five-coordinate species by π bonding from the *trans* ligand. The high *trans* effect of many ligands is thus explained and, properly, the model is a kinetic one, focusing on the five-coordinate species. Even so, the model is not without its problems. In particular, crystal structures of genuine five-coordinate platinum(II) complexes tend to show them to be square pyramidal rather than trigonal bipyramidal, although the model requires the latter. This is because in the π bonding model the *trans* group, the entering (Y) and leaving group (X), occupy the trigonal planar portions of the trigonal bipyramid. So, for the three ligands in this plane, the out-of-plane π orbitals of the *trans* ligand have a direct interaction with the π orbitals of both X and Y (and only X and Y). There is no comparable pattern possible in a square pyramidal complex.

A particularly unsatisfactory feature of the two models for the *trans* effect is that there is no way of combining them. They are introduced to the extent that they are needed to explain the experimental results. There are other, minority view, explanations of the *trans* effect and it is unlikely that the last word has been said. For instance, as we have mentioned, the effects of strong magnetic fields have shown that the reaction intermediates in electron-transfer reactions (Section 14.2) generally have magnetic properties rather different from those of both paramagnetic reactant and product species. Perhaps here, too, it could be that in going to the reaction inter-mediate, the distortion mixes higher electronic states into the ground state, thus introducing the possibility, for example, of significant relativistic effects, effects which are known to be generally large for platinum.[5] One final word, on $[Pt(NH_3)_nCl_{4-n}]^{(n-2)+}$ species. For these, systems in which the *trans* effect operates, it has been found that, when isomers exist, the isomer formed most rapidly is the thermodynamically unstable one. That is, for these systems at least, synthetic application of the *trans* effect depends on a careful choice of time at which the products are separated. Leave the reaction mixture too long and it appears that the *trans* effect is no more! The relevance of such observations—which are some 30 years old—to the explanations offered for the *trans* effect awaits full discussion, although it has been recognized that successive displacements can lead to isomerization, as shown in Fig. 14.8. A relevant recent discovery is that the cyanide anion

[5] Note that a (relativistic) σ orbital automatically contains a (rotating) π component, suggesting a simultaneous explanation for σ and π *trans* effects.

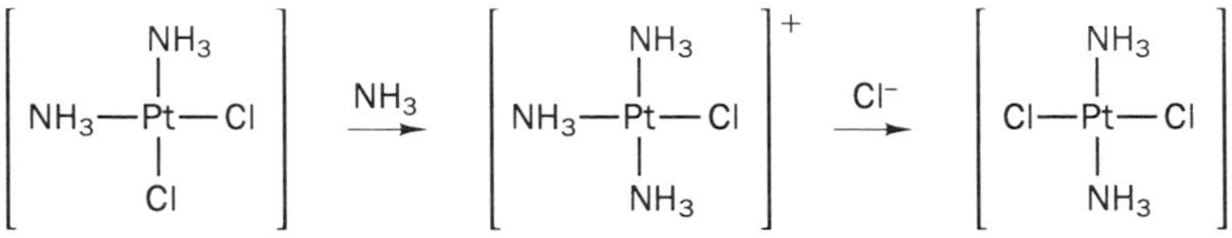

Fig. 14.8 A mechanism for the slow conversion of *cis*-diamminodichloroplatinum(II) into the *trans* isomer.

$$
\left[\begin{array}{c} NH_3 \\ | \\ NH_3-Pt-Cl \\ | \\ Cl \end{array}\right] \xrightarrow{NH_3} \left[\begin{array}{c} NH_3 \\ | \\ NH_3-Pt-Cl \\ | \\ NH_3 \end{array}\right]^{+} \xrightarrow{Cl^-} \left[\begin{array}{c} NH_3 \\ | \\ Cl-Pt-Cl \\ | \\ NH_3 \end{array}\right]
$$

not only exerts a *trans* effect but can exert a *cis* effect also. One is left with the distinct feeling that the *trans* effect, long a cornerstone in synthetic inorganic chemistry, merits more detailed investigation into its real nature and origin.

14.5 Substitution reactions of octahedral complexes

As implied in Section 14.3, the substitution reactions of octahedral transition metal complexes are in large measure dissociation-controlled (I_d and D) and so reaction rates are very dependent on the ligand displaced but not on the entering ligand. However, the field covered by the title of this section is a wide one, not confined to transition metals, and includes an enormous diversity of behaviours. Another aspect of this complexity is that caution is always necessary because simple kinetic data cannot reliably distinguish between A, I_a, I_d and D mechanisms. Our discussion will be restricted to areas which have been the subject of much work and for which some considerable insight has been gained. An important general theme will be the importance of—almost invisible—pre-equilibria and of solvent participation. For the former, work in non-aqueous solvents has shown that a complex which, formally, is positively charged may have closely associated with it so many anions that, if these are included, the overall charge changes sign, and with it our expectations! In contrast to the statement at the beginning of this section, the observation that the rate of substitution of transition metal ion complexes is often proportional both to the complex concentration and to the concentration of the incoming ligand seems to point to an associative mechanism:

$$
ML_6 + L \xrightarrow{\text{slow}} [\text{intermediate}] \xrightarrow{\text{fast}} \text{products}
$$

However, the actual rates show surprisingly little dependence on the chemical nature of L and other quantities—entropies, enthalpies and, to a lesser extent, volumes of activation—are also surprisingly constant. This sort of inconsistency is typical in the field, pointing to a hidden complication. The data can best be understood by a mechanism proposed by Eigen, Tamon and Wilkins and nowadays referred to as the *Eigen–Wilkins mechanism*. This mechanism is one in which the complex C, and the incoming ligand Y, diffuse together to form a weakly bonded *encounter complex*, a rapidly established equilibrium existing between this encounter complex and the free components:

$$
C + Y \underset{}{\overset{K_E}{\rightleftharpoons}} CY; \quad K_E = \frac{[CY]}{[C][Y]} \tag{14.1}
$$

Values of K_E can seldom be measured, but, perhaps surprisingly, they can be calculated, at least approximately. Two rather different models of the situation give rise to the same final mathematical equation—and this model-independence adds confidence in the result. The general conclusion is not surprising: large ions bump into each other more often than do small ones (and so give rise to larger K_E values), ions of opposite charge bond more strongly than do those of the same charge. In the second step of the Eigen–Wilkins mechanism, the encounter complex occasionally rearranges to give the final products in the rate-determining step

$$CY \xrightarrow[k]{slow} products$$

The observed rate of reaction is, then

$$rate = k[CY] \tag{14.2}$$

From Equation 14.1 and rearranging,

$$[CY] = K_E[C][Y] \tag{14.3}$$

Now the concentration of the total complex, $[C]_T$, is the sum of that which is free, $[C]$, and that which is in the encounter complex, $[CY]$:

$$[C]_T = [C] + [CY]$$
$$= [C] + K_E[C][Y]$$

Rearranging,

$$[C] = \frac{[C]_T}{1 + K_E[Y]} \tag{14.4}$$

Combining Equations 14.2 and 14.3,

$$rate = kK_E[C][Y] \tag{14.5}$$

and, from Equations 14.4 and 14.5, we have the final result:

$$rate = \frac{kK_E[C]_T[Y]}{1 + K_E[Y]}$$

It can immediately be seen that when the product $K_E[Y]$ is small compared to 1 then the commonly observed proportionality of the rate to both $[C]_T$ and $[Y]$ is explained. It transpires that the—often unexpectedly small—variations in rate constant mentioned above are the consequence of changes in K_E, not of the intrinsic rate k. So, for nickel(II) complexes the observed formation rates with respect to $[Ni(H_2O)_6]^{2+}$ vary from about 200 to 7 000 000 $M^{-1} s^{-1}$, a factor of over 30 000, but the corresponding k values only vary by a factor of about 2. This means that the observed reaction rate is controlled by the amount of encounter complex in solution. Relatively constant is the dissociation step, the rate at which a ligand (here, H_2O) leaves the coordination sphere, to be replaced by the ligand in the encounter complex—a ligand which can be thought of as lurking, waiting for the opportunity to insert. This interpretation of the reaction sequence is entirely in accord with the statement at the beginning of this section—and is a warning not to jump too quickly to mechanistic conclusions from rate expressions alone.

Another general aspect of importance is the fact that it is almost invariably found that in aqueous solution substitution of one ligand by another proceeds through the intermediate formation of an aqua complex (exceptions are found in platinum(II) chemistry, where, it will be recalled, a direct substitution may occur). In fact, this generalization can itself be generalized—solvent participation is more the norm than the exception. Even if there is no directly observable *solvento complex*, one may well be kinetically important. The reason is not difficult to see. First, the solvent is far more abundant than any other species present. Secondly, the majority of solvents do not form strong complexes so that a solvento complex, once formed, is a reactive species. A name has been introduced to cover this situation: *cryptosolvolysis*, solvent-mediated substitution in a situation in which no measurable amount of the solvento species is present. The phenomenon of cryptosolvolysis dovetails with the Eigen–Wilkins mechanism described above. In the language of that discussion, the species CY is saturated; that is, because here Y is a solvent molecule, all of the complex species C is associated with—at least—one solvent molecule. The first product species is, then, that in which the leaving group has been replaced by a solvent molecule. The leaving group may manage to regain the coordination site by displacing the solvent molecule and thus reform the starting material. Alternatively, the (new) incoming ligand may displace the solvent molecule to form the final product. One can at once see why there should be but little sensitivity to the incoming ligand but much to the ligand displaced. Once the latter has been displaced a reactive, low concentration, species is produced. It is obtaining this reactive species which is the difficult step.

As has been seen, this sequence is compatible with the Eigen–Wilkins mechanism but, nonetheless, has a different emphasis which finds expression in a different mathematical treatment—the steady-state approximation is applied to the concentration of the solvento species. However, evidence for the presence of the solvento species is usually indirect, frequently dependent upon seeking a rationalization, for example, for the pattern obtained from the study of a variety of closely related reactions in a series of closely related solvents. Not surprisingly, the role of water as a solvent has been the object of particular study; cations are commonly commercially available as their hydrates and so the interchange of coordinated and solvent water is a topic of great interest. Focusing on water at this point in the text has the advantage of bringing in ions other than transition metal species and so widening the scope of our discussion. Exchange of water between a coordination sphere and bulk at room temperature can be extremely fast through to extremely slow; the half-life of a coordinated molecule can be anything from 10^{-9} s through to 10^5 s. The geometry of the coordination sphere does not seem to have any great effect. At the fast end are ions such as Cs^+, Li^+, Pb^{2+}, Cu^{2+}, Ca^{2+}, Cd^{2+}, Ba^{2+}, Tl^{3+} and Gd^{3+}. At the slow end (half-lives greater than 1 s) are Al^{3+}, Pt^{2+}, Cr^{3+}, Ru^{3+} and Rh^{3+}. One would expect to find Co^{3+} in this region, too, but, of course, the ion is not stable in water. Across the first transition series volume of activation (pressure dependence of rates) measurements show a smooth change from an associative mechanism at the beginning (Ti^3, V^{3+}) to a dissociative mechanism towards the end (Co^{2+}, Ni^{2+}). The change-over point occurs between d^5 (Mn^{2+}, Fe^{3+}) and d^6 (Fe^{2+}) configurations. These data do not correlate with the rates of reaction and, again, one is warned of the danger of predicting the behaviour of one species from a knowledge of that of another. Nonetheless, quite simple ideas do seem to have some validity. Thus, as would be expected, for a given formal charge, larger ions exchange more rapidly than smaller. So, the rate constants vary: $Cs^+ > Rb^+ > K^+ > Na^+ > Li^+ \approx Ba^{2+} > Sr^{2+} > Ca^{2+} \gg Mg^{2+} \gg Be^{2+}$ and so on; it is also usually true that for a given ionic size an increase in formal charge is associated with a decrease in reaction rate. The observed tendency of a change from dissociative to associative activation on moving down a Group in the Periodic Table becomes, therefore, rather more understandable.

It is not always a simple matter to follow solvent exchange—the overall reaction is one in which there is no apparent change. Somehow, a label has to be introduced by means of which exchange can be followed. The obvious technique is that of isotopic exchange—for instance, preparing a complex containing coordinated $H_2^{17}O$ and following the exchange with solvent $H_2^{16}O$, perhaps by mass spectrometry. Clearly, such a method is only applicable to slow exchanges for otherwise exchange will be complete before measurements begin. As we have seen, another, much more widely used, set of techniques is based on NMR measurements, where, effectively, a spin label is used. A nucleus which exchanges between bulk and coordination slowly relative to the instrumental timescale will give two basic peaks (ignoring fine structure), the relative intensity of which is dependent on the ratio of bulk to coordinated solvent molecules (and so a ratio which can be used to obtain the coordination number). A nucleus which exchanges rapidly relative to the instrumental timescale gives a single peak at an average position; when exchange and instrumental timescales are similar, broadened peaks result. The NMR spectra can be studied as a function of temperature, pressure and time. The latter—by watching the evolution of an NMR signal over a period of time—enables the technique to be used for exchanges which are too slow to be studied by the line-broadening technique. These same techniques find applicability in the study of fluxional systems, a topic which forms the subject of a subsequent section of this chapter.

First however, a brief discussion of reaction intermediates. As has been emphasized, the majority of the reactions of octahedral transition metal complexes are I_d (the most common) or D in type. Let us confine our discussion to the latter, for this is the more clear-cut. A D-type intermediate must be five-coordinate, which means that, just as for platinum(II) reactions, square pyramidal and trigonal bipyramidal intermediates have to be considered, together with a possible interchange between them. Here, however, our concern is that of adding a ligand to the five-coordinate intermediate, not of losing one from it, as in platinum(II) chemistry. Is it possible to say anything about the geometry of the intermediate? What influence do the other ligands have? Is there a *trans* effect? It is clear that ligands other than that being replaced are important. Indeed, steric effects arising from these ligands are increasingly being invoked as a determining factor. In general, ligands *trans* to that being replaced seem to be no more important than those *cis*, although there are exceptions—in rhodium(III) complexes, for example, although the data are limited.

The series *cis* and *trans* $[CoLX(en)_2]^{2+}$, where X is the leaving group and L the ligand which is either *cis* or *trans* to it ($X = Cl^-, Br^-$; $L = Cl^-, OH^-$,

NCS$^-$, NO$^-$), provide a good example of another phenomenon. This is that the *cis* complexes show no tendency to isomerize to the *trans* on substitution whereas the *trans* give some *cis* product. A possible explanation which has been put forward is that the *cis* compounds dissociatively form a five-coordinate intermediate which is close to a square pyramid. The vacant position which accommodates the incoming ligand is unique and so offers no opportunity for rearrangement. If, on the other hand, the *trans* species dissociatively forms a near-trigonal bipyramidal intermediate then several non-equivalent insertion sites would exist, offering the possibility of rearrangement. This is a seductive argument, the more so when a possible π bonding stabilization of the trigonal bipyramid by the *trans* ligand L is added (see above). However, the comment at the end of Section 14.4 has relevance: it is unlikely that there is no free energy difference between *cis* and *trans* products. If the *cis* were inherently the more stable, perhaps for steric reasons, then an alternative interpretation becomes available—the *trans* form tends to give some of the more stable *cis* whereas the *cis* does not give the less stable *trans*. If this alternative were valid, then it is by no means evident that different geometry intermediates need be postulated.

14.6 Base-catalysed hydrolysis of cobalt(III) ammine complexes

It is common for studies of substitution reactions of cobalt(III) ammine complexes in aqueous solutions to be carried out under slightly acid conditions, for then protonation of the displaced ligand can occur, inhibiting the back-reaction and simplifying the kinetics. The usual observation under these conditions is of a rate which is proportional to the complex ion concentration and to nothing else. In contrast, when alkaline solutions are used, relatively rapid substitution occurs and reaction rates are usually proportional to the hydroxide ion concentration as well as to the concentration of the complex. The explanation for these observations was long disputed but is now resolved for cobalt(III) ammines and related species. What remains unclear is why the explanation is not more general—why other metal ammine species do not follow the same pattern (a few do, but only with complications). For the cobalt(III) ammines, the OH$^-$ removes a proton from a NH$_3$ (or, more generally, a NR$_2$H) ligand to give the conjugate base. The reaction is of the type

$$[Co(NH_3)_5Cl]^{2+} + OH^- \rightleftharpoons [Co(NH_3)_4(NH_2)Cl]^+ + H_2O$$

This pre-equilibrium is well over to the left-hand side and is rapidly established. However, the deprotonated species is up to ca. 10^{13} times as reactive as the species from which it is derived.

There is now a wide variety of evidence supporting this *conjugate base mechanism*, of which just three pieces may be mentioned. First, there is the unique character of the hydroxyl anion. In aqueous solution there are few cases where another substituting anion appears in a rate expression and so a unique mechanism involving the hydroxyl ion is indicated. Secondly, supporting evidence comes from the fact that exchange of the protons in $[Co(NH_3)_5Cl]^{2+}$, for example, with those of the solvent water is several orders of magnitude faster than the rate of base hydrolysis. Finally, the failure of complexes without such protons, bipyridyl complexes, for example, to undergo rapid base hydrolysis supports the conjugate base mechanism. When the stereochemistry is studied, for example by working with *cis* and also with *trans* disubstituted cobalt(III) ammines, then it is found that both scramble; stereochemistry is not retained. This is taken to mean that the five-coordinate intermediate is close to a trigonal bipyramid. A major effect of π bonding involving the deprotonated ammine on the behaviour of this intermediate has been postulated. However, although plausible, the model is not proven and it does not easily explain all of the available data.

This, then, is another example in which a pre-equilibrium has a major effect on a rate law—in this case, introducing a proportionality to the hydroxyl ion concentration. It is not difficult to see how this arises. In the example given above, let the forward rate be k_1 and the backward k_{-1}:

$$[Co(NH_3)_5Cl]^{2+} + OH^- \underset{k_{-1}}{\overset{k_1}{\rightleftharpoons}} [Co(NH_3)_4(NH_2)Cl]^+ + H_2O$$

Suppose the rate-determining step involves the expulsion of a chloride ion from this amide complex to give a five-coordinate intermediate, followed by rapid addition of water to give the final product. That is:

$$[Co(NH_3)_4(NH_2)Cl]^+ \xrightarrow[k_2]{slow} [Co(NH_3)_4(NH_2)]^{2+} + Cl^-$$

followed by

$$[Co(NH_3)_4(NH_2)]^{2+} + H_2O \xrightarrow{fast} [Co(NH_3)_5OH]^{2+}$$

Because the $[Co(NH_3)_4(NH_2)]^{2+}$ aquates as soon as it is formed, the reaction rate is equal to

$$k_2[Co(NH_3)_4(NH_2)Cl] \tag{14.6}$$

Now,

$$\frac{d[Co(NH_3)_4(NH_2)Cl]}{dt}$$

is equal to

$$k_1[Co(NH_3)_5Cl][OH] - k_{-1}[Co(NH_3)_4(NH_2)Cl][H_2O] - k_2[Co(NH_3)_4(NH_2)Cl] \tag{14.7}$$

Because the deprotonated species is present in such a small amount, its concentration will remain approximately constant (approximately zero!) and so

$$\frac{d[Co(NH_3)_4(NH_2)Cl]}{dt} \approx 0$$

It follows from Equation 14.7, then, that

$$[Co(NH_3)_4(NH_2)Cl] = \frac{k_1[Co(NH_3)_5Cl][OH]}{k_{-1}[H_2O] + k_2} \tag{14.8}$$

For convenience, the $[H_2O]$ term—a constant since we would be working in a very dilute aqueous solution—can be included in the k_{-1}.

Putting Equation 14.8 into 14.6 gives the final result:

$$\text{rate} = \frac{k_1 k_2}{k_{-1} + k_2} [Co(NH_3)_5Cl][OH]$$

where the experimentally observed rate constant is seen to be equal to

$$\frac{k_1 k_2}{k_{-1} + k_2}$$

and the observed rate law explained.

14.7 Mechanisms of ligand substitution reactions: postscript

In the preceding sections a wide variety of reaction mechanisms and kinetic patterns have been met. It has been emphasized that it is difficult to draw firm conclusions which cover more than one coordination geometry of one metal ion in one valence state. Can *any* general statements be made? Are there any general theories? The answer is yes to both questions.

It is found that inert and labile transition metal complexes are associated with quite specific d orbital occupancies. So, inert complexes are formed by ions with the configurations t_{2g}^3, t_{2g}^4, t_{2g}^5 and t_{2g}^6. Labile complexes are those in which the metal ion has electron configurations in which the e_g orbitals are occupied by one or more electrons, together with those with the configurations t_{2g}^1 and t_{2g}^2. The discussion of Section 13.5 is to be recalled—site preference energies are related to crystal field stabilization energies. It seems reasonable to expect that, provided that their reactions are mechanistically similar and that other energies vary smoothly along the series, the differences in kinetic behaviour associated with different d electron configurations may be explicable in terms of crystal field stabilization energies. That is, in going from the ground state of the reactants to the activated complex, the contribution made by the change in crystal field stabilization energy to the potential hill may be of importance. To test this theory one has to know the detailed geometry of the activated complex, not only its shape, but also the metal–ligand bond lengths (for these may not be the same as in the ground-state complex), so that the crystal field stabilization energy of the activated complex may be calculated. Unfortunately, our ignorance of the activated complex is profound. Calculations have been carried out assuming idealized geometries for activated complexes in which the metal ion is seven- or five-coordinate and give reasonable agreement with experiment. It is predicted that t_{2g}^1, t_{2g}^2 and $t_{2g}^3e_g^2$ ions will always react rapidly (that is, they do not lose crystal field stabilization energy in forming the activated complex). For t_{2g}^3, t_{2g}^4, t_{2g}^5 and t_{2g}^6 configurations a loss of crystal field stabilization energy in forming the activated complex predicts, as observed, relatively slow reaction. For other configurations, the predictions usually depend on the geometry assumed for the activated complex.

14.8 Fluxional molecules

So far in this chapter organometallic and low metal-valence state systems have not been mentioned, although such systems have been much studied. Early work had a parallel with some sections of this chapter—there was interest in the ligand substitution reactions of transition metal carbonyls, for instance. It was soon found that the kinetics of reactions in the light could be very different from those in the dark; in the next section the kinetics of photochemical reactions of complexes will be covered, although the metal carbonyls have not been the main focus of these studies. Within the organometallic field there has been much work stimulated by the recognition that many of these molecules are internally rather floppy—that they are *fluxional*, stereochemically non-rigid. Of course, this is not a phenomenon confined to this field—the turning-inside-out, the umbrella, motion of ammonia and pseudorotation in the trigonal bipyramidal PF_5, which interchanges axial and equatorial fluorines, are two well-known examples from main group chemistry. In the organometallic field there are many examples but we will concern ourselves with just one, very important, class. This class parallels the example of PF_5: those compounds in which atoms which might reasonably be considered equivalent, are, in fact, not equivalently bonded. Examples are σ-cyclopentadiene complexes, of both main group and transition elements, which contain a

unit. If the C_5H_5 ring moves round by one or two steps then the final molecule is equivalent to the starting one; if the interchange occurs readily, as it does, then it indicates that the barrier presented by an intermediate bonding position, such as

is not high. Not surprisingly, there has been much discussion on the nature of this intermediate (it could, for example, be one in which the C_5H_5 ring is η^5, π-bonded with all CH groups equivalent). In fact, it seems that 1,2 shifts provide a general mechanism. Examples of this behaviour are provided by $Hg(C_5H_5)_2$ $(=Hg(\eta^1\text{-}C_5H_5)_2)$, $Fe(CO)_2(\eta^5\text{-}C_5H_5)(\eta^1\text{-}C_5H_5)$ and $Cu(PEt_3)_3(\eta^1\text{-}C_5H_5)$.

A second general example is provided by CO ligands, in complexes containing this ligand bonded in a variety of ways. Crystal structure measurements have revealed that there is an almost continuous range of

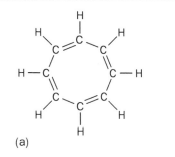

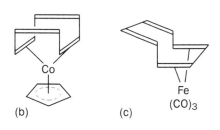

Fig. 14.9 (a) cyclooctatetraene, drawn as a planar molecule with four localized double bonds as might be appropriate for it acting as a symmetrically bonded ligand. (Actually, the molecule is non-planar; if it were planar the double bonds would be delocalized.) (b) And (c) are two complexes containing cyclooctatetraene as a ligand.

bonding positions for CO, from bonding uniquely to one metal atom (terminal CO), increasingly being bonded to a second until it is equivalently bonded to two (bridged CO), and then an increasing interaction with a third until it is equivalently bonded to three (face-bonded CO). This adaptability has been recognized in some models of metal cluster carbonyls, such as $Ir_4(CO)_{12}$, where a central metal cluster (here a tetrahedron) is surrounded by a polyhedron of CO groups (here a cube-octahedron), the arrangement within the polyhedron apparently as much being determined by packing constraints within the polyhedron as by metal–CO bonding considerations. From such models it is but a short step to expect that when not all of the CO groups are equivalently bonded, as in $Co_4(CO)_{12}$ and $Rh_4(CO)_{12}$—molecules which contain three bridging CO groups—then the CO positions will scramble rather rapidly. Indeed this is just what is found.

Finally, a ligand such as the (cyclic) cyclooctatetraene is too big for all of the carbons to be equivalently bonded to most metal atoms and in complexes the ligand commonly adopts an asymmetric position (Fig. 14.9). Again, not surprisingly now, the particular carbons bonded to the metal often change rapidly. The phenomenon has been referred to, rather colloquially but warmly, as 'ring whizzers'. There is one question which may be asked—does the ring rotate or does the metal hop?—which is less meaningful than it appears. We are not concerned with the translations and rotations of the entire molecule. It follows that any internal motion must have zero linear momentum and zero angular momentum. Consequently, the motion involved in ring whizzers must involve both ring rotation *and* metal hopping, in opposite directions—ring rotation on its own would mean that the motion had angular momentum. The study of such phenomena parallels that discussed in Section 14.5 for metal hydrates. Nuclear magnetic resonance spectroscopy is used and the broadening and coalescence behaviour of a signal as a function of temperature studied. Commonly, the experimental data are compared with computer-generated model spectra; in many cases the experimental and theoretical spectra are distinguished only by the noise in the experimental! In such cases the confidence in a correct interpretation is high, although this confidence is dependent on the correct assignment of peaks in the slow exchange limit spectrum. The ability to make such studies has depended critically on developments in NMR spectroscopy, particularly Fourier transform techniques which have enabled low-abundance nuclei, such as ^{13}C and ^{17}O, to be studied. Nowadays it is not uncommon for several different nuclei in one molecule to be accessible for study. Such work has made it clear that there is often more than one fluxional process occurring simultaneously in a molecule. We are left with the pattern presented at the beginning of this section—of molecules that are very floppy, perhaps to be pictured as rounded lumps of jelly stuck together, as if by surface tension, but quite free for the various bits to wobble, slide and rotate, rather than as collections of spheres of various sizes, rigidly locked together.

14.9 Photokinetics of inorganic complexes

Inorganic pigments have been used since antiquity, most of them having the attribute that their colours do not change with time, any deterioration in the paint resulting from changes in the medium supporting the pigment. In contrast to this stability there are other inorganic systems which are very sensitive to the action of light. Although the silver-halide photographic process is outside the scope of the present book, other photographic processes fall within it. For instance, iron(III) salts are readily reduced to iron(II) and the oxalate anion is a useful reducing agent. So, although the complex anion $[Fe(ox)_3]^{3-}$ is easy to prepare—concentrated aqueous solutions of ammonium oxalate and almost any soluble iron(III) salt give pale green crystals containing the anion—it is not surprising to learn that it is not very stable, decomposing under the action of light to give iron(II) oxalate and carbon dioxide. This behaviour is readily explained by the light exciting an electron in a ligand-to-metal charge transfer transition (LCMT, see Section 8.10.2). This electron transfer oxidizes the ligand and reduces the metal. The phenomenon is exploited in two ways. First, by impregnating paper with a mixture of $K_3[Fe(ox)_3]$, or a similar species, and $K_3[Fe(CN)_6]$. When the paper is placed under a mask and then exposed to light—ultraviolet light is best—and the paper sprayed with water, the areas not protected by the mask turn blue because of reaction between the photo-produced Fe(II) and $[Fe(CN)_6]^{3-}$ to give the blue pigment Turnbull's blue, $KFe(II)[Fe(CN)_6]$. This simple technique for producing blueprints is used up to the present day for engineering drawings and the like because it can easily be applied to sheets of paper much larger than can be accommodated by ordinary photocopying machines. The salt $K_3[Fe(ox)_3]$ is also used in chemical actinometry—when an aqueous solution of the salt is exposed to light under standard conditions, the amount of Fe^{II} produced is a measure of the total amount of light that has passed through the solution. The quantum yield of Fe^{II}—the number of Fe^{II} produced for each quantum absorbed—is almost independent of the intensity of the incident light and of the concentrations of Fe^{III} and Fe^{II} species. However, the system is only useful for light of shorter wavelength than yellow. Photolysis of $[Cr(NH_3)_2(NCS)_4]^-$, which causes release of NCS^-, is effected by all visible light except deep red and is also used in actinometry. For simplicity, all of the above examples concern complexes containing just one metal atom. A recent example of a more complicated photochemical system is the anion $[Os_{18}Hg_3C_2(CO)_{42}]^{2-}$, a molecule in which, essentially, two Os_9 clusters are joined by a triangle of mercury atoms. Photolysis leads to the expulsion of a mercury atom, giving $[Os_{18}Hg_2C_2(CO)_{42}]^{2-}$ where the Os_9 clusters are joined by just two mercury atoms. In the dark and the presence of mercury this latter complex adds a mercury atom to regenerate the original compound.

In recent years, inorganic photochemistry has received much attention and changed considerably from that typified by the historically important examples just given. There have been three main directions of research. First, the study of shorter and shorter timescales. This means using short time-pulse lasers and exploring their immediate (or, if of interest, their not-so-immediate) consequences. In this way, the sequence of steps following absorption of radiation can, hopefully, be followed. In practice, there are such intricate networks of finely separated energy levels, electronic, vibrational and rotational, that the energy absorbed when an electron is excited usually very quickly reappears as heat, the system rapidly changing levels when they cross, until the excitation is dissipated.

The second line of research is in the opposite direction. If energy can be pumped into a molecule by shining light on it and that energy, in large measure at least, retained in the molecule, then there is the possibility of using the energy to carry out chemistry. So, for example, there has been a persistent search for systems, perhaps several combined, that will photo-decompose water. If such a process could be made economically viable it could be an excellent way of harnessing solar energy. It is scarcely likely that a system could be found in which absorbed photoenergy does not degrade—indeed, there are good theoretical arguments that such a system could not exist. So the search is for a system which, as energy is lost and the energy ladder descended, eventually finds itself in an excited state from which there is no easy escape route. If, for instance, the excited state trap had a different spin multiplicity from the ground state then photoemission, at any rate, would be spin forbidden. Similarly, if the excited state were of the same parity (g or u) as the ground state then the transition used would be orbitally forbidden also. If it were a relatively simple system the number of vibrational modes, and so the possibility of vibrational deactivation, would also be reduced. Clearly, these requirements could well be met by simple, high-symmetry transition metal complexes. Of course, there is still the non-trivial problem of subsequently carrying out the required chemistry—but our first task must be to find a molecule that remains excited long enough for it to make contact with any molecule with which we would wish it to interact. It has been estimated that for an excited state to have a chemistry it must live for at least ca. 10^{-9} s. Such systems are known. The most studied species is $[\mathrm{Ru(bpy)_3}]^{2+}$ (bpy = 2,2′-bipyridine) which has an excited state lifetime in solution at room temperature of about 10^{-8} s and, at lower temperatures, 10^{-6} s—or 10^{-5} s in some similar species containing ligands with substituents on the pyridine rings.[6] An unexpected result is that in the excited state of the molecule the excitation is localized in a single bpy ligand molecule (all the evidence is that the excited state has a lower symmetry than the D_3 of the ground state), that ligand being reduced (and the ruthenium oxidized to ruthenium(III)). Such electron transfer is just what is involved in metal-to-ligand charge transfer (MLCT) transitions; they were first met in Section 8.10.3, although the idea that ligands equivalent in the ground state may not be so in the excited state was not mentioned there. There is no doubt that excited $[\mathrm{Ru(bpy)_3}]^{2+}$ is long-lived enough for it to react with other species in solution. However, as yet no viable system for obtaining H_2 and O_2 from water containing $[\mathrm{Ru(bpy)_3}]^{2+}$ has been discovered, although the energy available in the photoexcited state is sufficient. Although there seems to be no connection with the behaviour of the ruthenium compound, the complex $[\mathrm{Ir(bpy)_3}]^{3+}$ undergoes an isomerization of one of its ligands on photolysis, as shown in Fig. 14.10.

As implied above, it is generally true that the reactions undergone by photoexcited species are different from those of the ground state species. Most work has been done on octahedral complexes of d^3 and d^6 configurations (half full and full t_{2g} shells) and species derived from them—$[\mathrm{Ru(bpy)_3}]^{2+}$ is a d^6 ion. Complexes of Cr^{III} and Co^{III} have been particularly studied. Of course, the aquation of $[\mathrm{Cr(NH_3)_6}]^{3+}$ involves NH_3 loss by both

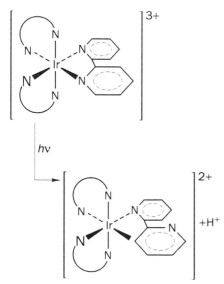

Fig. 14.10 The photoisomerization of $[\mathrm{Ir(bpy)_3}]^{3+}$. Only one bpy ligand is shown in detail in the starting material.

[6] This explains the interest in Creutz–Taube type Ru^{II}–Ru^{III} species mentioned in Section 14.2.

thermal (normal) and photochemical routes. However, for species of general formula $[Cr(NH_3)_5X]^{3+}$, where X is a halide or pseudohalide, thermal aquation leads predominantly to loss of X but photochemical aquation largely leads to loss of NH_3. Similarly, whereas $[Cr(NH_3)_6]^{3+}$ shows little photosensitivity, not only are the species $[Cr(NH_3)_5X]^{2+}$ much more sensitive, but the favoured process depends on the wavelength of the radiation. Excitation of spin-allowed transitions (particularly charge-transfer) preferentially leads to aquation by replacement of NH_3, whereas Cl^- tends to be replaced when the excitation is into a band that is spin forbidden (usually, d → d) from the ground state.

Although transient excited states living ca. 10^{-12} s have been observed in some cases, relatively little is known about the details of the processes involved in the photochemistry outlined above. Although they are the only ones we have so far mentioned, internal deactivation and chemical reaction are not the only processes that occur. Excited states may re-emit radiation, essentially, immediately. This is fluorescence and involves a transition from an excited level from which emission to give the ground state is a spin-allowed process. If there is deactivation to give an excited state from which emission to the ground state is spin-forbidden, then *phosphorescence* occurs with lifetimes of from milliseconds to minutes for solid-state species. In solution, the corresponding lifetimes are much shorter. When there is only a small energy difference between the two (spin-allowed and spin-forbidden) transitions just discussed then there is the possibility that the thermal energy of the system may be sufficient, for example, to promote molecules from a lower, spin-forbidden transition, level to an upper spin-allowed. When this happens fluorescence with a long lifetime is observed, a lifetime which is characteristic more of a spin-forbidden than a spin-allowed process (but the nature of which is usually evident from the absorption spectra). It is now easy to understand why t_{2g}^3 and t_{2g}^6 ions have proved of photochemical interest. For the t_{2g}^3 configuration there are in addition to the ground state, $^4A_{2g}$, term a host of spin doublets (^2E_g, $^2T_{1g}$, in particular, see Table 7.5 and Appendix 7; these have energies that have approximately the same Δ dependence as the ground state; the former is involved in the light emission process of the ruby laser). Similarly, cobalt(III) complexes, with a $^1A_{1g}$ ground state, have low-lying spin triplets, in particular a $^3T_{1g}$, which is the one usually implicated (again, see Table 7.5 and Appendix A.7). A diagram which is appropriate to $[Cr(NH_3)_6]^{3+}$ is given in Fig. 14.11. It shows at the top a (spin quartet) electronic excited state derived from the configuration $t_{2g}^2 e_g^1$ (and so, remembering that e_g electrons are weakly antibonding, with slightly longer Cr–N bond lengths than the ground state). Associated with this level is a host of vibrational energy levels. Intersystem crossing from the spin quartet to a spin doublet derived from the t_{2g}^3 configuration occurs (because of the spin change, spin–orbit coupling is implicated in this process). Again, vibrational deactivation occurs until some process to give the ground state is all that remains. The possibilities for this are discussed above.

A final example of how the relative stability of t_{2g}^6 systems in their lowest excited state can be exploited is provided by the $[Fe(CN)_5NO]^{2-}$ anion. A neutron diffraction study has been carried out on a salt of this anion whilst the crystal was under intense laser irradiation. The intensity of the light and the lifetime of the lowest excited state (at liquid helium temperature) were

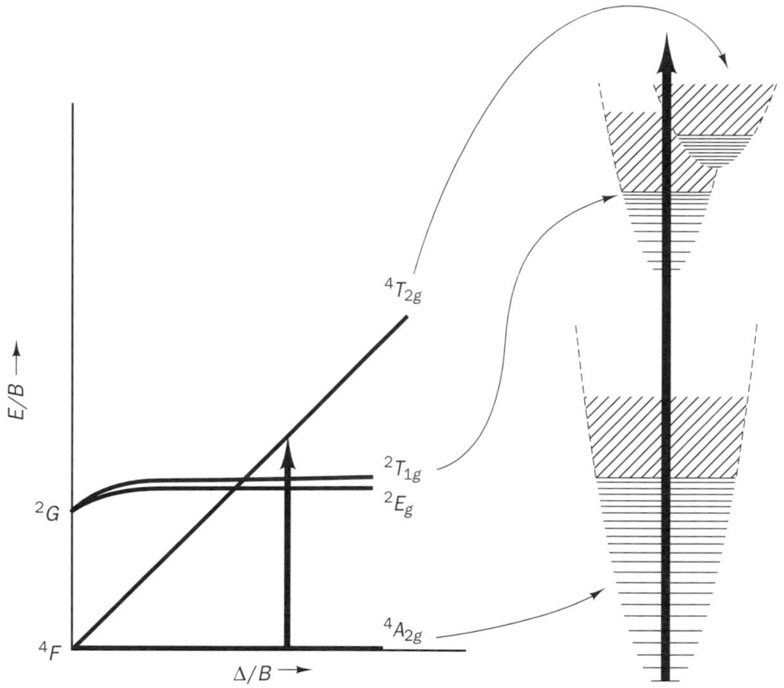

Fig. 14.11 On the left is shown a simplified Tanabe–Sugano diagram for the octahedral d³ case. The spin-allowed $^4T_{2g} \leftarrow {}^4A_{2g}$ transition is detailed on the right-hand side, where vibrational levels are included. There is intersystem crossing between $^4T_{2g}$ and $^2T_{1g}/{}^2E_g$ (only one of the latter is represented). In contrast, the spin-forbidden step from the latter to the ground state is more difficult.

such that approximately half of the anions in the crystal were in the excited state throughout the diffraction experiment. After measurements had been completed, the laser was turned off and the measurements repeated, this time with all anions in their ground state. Comparison of the two sets of measurements revealed that in the excited state the Fe–N bond (of Fe–NO) was lengthened relative to the ground state, all other bond lengths being unchanged within error. Clearly, the result requires an excited state localization which is not consistent with the octahedral t_{2g}^6 configuration (just as does the ground state structure), but the essential point—that the lowest excited state involves an electron in a metal–ligand antibonding orbital—remains valid.

Further reading

An older book which is both so easy to read and so forward-looking that it deserves mention is *Inorganic Reaction Mechanisms* by M. L. Tobe, Nelson, London, 1972. A good contemporary text is *Kinetics and Mechanisms of Reactions of Transition Metal Complexes* by R. G. Wilkins, VCH, New York, 1991.

Up-to-date accounts will be found in *Comprehensive Coordination Chemistry* G. Wilkinson, R. D. Gillard and J. A. McCleverty (eds.), Pergamon Press, Oxford, 1987, Vol. 1: Chapter 7.1 by M. L. Tobe 'Substitution Reactions'; Chapter 7.2 by T. J. Meyer and H. Taube 'Electron Transfer Reactions'; Chapter 7.3 by C. Kutal and A. W. Adamson 'Photochemical Processes'.

Other references worth mentioning are 'An appraisal of square-planar substitution reactions' R. J. Cross, *Adv. Inorg.*

Chem. (1989) *34*, 219; *Electron Transfer Reactions* by R. D. Cannon, Butterworth, London, 1980.

A Nobel Lecture gives both a fascinating and enlightening account: 'Electron Transfer Reactions in Chemistry: Theory and Experiment' R. A. Marcus, *Angew. Chem. Int.* (1993) *32*, 1111.

For a touch of the unexpected: 'Dissociation Pathways in Platinum(II) Chemistry' R. Romeo, *Comments Inorg. Chem.* (1990) *11*, 21.

An up-to-date and easy-to-read overview is provided by *Reaction Mechanisms of Inorganic and Organometallic Systems* R. B. Jordan, Oxford University Press, Oxford, 1991.

A wide-ranging review is 'Ru(II) Polypyridine Complexes: Photophysics, Photochemistry, Electrochemistry and Chemiluminescence' A. Juris, V. Balzani, F. Barigelletti, S. Campagna,

P. Belser and A. von Zelewsky, *Coord. Chem. Rev.* (1988) *84*, 85.

Dovetailing well with the general level of presentation in the present book is 'The Photophysics of Chromium(III) Complexes', L.S. Forster, *Chem. Rev.* (1990) *90*, 331.

'Fluxionality of Polyene and Polyenyl Metal Complexes' B.E. Mann, *Chem. Soc. Rev.* (1986) *15*, 167.

Questions

14.1 Distinguish between the order and molecularity of a reaction. Give examples of how pseudofirst- and pseudosecond-order kinetics may arise in inorganic substitution reactions.

14.2 $[Fe(CN)_6]^{2-}$ and $[Fe(CN)_6]^{3-}$ are inert complexes yet they react with each other rapidly. $[Co(NH_3)_6]^{3+}$ is inert and $[Co(H_2O)_6]^{2+}$ is labile yet these two react with each other slowly. Detail the current explanation for this different behaviour.

14.3 A useful rule of thumb is that for suitable systems (usually d^3 or d^6) light emission is the more favoured the greater the number of vibrational steps separating the low-lying excited state from the bottom of the ground state. This means that simple anions such as $[MI_6]^{3-}$ would be favoured for emission studies. Yet they have not been discussed in this chapter. Suggest plausible reasons for this omission.

14.4 Detail the experimental distinctions between A, I_a, I_d and D mechanisms. Suggest the tactic you might employ in choosing ligands to force an ion such as Pt^{II}, for which A-type kinetics are typical, to follow a D-type mechanism.

15

Bonding in cluster compounds

15.1 Introduction

The study of cluster compounds is an active area of current research. The scope is widening and systematic methods of synthesis beginning to appear based, for example, on the isolobal principle first mentioned in Chapter 10 and which will be met again later in this chapter. It seems likely that it will eventually be found that such compounds may involve almost any element, perhaps with the exception of the most electropositive. Some representative examples are given in Fig. 15.1. Despite all this activity, our understanding of the bonding in cluster compounds is relatively primitive. Most of them are beyond the scope of detailed *ab initio* calculations and resource has to be made to more approximate methods, the $X\alpha$ and its refinements probably being the most reliable (see Section 10.3.1). However, it seems that even the simplest of methods, in particular the extended Hückel (Section 10.5), can give the highest occupied orbitals with a tolerable accuracy, though being far less reliable for more deeply lying orbitals. Unfortunately, cluster molecules have such a plethora of bonding molecular orbitals that any photoelectron spectroscopic data (Section 12.7) can only be interpreted with the greatest difficulty. So, it is difficult to find any reliable external check on the available calculations.

Before commencing a detailed study it will be found useful to consider a well known and simple molecule which may be regarded as a cluster—the tetrahedral molecule P_4. It will prove helpful to compare this with another tetrahedral molecule which has been the subject of study—the boron chloride B_4Cl_4, which consists of a B_4 tetrahedron with a chlorine pointing radially outwards from each boron; it is made from gaseous BCl_3 passed through an electrical discharge. The argument developed will be symmetry-based and so in Table 15.1 is given the character table for the T_d point group. Three different models of the bonding in P_4 will be compared, the object of the exercise being to show the relationship between the three models. Rather similar relationships occur between the extant descriptions of the bonding

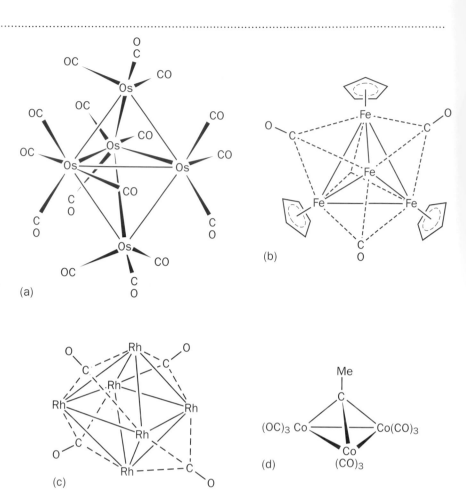

Fig. 15.1 Some examples of cluster compounds. (a) $Os_5(CO)_{16}$—at first sight this appears a high-symmetry cluster. It is not. It contains one $Os(CO)_4$ group and four $Os(CO)_3$. At the moment, all that can be hoped for with such structures is a retrospective rationalization. (b) $[\eta^5\text{-}C_5H_5Fe(CO)]_4$—a tetrahedral cluster with a CO group above each tetrahedron face. The $\eta^5\text{-}C_5H_5$ group attached to the closest iron atom has been omitted, as has the most distant CO. (c) $Rh_6(CO)_{16}$—an octahedron of rhodiums but four face-centering CO ligands reduce the symmetry of the cluster to T_d. To each Rh are attached two terminal CO groups (not shown) (d) $Co_3(CO)_9CCH_3$—this cluster is typical of those taken to model the attachment of a reactive organic molecule to a bare metal surface.

in more exotic, more complicated, clusters although these relationships tend to be more difficult to see than in the cases now to be considered.

15.2 Bonding in P_4 (and B_4Cl_4)

The three models that follow all lead to the same qualitative result but they do so by such different routes that one might reasonably conclude that they are different. As will become evident, these apparent differences are not fundamental. Although minor differences may persist, in the things of importance, the three models are equivalent. The P_4 molecule has been chosen because just s and p valence shell atomic orbitals are involved; it is the additional involvement of d orbitals that leads to the more complicated situation in clusters involving transition metals.

15.2.1 'Simple ammonia' model for P_4

In NH_3, one may think of three equivalent N–H bonding orbitals and of an additional lone pair of electrons on the N atom (Appendix 2). The simplest picture of the bonding in P_4 is to think of the N of NH_3 being replaced by P and each H also being replaced by a P, the resulting overall

Table 15.1 The T_d character table

T_d	E	$8C_3$	$3C_2$	$6S_4$	$6\sigma_d$
A_1	1	1	1	1	1
A_2	1	1	1	−1	−1
E	2	−1	2	0	0
T_1	3	0	−1	1	−1
T_2	3	0	−1	−1	1

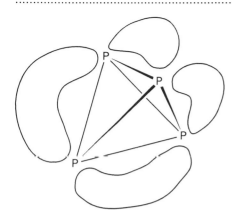

Fig. 15.2 A symbolic representation of four of the six P–P edge bonds in P_4. The exact orbital makeup and detailed orientation in space of these edge bonds are not important as they do not change the symmetry properties of members of the set.

Table 15.2 The transformation properties of the tetrahedron edge orbitals of P_4. Characters generated by the six orbitals under the operations of the T_d point group (see Fig. 15.2) which is the sum of the characters of the irreducible representations $A_1 + E + T_2$

E	$8C_3$	$3C_2$	$6S_4$	$6\sigma_d$
6	0	2	0	2

bonding network leading to the P_4 molecule. In this model, each P has a lone pair of electrons, just like the N in NH_3, and is also involved in three P–P bonds, just as N in NH_3 is involved in three N–H bonds. In this model each of the six edges of the P_4 tetrahedron is associated with a P–P bond. Such a bond need not lie along the edge, perhaps a pattern such as that in Fig. 15.2 is more probable. The symmetries of the (delocalized) bonding molecular orbitals in P_4 can be obtained from the local orbitals in Fig. 15.2 by using them as basis functions for the generation of a reducible representation in the T_d point group. This development is detailed in Table 15.2 where the conclusion is reached that they have $A_1(0) + T_2(1) + E(2)$ symmetries. The numbers in brackets refer to the number of nodal planes inherent in each molecular orbital. To obtain these numbers explicit mathematical expressions for the orbitals are needed; they are readily obtained by the methods described in Appendix 6, and representative examples of the molecular orbitals are shown in Fig. 15.3. The more nodes, the higher the energy, and so we conclude that the stability sequence is $A_1 > T_2 > E$. These are stabilities relative to the stability of the P–P bonds that we started with; as interactions between these bonds will, as a first approximation, lead to a splitting—and not an overall shift in energy—we conclude that the A_1 orbital is stabilized and the E pair destabilized relative to the P–P bonding orbitals with which we started. The overall picture is that given in Fig. 15.4, where a small displacement of the T_2 set, given by a more detailed analysis, has been included. Feeding electrons into the bonding molecular orbitals gives the scarcely surprising result that all are filled. Of course, had the E orbitals in Fig. 15.4 been much higher in energy then they would perhaps have been too high in energy to be occupied. But that is another story.

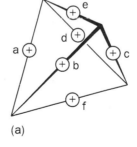

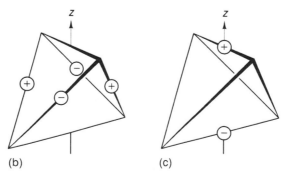

Fig. 15.3 Representative examples of the edge group orbitals of P_4. The signs indicate relative phases. So, (a) the A_1 is $1/\sqrt{6}(a + b + c + d + e + f)$; (b) one of the E is $1/2(a - b + c - d)$; and (c) one of the T_2 is $1/\sqrt{2}(e - f)$.

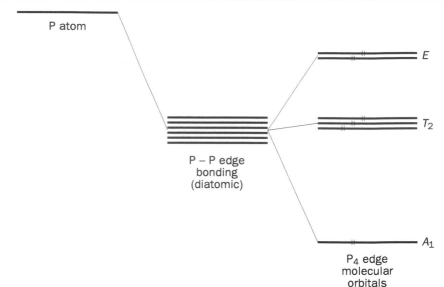

Fig. 15.4 A schematic energy level diagram for the six edge molecular orbitals of P_4.

15.2.2 'Twisted ammonia' model for P_4

One sometimes encounters the statement that a rotational axis which is threefold, C_3, or above (C_4, C_5 etc.) shows cylindrical symmetry. Now, the view down the C_3 axis of a molecule such as ammonia shows something more bumpy than cylindrical, so just what does cylindrical symmetry mean? It means that, as long as they remain mutually perpendicular, the x and y coordinate axes can be placed *anywhere* perpendicular to the C_3, C_4, C_5 etc. axis (this axis is conventionally chosen as z). Of course, some choice of x, y directions may be more convenient—they lead to simpler mathematics and simpler pictures, but any other choice would be just as acceptable and would in no way change the final result. In a cylinder, x and y can be chosen anywhere perpendicular to the principal rotational axis; so, too, here. What applies to x and y coordinate axes also applies to p_x and p_y atomic orbitals, and, for us, this is the important aspect of cylindrical symmetry.

When, in Section 15.2.1, we talked of the N–H bonding orbitals of ammonia, we did not enquire into the nitrogen orbitals involved in this bonding. Had we done so, they would have been mixtures of nitrogen 2s, $2p_z$, $2p_x$ and $2p_y$ orbitals and would have looked something like those shown in Fig. 15.5(a). But nitrogen 2s and $2p_z$ are axially symmetric and, as we have seen, $2p_x$ and $2p_y$ can be chosen to point anywhere perpendicular to z. This means that as an alternative to Fig. 15.5(a) we could have worked with the orbitals of Fig. 15.5(b). They look different but, remembering that they show contour diagrams, the actual electron density distributions implied by Figs. 15.5(a) and (b) are identical.

Return now to the P_4 molecule, and regard the contribution to the bonding from each phosphorus to arise from orbitals such as those of Fig. 15.5(b). Because they point towards the centres of faces of the P_4 tetrahedron they combine to give three-centred bonding orbitals of the sort shown in

(a)

(b)

Fig. 15.5 A set of sp^2 or sp^3 hybrids viewed down a threefold axis of P_4. They may be orientated towards (a) the tetrahedron edges or (b) faces. Although these alternatives appear rather different, they both correspond to cylindrical symmetry of electron density distribution around the threefold axis.

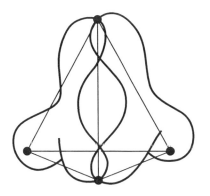

Fig. 15.6 An approximate picture of two of the four tetrahedron face bonding orbitals of P_4. Each of these orbitals is totally symmetric with respect to rotation about the threefold axis on which it lies.

Table 15.3 The transformation of the tetrahedron face orbitals of P_4. Characters generated by the four orbitals under the operations of the T_d point group (see Fig. 15.6) which is the sum of the characters of the irreducible representations $A_1 + T_2$

E	$8C_3$	$3C_2$	$6S_4$	$6\sigma_d$
4	1	0	0	2

Fig. 15.6. There are four faces of the tetrahedron and so just four orbitals like that in Fig. 15.6. These orbitals can be used to obtain delocalized molecular orbitals; the steps in the sequence are detailed in Table 15.3. The orbitals that result are of $A_1(0)$ and $T_2(1)$ symmetries; again, their explicit forms may be derived using the methods of Appendix 6 and the numbers in brackets refer to the number of inherent nodal planes; A_1 is more stable than T_2. The final molecular orbital energy level diagram that results is shown in Fig. 15.7. It is immediately clear that this figure poses problems for P_4. Placing a lone pair of electrons on each P leaves six bonding electron pairs to be fed into the four orbitals of Fig. 15.7. It cannot be done. So what of the statement made earlier that 'the actual electron density distributions implied by Figs. 15.5(a) and (b) are identical'—if this is so, why do the different choices lead to different results? Clearly, this is a point to which we must return. Although the present model seems to have failed for P_4, we end this section on a more encouraging note: the model works for B_4Cl_4, for which it is a simple matter to see that there are just eight valence shell electrons available for the cage bonding. Three-centred face bonding orbitals of the type just encountered are commonly used in discussions of the bonding in the boron hydrides and related compounds. Such species are cage (or cluster) molecules and their apparent electron deficiency is bypassed by using such three-centred orbitals. Allocating two electrons to each terminal B–Cl bond, in B_4Cl_4 there are two electrons from each boron to allocate to cage bonding molecular orbitals, a total of eight. Figure 15.7 is, now, clearly appropriate. But, conversely, Fig. 15.4, which requires 10 electrons, is not. Or is it? At the end of Section 15.2.1, it was suggested that if the highest E orbitals are sufficiently destabilized, then they will be unoccupied. It seems that this is the case in B_4Cl_4, leading the three-centred bonding model to be preferred.

Fig. 15.7 A schematic energy level diagram for the four face molecular orbitals of P_4.

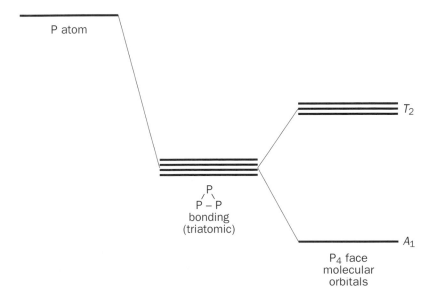

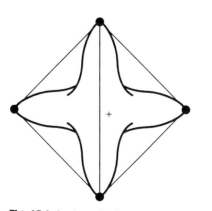

Fig. 15.8 The $3p_z$ orbitals of the P atoms in P_4. Each local z axis is directed radially outward, away from the centre of gravity of the tetrahedron.

Table 15.4 The transformation properties of the phosphorus σ orbitals (p_z, pointing towards the centre of the tetrahedron) in P_4. Characters generated by the four orbitals under the operations of the T_d point group (see Fig. 15.8) which is the sum of the characters of the irreducible representations $A_1 + T_2$

E	$8C_3$	$3C_2$	$6S_4$	$6\sigma_d$
4	1	0	0	2

Fig. 15.9 A schematic picture of the totally symmetric (A_1) combination of the four $3p_z$ orbitals of Fig. 15.8.

15.2.3 Atomic orbital model for P_4

If one were to wish to carry out a detailed molecular orbital calculation on P_4 one would consider, at least, all of the valence shell atomic orbitals on each phosphorus. The local C_3 axis would be chosen as z, and so defining p_z, whilst, as the previous section indicated, there would be considerable freedom about the choice of local x and y. To simplify the problem we shall ignore the phosphorus 3s orbitals. This is not too bad an approximation. As in ammonia, each phosphorus has a lone pair of electrons and the major contributor to this lone pair orbital is 3s; we are simply ignoring the small $3p_z$ contribution. In this model, each phosphorus $3p_z$ orbital in P_4 points towards the centre of the tetrahedron (Fig. 15.8); simple group theory, detailed in Table 15.4, shows that they combine to give $A_1(0)$ and $T_2(1)$ orbitals, the former being bonding and the latter, correspondingly, antibonding. The A_1 orbital is schematically illustrated in Fig. 15.9. We now have to consider the $3p_x$ and $3p_y$ orbitals on each phosphorus atom; here the group theory is not so simple but is outlined in Table 15.5. It is found that T_2(most stable) + E(intermediate) and T_1(least stable) orbitals result. Even with the detailed form of the orbitals (and these are given in the reference given in Table 15.5), it is no trivial matter to obtain the relative energies of these T_1, E and T_2 sets, although that just given seems to be the most general. The detailed pattern is rather dependent on the actual element forming the tetrahedral cluster. These three levels are combined with the $A_1 + T_2$, arising from the orbitals of Fig. 15.8, in Fig. 15.10(a). This diagram is adapted to P_4 and B_4Cl_4 in Figs. 15.10(b) and (c), respectively, where the results of photoelectron spectroscopic measurements on the former species are included (see Section 12.7). In Figs. 15.10(b) and (c) the electron pairs excluded in the cluster bonding discussion have been added (phosphorus lone pairs and B–Cl bonding pairs, respectively).

Table 15.5 The transformation properties of the p_x and p_y orbitals of each phosphorus atom in P_4. Characters generated by the eight orbitals under the operations of the T_d point group which is the sum of the characters of the irreducible representations $T_1 + T_2 + E$

E	$8C_3$	$3C_2$	$6S_4$	$6\sigma_d$
8	-1	0	0	0

Note: it is not a trivial task to generate the above reducible representation. First, the character of -1 under $8C_3$. The -1 is a sum of $(-\frac{1}{2}) + (-\frac{1}{2})$; the calculation is very similar to that detailed in Appendix 4. Secondly, the character of 0 under $6S_4$. To obtain this, it is helpful to remember, as discussed in Section 15.2.2, that there is considerable freedom of choice of x and y axes at any phosphorus atom. The value of 0 ($= +1 - 1$) is most readily obtained if x and y are chosen one to be symmetric and one antisymmetric with respect to reflection in the local σ_d mirror plane. See Appendix 4 of *Symmetry and Structure*, S. F. A. Kettle, Wiley, Chichester, 1995 for a detailed derivation.

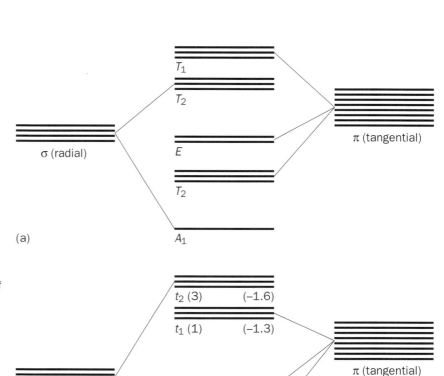

Fig. 15.10 (a) A combined schematic molecular orbital energy level diagram arising from the radially directed σ orbitals (σ(radial)) of a tetrahedron together with the π that are tangential to the local z axes (π(tangential)). (b) Fig. 15.10(a) adapted to P_4. The results of calculations/photoelectron measurements on P_4 are indicated in units of eV. Values in brackets are calculated.

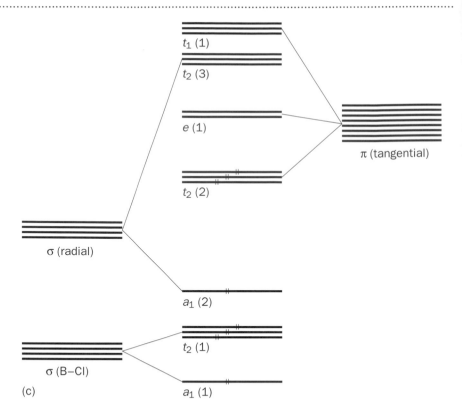

Fig. 15.10 (*continued*) (c) Figure 15.10(a) adapted to B_4Cl_4, where the e(1) level is empty.

15.2.4 Unity of the three models of P_4 bonding

Although they have features in common, Figs. 15.4, 15.7 and 15.10(a), all of which purport to describe the same molecule, P_4, are rather different—why? In Section 15.2.3, the atomic orbital model, all the valence shell atomic orbitals of the phosphorus atoms were included. In Sections 15.2.1 and 15.2.2, describing localized bonding orbitals, just the face and edge bonding molecular orbitals, respectively, were included: their antibonding counterparts were ignored. We shall not discuss the derivations associated with these antibonding orbitals (they involve complications very similar to those mentioned in Section 15.2.3 in connection with the phosphorus $3p_x$, $3p_y$ orbitals) and merely quote the results. These are given in Fig. 15.11. In Fig. 15.11(a) are given approximate energy level diagrams arising from edge and face antibonding orbitals. Note particularly that the stabilization resulting from in-phase interactions between these orbitals can be such as to more than compensate for their original antibondingness and so give overall bonding. In Fig. 15.11(b) the levels are combined with their bonding counterparts. The resulting face-basis and edge-basis energy level diagrams are very similar, not only with each other but also with Fig. 15.10(b). At long last, all three models agree—at least, superficially. In particular, we can now use the three-centred-bond model to explain why P_4 is stable. An *E* molecular orbital, derived from what was originally an antibonding orbital, becomes sufficiently stabilized for it to become bonding (Fig. 15.11(b)). Comparison of the two models represented in Figs. 15.11(b) make understandable the fact that sometimes this orbital is occupied (P_4) and sometimes

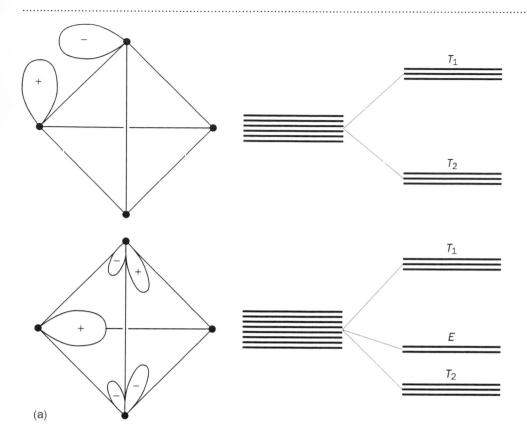

Fig. 15.11 (a) Representative edge (upper) and face (lower) antibonding orbitals of a tetrahedron. The face antibonding orbitals are doubly degenerate and the figure shows both components, one in the left-hand face and the other in the right. If both components are to refer to a single face one component (either) should, mentally, be reflected in a vertical mirror plane containing the vertically drawn edge in the lower diagram.

unoccupied (B_4Cl_4); in Section 15.5 further insights into this E orbital will be gained.

One final point. All three models agree that the lowest lying orbital is of A_1 symmetry but picture it rather differently. In the edge-bond model it is composed of edge bonding orbitals (Fig. 15.3), in the face-bond model it is an in-phase combination of all four of the face-bonding orbitals (we have not attempted to picture it because such a picture would be rather convoluted and complicated) whilst in the atomic-orbital model it is as shown in Fig. 15.9. Yet all three are attempting to describe the same molecular orbital! Are the differences real? No, not if all three models are taken to the same limit—the models differ in the sequence in which interactions are 'switched on', not in the final result. Our conclusion is clear; different models of cluster bonding may produce results which, superficially, are very different but these differences may be illusory.

15.3 Wade's rules

A focal point of interest of clusters such as those shown in Fig. 15.1 is the answer to the question: how do they hold together and what is the bonding? This question seems almost impossibly difficult. One might reasonably assume that the 18-electron rule holds for each transition metal atom but how many electrons can be regarded as metal–metal bonding and how many

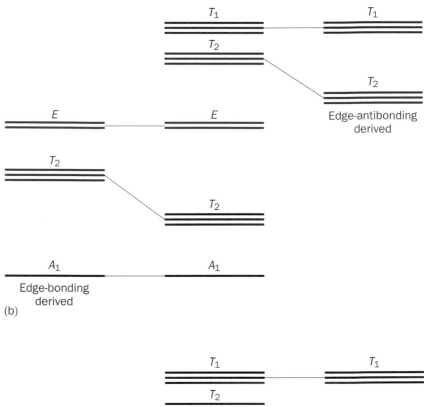

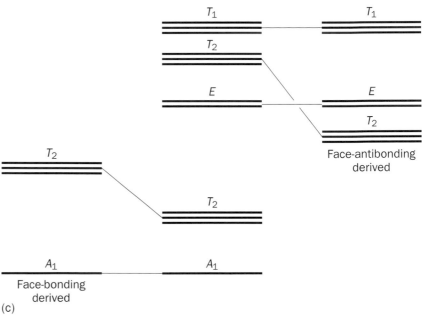

Fig. 15.11 (*continued*) (b) A qualitative energy level diagram which indicates the result of combining edge bonding and antibonding orbitals to give (tetrahedral) molecular orbitals. (c) The corresponding diagram for 'face' bonding and antibonding orbitals, which combine to give (tetrahedral) molecular orbitals. In both upper and lower diagrams the final result is akin to Fig. 15.10(a).

metal–ligand? There might even be orbitals significantly involved both in metal–metal and in metal–ligand bonding. In the event, significant progress has been made. The first step is to decide how many electrons are to be associated with metal–metal bonding, at least as a first approximation.

The insight that provides an answer to this question comes from boron chemistry. There are several hundred boron hydrides and their derivatives; structurally they are based on polyhedra with boron atoms at their apices.

Characteristically, each boron has a terminal hydrogen atom bonded to it, that is, one in which the boron–hydrogen bond points away from the centre of the polyhedron. Some examples are shown in Fig. 15.12. The existence of such a large number of closely related molecules based on a simple common unit, B–H, invites attempts to provide a description of their bonding, applicable to all. Apart from commenting that, just as for B_4Cl_4 in Section 15.2, empty orbitals are important, we shall content ourselves with the simpler question—how many electrons are associated with boron–boron bonding? The number is easy to calculate. Each B–H unit is associated with four electrons, three from each boron and one from each hydrogen; two of these electrons we allocate to the terminal B–H bond. This leaves two electrons per B–H unit for cluster bonding, to which have to be added any formal charge on the molecule (many boron hydrides are anionic). Making use of the experimental observation that anions of general formula $B_nH_n^{2-}$ are always closed complete polyhedra with triangular faces, we conclude that such polyhedra have $(n + 1)$ electron pairs involved in skeletal bonding. Such species are termed *closo* (closed cage). If a molecule differs from a *closo* species by a single B–H unit being absent, it is called a *nido* (nest-like) species. They all contain bridging hydrogens which 'sew up' the edges of the hole caused by the omitted B–H. Such molecules are usually neutral species, of general formula B_nH_{n+4}, but are conveniently thought of as the composite ($B_nH_n^{4-}$ + 4 protons). As the sewn up comment above implies, the four protons are associated with orbitals also involved in cage bonding; we conclude that *nido* species are characterized by $(n + 2)$ skeletal electron pairs. If two cage B–H units are missing from a close polyhedron, one has the *arachno* (web-like) series, of general formula BnH_{n+6} and $(n + 3)$ skeletal electron pairs. A rare series is the *hypho* (net-like) with three B–H units missing to give B_nH_{n+8} and $(n + 4)$ skeletal electron pairs. Clearly, this can be worked backwards; given the formula of a boron hydride species the structure can be deduced with some reliability. Thus, B_5H_9 is B_nH_{n+4} with $n = 5$ and so a *nido* species with one B–H unit missing from the basic polyhedron. The basic polyhedron is therefore an octahedron; we correctly deduce that B_5H_9 has C_{4v} symmetry. One important aspect of these rules (called *Wade's rules*, after the person who first formulated them) is that they may also be applied to metal clusters.

Important in this extension is the recognition that, in the boron hydride discussion above, each B–H unit supplied not only two electrons to skeletal bonding but also three atomic orbitals from the boron. We could think of these as three sp^3 hybrids (the fourth being involved in B–H bonding). However, it will prove more convenient to think of the three orbitals as one sp hybrid (the other sp hybrid being involved in B–H bonding) and the (p_x, p_y) pair. The key point is that the set of three should transform as $A_1 + E$ under local C_{3v} or C_{4v} symmetry (other cases are actually similar—but different symmetry labels apply). Provided that this condition is met, a wide variety of units, with varying electron count, can substitute for B–H in boron hydrides. Indeed, and as is usually the case in transition metal cluster compounds, there need be no B–H units in the cage molecule. Nonetheless, recognition of the boron hydride pattern reveals the general cluster pattern also. So, a C–H unit is isoelectronic with a $(B–H)^-$ unit and has a similar orbital pattern (and, perhaps more important, $C–H^+$ and B–H are related

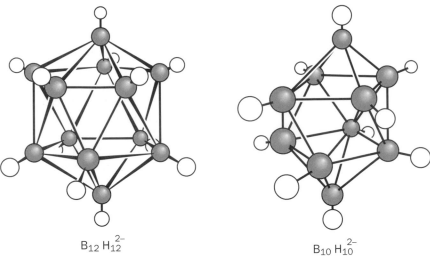

$B_{12}H_{12}^{2-}$

$B_{10}H_{10}^{2-}$

Fig. 15.12 Representative boron hybride molecules. Three are *closo* and two are *nido*.

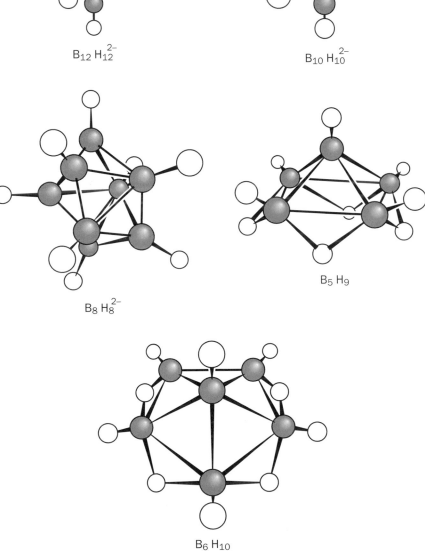

$B_8H_8^{2-}$

B_5H_9

B_6H_{10}

in the same way but with one electron fewer). Following the argument above, it would be expected that $B_5H_5^{2-}$ is a *closo* molecule, and so a trigonal bipyramid. In fact, this anion has not been prepared. Writing it as $B_3H_3(BH)_2^{2-}$, and substituting two BH^- by CH, one concludes that $B_3H_3(CH)_2$, $C_2B_3H_5$, which has been prepared, is also expected to be a trigonal bipyramid, as indeed it is.

Application to transition metal cluster compounds starts with the 18-electron rule, in the sense that we take nine as the number of valence shell metal orbitals of relevance. We reserve three of these, satisfying the $(A_1 + E)$ symmetry requirement given above, for metal–metal bonding, leaving six for bonding of ligands, back-bonding, non-bonding—or anything else. Not surprisingly, we do not attempt to enquire into the detailed characteristics of the six orbitals! All we do is to allocate 12 electrons to them. Any electrons in excess of this number are to be associated with the three orbitals we have reserved for cluster bonding. Consider the $Fe(CO)_3$ unit as an example. The neutral iron atom has a ... $3d^6 4s^2$ configuration—that is, eight valence shell electrons. In applying the 18-electron rule to the iron atom we add two (σ) electrons from each carbonyl ligand to give a total of 14 electrons. Of these, 12 are allocated to the six non-cluster orbitals, leaving two electrons to be associated with the three, $A_1 + E$, cluster orbitals. We conclude that the $Fe(CO)_3$ unit, like B–H, has two electrons and three orbitals associated with cluster bonding. This relationship between B–H and $Fe(CO)_3$ is expressed in the statement that the two units are *isolobal*, a term already met in Section 10.4.3. Ideally, isolobal units have the same number of orbitals of the same effective symmetries, of similar size and energies and, formally, are associated with the same number of electrons. So, because $B_nH_n^{2-}$ and $C_2B_{n-2}H_n$ have *closo* structures, it follows that if we were to add another BH, another CH^+, or another $Fe(CO)_3$ (all three are isolobal) that the resulting molecule would also have a *closo* structure. In particular, it is to be expected that species of general formula $C_2B_nH_{n+2}Fe(CO)_3$ will be *closo*. Such molecules have been prepared and, indeed, are *closo*, the C, B and Fe atoms lying at the apices of a (slightly distorted) polyhedron with $(n + 3)$ vertices.

As an example of the application of Wade's rules to a purely transition metal cluster consider the tetrahedral molecule $Ir_4(CO)_{12}$, where each Ir atom is part of an $Ir(CO)_3$ unit (Fig. 15.13); the preparation of this compound was briefly described in Chapter 4 and, in more detail, in Question 4.4. Because the electron configuration of an Ir atom is ... $5d^7 6s^2$, there are $9 + 6 = 15$ valence electrons associated with each $Ir(CO)_3$ group (the six coming from CO σ orbitals). Placing 12 in the six non-cluster orbitals, three electrons are left to be associated with the three Ir orbitals allocated to cluster bonding. This gives a total of 12 electrons for bonding within the tetrahedron, exactly the same number as for P_4 in Section 15.2. Clearly, the general arguments used in Section 15.2 can be applied to $Ir_4(CO)_{12}$, although the composition of the various orbitals involved will be less clear, for each may have a 4d orbital component. In summary, then, there are 60 $(4 \times 12 + 12)$ electrons associated with the cluster $Ir_4(CO)_{12}$, although only a few of them are formally involved in metal–metal bonding. It is found that the number 60 is a characteristic of tetrahedral transition metal clusters. The recognition of this fact, together with the generalizations implied by Wade's rules, leads us to take an interest in the number of cluster valence electrons (often referred

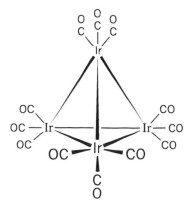

Fig. 15.13 The tetrahedral structure of $Ir_4(CO)_{12}$.

Table 15.6 Cluster valence electron (CVE) counts for some common polyhedra

Geometry	CVE	Example
Tetrahedron	60	$Ir_4(CO)_{12}$
Butterfly (a tetrahedron with one edge open)	62	$[Re_4(CO)_{16}]^{2-}$
Square pyramid	74	$Fe_5(CO)_{15}C$
Trigonal bipyramid	72	$Os_5(CO)_{16}$
Octahedron	86	$Ru_6(CO)_{17}C$
Trigonal prism	90	$[Rh_6(CO)_{15}C]^{2-}$

to as the CVE; an alternative name is skeletal electron-pair count) in a molecule. Indeed, it transpires that this is a useful exercise—different geometries are associated with different CVE counts. Some examples are given in Table 15.6. In all cases it is possible to develop arguments similar to those given above for $Ir_4(CO)_{12}$ to justify the CVE count. So, in the case of 'butterfly' clusters, with two electrons more than the 60 required for a tetrahedron, the additional two are taken to fill an antibonding edge orbital, such as one of those shown in Fig. 15.11(a), thus breaking the edge bond associated with this edge and giving the observed structure.

In this section, Wade's rules have been introduced and the bonding in some tetrahedral cluster molecules considered in outline. In a sense, these two topics are incompatible. Wade's rules do not work for tetrahedral clusters! The tetrahedral molecules we have considered are *closo*, and so, according to Wade's rules, should be derived from the unknown anion $B_4H_4^{2-}$ or, equivalently, $B_4Cl_4^{2-}$. But, of course, we have found that such tetrahedral molecules actually have eight (B_4Cl_4) or 12 (P_4, $Ir_4(CO)_{12}$) cluster bonding electrons, not the 10 indicated by formulae such as $B_4H_4^{2-}$ and $B_4Cl_4^{2-}$. Both the strength and limitations of Wade's rules are at once evident. They are powerful, enabling the majority of simple clusters to be given a common rationalization. However, they do not attempt to provide detailed insights into the bonding in individual molecules and so exceptions to the rules must be expected. Not surprisingly, one of the objects of detailed calculations has been to discover just why Wade's rules work and what are their limitations.

In Table 15.7 are listed the number of cluster electrons associated with units which may be regarded as building blocks for cluster compounds. In each case, three orbitals are allocated for cluster bonding; *closo* structures have $(n + 1)$ skeletal bonding pairs associated with n building units. Similarly, following Wade's rules, *nido* structures have $(n + 2)$ skeletal electron pairs and *arachno*, $(n + 3)$ skeletal pairs, all associated with n cage atoms. An advantage of the use of Wade's rules is evident when a cluster contains bridging ligands—one simply includes in the cluster electron count an appropriate number of electrons per bridging ligand—two for a bridging CO, for example. The calculation is otherwise unchanged.

Wade's rules only apply to clusters which may be regarded as derived from a cage, *closo*, molecule. However, many exotic clusters exist which are much more complicated. When such molecules may be regarded as resulting from the fusion of molecules which, separately, are derived from *closo*

Table 15.7 Transition metal cluster building blocks (Cp $= \eta^5 - C_5H_5$)

Molecular fragments associated with -1 skeletal electrons per unit
$Mn/Tc/Re(CO)_2$; $Cr/Mo/WCp$

Molecular fragments associated with 0 skeletal electrons per unit
$Fe/Ru/Os(CO)_2$; $Mn/Tc/ReCp$; $Cp/Mo/W(CO)_3$

Molecular fragments associated with 1 skeletal electron per unit
$Co/Rh/Ir(CO)_2$; $Fe/Ru/OsCp$; $Mn/Tc/Re(CO)_3$

Molecular fragments associated with 2 skeletal electrons per unit
$Ni/Pd/Pt(CO)_2$; $Co/Rh/IrCp$; $Fe/Ru/Os(CO)_3$; $Cr/Mo/W(CO)_4$

Molecular fragments associated with 3 skeletal electrons per unit
$Ni/Pd/PtCp$; $Co/Rh/Ir(CO)_3$; $Mn/Tc/Re(CO)_4$

Molecular fragments associated with 4 skeletal electrons per unit
$Ni/Pd/Pt(CO)_3$; $Co/Rh/Ir(CO)_4$

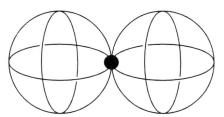

Fig. 15.14 A schematic representation of single-atom fusion between two cages. Only the fusion atom is shown explicitly.

structures then, again, electron counting rules may be applied. Suppose the fusion occurs because one transition metal atom is common to two cages (an example is given in Fig. 15.14). Remembering that the 18-electron rule (Section 10.1) is most probably applicable, the 18 electrons associated with the common atom have been counted twice—once in each cage. The electron count for the entire cluster is, then, the sum of those for the component, separate, cages less 18. When fusion occurs between two transition metal atoms which are bonded to each other and common to both cages then the count is the sum of those for the component, separate, cages less 34 (34 = twice 18 less 2 electrons for the bond between the two atoms common to both cages). Similar counts are applicable to more complicated fusion patterns but will not be discussed—the above two examples serve to establish the pattern.

15.4 Topological models

Surprising as it may seem, aspects of topology and chemistry are closely related. When one draws a line between the carbon and a hydrogen atom in methane, one is indicating a 1:1 correspondence between a topology—a connectedness, a line—and a bonding interaction. This goes even deeper. It is a simple application of group theory to use Table 5.1 to show that the four C–H bonds in methane, or rather the four lines drawn from carbon to hydrogens, span the $A_1 + T_2$ irreducible representations of the T_d point group. The four topological lines lead us to conclude that there is a singly degenerate (A_1) C–H bonding molecular orbital in methane and a triply degenerate set (T_2)—and that there are no others. Similarly, for P_4, in the simple ammonia model, Section 15.2.1, we effectively used the six edges of a tetrahedron to conclude that the interactions responsible for the bonding were of $A_1 + T_2 + E$ symmetries. Not surprisingly, this rather simple and apparently straightforward way of obtaining insights into cluster bonding has been the subject of many enquiries. Unfortunately, the model is not always quite as simple as in the examples above. When discussing P_4 we

Table 15.8 The octahedral character table and octahedral clusters

O_h	E	$6C_4$	$3C_2$	$6C_2'$	$8C_3$	i	$6S_4$	$3\sigma_h$	$6\sigma_d$	8_6
A_{1g}	1	1	1	1	1	1	1	1	1	1
A_{2g}	1	−1	1	−1	1	1	−1	1	−1	1
E_g	2	0	2	0	−1	2	0	2	0	−1
T_{1g}	3	1	−1	−1	0	3	1	−1	−1	0
T_{2g}	3	−1	−1	1	0	3	−1	−1	1	0
A_{1u}	1	1	1	1	1	−1	−1	−1	−1	−1
A_{2u}	1	−1	1	−1	1	−1	1	−1	1	−1
E_u	2	0	2	0	−1	−2	0	−2	0	1
T_{1u}	3	1	−1	−1	0	−3	−1	1	1	0
T_{2u}	3	−1	−1	1	0	−3	1	1	−1	0

Each threefold axis of an octahedron passes through the centre of a pair of opposite faces and each σ_d mirror plane bisects four faces, facts that are apparent in the characters generated by the transformations of the eight faces of an octahedron under the 48 operations of the O_h point group:

$$
\begin{array}{cccccccccc}
E & 6C_4 & 3C_2 & 6C_2' & 8C_3 & i & 6S_4 & 3\sigma_h & 6\sigma_d & 8S_6 \\
8 & 0 & 0 & 0 & 2 & 0 & 0 & 0 & 4 & 0
\end{array}
$$

$$= A_{1g} + A_{2u} + T_{1u} + T_{2g}$$

Similarly, each pair of opposite octahedral edges are bisected by a C_2 axis; these edges are contained in a σ_h mirror plane and are perpendicular to a σ_d, leading to the reducible representation

$$
\begin{array}{cccccccccc}
E & 6C_4 & 3C_2 & 6C_2' & 8C_3 & i & 6S_4 & 3\sigma_h & 6\sigma_d & 8S_6 \\
12 & 0 & 0 & 2 & 0 & 0 & 0 & 0 & 4 & 2 & 0
\end{array}
$$

$$= A_{1g} + E_g + T_{1u} + T_{2g} + T_{2u}$$

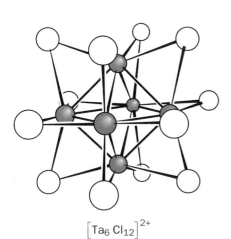

$$\left[\text{Mo}_6\,\text{Cl}_8\right]^{4+}$$

$$\left[\text{Ta}_6\,\text{Cl}_{12}\right]^{2+}$$

Fig. 15.15 The octahedral clusters $[\text{Mo}_6\text{Cl}_8]^{4+}$ and $[\text{Ta}_6\text{Cl}_{12}]^{2+}$ (there are several examples known of each type). In both cases metal atoms are black and chlorines are white circles.

were able to relate bonding orbitals either to tetrahedron edges (Section 15.2.1) or to tetrahedral faces (Section 15.2.2), but not both. For transition metal clusters it is necessary to invoke both simultaneously. Further, as we saw in Section 15.2.2, there is the possibility of delocalization stabilizing what, locally, are antibonding orbitals—so that they may become occupied. This general approach is commonly referred to as the topological equivalent orbital (TEO) model. As a further example of its application we consider the two cluster halide species $[\text{Mo}_6\text{Cl}_8]^{4+}$ and $[\text{Ta}_6\text{Cl}_{12}]^{2+}$, shown in Fig. 15.15, both occurring in solids made by reduction of a higher chloride (so, $[\text{Mo}_6\text{Cl}_8]^{4+}$ occurs in MoCl_2, a compound which is actually $[\text{Mo}_6\text{Cl}_8]^{4+} \cdot 4\text{Cl}^-$). In $[\text{Mo}_6\text{Cl}_8]^{4+}$ the chloride ligands are centred one above each octahedral face—and so each is involved in σ bonding with a face localized orbital. The metal–metal bonding involves localized orbitals which are in a 1:1 correspondence with octahedron edges. It is a simple matter to show (Table 15.8) that the 12 octahedral edges span the $A_{1g} + T_{1u} + E_g + T_{2g} + T_{2u}$ irreducible representations of O_h and so these are the symmetries of the metal–metal bonding molecular orbitals (a conclusion generally supported by more detailed calculations). Use of Table 15.8 also shows that the octahedral faces span the $A_{1g} + T_{1u} + T_{2g} + A_{2u}$ irreducible representations, so these are the symmetries of the chlorine-cluster σ delocalized molecular orbitals, a conclusion similarly supported by detailed calculations. In $[Ta_6Cl_{12}]^{2+}$, where the chlorines are located above octahedron edges, the pattern is reversed. The metal–metal delocalized bonding molecular orbitals, associated with octahedron faces, are of $A_{1g} + T_{1u} + T_{2g} + A_{2u}$ symmetries and the chlorine-cluster σ delocalized molecular orbitals of $A_{1g} + T_{1u} + E_g + T_{2g} + T_{2u}$ symmetries. Electron counts on the two molecules indicate that they are both, effectively, 40-electron systems with all of the orbitals listed above—$2A_{1g} + 2T_{1u} + E_g + 2T_{2g} + T_{2u} + A_{2u}$—being filled with either metal electrons or chlorine.

Although a simple topological model works for the cases just considered, it has to be recognized that these are high-symmetry cases and, consequently, contain symmetry-enforced degeneracies. For low-symmetry molecules (which, for comparison, it is often simplest to think of as an isomer of a high-symmetry case whenever possible) the degeneracies will be split and a scatter of energy levels will occur. In such cases, a component of a (high-symmetry) weakly bonding orbital may become non-bonding or antibonding; similarly, a low-lying (high-symmetry) antibonding orbital may become non-bonding or bonding. Equally, the basis becomes unclear—just how far apart do two metal atoms have to be before one no longer associates an edge bond with them? So, ambiguities weaken the immediate applicability of the model. Nonetheless, the approach remains useful and, in particular, enables face- and edge-bonded ligands to be incorporated within a general discussion of metal–metal bonding.

Another aspect of topology which has been studied and which is not too different, at the beginning at least, from that which has just been discussed, is graph theoretical. Essentially, one draws up a table, a matrix, in which connectivities are listed. Each cluster vertex atom is listed across the top and down the side of the matrix; in the matrix itself a 1 is entered whenever atoms are connected and 0 when they are not; the mathematical properties of the matrix are then considered. If, for example, an additional atom, an additional vertex, is added to a cluster then the size of the matrix increases and a modified connectivity pattern is represented. Comparison of the 'before' and 'after' patterns potentially provides insights into whether such an addition is likely. Although much work has been done, the present impact of the method is not great. This is largely because the precise link between the properties of a matrix and properties of the corresponding molecule is not always clear. However, it is likely that the situation will improve and that the method will assume a greater importance.

One final general conclusion which results from simple topological models—but which reappears in the more complicated models considered in the next section—is to be noted. In a topological model, similar bonds are regarded as equivalent. So, if molecular stability is to be maximized, the more bonds the better. Consider an array of three atoms. If they form a linear chain, M–M–M, there are just two bonds. If they are arranged at the corners of a triangle, then there are three. So, we would expect to find that molecular cages will be built of triangular faces, and, indeed, triangular faces are found to be by far the most common—square faces, for example, are relatively uncommon.

15.5 Free-electron models

One example of the use of something rather akin to a free-electron model (a model in which the electrons are free to wander over some surface, a surface which is free from nuclei and the complications of their associated point charges), has already been met in Section 10.2, where metal–fullerene complexes were considered. This partial duplication of the discussion makes available to the reader an alternative (and simpler) treatment of much the same material as that which follows. It was felt that this would be helpful since the theory to be developed can pose conceptual difficulties for the student.

One of the problems encountered by the newcomer to bonding in cluster molecules is a change in language. Although the labels 'σ' and 'π' are encountered—the former was used in the last section—they are rather rare. The reason is that they are often not helpful. Figure 15.16 shows an octahedron of metal atoms, together with p orbitals on three of them. Between one pair of p orbitals the interaction is largely σ, although it also has a π component, between another pair it is pure π and between the third pair it is non-bonding! Although in an octahedron, application of group theory would help simplify the problem (in the symmetry-adapted linear combinations that result from its use, the non-bonding pattern is not found), the basic ambiguity problem remains, and becomes worse in low-symmetry molecules. Bonding interactions often cannot be spoken of as simple σ or π. In the present section, in particular, such labels will be little used to describe bonding interactions between atoms, although they will be much used in a different way, to differentiate between the orbital types of any one atom.

In *closo*, complete cage, structures all of the atoms of the cluster that are bonded to each other lie approximately, and often accurately, equidistant from the centre of the molecule. That is, they lie on the surface of a sphere. Electrons which are responsible for cluster bonding will lie between the bonded atoms and, so, approximately, will be largely located close to the surface of the sphere. Now, if the sort of orbital which can lie on the surface of a sphere has any particular characteristics, these characteristics are likely to be reflected in the bonding of the cluster atoms. Indeed, as a first

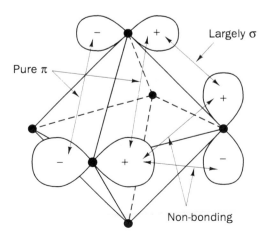

Fig. 15.16 Examples of the variety of interactions that can occur between adjacent π(tangential) orbitals in a tetrahedron.

Largely σ

Pure π

Non-bonding

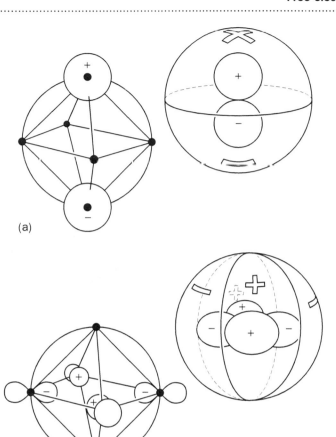

Fig. 15.17 The projection of the atomic orbitals at the centre of a sphere onto the surface of the sphere: (a) a p orbital, (b) a d orbital. In each case a corresponding orbital of an octahedral cluster is shown: (a) a t_{1u}, (b) an e_g.

approximation, we might ignore the cluster atoms altogether and look at the wavefunctions of a free electron on a sphere. Later, of course, we would have to add the cluster atoms, whereupon the symmetry would drop from spherical to that of the cluster and there would be a splitting of some of the high degeneracies which are associated with spherical symmetry. This is the approach which will now be developed. Now, an isolated hydrogen atom has spherical symmetry and so we would expect some sort of connection between the possible wavefunctions of the electrons of a hydrogen atom—orbitals—and the wavefunctions of an electron free to move over the surface of a sphere. Our expectation is justified. There is a 1:1 relationship between the two. In Fig. 15.17 is shown, schematically, how the projection of an atomic orbital onto the surface of a sphere enables one to obtain 'free electron on a sphere' functions from those of the corresponding atomic orbital. The degeneracies of the hydrogen orbital functions, 1(s), 3(p), 5(d), 7(f) etc. are also the degeneracies of the free electron on a sphere functions. To determine how these degeneracies are reduced by the presence of the cluster atoms data such as that in Table 7.6 can be used (replacing upper case symbols by lower case). However, it has been found possible to make considerable progress even if this apparently essential step is omitted!

This progress is dependent on the recognition that there is an enormous weakness in the free electron on a sphere model that has so far been developed. To see this weakness, readers are asked to imagine themselves at the centre of the sphere, looking out at the atoms lying on the sphere which form the cluster molecule. The atomic orbitals of each atom are classified from this viewpoint. A common classification is thus used for all of the cluster atoms, which is clearly an advantage. Further, at each atom, the surface of the sphere is perpendicular to our direction of vision. This means that our classification is particularly appropriate to a 'surface of a sphere' model. From this viewpoint, the atomic s, p and d orbitals are either σ in type (s, p_z or d_{z^2}, where we take our axis of view as the z axis appropriate to the atom at which we are looking), π (p_x, p_y; d_{xz}, d_{yz}) or δ (d_{xy}, $d_{x^2-y^2}$)—and this is the meaning that will be attached to the labels σ, π and δ for the remainder of this section. It is for the latter two types that our model is deficient. They are characterized, from our viewpoint, by having nodal planes which contain the atom. Now, in our model, the cage atoms are scattered, usually in a fairly regular way, over the surface of the sphere. This relationship provides no guarantee that the electron-on-a-sphere wavefunctions such as those shown in Fig. 15.17 will have the right number of nodal planes in the right places to conform with the inherent nodal plane requirements of π and δ type orbitals. The wavefunctions of Fig. 15.17 are only applicable to σ orbitals, with no inherent nodal requirements perpendicular to the surface of the sphere. To be able to proceed we need functions akin to those of Fig. 15.17 but which conform to the nodal requirements of π and δ orbitals on atoms placed at any point on the surface of the sphere. Such functions have been calculated by Stone, and carry the rather off-putting name of *tensor surface harmonics*. We shall give pictures of some of them shortly, but first turn to a point which will become of key importance.

The point is the rather obvious one that, on each atom, each π set has two members—p_x and p_y, for example, as too does each δ set. Members of a π set are related to each other by a 90° rotation about the z axis (centre-of-sphere to metal atom); members of a δ set by a 45° rotation about the same axis. To see the significance of this it is simplest to consider a particular example. It turns out that there is one that we have already considered in sufficient detail. In Chapter 6 (Section 6.2.2) we discussed the ligand π orbitals in an octahedral metal complex, ML_6. If we remove the metal we are left with the π orbitals, defined appropriately for the present application, that of a M_6 octahedral cage. There are two π orbitals on each M, a total of twelve. They transform as $T_{1g} + T_{1u} + T_{2g} + T_{2u}$ under the operations of the O_h point group and representative examples of the first three were given in Figs. 6.13(a), 6.13(b) and 6.16 (the T_{2u} can be obtained from the T_{1u} by changing the phases of two *trans* p_π orbitals). If the reader refers to these figures, reproduced and modified as Fig. 15.18, he or she will be able to see an interesting point related to the 90° rotation that interconverts the two separate p_π orbitals. Consider the t_{1u} combination of Fig. 15.18. If each p_π orbital in it is rotated 90° in the same sense (clockwise or anticlockwise, about the centre-of-sphere to metal axis) then the t_{1g} combination results (possibly with all signs changed). The t_{2g} and t_{2u} combinations are similarly interconverted. The symmetry-adapted combinations of p_π orbitals all occur in pairs, $t_{1u} \leftrightarrow t_{1g}$ and $t_{2u} \leftrightarrow t_{2g}$. Now, the interactions between adjacent orbitals in the t_{1u} combination is in-phase and so bonding. Because of this, the interactions between adjacent orbitals in the paired, t_{1g},

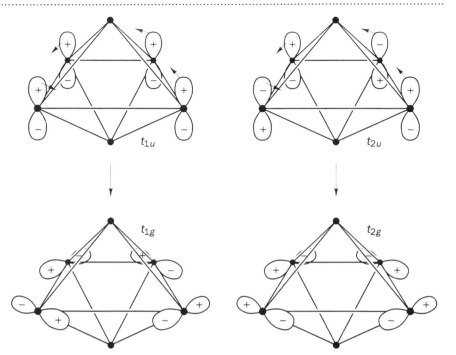

Fig. 15.18 Rotation in the same sense by 90° (indicated by small arrows in the upper diagrams) converts a (bonding) t_{1u} combination into an (antibonding) t_{1g}. Conversely, an (antibonding) t_{2u} is rotated into a (bonding) t_{2g}.

combination are antibonding. Similarly, t_{2g} is bonding but, as a consequence, t_{2u} is antibonding. Not only do the symmetry-adapted combinations appear in pairs, but one member is bonding and the other antibonding. One should not argue too much from a single example—and, after all, we are placing no restrictions on the position of atoms on the surface of the sphere—but it transpires that this conclusion is general, applying to π, δ and all higher functions. There is a pairing, such that for every symmetry-adapted combination that is bonding there is another which is antibonding, unless rather special circumstances—which will immediately become apparent—exist.

To illustrate the insight that this pairing pattern provides we consider the tetrahedron. Here, as has been seen (Section 15.2.3) the metal π orbitals transform as $T_1 + T_2 + E$. Clearly, the $T_1 \leftrightarrow T_2$ sets can be (and are) interrelated by 90° rotations, but what of the E? Here, it is the two separate E functions that are related one to another by a set of 90° rotations, as shown in Fig. 15.19. But, since the two E functions are symmetry-required to be degenerate, it is ridiculous to think of one as bonding and the other as its antibonding counterpart. The only way out is for them, as a pair, to be non-bonding. In a real-life molecule the E functions are unlikely to be exactly non-bonding but, rather, slightly bonding or slightly antibonding. We can at once understand why sometimes the E set is empty (as in B_4Cl_4) and sometimes full (as in P_4 and $Ir_4(CO)_{12}$).

So far we have proceeded by means of examples. What of the general case? As far as the 'from the centre of the sphere defined' σ orbitals are concerned, the 'free electron on a sphere' functions are simply projections of atomic orbitals, like those shown in Fig. 15.17. In energy order sequence, which is just the order of nodal counts, they are

$$S_\sigma(0), P_\sigma(1), D_\sigma(2), F_\sigma(3), \ldots$$

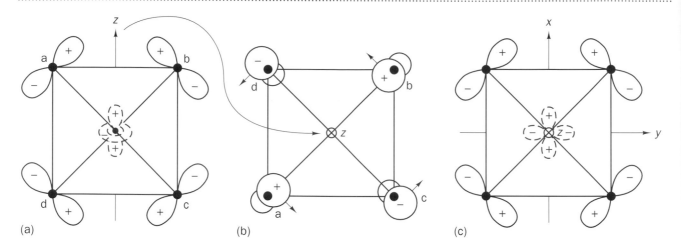

(a)   (b)   (c)

Fig. 15.19 (a) An e orbital combination which transforms like the d_{z^2} of a (hypothetical) atom at the centre of the tetrahedron. It is viewed perpendicular to the z axis. (b) The same conbination but viewed along z. The direction of rotation (see the text) is indicated by small arrows. This rotation, by 90°, gives (c), a combination which transforms as $d_{x^2-y^2}$ of an atom at the centre of the tetrahedron.

Here, S, P and D etc. have their usual meanings but (hopefully), for clarity, the number of inherent nodal planes have been added in brackets (the number is of those that are perpendicular to the surface of the sphere). Again, as usual, the degeneracies are given by $(2L + 1)$, where L is the total number of nodal planes. For each orbital on a cluster atom that is σ-like when viewed from the centre of the sphere, the above sequence holds. So, for each cluster atom, which we shall take to be a first row transition metal, it holds separately for 4s, $4p_z$ and $3d_{z^2}$ (z being the axis from the centre of the polyhedron to the atom). As we have seen, the π orbitals ($4p_x$, $4p_y$ and $3d_{xz}$, $3d_{yz}$) are both more complicated and interesting. In particular, they have the pairing properties discussed above. Because the π orbitals are inherently nodal they can never give a node-free pattern. Nodes can be added but not lost. Now, π orbitals are rather like vectors (a clear analogy can be seen between a $\oplus - \ominus$ pattern and the symbol $\leftarrow$) and so we shall use a vector arrow representation in our diagrams. It is important to recognize that on the surface of a sphere, as on any other surface, the vectors in an array do not have to be either colinear or parallel; they can be rotated relative to each other. In our case this freedom allows the members of a vector set to turn round—for one member to point in a slightly different direction to another, to rotate about an axis—without the intervention of a nodal plane. Simultaneously, the relative amplitudes of adjacent vectors can vary, reflected in the size of the arrows, so that pictures of the tensor surface harmonics tend to appear frighteningly complicated (but are not, unless the reader panics). Taking just one member of each pair of p_π orbitals the following are spanned (again including the number of nodal planes); note the way that, because we are dealing with π functions, the label S does not appear

$$P_\pi(1), D_\pi(2), F_\pi(3), \ldots$$

where one node is inherent in each case (we *are* dealing with p_π orbitals) and the others are global. The degeneracy is again $(2L + 1)$, where L includes all nodes, be they inherent or not. In Fig. 15.20(a) we show the three members of the $P_\pi(1)$ surface tensor harmonic. To make these more understandable, against each left-hand diagram (which applies to all polyhedral clusters, no

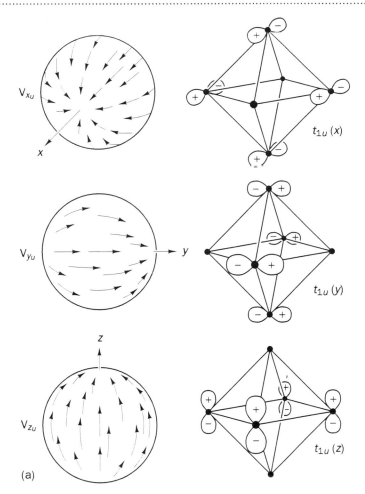

Fig. 15.20 (a) The $\bar{P}_\pi(1)$ set of tensor surface harmonics together with (right) their application to an octahedral cluster.

matter how many atoms they contain) is, on the right, given its application to the octahedral case. We shall discuss this latter case in more detail shortly. For the moment, before continuing further in this chapter, the reader should carefully study the relationship between corresponding left- and right-hand diagrams in Fig. 15.20(a); it may be helpful to look back at Fig. 15.17.

The corresponding, paired, rotated π tensor surface harmonics are conventionally distinguished from the original by the addition of a bar above each symbol and so are

$$\bar{P}_\pi(1); \bar{D}_\pi(2); \bar{F}_\pi(3), \ldots$$

Again, degeneracies are $(2L + 1)$. In Fig. 15.20(b) is shown the $\bar{P}_\pi(1)$ set and, again, at the right, the corresponding octahedral orbitals. For both parts of Fig. 5.20 on the right-hand side it is p_π orbitals of the octahedral cluster which have been shown—d_π could have been used as they cover similar sets, but the pictures would have been more complicated. In Fig. 15.20 note the way that for each vector (arrow) diagram in Fig. 15.20(a), in the corresponding diagram in Fig. 15.20(b) the vectors (arrows) are rotated by 90°. As has been noted above, concomitantly, the octahedral t_{1u} set becomes t_{1g}. A similar pattern is shown in Figs. 15.21(a) and (b), which show the $D_\pi(2)$

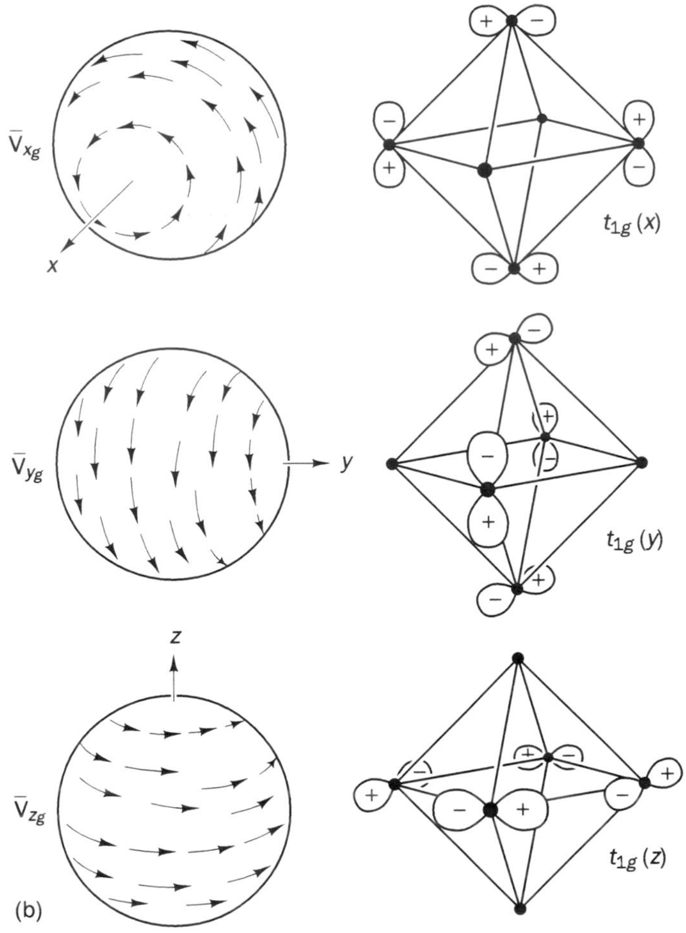

Fig. 15.20 (continued) (b) The $\bar{P}_\pi(1)$ set of tensor surface harmonics. Notice that each arrow is perpendicular to that in the corresponding $P_\pi(1)$ function. To the right is shown the octahedral cluster application. Again, orbitals are 90° rotations of those in (a).

and the $\bar{D}_\pi(2)$ functions. In both Figs. 15.20 and 15.21 approximate Cartesian labels for the individual surface tensor harmonics are given, although there is no accepted convention for these labels.

To complete the picture, δ orbitals ($d_{x^2-y^2}$, d_{xy}) on the cluster atoms have to be included. Again, there is an original-rotated set separation (but, for these functions the rotation needed to interconvert the two sets is 45°). Because δ orbitals are characterized by two nodal planes perpendicular to the surface of the sphere, we cannot have combinations with fewer. The combinations are therefore

$$D_\delta(2), F_\delta(3), G_\delta(4), \ldots$$

and

$$\bar{D}_\delta(2), \bar{F}_\delta(3), \bar{G}_\delta(4), \ldots$$

with degeneracies of $(2L + 1)$, all nodes, local and global, being included in the count. Of the above sets, in Fig. 15.22 only representative $D_\delta(2)$ and $\bar{D}_\delta(2)$ combinations are shown, along with corresponding orbitals for an octahedral cluster. In Fig. 15.22 the initial T used in the labelling of the functions may be thought of as a distorted form of + (the lobe pattern of a δ orbital) but

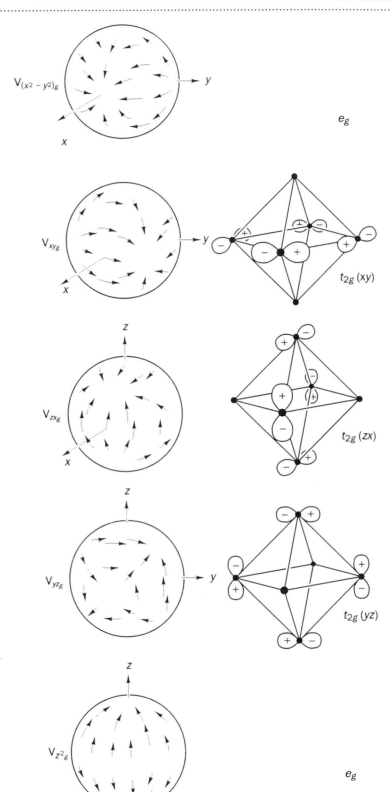

Fig. 15.21 (a) The $D_\pi(1)$ set of tensor surface harmonics together with, where appropriate (right) their application to an octahedral cluster.

$V_{(x^2-y^2)_g}$

e_g

V_{xy_g}

$t_{2g}(xy)$

V_{zx_g}

$t_{2g}(zx)$

V_{yz_g}

$t_{2g}(yz)$

$V_{z^2_g}$

e_g

(a)

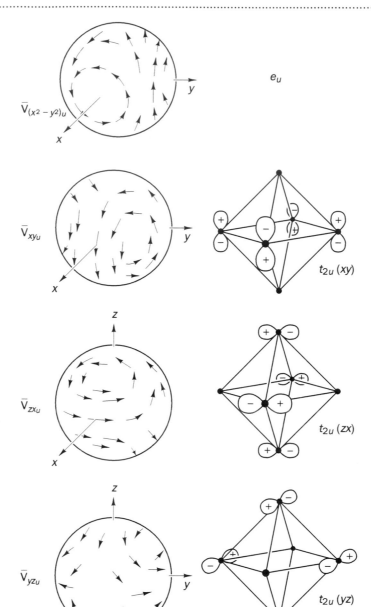

Fig. 15.21 (*continued*) (b) The corresponding $\bar{D}_\pi(1)$ set of surface tensor harmonics and (right) their application to an octahedral cluster. Note the 90° rotation patterns between corresponding $D_\pi(1)$ and $\bar{D}_\pi(1)$ functions.

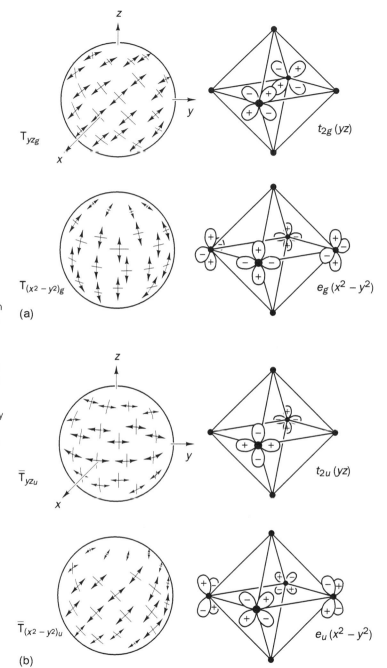

Fig. 15.22 Representative (a) $D_\delta(2)$ and (b) $\bar{D}_\delta(2)$ tensor surface harmonics together with (right) their application to an octahedral cluster. The Cartesian labels given refer to the axes of the first diagram. However, the orbitals at each cluster atom refer to local axes. In each case, local z is radial (i.e. directed outwards from the centre of gravity of the octahedron) and local x and y perpendicular to z. These local axes are taken to be parallel to the axes in the first diagram, although axis labels are not necessarily maintained. So, a local x may be parallel to global z. This pattern means that the e_g and t_{2u} octahedral functions are based on local $d_{x^2-y^2}$ orbitals, the e_u and t_{2g} on local d_{xy} (see also Table 15.10).

actually is a shorthand for tensor, just as V was for vector in Figs. 15.20 and 15.21. Beware of confusion: all the functions under discussion in this section are collectively called 'tensor surface harmonics' but the name 'tensor' also reappears to describe the specific functions with the T label.

Although, within this model, all functions of a given type, the seven $F_\delta(3)$ for example, are treated as degenerate, not all of them may exist for a given cluster. The actual symmetries spanned by the σ, π and δ orbitals on the cluster atoms has to be worked out for each case. Correlations then have to be made with the tensor surface harmonic functions given above, or rather their components, in the symmetry of the actual complex. Let us look at a specific example. In Figs. 15.20 and 15.21 an octahedral cluster was used as an example and we will continue with this choice. In doing so we will tidy up some loose ends left at the end of our topological equivalent orbital description of $[Mo_6Cl_8]^{4+}$ and $[Ta_6Cl_{12}]^{2+}$. To this end two things have to be done. First, the general discussion has to be applied to the octahedral case. Second, the symmetry properties of the σ, π and δ orbitals on the transition metal atoms have to be determined. The first of these tasks is not difficult. The data in Table 7.6 has to be applied to the spherical tensor harmonic functions listed above, although our notation must first be extended to distinguish between the different sets arising from the various metal orbitals of π symmetry, for instance. This is easy. Instead of talking about $D_\pi(2)$, for example, we will now call this D_π^p, D_π^d or D_π^f, where we have added p, d or f to indicate the sort of orbitals on the metal atoms which are involved (although not much use of D_π^f is to be expected!) and compensated by dropping the (2). The results are given in Table 15.9, which is quite horrendous in appearance. This is because it applies to all octahedral clusters, including octahedral clusters with dozens of metal atoms. We are only concerned with the simplest octahedral cluster and so only a small amount of the information in Table 15.9 is relevant. However, the complete Table 15.9 has a basic pattern which both makes it worthy of inclusion and of some study. The second thing which we have to do, the determination of the symmetry pattern of the σ, π and δ orbitals, is not too difficult either. We have already discussed in Sections 6.2.1 and 6.2.2 the σ and π bonding associated with an octahedral array of atoms. The fact that it is ligand orbitals which are discussed in these sections is irrelevant—transformation properties do not depend on the chemical nature of an atom and so the results needed are those given there. The discussion has, however, to be extended to include δ functions. There are two of these on each atom, $d_{x^2-y^2}$ and d_{xy}, using the coordinate axes defined in the caption to Fig. 15.22. No

---

Table 15.9 The symmetry properties of the surface tensor harmonics in the octahedral group O_h

Tensor surface harmonic	Corresponding rotated-basis tensor surface harmonic
$S_\sigma(0)$ $[S_\sigma^s, S_\sigma^p, S_\sigma^d]$: A_{1g}	–
$P_\sigma(1)$ $[P_\sigma^s, P_\sigma^p, P_\sigma^d]$: T_{1u}	–
$D_\sigma(2)$ $[D_\sigma^s, D_\sigma^p, D_\sigma^d]$: $T_{2g} + E_g$	–
$P_\pi(1)$ $[P_\pi^p, P_\pi^d, P_\pi^f]$: T_{1u}	$\bar{P}_\pi(1)$ $[\bar{P}_\pi^p, \bar{P}_\pi^d, \bar{P}_\pi^f]$: T_{1g}
$D_\pi(2)$ $[D_\pi^p, D_\pi^d, D_\pi^f]$: $T_{2g} + E_g$	$\bar{D}_\pi(2)$ $[\bar{D}_\pi^p, \bar{D}_\pi^d, \bar{D}_\pi^f]$: $T_{2u} + E_u$
$F_\pi(3)$ $[F_\pi^p, F_\pi^d, F_\pi^f]$: $A_{2u} + T_{1u} + T_{2u}$	$\bar{F}_\pi(3)$ $[\bar{F}_\pi^p, \bar{F}_\pi^d, \bar{F}_\pi^f]$: $A_{2g} + T_{1g} + T_{2g}$
$D_\delta(2)$ $[D_\delta^d, D_\delta^f, \ldots]$: $T_{2g} + E_g$	$\bar{D}_\delta(2)$ $[\bar{D}_\delta^d, \bar{D}_\delta^f, \ldots]$: $T_{2u} + E_u$
$F_\delta(3)$ $[F_\delta^d, F_\delta^f, \ldots]$: $A_{2u} + T_{1u} + T_{2u}$	$\bar{F}_\delta(3)$ $[\bar{F}_\delta^d, \bar{F}_\delta^f, \ldots]$: $A_{2g} + T_{1g} + T_{2g}$
$G_\delta(4)$ $[G_\delta^d, G_\delta^f, \ldots]$: $A_{1g} + E_g + T_{1g} + T_{2g}$	$\bar{G}_\delta(4)$ $[\bar{G}_\delta^d, \bar{G}_\delta^f, \ldots]$: $A_{1u} + E_u + T_{1u} + T_{2u}$

Table 15.10 Transformation properties of metal orbitals in an M_6 octahedron (see the caption to Fig. 15.22 for a definition of x and y axes)

Symmetry relative to an axis from the octahedron centre	Symmetry species spanned
σ	$A_{1g} + F_g + T_{1u}$
π	$T_{1g} + T_{1u} + T_{2g} + T_{2u}$
$\delta(x^2 - y^2)$	$A_{2g} + E_g + T_{2u}$
$\delta(xy)$	$A_{2u} + E_u + T_{2g}$

symmetry operation relates members of the $d_{x^2-y^2}$ set to members of the d_{xy}, so they can be considered separately. The collected results are given in Table 15.10.

Some progress can be made by bringing Tables 15.9 and 15.10 together. Only entries in Table 15.9 which appear in Table 15.10 are to be considered. Thus, because in Table 15.10 the σ basis contains no T_{2g} entry we can ignore the T_{2g} entry against $D_\sigma(2)$ in Table 15.9. Further, because of the nodal pattern sequence of these entries in Table 15.9, it can be concluded that the energy level stability pattern of the orbitals in Table 15.10 is $A_{1g} > T_{1u} > E_g$. We are making progress! To make more progress, only the d orbitals on the six metal atoms will be considered. This simplifies the problem and is actually much less restrictive than it seems. Inclusion of metal s and p orbitals would simply repeat (at higher energies) symmetry species that we shall meet with a d orbital-only basis. The way that we will handle the problem with this latter basis is the way that we would have handled it also for s and p functions.

Of the two species $[Mo_6Cl_8]^{4+}$ and $[Ta_6Cl_{12}]^{2+}$ shown in Fig. 15.15 only the former will be considered. Surprisingly, it is the more complicated and the latter, the simpler, is set as a problem at the end of this chapter. Actually, the discussion is the same as far as d_σ and d_π orbitals are concerned—both correspond to cylindrical electrical density distributions about the local C_4 axis at each metal atom. Unlike the d_π, the d_δ functions on each atom do not transform as a pair and so it is necessary to decide which of the pair it is appropriate to use in a particular discussion. For $[Mo_6Cl_8]^{4+}$ the d_{xy} (pointing towards the faces of the Mo_6 octahedron) are clearly involved in Mo–Cl bonding (the Cl atoms lie above these octahedron faces). So, as our concern is the direct metal–metal bonding we shall only include the $d_{x^2-y^2}$ functions in our discussion. This discussion is conveniently developed from Fig. 15.23. To the left in this figure are indicated the six d_σ, 12 d_π and six $d_\delta(x^2 - y^2)$, in order of increasing nodality. To their right are drawn the levels arising (Table 15.8) modified by the requirement that only entries appearing in Table 15.9 are acceptable (the unacceptable are indicated in brackets). The pictures of Figs. 15.20 and 15.21 have been used as a guide in placing the bonding orbitals below their antibonding counterparts, the greatest stability coming from interactions with a σ component. So, for the d_π orbitals the T_{2g} level is placed below the T_{1u} because the former involves mixed σ–π bonding between adjacent cluster atoms, whereas the latter is π-only bonding (see Fig. 15.16). Finally, to the right of Fig. 15.23, levels of the same symmetry are allowed to interact with each other. An example of this is shown in Fig. 15.24 where the $T_{2u}(\pi)$–$T_{2u}(\delta)$ interacting orbitals are shown. It is clear that they overlap and that a bonding and an antibonding combination will result. This conclusion is general—interaction between levels leads to the lower being stabilized and the higher being destabilized. The energy level pattern, and orbital occupancy, for the metal–metal bonding in the $[Mo_6Cl_8]^{4+}$ cation shown at the right of Fig. 15.23 results. This pattern contains one point which we could not have predicted—a crossing over between T_{2u} and T_{1u} levels such that the former is the HOMO and the latter is the LUMO. Although such a crossing is entirely consistent with the argument that has been developed, the strong argument in its favour, confirmed by detailed calculations, is the fact that it

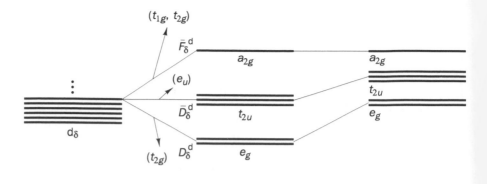

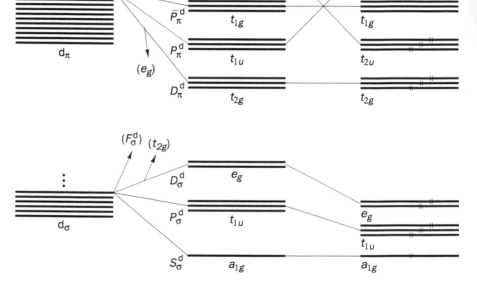

Fig. 15.23 Application of the tensor surface harmonic model to the metal–metal bonding in $[Mo_6Cl_8]^{4+}$. The d_σ, d_π and d_δ sets (left) are infinite in extent in this model but only the number of them needed for discussion of the bonding in the Mo_6 cluster is shown. Similarly, although in the model, levels such as S_σ^d, P_σ^d, D_σ^d, ... are degenerate, they are split apart in the central part of the diagram, a greater number of nodes being taken to indicate a higher energy. The central diagram, then, shows a first approximation to the orbital energy levels in the octahedral Mo_6 cluster. Interaction between levels of the same symmetry leads to the final orbital energy level sequence shown at the right.

is predicted by the topological model developed in the previous section. Not for the first time we find that different models complement each other. The advantage of the spherical tensor model is its general applicability—it can be expected to give insights into the cage bonding in any near-spherical cluster species. Its disadvantage is the fact that it includes no atoms! It seems that this disadvantage is surprisingly unimportant and that fine-tuning of the sort used at the end of our discussion of the bonding in $[Mo_6Cl_8]^{4+}$ adequately compensates.

We conclude this section by using the results obtained to justify the Wade's rule count of $(n + 1)$ skeletal electron pairs needed to bond a *closo* cluster of n atoms. The justification rests on two features. First, that the π orbital interactions give rise to two sets (we initially called them $P_\pi(1)$, $D_\pi(2)$, ... and $\bar{P}_\pi(1)$, $\bar{D}_\pi(2)$, ...); when the members of one set are bonding, the corresponding members of the other set are antibonding. Now the total number of π orbitals is $2n$, where n is the number of cluster atoms,

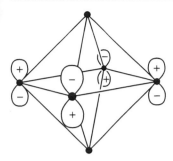

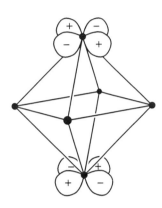

Fig. 15.24 The T_{2u} combinations of π- and δ-type functions that are shown as interacting in Fig. 15.23. Note that, for simplicity, the π-type function shown is p_π; in fact, a d_π is the one involved in Fig. 15.23.

so we may reasonably anticipate that we have just n potentially bonding orbitals derived from the π basis (these orbitals spanning subsets within the general P_π, D_π etc. functions available for the geometry of the particular cluster). Comparison with the transformations of the σ (which give rise to S_σ, P_σ, D_σ, . . .) shows that the only symmetry type within the latter which is not present from the π orbitals is S_σ. Because this is spherically symmetric it is an in-phase combination of all of the σ basis functions and so is certainly bonding and therefore occupied. All of the other symmetries arising from the π set are found in the σ; within each pair with the same symmetry label one will be bonding and one antibonding; similar arguments hold for the δ. We have then, $(n + 1)$ bonding pairs—the number required by Wade's rules. It is worth dwelling on the reason why the P_π^s, D_δ^d etc. are unoccupied. As just indicated, the answer is something which we recognized earlier when talking of $[Mo_6Cl_8]^{4+}$—that when interaction occurs between two orbitals of the same symmetry species, the result is that one becomes more bonding and one more antibonding. So, what we have so far called P_σ bonding orbitals will contain some P_π component (and also D_π) but the actual number of bonding molecular orbitals is unchanged—it is $(n + 1)$, and this is just Wade's rule applied to a closed cage molecule. It is to be recognized that the argument that has just been used is one that enables bonding orbitals to be counted. It is not really an argument about bonding.

A partial check on the correctness of the approach used in obtaining the $(n + 1)$ bonding electron pair count, Wade's rule, is provided by the tetrahedral and octahedral molecules that have been considered in some detail in this chapter. If consideration is confined to cage-bonding orbitals, in no model was the same symmetry species spanned twice. This is an expression of the argument used above. Of course, in lower symmetry molecules a similar statement could probably not be made but, even here, if the relevant orbitals are correlated back to their counterparts in Figs. 15.18 and 15.22, it will almost always be found that the exclusion we have invoked holds at the low-symmetry level. Further illustrations of the way that levels of like symmetry interact so as to lead to only one bonding combination will be found in the way that figures such as Fig. 15.4 combine with Fig. 15.11(a) to lead to a final pattern such as that in Fig. 15.10(b); similarly Figs. 15.7 and 15.11(a) also combine to give the second pattern in Fig. 15.11(b).

15.6 Detailed calculations

As has been mentioned already, the task of carrying out detailed molecular orbital calculations on cluster compounds is difficult in the extreme. They contain far too many atoms and far too many electrons! The common way that has been followed to attempt to circumvent the problem is that of carrying out a calculation on the metal cage (with no ligands) and also one on the ligand cage (with no metals) and then bringing the two together. It is interesting that two distinct approaches along these lines have developed— one coming from physics and the other from chemistry (that of this chapter). The physicists have been interested in the answer to questions such as: how small can a cluster of metal atoms be and yet show metallic properties? For instance, how many iron atoms are needed before they show magnetic properties akin to those of bulk iron? The answer to such questions seems to

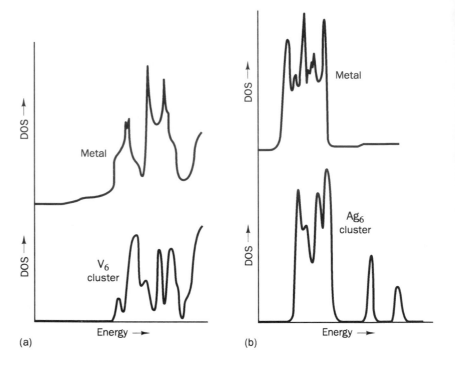

Fig. 15.25 Density of state (DOS) energy patterns for (a) vanadium metal and V_6 (b) silver metal and Ag_6. Adapted and reproduced with permission from G. Seifert and H. Eschrig *Phys. Stat. Sol.* (b) (1985) *127*, 573.

depend on the quantity used as a criterion—anything from four (electrical conductivity) to a hundred (ionization energy). It has been commented that 'the solid does not emerge all at once, but like the Cheshire cat, it fades into view slowly, with the smile appearing first'. The physicists have used models rejoicing in names such as 'the droplet model' and 'the jellium model'— names which are reasonably self-explanatory and which find experimental justification in the metal–atom mobility within some clusters observed by (metal atom) NMR. For the chemist, there is one problem with this approach. Because in bulk metals there are so many energy levels one gives up all attempts to list them and talks instead of density of states (DOS); so, for small metal clusters the same language is often adopted by physicists. As a result, the chemist may well find when reading the physics literature that no use is made of the symmetry labels used throughout this chapter. Figure 15.25 shows a comparison of DOS patterns for bulk metal and octahedral clusters, calculated using the Xα method. Although there are clear differences, the gross similarities are perhaps more evident.

Assuming that most bare metal clusters can be taken to have some metal-like properties, the next question is the extent to which these properties persist in metal cluster compounds. We are immediately plunged into an area of controversy. Whilst no ferromagnetic cluster compounds of iron are known (although some osmium cluster complexes display paramagnetism) this is temperature independent paramagnetism (TIP) in origin— see Section 9.3); this is scarcely relevant to the real issue. In bulk transition metals almost all of the cohesive energy—the bonding energy—comes from interaction between d orbitals. The s and p have little direct involvement. Is this also true of cluster compounds? Extended Hückel calculations tend to say no while Xα (and the related, but better, density functional) method tends

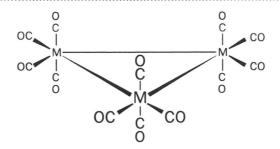

Fig. 15.26 The D_{3h} structure of $M_3(CO)_{12}$, M = Ru or Os (the iron compound has a different, but related, structure which is discussed in Section 12.5.3).

towards yes. Both, speaking from their own corner, offer explanations of Wade's rules. Does this matter? Quite possibly not; it depends on the detail one wishes to explain. So, for the compounds $Os_3(CO)_{12}$ and $Ru_3(CO)_{12}$ (shown in Fig. 15.26) the Xα method gives the sequence of highest occupied molecular orbitals (in D_{3h} symmetry) as

$$a_1' < a_2'' < e'' < a_1'' < e'' < e' < e' < a_1'$$

whereas extended Hückel gives the same set in a different order

$$a_1' < a_2'' < e' < e'' < a_1'' < e' < e'' < a_1'$$

Of these, the highest e' and a_1' are largely involved in metal–metal bonding (note that these are also the irreducible representations generated by the M–M bonds between the three $M(CO)_4$ groups in $M_3(CO)_{12}$). In the photoelectron spectra (that of $Os_3(CO)_{12}$ is shown in Fig. 15.27) ionization from the metal–metal bonding orbitals is associated with the lowest energy peak, arrowed in Fig. 15.27. Ionization from the other levels listed above accounts for the two peaks on the immediate right of that arrowed; the orbitals involved are predominantly d, with significant CO bonding involvement. The intense peaks with energies greater than 12 eV in Fig. 15.27 are largely associated with ionizations from the CO groups. Looking at spectra such as that in Fig. 15.27, which is for a small, high-symmetry cluster, together with the orbital energy sequences listed above, one can understand the attractions of a density-of-states approach! However, if one wishes to interpret electronic spectra then a detailed knowledge of the occupied and lowest unoccupied orbitals becomes essential.

Because of the inherent problems associated with calculations on large clusters, most of the detailed calculations that have been performed concern diatomic systems such as the $[Re_2Cl_8]^{2-}$ anion. This anion has a short metal–metal bond and the two $(ReCl_4)$ units are eclipsed (Fig. 15.28(a)). Since the short metal–metal bond can only increase steric interactions between the two sets of Cl_4 ligands, a staggered configuration might confidently be predicted! The only evident explanation for the observed eclipsed arrangement is the presence of significant δ bonding between the two rhenium atoms (Fig. 15.28(b)). Does it occur? The answer given by the best calculations (usually Xα) is yes ... but! The 'but' occurs because of a problem we have met several times in this book—electron correlation. Indeed, one of the workers in the field has written an article entitled 'Problems in the theoretical description of metal–metal multiple bonds or

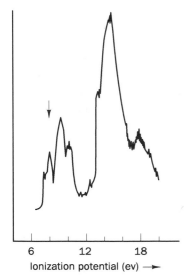

6   12   18

Ionization potential (ev) ⟶

Fig. 15.27 The photoelectron spectrum of $Os_3(CO)_{12}$. Adapted and reproduced with permission from J. C. Green, D. M. P. Mingos and E. A. Seddon, *Inorg. Chem.* (1981) **20**, 2595.

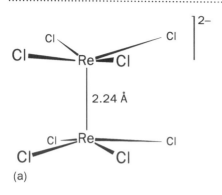

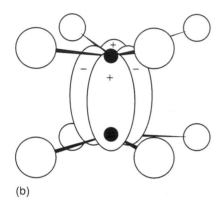

Fig. 15.28 (a) The eclipsed structure of $[Re_2Cl_8]^{2-}$. (b) The Re–Re δ bond which would be broken were the anion to adopt the sterically-preferred staggered configuration. In this case the Re–Re bond is short (2.24 Å) and so electron correlation relatively unimportant (see text).

how I learned to hate the electron correlation problem' (see the Further Reading at the end of the chapter). Essentially, electron repulsion forces electron density from bonding orbitals, which may be σ, π or δ, into low-lying antibonding orbitals, which may be σ^*, π^* or δ^*. The bonding orbital occupancy is less than given by the simple model and the antibonding orbital occupancy greater. Both vary with change in metal–metal internuclear distance. Larger bond lengths usually mean a greater occupation of antibonding orbitals (in the limit, where the bond breaks, bonding and antibonding orbitals are equally occupied). This can be placed in some perspective by the outcome of some calculations that have been carried out on the bare Mo_2 molecule. It might reasonably be assumed that the 12 valence electrons in the molecule might be distributed over the two σ bonding molecular orbitals (one originating in the s and one in the d_{z^2} orbital on each atom), the two (degenerate) π, d orbital in origin and the two δ (degenerate). The observed bond length in the gaseous molecule is 1.93 Å; in principle a sextuple bond could be involved! In fact, it was only with considerable difficulty that the calculations could be induced to show any bonding. They did so only after inclusion of considerable electron correlation. The absence of ligands does not seem to be the vital factor; in calculations on the dimeric diamagnetic molecule chromium(II) acetate the presence of the ligands did not give results greatly different from those on Cr_2^{4+}; again, a correlation energy contribution was required to give bonding. Note particularly that, as has been indicated above, the correlation energy contribution is expected to *increase* with increase in metal–metal distance. Yet further perspective can be added to this by the recognition that the metal–metal π bonding overlap seems invariably less than the carbon–carbon π overlap in benzene—and the metal–metal π and δ are presumably yet smaller. It has even been claimed, on theoretical grounds, that a Cr–Cr quadruple bond is to be expected to be as weak as a Cr–Cr single bond if there are no bridging ligands in the molecule (only one compound with a Cr–Cr bond but without bridging ligands is known). The spectroscopic evidence is unambiguous.

In $[Re_2Cl_8]^{2-}$ and related anions the electronic transition from the δ bonding to the δ antibonding orbital lies at about 7000 Å, an energy corresponding to a fairly typical crystal field splitting (Chapter 8). So, returning to larger clusters, where the bond lengths are much greater than in species such as $[Re_2Cl_8]^{2-}$, must we expect electron correlation to be of any importance? It seems that the answer is yes, although the delocalized nature of most orbitals, a phenomenon which means that electron density is spread out and so the effects of repulsion reduced, offers hope that the importance is limited. Nonetheless, it is the difficulty of including electron correlation which is a limiting factor in carrying out accurate calculations on cluster compounds. However, one must proceed with extreme caution. Not only is the problem a difficult one but it has been found that agreement with experiment is not invariably a reliable guide to the accuracy of a calculation. So, as has been mentioned in Chapter 10, there are cases known in which a calculation gives good apparent agreement with experiment—a bond length is accurately predicted, for instance—but, when the calculation is improved, perhaps by the inclusion of more orbitals on some atoms or the inclusion of more excited states (which can interact with and thus modify, the ground state) the apparent good agreement is lost. This is not all. There is the problem of relativistic effects. These have briefly been mentioned before (Section 11.12, their origin is outlined in the footnote to page 243); they are particularly important for the heavier elements and particularly for Pt, Au and Hg. So, it has been calculated that in the bare Pt_3 equilateral triangle the non-relativistic bond energy is $7.1 \, kJ \, mol^{-1}$ ($1.7 \, kcal \, mol^{-1}$) but that the relativistic is $36.4 \, kJ \, mol^{-1}$ ($8.7 \, kcal \, mol^{-1}$); where the general pattern is more to be noted than the particular energies. The origin of the particular strength of the Hg–Hg bond in $[Hg_2]^{2+}$ compounds is almost certainly relativistic, particularly when electronegative atoms, such as halogens, are attached. It seems that advances in our understanding of the bonding in cluster compounds may involve the introduction of yet more novel ideas.

15.7 Clusters and catalysis, a comment

Transition metals, either pure or in mixtures, are of wide industrial use in catalytic systems. They are used as wire meshes or finely divided on a support, the object being that of increasing surface area. Metal cluster compounds, including many of those shown in Fig. 15.1, may often be regarded, structurally, as a fragment of the bulk metal to which ligands are attached. Further, the majority of the metal atoms are on the surface of the cluster. Could not these atoms chemically resemble those on the surface of the bulk metal? If so, could not cluster compounds be a rich source of catalytic molecules? In that they also offer the possibility of fine-tuning by appropriate choice of ligands—both the steric and the electronic properties of the ligands can be varied—the prospect is attractive indeed. In practice, whilst important cluster catalysts certainly exist, the high hopes have been little fulfilled. Although the vibrational characteristics of ligands on clusters have proved very similar to those of the same ligands absorbed on bulk metal, catalytically, clusters have not modelled metal surfaces particularly closely. Why? There are, no doubt, many reasons. The surfaces of (chemists') clusters are usually saturated with ligands, metal surfaces often are not. Metal surfaces have surfaces, edges, defects and corners which are not matched by small clusters. However, we can also speculate on an electronic difference affecting reaction between two coordinated molecules. We have seen that, although the details remain a matter for debate, ligands significantly modify the electronic properties of bare metal clusters. There are no known paramagnetic cluster compounds of iron! In contrast, metallic iron, perhaps dispersed on an alumina surface, remains paramagnetic despite the presence of ligands. It seems possible that in the latter case, the reservoir of electrons and electron energy levels associated with the bulk metal provides a damping which enables individual ligands to undergo the gymnastics which are an essential part of most catalytic reactions, relatively independently of one another. For a small metal cluster, the gymnastics of one ligand have a sufficient influence on the orbitals and energy levels of the entire cluster for the gymnastics of a second ligand to be constrained. If this explanation is correct, then complexes of a type which are just beginning to be made— consisting of two or more concentric cages of metal atoms—are likely to be more catalytically interesting, for, if it is meaningful to make the distinction, the inner cages will surely have properties not far removed from those of the bulk metal.

Further reading

A brief overview of cluster structure and, to some extent, bonding, is contained in *Metal clusters revisited*, by J. Lewis, *Chemistry in Britain* (1988) 795. A more complete picture is given in *The Chemistry of Metal Cluster Complexes* by D. F. Shriver, H. D. Kaesz and R. D. Adams (Eds.), VCH, New York, 1990.

A review of bonding theories, including those mentioned in the early part of this chapter, is contained in 'Theoretical Models of Cluster Bonding' by D. M. P. Mingos and R. L. Johnson, *Struct. Bonding* (1987) *68*, 29. See also 'Bonding Models for Ligated and Bare Clusters' by D. M. P. Mingos, T. Slee and L. Zhenyang, *Chem. Rev.* (1990) *90*, 383.

Wade theory is detailed in *Transition Metal Clusters* B. F. G. Johnson (Ed.), Wiley, 1980, Chapter 3 'Some Bonding Considerations', by K. Wade.

Stone theory is described in 'New Approach to Bonding in Transition-Metal Clusters and Related Compounds' by A. J. Stone, *Inorg. Chem.* (1981) *20*, 563.

The problem of electron correlation is discussed in 'Problems in the Theoretical Description of Metal-Metal Multiple Bonds or How I Learned to Hate the Electron Correlation Problem', M. B. Hall. *Polyhedron* (1987) *6*, 679.

A brief review of graph-theoretical ideas is to be found in 'Mathematical methods in coordination chemistry: topological and graph-theoretical ideas in the study of metal clusters and polyhedral isomerizations' by R. B. King, *Coord. Chem. Rev.* (1993) *122*, 91.

A book that covers both experiment and theory is *Multiple Bonds Between Metal Atoms*, 2nd edn., by F. A. Cotton and R. A. Walton, Oxford University Press, Oxford, 1993.

A useful source is Volume 1 of *Comprehensive Coordination Chemistry* G. Wilkinson, R. D. Gillard and J. A. McCleverty (Eds.), Pergamon Press, Oxford, 1987, Chapter 4 'Clusters and Cages', by I. G. Dance).

Questions

15.1 'Insights into cluster bonding come more readily from an approach in which the details of atomic structure are not given prominence.' Comment on the validity of this statement.

15.2 In the compound $Rh_6(CO)_{16}$ (Fig. 15.1(c)) the rhodium atoms have O_h (octahedral) symmetry but the CO groups reduce the symmetry to T_d (tetrahedral). Outline and compare the pictures of the cluster bonding in the O_h and T_d point groups.

15.3 The molecule B_4Cl_4 exists but Al_4Cl_4 does not. The molecule N_4 does not exist but P_4 does. Use these examples to highlight the limitations of our present qualitative understanding of bonding in cluster species.

15.4 To what extent do the molecules shown in Fig. 15.12 conform to the predictions of Wade's rules?

15.5 Using $[Mo_6Cl_8]^{4+}$ as an example, apply Stone's surface tensor harmonic model to the species $[Ta_6Cl_{12}]^{2+}$. To what extent does the topological equivalent orbital model facilitate the application?

15.6 Review those areas in coordination chemistry in which it seems electron correlation is important for a detailed understanding (use the index at the end of the book).

15.7 In Section 15.4, $[Mo_6Cl_8]^{4+}$ and $[Ta_6Cl_{12}]^{2+}$ were presented as 40-electron systems although in Table 15.6 such octahedral species were associated with 86-electron counts. Explore the rapprochement between these two numbers. (Hint: consider the symmetry implications of the 86-electron count.)

16

Some aspects of bioinorganic chemistry

16.1 Introduction

Coordination compounds are central to living processes. Although the cations of Group 1 (Na^+, K^+) are usually thought of as very mobile, as Section 5.7 shows, it is possible to have relatively stable complexes of them when there is a matching between their size and that of a cavity in a ligand; change the cavity size and the stability changes. Perhaps, in part, this lies behind the function of these mobile cations in biology, for example in the transmission of nerve impulses. Group 2 cations are also important, presumably also through complexes that they form—Ca^{2+} in physical structures such as teeth, skeleton and shell, as well as functioning as a trigger in neurotransmission, as a messenger in initiating hormone action and in initiating blood clotting. The presence of Mg^{2+} in chlorophyll is well known. Zn^{2+} is present in many enzymes; sometimes it seems to have a structural role, maintaining a protein structure in a particular conformation for example, in others it is intimately involved in the chemistry. Complexes of transition metal ions are of vital biological importance. They provide mechanisms for electron storage, for electron transport and for catalysis—in particular involving small molecules and ions such as O_2, N_2, NO, NO_3^- and SO_4^{2-}. Is it understandable that bioinorganic chemistry is currently one of the fastest growing parts of inorganic chemistry.[1] In the present chapter this growth is not the focus of attention. Rather, attention is focused on the way that the subject throws up fascinating problems which offer a great challenge and require careful and, often, great ingenuity and subtle methodology to solve. This, then, is a problem-solving chapter. We start with a general overview of the subject which will provide a background for the specific problems that will be considered.

---

[1] A very readable paper which develops and enlarges the theme of this paragraph is 'The Chemical Elements of Life' by R. J. P. Williams, *J. Chem. Soc., Dalton Trans.* (1991) 539.

Bioinorganic chemistry is in large measure concerned with the study of metal complexes in which one ligand (perhaps, the only ligand) is a large, naturally occurring organic molecule—a protein, a sugar or a heme, for example. There are many factors contributing to its growth in recent years. First, X-ray crystallographic and related techniques (not least, the art of obtaining single crystals of complicated bioinorganic molecules) have developed to the point at which it has been possible to obtain accurate crystal structure data on many, if not all, species. This means that the local structure, composition and geometry around the metal ion is known for many compounds. Secondly, spectroscopic methods of study have been developed which are both sensitive and selective, probing the properties of, for instance, the metal ion and its immediate environment, notwithstanding the low concentration of this ion—usually much less than 1% on a molar basis. The present chapter will place some emphasis on a selection of these techniques. Thirdly, it has been recognized that bioinorganic species play a key role in the most important of biological reactions—respiration, energy transfer, photosynthesis and nitrogen fixation to name but a few. Our understanding of such processes is dependent on our first understanding the bioinorganic chemistry involved. Fourthly, bioinorganic species are often remarkably effective catalysts. Enzymes, and enzymatic reactions, are at the heart of biochemistry. As an illustration of this area, consider the fact that at the present time, the nitrogen fixation ability associated with bacteria present on the roots of some plants—clovers, peas and beans, for example—cannot be matched or mimicked in the laboratory. Perhaps one day it will be, threatening Haber ammonia plants with closure and possibly[2] changing the basis of agricultural production. Fifthly, it often turns out that in bioinorganic species the metal ions have unusual coordination numbers and/or geometries. One of the best known is the square planar coordination of the Mg^{2+} ion in chlorophyll. This point is surely connected with the fourth; it has stimulated attempts to make relatively simple inorganic complexes having coordination patterns similar to those found in nature. Such *model compounds*, when prepared, are likely to be much more readily available in quantity and much cheaper than their naturally occurring counterparts. However, they may well display some of the chemical properties of the naturally occurring materials, catalytic properties, in particular. This line of research has stimulated the synthesis of exotic ligands, mimicking naturally occurring ligands in the local environment that they provide the metal atom. Some of these ligands were met in Chapter 2. Sixthly, some metal species have proved to be effective therapeutic agents. Whilst the treatment of deficiency diseases (for instance treating anaemia, lack of iron, with complexes of Fe^{II} or Fe^{III}) are important, perhaps best known is the use of *cisplatin*, *cis*-$[Pt(NH_3)_2Cl_2]$, in the treatment of some cancers. Similarly, gold(III) complexes, often with ligands bonded through sulfur, are used in the treatment of some arthritic conditions. However, these drugs can have bad side-effects—cisplatin, used alone, can have a destructive effect on kidneys. Although these side-effects can be mitigated by administering the metallic compound along with suitable additional, compensating,

---

[2] 'Possibly', because an alternative line of research is to incorporate into wheat, for instance, the gene involved with mediating nitrogen fixation, thus making the species self-sufficient and, ultimately, removing the need for nitrogen-containing fertilizers in agriculture.

compounds and much water, these are only temporary expedients. Much better would be the use of metal complexes without the side-effects. Although the buck-shot approach (synthesize a lot of likely-looking compounds and try them) can be used, the desired goal is more likely to be reached by a detailed understanding of the mode of action of a drug, the *in vivo* transformations that it undergoes and thus, ultimately, the nature of the species that is actually leading to the desired effect. This calls for all the apparatus of bioinorganic chemistry—the identification, isolation, handling and study of biologically active species, the preparation of model compounds and so on.

As already indicated, it is not the objective of the present chapter to attempt to give either a comprehensive account of, or even an overview of, bioinorganic chemistry. Rather, we shall explore the application of subject matter in earlier chapters to the topic. From this viewpoint, bioinorganic chemistry provides a series of novel applications of this subject matter. Further, some advances in bioinorganic chemistry have involved specialized techniques and ideas which, to date, have been less applied to more traditional areas of coordination chemistry; we shall look at some of these also. In that such applications will surely become more commonplace, bioinorganic systems provide an indication of the results that will be obtained and of the information to be learned from them.

We start with four important points, some of which contain lessons in their own right. First, although it seems that seldom, if ever, does a metal ion occur in a biological molecule and not have a function, it should not be thought that the sole function available is a catalytic one, although this aspect is often emphasized. Quite the contrary, for instance, a number of cases are known where the function of Zn^{2+} is classified as structural. In some of these, Zn^{2+} forms rather strong bonds to sulfur ligands (usually from the amino acid cysteine, $HS–CH_2–CH(NH_2)–COOH$) which are part of a protein amino acid sequence. By so doing, the protein chain, or several chains together, are locked into a particular geometrical arrangement. It seems that the resulting global, quaternary, structures are favourable to the action of the protein (and this action could be a catalytic one) and are thus stabilized. Other non-catalytic functions of metal ions include the transmission of impulses along nerve fibres (each impulse is associated with a sudden change from a 'high K^+ concentration inside the fibre, high Na^+ outside' pattern. Immediately after the impulse, a pump mechanism restores the original concentrations). Another function is that played by Ca^{2+} as a messenger in nerve action (from one nerve cell to the next) and in blood clotting. It seems that Ca^{2+} serves such functions because the complexes it forms in these situations are rather balanced—not too weak, not too strong; not too labile, not too inert.

The second point is related to the first and echoes one that has already been mentioned. Metal ions in biological systems are often in rather unusual environments. This is not just because of the geometrical arrangements of the coordinating atoms but because of the influence of what one might think are the more distant parts of the coordinating molecule. Typically, these are convoluted around the metal ion and its immediate ligating atoms and, effectively, play the role of the solvent in simpler systems. Except, of course, that the solvent is a rather rigid one, one that severely limits access of molecules to (and egress from) the metal ion; seldom, if ever, is it as innocent as it may appear! Further, the physical bulk of the main ligating molecule both limits access to, and the mobility of, ligands attached to the metal ion. By the hydrophobic or hydrophillic surfaces that it presents, it to some extent selects the species that can approach the metal ion. These features combine to lead to a rather different chemistry. A link with classical coordination chemistry is forged by the observation that the coordination equilibria of complexes of polyfunctional ligands are very sensitive to changes in solvent composition.

Thirdly, we note that several of the ions mentioned in this paragraph are not too easy to study. Zn^{2+} and Ca^{2+}, in particular, lack convenient and sensitive spectroscopic 'handles'. They have no low-lying electronic transitions by which they can be conveniently characterized; they have no high-abundance isotope which can be studied by NMR; they do not give Mössbauer spectra, and so on. The problem is addressed by a technique which would be almost unthinkable in conventional coordination chemistry. Such ions are studied by replacing them with a similar ion which *does* have suitable spectroscopic characteristics. The reason for this that, although the biological macromolecule may distort a little when a look-alike ion is substituted, the distortion is small and good insight into the environment of the original metal ion may be obtained. So, if bioinorganic species containing Zn^{2+} are treated with an aqueous solution of 1,10-phenanthroline (phen, Table 2.3) the zinc is extracted because it forms a very strong complex with phen. If the zinc-free macromolecule is separated it is then readily coordinated to another cation. Co^{2+} is a popular replacement ion, but Mn^{2+}, Cu^{2+} and Cd^{2+} have also been used. Replacements which have been used for Ca^{2+} are lanthanides, particularly Nd^{3+} and Eu^{3+}.

The fourth and final point is that species containing two or more metal ions, linked by bridging atoms, are fairly common in bioinorganic chemistry. The two ions can be of different metals, so that one is immediately faced with the question of whether chemistry takes place at both of them, perhaps simultaneously, or whether one fine-tunes the chemistry of the other. Commonly, the ions are of transition metals and it is then usual to find that interaction occurs between what would otherwise be unpaired electrons at each centre. We shall not develop this particular topic here because all the important points have already been covered in Section 9.11.

16.2 Myoglobin and hemoglobin

In this and the following two sections a particular emphasis will be placed on iron-containing species. This is a distortion of the subject of bioinorganic chemistry. However, these systems are rich in the spectroscopic methods which have been applied to them, making them particularly suitable for a study of such techniques in bioinorganic chemistry.

So common is talk about 'the hemoglobin in the blood' that one is tempted to believe that it is the only species involved in oxygen transport from the lungs to the place where it is needed. It is not, myoglobin is also involved. One can perhaps see a reason for a two-component process, because oxygen transport is, in a sense, self-contradictory. One needs a species with a high affinity for oxygen in order to extract it from the air but

(a)

Fig. 16.1 The structure of myoglobin. In (a) the protein sheath is black and the heme, iron and histidine indicated. (b) The Fe–heme group and (c) the attachment of the iron to the protein through a histidine group.

(b)

(c)

one with a low affinity in order that the oxygen can readily be made available where it is needed, within a muscle, for example. Hemoglobin has such properties, but the final transport is carried out by myoglobin. Myoglobin consists of a protein with a molecular weight of just under 18 000 and has just one iron–porphyrin unit, a heme (Fig. 16.1). Hemoglobin, however, has four heme units, each with its own protein chain, the whole intermeshed to give a single molecule (Fig. 16.2). The unusual oxygen-affinity properties of hemoglobin arise from a cooperative behaviour between the four heme groups, a cooperation that must surely be mediated by the protein chains, for the heme units are from 25 to 40 Å apart. The heme cooperation in hemoglobin means that when one or two O_2 molecules are bonded, the oxygen affinity of the other (O_2 free) heme sites is enhanced. The more oxygen it has, the more it wants! Clearly, this is a useful property to have when scavenging for oxygen in the lungs. Conversely, when transfer of oxygen to myoglobin is under way, the less oxygen the hemoglobin has, the less it wants. Just the necessary characteristics! Oxygen is absorbed by attachment at iron atoms in myoglobin and hemoglobin (Fig. 16.3), interestingly, neither uniquely σ nor π bonded, but rather at an intermediate, angular orientation

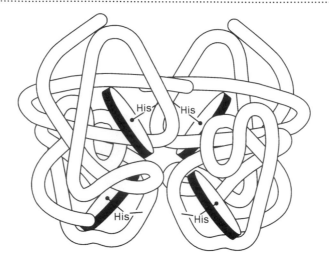

Fig. 16.2 The hemoglobin molecule; the attachment of the iron (black dots) to the protein through histidines is indicated.

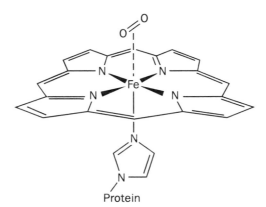

Fig. 16.3 The heme unit in myoglobin (substituents on the porphyrin ring have been omitted). In the absence of O_2 the Fe is about 0.4 Å below the plane of the ring, as shown. When coordinated to O_2 (shown dotted) the Fe moves into a position almost coplanar with the ring.

(when CO is absorbed in place of oxygen it bonds end-on through carbon to the iron, just as in most metal carbonyls).

In this chapter the use of model compounds (which are cheaper and available in much greater quantity than most naturally occurring molecules of interest) to explore the properties of naturally occurring species will often be mentioned, but here we see a limitation. We will see models for myoglobin but a model for hemoglobin which begins to mimic its cooperative behaviour is another matter; its design and synthesis will require enormous chemical insight and ingenuity. Perhaps it is as well that hemoglobin is a relatively easy material to obtain and to work with—it was first crystallized in 1849 and its oxygen-transport properties recognized in 1864 (only 40 years later did its CO_2 transport ability become known). Just to complicate things, oxygen release from hemoglobin is not solely determined by cooperative factors. It is also pH-dependent (protonation and deprotonation of amino acid residues may well change the preferred geometry of the peptide chains). Low pH leads to a more ready release of oxygen, so muscle tissue, where CO_2 and lactic acid are generated, is an especially favoured site for oxygen release.

The salts of iron(II) are susceptible to oxidation by the oxygen of the air—iron(II) sulfate (ferrous sulfate), pale green crystals when freshly prepared, tend to become brown after having been stored for years; the iron(II) salt used in volumetric analysis as a reducing agent, ferrous ammonium sulfate is so used, in part, because it is much less prone to oxidation than most iron(II) alternatives. So, too, with iron(II) porphyrins (and the iron(II) found in myoglobin and hemoglobin is in a porphyrin ring, Figs. 16.1–16.3). They are rapidly oxidized to their corresponding iron(III) compounds by oxygen. Such oxidation can occur, although much less readily, with myoglobin and hemoglobin, when met-myoglobin and met-hemoglobin are formed (these names should not be written with a hyphen; one is included here solely to facilitate a correct pronunciation). Metmyoglobin and methemoglobin do not transport oxygen! About 3% of the total hemoglobin in the blood is oxidized to methemoglobin each day, but mechanisms exist within the red blood cells for regenerating hemoglobin from methemoglobin. The reason that the loss is so low—and 3% is low—is reasonably well understood, as follows. Kinetic studies show that the rate of oxidation of Fe^{II} porphyrins is second order in the iron(II) species and first order in oxygen. The most probable explanation is that a 1:1 complex is first formed between an iron(II) porphyrin molecule and O_2 which subsequently reacts with another iron(II) porphyrin to give an O_2 bridged complex—an example of such a compound of cobalt is given in Table 2.2. The decomposition of the bridged dimer leads to the oxidation products. In myoglobin and hemoglobin the protein provides a protective shroud for the O_2 molecule such that formation of a dimeric species is scarcely possible, whilst remaining sufficiently open to enable ready O_2 access and egress. The low rate of formation of the met derivative is thus explained.

The object of models of the myoglobin molecule has been to provide a similar environment to that found in the natural material; some examples are shown in Fig. 16.4. One of these ligands, the picket fence molecule, was given in Table 2.4 and it will be given again in detail in Fig. 16.5. What characteristics do the models have to attempt to emulate? In myoglobin and hemoglobin the iron(II) is in a square pyramidal environment, the four square-planar coordination sites being associated with the porphyrin ring and the axial with a histidine (see Fig. 16.1(c)) from the protein chain. Oxygen enters the sixth site, completing an octahedral coordination around the iron(II). Whereas the five-coordinate iron is high-spin—four unpaired electrons—the O_2 complex, oxyhemoglobin, is low-spin, no unpaired electrons. The reduction in ionic radius of iron(II), high-spin, to low-spin (Section 13.2) is sufficient to enable the iron(II), initially a bit too big to fit in the porphyrin hole and so sitting slightly out of the ring and towards the histidine ligand, to drop more or less into the plane of the ring in oxyhemoglobin. The consequent tug on the histidine connecting the iron(II) to the porphyrin presumably contributes to the cooperative oxygenation effect seen in hemoglobin. How well is this behaviour of myoglobin mimicked by the model compounds? To those shown in Fig. 16.4 we have to provide a ligand, for obvious reasons a histidine derivative, to occupy the open coordination site at the bottom of each species shown. With this addition, comparison is possible. Of course, the models can offer no information on the detailed involvement of the protein chain—this problem is being explored by site-directed mutagenesis (creating mutants in which specific amino acids are changed—and this is more than just an academic exercise; a number of

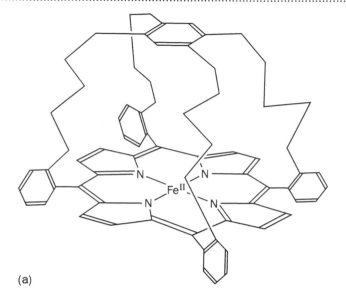

(a)

Fig. 16.4 Some examples of models of myoglobin. Only schematic structures are shown—there are usually other heteroatoms (O, N) within the molecules but not shown here: (a) a capped porphyrin, (b) a crowned porphyrin (c) picket fence porphyrin.

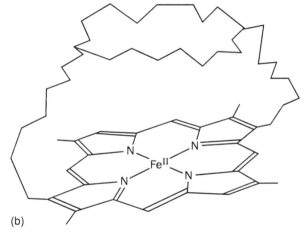

(b)

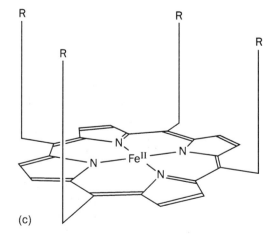

(c)

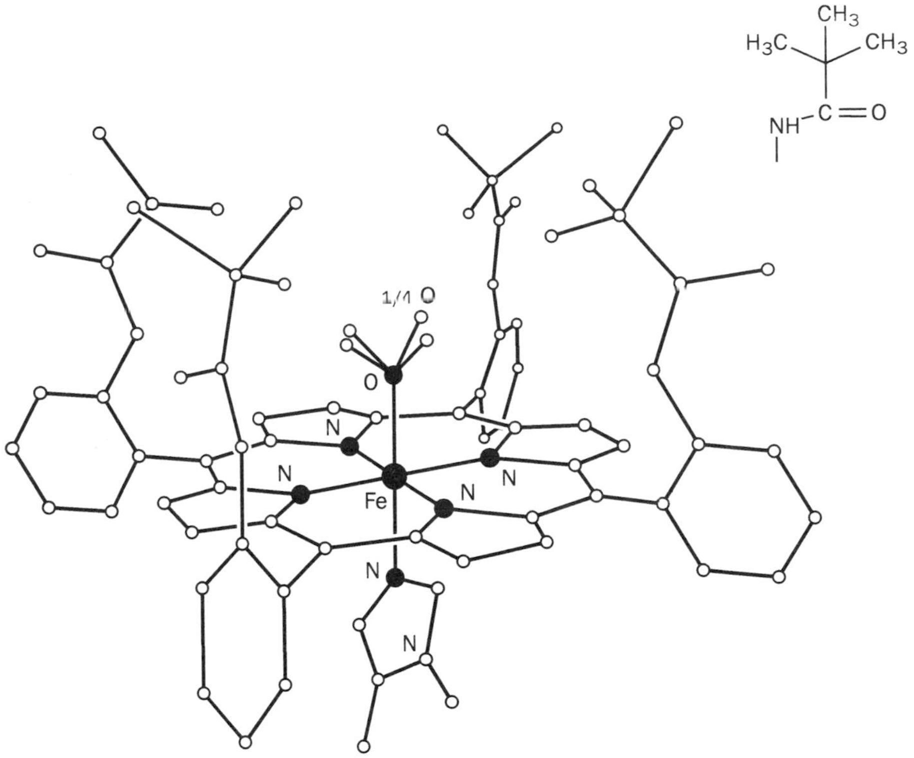

Fig. 16.5 The crystal structure of picket fence porphyrin with a histidine coordinated to the iron. The O_2 molecule also coordinated is disordered over four equivalent sites. The detail of the 'fence posts' is shown as an insert.

diseases are known that arise from hemoglobin mutants). We concentrate, then, on probes that investigate the iron(II) and its immediate environment, particularly those discussed in Chapter 12.

Although the crystal structures of myoglobin and oxymyoglobin are available—and their realization was a considerable achievement—they are not as well resolved as those of model compounds, although the iron-atom slip and tilted O_2 binding, mentioned above for the natural species, are both clear. In Fig. 16.5 is shown the crystal structure of Colman's picket fence porphyrin model of hemoglobin, one that will be considered in some detail. Notice that although the bottom of the O_2 molecule is anchored to the Fe along the latter's fourfold axis, the top of the O_2, as drawn, is tilted to one side and disordered over four sites. O_2 is a diradical, two unpaired electrons, one in each of the degenerate π^* orbitals of O_2. There is theoretical evidence that it is reasonable to think of the two π^* electrons pairing up in just one of the two π^* orbitals and as being involved, along with the σ electrons on the O_2, in σ donation to the Fe^{II} d_{z^2} orbital (so that the latter receives electron density from two sources). For the d_{z^2} to be empty in order to receive them, spin pairing occurs on the iron(II), $t_{2g}^4 e_g^2 \rightarrow t_{2g}^6 e_6^0$, leading to a diamagnetic complex with an obliquely coordinated O_2.

How strong is the Fe–O bond? Is its strength in the model comparable with that in oxyhemoglobin? How accurate is the bonding description that we have just presented—in that it invokes π^* electron density being removed from O_2 it might reasonably be expected that the O–O bond would be strengthened by complexation. The available crystallographic work provides a partial answer; in Table 16.1 are listed some relevant comparisons.

Table 16.1 Geometrical features of oxyhemoglobin and model compound

Feature	Picket fence model (Fig. 16.5)	Oxyhemoglobin
Fe–(O–O) distance	1.75 Å	1.83 Å
Fe–O–O angle	136°	ca. 130°
O–O[a] distance	1.25 Å	

[a] In O_2, 1.21 Å; O_2^- 1.28 Å.

Reasonable accord is found, although contrary to the model of the bonding presented, the O–O bond length increases on complexation. This discrepancy is easily explained by adding π back-donation from iron to the empty π^* orbital of the oxygen (and the empty orbital included is purely π^* in character) to the picture. In this way, the Fe–O bonding is strengthened and the O–O weakened.

Another way of answering questions about bond strengths is provided by vibrational spectroscopy, provided that the vibrational spectra can be sufficiently well understood. Not an easy task for complicated molecules! In the present case two ameliorating features exist. First, the Fe–O_2 unit is simple. Secondly, resonance Raman spectroscopy (Section 12.3) enables small parts of a molecule to be studied. With any luck, they are the parts one is interested in! However, before running a resonance Raman spectrum one would invariably first run an infrared. Unfortunately, at first it proved impossible to find any infrared feature associated with the O_2 unit in the model compound; spectra of $^{16}O_2$ and $^{18}O_2$ species looked the same. However, low-temperature work eventually led to the discovery of a sharp peak at 1385 cm^{-1}, which broadened almost out of sight at room temperature (perhaps this is associated with the disorder seen in the crystal structure). For comparison, a peak at 1103 cm^{-1} has been reported for oxymyoglobin, quite close to the value of 1145 cm^{-1} reported for the O^{2-} anion. Although 1385 cm^{-1} is well below the frequency reported for the ν(O–O) stretching vibration in gaseous oxygen (1556 cm^{-1})—and this difference is in accord with the modified picture of the Fe–O_2 bonding just presented—the difference between it and the oxyhemoglobin value is more than one might wish. However, by changing the *trans* ligand (to a dimethylimidazole) it proved possible to observe the ν(O–O) band in solution, where it proved to have a value of 1159 cm^{-1}, in much better accord with the hemoglobin value. It has been found that vibrational coupling, and consequent frequency shifts, can occur between coordinated O_2 and the *trans* ligand. Even so, the shift from 1385 to 1159 cm^{-1} seems too large to be explained in this way alone. The problem remains unresolved. As anticipated above, it has proved possible to obtain data from resonance Raman measurements. In addition to providing a (solution) value for ν(O–O) of 1140 cm^{-1}, the method has also shown ν(Fe–O) to be at 568 cm^{-1} in the picket fence model, in excellent accord with the value of 567 cm^{-1} found for oxymyoglobin.

What other measurements can be made which offer a comparison between model and biological compounds? Iron is an element which can be studied by Mössbauer spectroscopy (Section 12.5.3). Mössbauer spectroscopy provides a way of determining whether, for instance, the electric field gradient at the iron is similar in model and biological species. The comparison is detailed in Table 16.2. Without going into a detailed discussion of the meaning of the quantities in this table (although Sections 12.5.2 and 12.5.3 provide an indication), it is evident that a reasonable accord exists and this conclusion is confirmed by a more detailed analysis—the iron atoms, indeed, are in similar environments.

Whereas Mössbauer spectroscopy probes the immediate environment of the iron atom, electronic spectroscopy is sensitive to both metal and ligands—all ligands, not just the O_2. It is a characteristic of porphyrin rings containing a metal atom that they show three bands in their visible and

Table 16.2 Comparison of Mössbauer spectroscopic data for oxyhemoglobin and model compound

Compound	Centre shift from metallic iron (mm sec^{-1})		Quadrupole splitting (mm sec^{-1})	
	4.2 K	195 K	4.2 K	195 K
Picket fence model	0.29	0.25	2.10	1.34
Oxyhemoglobin	0.24	0.20	2.24	1.89

Table 16.3 Electonic absorption specta of oxyhemoglobin and a picket fence model compound

Compound	Absorption maxima (nm)		
	α	β	Soret
Oxyhemoglobin (human)[a]	582	544	418
Picket fence model		548	429

[a] The data vary little with source of oxyhemoglobin.

Table 16.4 Oxygen uptake of myglobin and a picket fence model compound

Compound	Fraction of saturation at equilibrium (20 °C)	ΔH^0 (kJ mol^{-1})	ΔS^0 (JK^{-1} mol^{-1})
Picket fence model	97%	−65.3	−15.9
Myoglobin	93%	−56.1	−13.4

near-ultraviolet spectra associated with $\pi^* \leftarrow \pi$ transitions in the rings. The strongest band (by a factor of about ten) is in the near-ultraviolet and is known as the *Soret* band. The other two, in the visible and of comparable intensities, are called the α and β bands. The α band is at the longer wavelength. In Table 16.3 data for the picket fence model are compared with those for oxymyoglobin. As might have been anticipated, although there is a general similarity, the differences in the porphyrin ring substituents, in particular, are evident. The picket fence data are for 1-methyl imidazole as *trans* ligand, but varying this has little effect on the numbers in Table 16.3. Finally, a simple, but important, question can be asked. How does the model compare with the biological when it comes to oxygen uptake? Again, not at all badly, as the data in Table 16.4 show.

16.3 Search for reaction intermediates

As has been seen, bioinorganic species tend to be complicated and one must always be wary that in concentrating on the metal ion and its immediate environment one becomes blind to other changes in the molecule. Even definitions can become difficult. If a sugar, peptide or other part of the molecule changes orientation or conformation by a series of steps, is each step to be regarded as a reaction intermediate? Such problems are resolved experimentally, if at all possible. Try to find the intermediate and study it 'on the fly'. Such searches have led to a wide use of time-resolved methods. Flash photolysis, in which the photolysis products are studied immediately after they are generated, was one of the first. More recently, laser pulse techniques, in which one laser pulse prepares—excites—the system and a second probes—studies—the excited molecules, have been extensively applied. As an example of the tricks that can be used, it is possible to study very short time intervals by deriving the probe pulse from the exciting pulse but delaying it very briefly by allowing it to travel away from the apparatus and then reflecting it back with a mirror. The delay, the time taken for the round trip, is varied by moving the mirror.

Many of the techniques discussed in Chapter 12 can be adapted to give such time-resolved information—and it seems that the complexity of many bioinorganic systems is such that they can only be properly understood once these sort of data are available. A typical example of the method, if not of the result, is provided by circular dichroism (CD; Section 12.4). As is well known, hemoglobin absorbs carbon monoxide. The complex formed, which is often written HbCO, can be photodecomposed by irradiation by light with a wavelength corresponding to an absorption band of hemoglobin. In Fig. 16.6 are shown the CD of HbCO, Hb and HbCO 10 µs after a laser pulse at 532 nm. The interpretation is clear—the data provide no evidence for an intermediate. If there is one, experiments with a shorter time scale will be needed to find it (to date, none has been found). One reason that the 10 µs and Hb traces in Fig. 16.6 are not quite identical becomes evident from Fig. 16.7, which shows that 20 µs after the laser pulse, HbCO is being reformed. Of the other methods described in Chapter 12, the time resolution of NMR methods is well known and has been discussed in Section 14.8. In general, spectroscopies in which Fourier transform methods may be used are particularly well adapted to time-resolved studies because they study the entire spectral range simultaneously. One scan/pulse gives the entire spectrum; with luck, effort, or both, the signal to noise ratio of the final spectrum will be good enough (although, in general, in Fourier transform

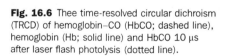

Fig. 16.6 Thee time-resolved circular dichroism (TRCD) of hemoglobin–CO (HbCO; dashed line), hemoglobin (Hb; solid line) and HbCO 10 µs after laser flash photolysis (dotted line).

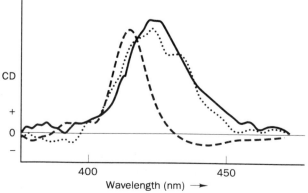

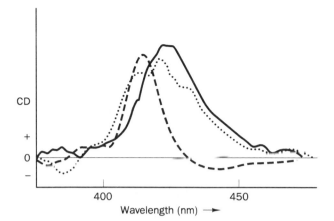

CD

+

0

−

400   450

Wavelength (nm) →

Fig. 16.7 As Fig. 16.6 except that the dotted line is now HbCO 20 µs after laser flash photolysis.

methods repetitive scans/pulses are accumulated to improve the final signal to noise ratio).

16.4 Peroxidases

The peroxidases are heme enzymes that catalyse the oxidation of organic substrates, the actual oxidant that is catalysed being hydrogen peroxide. In a typical peroxidase—horseradish peroxidase is one of the best studied—it seems that iron(III) is coordinated by a heme ligand, a fifth coordination site being occupied by the nitrogen of a histidine. The action of hydrogen peroxide on this enzyme gives an intermediate, usually called compound I, which carries out part of the oxidation of the organic substrate, itself being reduced to compound II. Compound II completes the oxidation of the substrate, the native enzyme being regenerated. The question is: 'what are compounds I and II?' Before seeking experimental evidence, it is sensible to make some intelligent guesses. First, it seems likely that they contain both heme and histidine. If the histidine were lost during the oxidation cycle then it would have to be replaced before the cycle could repeat. Such a need to scavenge for histidine would introduce a bottleneck which is scarcely compatible with an effective catalyst. Secondly, they are likely to contain the iron and/or heme in an oxidized form. For the iron this means Fe^{IV} or perhaps Fe^{V}, both unusual and therefore interesting valence states. Finally, since hydrogen peroxide is the ultimate oxidant, it may be that some fragment of the H_2O_2 molecule is present in either or both of compounds I and II.

The sixth coordination site around the iron is formally vacant in the native enzyme and seems an obvious place for interaction between the enzyme and H_2O_2 to take place. Of particular interest in this field have been studies of the absorption of white X-rays. In this context X-rays may best be thought of in the sequence

visible light → ultraviolet → vacuum ultraviolet → X-rays

All excite electrons, the X-ray excitation being the most energetic. As is well known, X-rays are emitted when an outer electron drops into a hole in a low-lying incompletely filled orbital (this is the basis of everyday X-ray

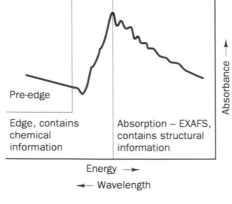

Fig. 16.8 The absorption of white X-rays at a wavelength chosen because it is one characteristic of the element one wishes to study. The fine details contain useful information.

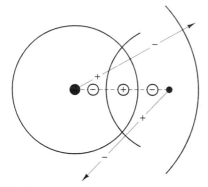

Fig. 16.9 An electron photoionized in an EXAFS experiment may be thought of as giving rise to a transient standing wave. In the upper diagram the wave from the emitted electron (ejected from the larger black, metal, atom) constructively interferes along the metal–ligand axis with the reflected wave from the ligand. The same happens along all other metal–ligand bonds of the same length. The phases of the waves from each centre are indicated along the arrow emanating from that centre. The phases of the overlaps along the internuclear axis is indicated in circles. The lower diagram similarly illustrates how, for a different wavelength, destructive interference can occur.

generators, the holes being produced by electron impact—thus, the domestic colour television is a potential source of soft X-rays). So, the reverse: X-rays of the right wavelength excite inner electrons into an outer orbital, the wavelength region being a characteristic of the element under study. Provided that access can be gained to one of the national or international synchrotron radiation facilities at which high intensity white X-rays are available, one can decide to study just the iron atom in compound I or compound II provided that the molecule of interest is available in sufficient quantity and sufficient purity. The details of the absorption observed indicate not only the valence state of the element studied (the higher the formal charge, the greater the energy needed to excite the inner electron) but also something of the symmetry of the ligand field.

A typical absorption spectrum is shown in Fig. 16.8, where it is contained within the (absorption) edge. For compounds I and II such data indicate that the iron is Fe^{IV} in the sort of tetragonal ligand field that we expect. This is not all, however. What of X-rays that posses enough energy to ionize the atomic species under study? An electron will be emitted. One might think that, once the incident X-ray energy is sufficient to cause ionization, higher energy X-rays would be no less effective. Figure 16.8 shows that it is not this simple, there are wiggles, fine structure, in the X-ray absorption. This is called extended X-ray absorption fine structure or, invariably *EXAFS*.

In EXAFS, an electron is photoionized by the incident X-rays. As is well known, the electron sometimes behaves like a particle (usually, when large distances are involved), sometimes like a wave (usually, when short distances are involved). The EXAFS wiggles arise because of the wave-like nature of the emitted electron. The surrounding atoms reflect the electron wave back to the atom from which it originates (Fig. 16.9). If there is constructive interference between reflected and original waves an enhanced probability of electron emission results. Destructive interference leads to reduced emission. Hence the wiggles, as the energy (and therefore wavelength) of the emitted electron varies. Evidently, if the distances between the central and reflecting atoms are changed, so too will the wiggle pattern. The wiggles contain structural information. We have a method by which we can probe the local geometry around an individual atom in a macromolecule! Whilst this is true, it is also true that extraction of the structural information from the experimental data is often far from trivial.

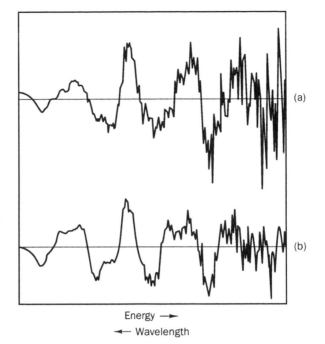

Fig. 16.10 The EXAFS of (a) horseradish peroxidase compound I and (b) that of the model compound shown in Fig. 16.11. Data processing has removed the gradient evident in Fig. 16.8 and by amplifying the right-hand side of the spectra to give the same signal level as the left-hand side has also increased the noise.

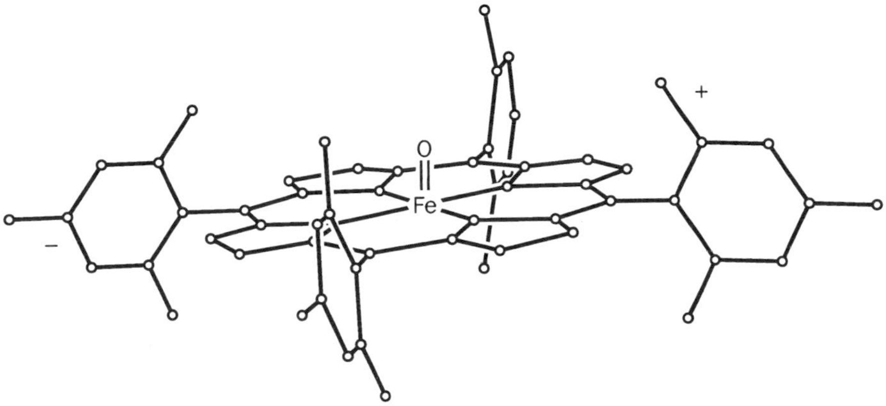

(a)

(b)

Energy ⟶
⟵ Wavelength

Fig. 16.11 The model compound referred to in Fig. 16.10.

Life is made much simpler if a good model compound is available. In Fig. 16.10(a) is given the EXAFS of compound I and (b) that of the model compound shown in Fig. 16.11. The similarity of the two EXAFS spectra is such that one has considerable confidence in the accuracy of the model compound's mimicry. It is seen that this model compound has an Fe=O bond. Detailed analysis of the EXAFS data gives an Fe=O bond length of 1.61–1.64 Å in compound I and 1.62–1.66 Å in the model; the Fe–N(porphyrin) bond length is found to be 1.99–2.00 Å in both compound I and the model. The agreement is excellent. Compound II has been similarly modelled and is believed to be simply compound I without a positive charge (i.e. with one additional electron). Several weaknesses of EXAFS should be noted, however. Although it gives bond lengths it only indirectly gives bond angles, so it is limited in what it can say about coordination geometries. Second, it cannot distinguish between elements which are close together in the periodic table—N and O; S and Cl. Finally, it is at its best when there is only one type of individual atom present, one sort of Fe, for example. When there are several chemically different Fe atoms, it will only yield an average structure.

Another technique that has been used to probe the structures of compounds I and II is *magnetic circular dichroism* (MCD). Circular dichroism (CD) has been met both earlier in this Chapter and also in Chapter 12. It has the disadvantage that it is a phenomenon confined to molecules which are optically active. MCD is a similar phenomenon but is applied to molecules which are not optically active. The trick is this. All molecules distort very slightly in a magnetic field and the distortion is always one that changes the optical activity of a molecule. A previously inactive molecule has no alternative, it becomes slightly optically active. The magnetic field distorts the molecule so as to destroy any centre of symmetry and any mirror planes. So, in the intense magnetic field of a superconducting magnet, optical activity is the norm. MCD is particularly useful when the field-free molecule has some symmetry, for then the magnetic field splits degenerate electronic energy levels. This means that it is particularly useful for exploring transition metal ions and their environments. In MCD there are two cases, A and C, of particular importance:

- case A: the electronic ground state is non-degenerate but the excited is degenerate in the absence of a field;
- case C: and the opposite pattern, in which the ground state is degenerate but the excited state non-degenerate in the absence of a magnetic field.

Because of the theory used to describe the phenomenon, the spectral patterns resulting (Fig. 16.12) are described as A term and C term, respectively. As the reader may have guessed, B terms also occur but are much less important (they relate to magnetic-field mixing of wavefunctions and so are akin to TIP, mentioned in Section 9.3). Figure 16.12 shows the energy level splittings which are associated with A and C terms. Because the ground-state splittings

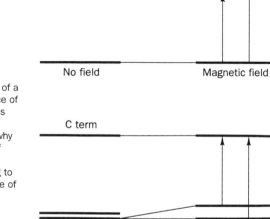

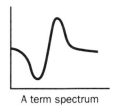

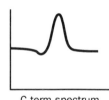

Fig. 16.12 The origin of A and C terms in MCD. A triplet, degenerate in the absence of a magnetic field, is split apart in the presence of such a field. Two transitions are allowed; as the A term case shows, they are equal in magnitude and opposite in sign (which is why no spectrum is observed in the absence of a field). In the C term case the thermal populations of the split levels differ leading to bands which, although of opposite sign, are of unequal intensity.

of C term spectra are small, the thermal populations of the split levels vary significantly with temperature; the lower the temperature, the stronger the spectral bands. In contrast, A terms are essentially temperature-independent.

MCD appears to be a somewhat esoteric technique but, in fact, it is of great value. A complementarity between CD and MCD should be noted. CD explores a static optical activity and this usually means exploring a property of the ligand system around, but relatively remote from, a metal ion; MCD offers the possibility of exploring the properties of the immediate environment of the metal ion for this is, in part, responsible for degeneracies. Further, it is not surprising that for paramagnetic species, this paramagnetism is reflected in their MCD spectra (the technique, after all, is concerned with the splitting of degeneracies and this is what Chapter 9 was all about, too). Indeed, MCD can be used to measure magnetic properties and, for this, has advantages. It is much less sensitive to impurities than are the methods described in Appendix 8. EPR (Section 12.6) also provides magnetic data but, really, only for systems with an odd number of electrons. MCD provides data for both even and odd. Just as a detailed understanding of the magnetic properties of a complex is none too simple (Appendix 9 gives the simplest case), so too with MCD. No attempt, therefore, will be made to give a detailed analysis, but simply to give in Fig. 16.13 the low-temperature MCD spectra of compound I and compound II. The differences, believed to result from the one extra electron in compound II, are obvious. Just as for CD, it is possible to carry out time-resolved MCD measurements and to use the technique to follow the effects of photolysis, for example. It is in the area of using spectroscopic techniques to follow such changes that many advances in bioinorganic chemistry are presently being made.

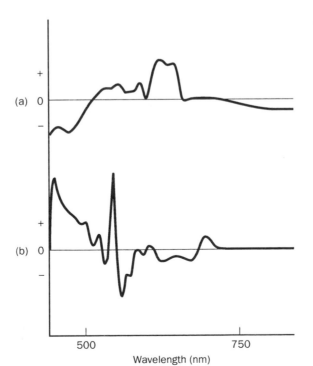

Fig. 16.13 (a) The MCD of horseradish peroxidase, compound I, and (b) horseradish peroxidase, compound II.

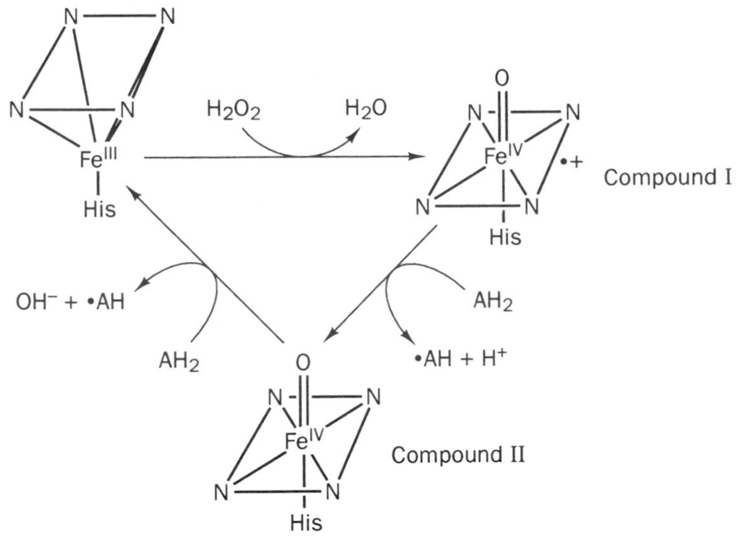

Fig. 16.14 The current view of the catalytic cycle in which peroxidase (top left-hand) is oxidized by hydrogen peroxide to give compound I. This oxidizes the substrate AH_2 to give $AH^\cdot$ (dots indicate free radicals), H^+ and compound II. This similarly oxidizes more substrate to regenerate the starting material (His = histidine).

This section will be concluded with a survey of other evidence on compounds I and II. First, in that it takes two one-electron reduction steps (the formation of compound II is the first) to regenerate the Fe^{III} compound, peroxidase, it would appear reasonable that compound I might be regarded as containing Fe^V. However, the evidence from Mössbauer and magnetic susceptibility measurements, as well as EXAFS and MCD, suggest otherwise. It appears that it is more accurately described as low-spin Fe^{IV}, spin = 1, coupled to a porphyrin π-radical cation. Both NMR and visible/ultraviolet spectroscopic results are compatible with the presence of a π-radical cation. This interpretation is not unreasonable. One must expect that the porphyrin does more than just provide a suitable ligand field and a set of geometric constraints. Indeed, there is evidence that the oxidation of organic substrates by compounds I and II is carried out through the edge of the heme group and not through the Fe=O unit. To complete the structural data, resonance Raman work and oxygen isotope labelling indicate that there is an Fe=O unit in both compounds I and II. Finally, in Fig. 16.14 is shown, schematically, the catalytic cycle as it is presently believed to occur. Although it leaves many questions unanswered, the cycle is remarkably simple, notwithstanding the relative complexity of the molecules involved. Simplicity in the midst of apparent complexity is not uncommon in bioinorganic chemistry and has acted as an additional attraction of the field.

16.5 Blue copper proteins

In the previous sections, the availability of model compounds often, although not always, provided important contributions to the understanding of the native material. What if it proves difficult to prepare suitable models? Such a situation is encountered for the blue copper proteins and provides an opportunity of demonstrating the use of some of the experimental methods and their interpretations which have been the subject of previous chapters. The blue copper proteins form a series of compounds which are involved in long-range electron transfer reactions, so that the interplay of Cu^{II} and Cu^{I} is of the essence. Although crystallographic work has confirmed the ultimate conclusion it is interesting to trace the non-crystallographic path leading to the structure of the site of the copper ion. The compounds are called blue because of their intense colour, about 400 times more intense than that of most copper(II) salts. The conclusion drawn was that the Cu^{II} in the protein is in a tetrahedral site, because as seen in Section 8.8, such a site makes more allowed d–d transitions because of mixing of d and p orbitals by the ligand field. The conclusion was right but the reasoning wrong. The blue colour is due to a charge transfer transition, not d–d, although the suggestion of a (very) distorted tetrahedral geometry was correct. Evidence for the charge-transfer assignment comes from a comparison of the electronic spectrum with the MCD spectrum (Fig. 16.15). The electronic spectrum can be decomposed into a sum of up to eight individual transitions, as many charge transfer as d–d (there is separate excitation from each of the components of d sets which would be degenerate in T_d symmetry), a spectrum which has reasonably well been interpreted by Xα calculations. The lower energy bands are the ones showing the greatest MCD activity and so are likely to be the most metal-localized. The blue band also persists when the Cu^{II} is replaced by Ni^{II} or Co^{II}, but it moves into the near-ultraviolet, consistent with a low energy ligand-to-metal charge-transfer assignment.

In almost 30 blue copper proteins, all of which contain about 100 amino acids, the amino acid sequence has been determined. About a quarter of the amino acid sequence is common to all and so this sequence almost certainly contains the ligands which bond to the copper. In this way, the amino acids

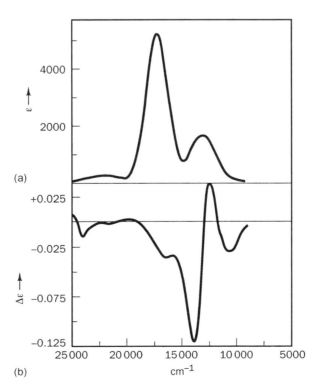

Fig. 16.15 A comparison of (a) the visible and (b) the MCD spectra of the blue copper protein plastocyanin.

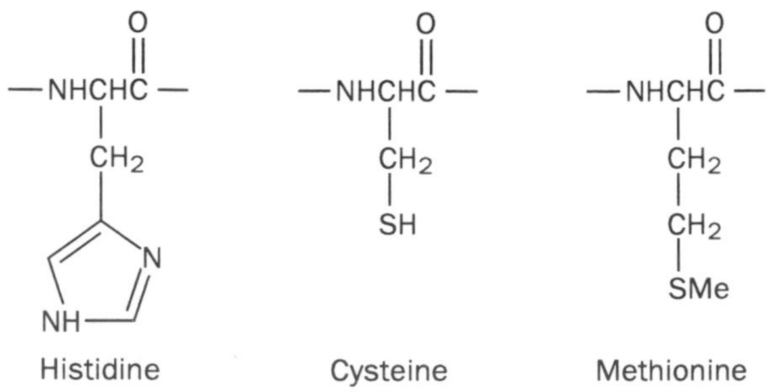

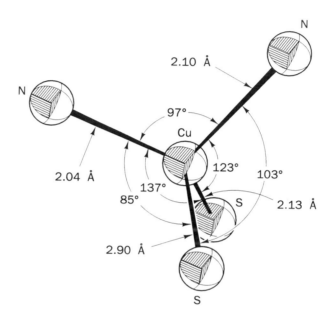

Fig. 16.16 Chemical structures of amino acids bonded to copper in blue copper proteins.

Histidine   Cysteine   Methionine

Fig. 16.17 The coordination environment of the copper in blue copper proteins.

histidine (two), cysteine and methionine (Fig. 16.16) have been implicated as involved in the coordination. This conclusion has been confirmed by X-ray studies, studies that have also shown that the site at which the copper is located changes little when the copper is removed. The copper is at a site with a ligand-imposed geometry (Fig. 16.17). This at once provides one reason why it has proved difficult to devise really good model compounds. Copper(II) is rapidly reduced by thiols

$$2Cu^{II} + 2RS^- \rightarrow 2Cu^I + RSSR$$

and it is necessary to arrange that this fate does not befall the cysteine, which contains an RSH unit. Some apparently promising model ligands form stable complexes with Co^{II} but react with Cu^{II}.

Another interesting characteristic of the blue copper proteins is the visibility of the histidine ligands in their EPR spectra. Like many copper(II) species, the blue copper proteins give strong EPR signals. When there is a nitrogen ligand bonded to Cu^{II} it is usual for its presence to be indicated by nitrogen hyperfine structure (see Section 12.6) and this is seen in the EPR spectra of blue copper proteins (Fig. 16.18). Finally, EXAFS measurements on the proteins provided further evidence of the environment of the copper

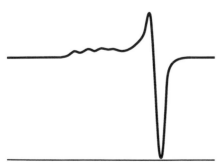

Fig. 16.18 The EPR spectrum of plastocyanin; the wiggle pattern at the centre-left of the spectrum is the nitrogen hyperfine structure.

ions at about the same time as the results of the first crystal structure determination became available. Information on the long-range electron-transfer properties of the blue copper proteins has come from several sources; the use of $[Ru(NH_3)_5(OH_2)]^{3+}$ is particularly interesting. It has been found that this ion binds to the surface of the protein surrounding the copper in at least one blue copper protein, coordinating to an exposed histidine (presumably losing the H_2O ligand) and thus being held almost 12 Å from the copper. Flash photolysis of $[Ru(bpy)_3]^{2+}$, added to a solution of the blue protein/rutheniumammine adduct gives $[Ru(bpy)_3]^{3+}$ and the $[Ru(NH_3)_5]^{2+}$–protein adduct. The $[Ru(NH_3)_5]^{3+}$ adduct is reformed by electron transfer from the Cu^{II}, which becomes Cu^{I}, the protein losing its blue colour, providing a convenient method of measuring the electron-transfer process. The observation of a rate constant (about $2 s^{-1}$) which is temperature-independent over the range studied indicates the absence of any significant activation energy although there is an entropic factor, probably associated with reorganization of the water molecules around the Ru site; the site of the copper is optimized for electron transfer.

16.6 Nitrogen fixation

Just as Cu^{II}–Cu^{I} oxidation–reduction is used in the blue copper proteins, so the Fe^{III}–Fe^{II} pair is used in a number of others, of which examples have already been met. Some particularly important systems contain the Fe^{III}–Fe^{II} system together with sulfur ligands (Fig. 16.19). In the *rubredoxins* a single iron atom is tetrahedrally surrounded by the sulfurs of four cysteine ligands. In plant *ferredoxins* there are two iron atoms, again tetrahedrally coordinated and associated with four cysteine sulfurs but bridged by two so-called 'inorganic' sulfide, S^{2-}, ligands. These sulfides can readily be displaced and so they are also called labile sulfides. Finally, in bacterial ferredoxins there are four iron atoms, again tetrahedrally coordinated and with four cysteine ligands, bound together into a tetramer by four triply bridging inorganic S ligands; a relationship with some of the species discussed in Chapter 15 may be noted. The rubredoxins and ferredoxins have been the subject of intense study using most of the methods described already and could well be used to provide further examples of these applications. However, the present section is more speculative.

The mystery of the mechanism by which biological nitrogen fixation is achieved has recently, to some extent, been removed by the publication of some structures based on crystallographic and related work (EXAFS, mutagenisis results, peptide sequencing and some spectroscopy).[3] Two different molecules are involved; one is a protein with two identical subunits bridged by a $[Fe_4S_4]$ cluster of the ferredoxin type, bonding to each protein subunit being through two cysteines. The other molecule contains two pairs of clusters, the members of one pair themselves each contain a pair of ferredoxin $[Fe_4S_4]$ clusters, bridged through two of the cysteines (Fig. 16.20(a)). The other pair of clusters in the second protein also each contain two bridged clusters, one a modified ferredoxin $[Fe_4S_4]$ cluster (not all of the sulfur-containing ligands attached to Fe are cysteine) and the other a

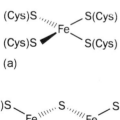

Fig. 16.19 The typical environment of Fe in (a) rubredoxins, (b) plant ferredoxins and (c) bacterial ferredoxins.

[3] J. Kim and D. C. Rees, *Science* (1992)257, 1677.

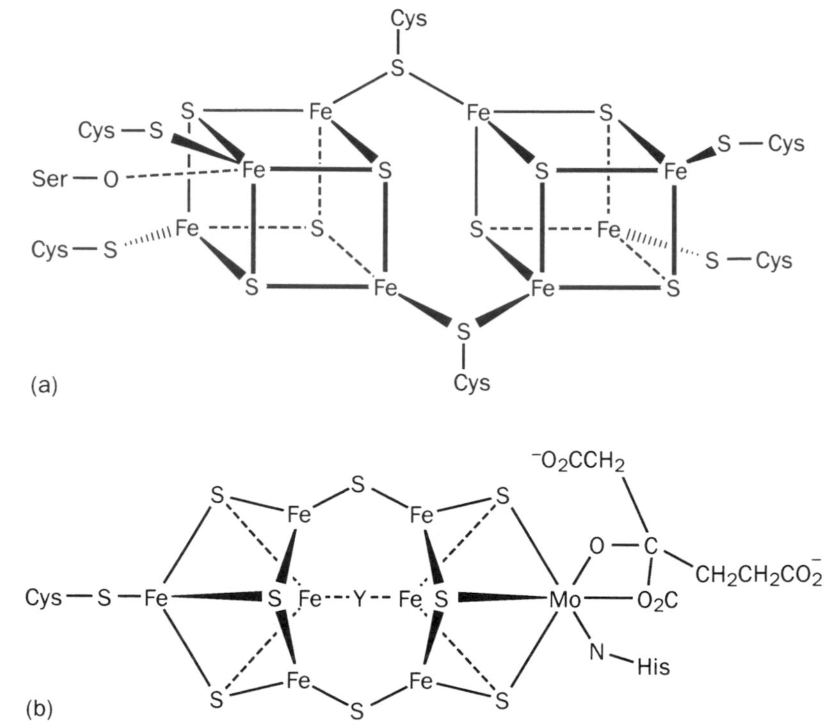

Fig. 16.20 (a) The structure of the FeMo cofactor Fe$_8$ cluster in nitrogenase. (b) The structure of the so-called P-cluster (Fe$_7$Mo) in nitrogenase (the identity of Y is not known).

similar cluster but with one of the Fe atoms replaced with a Mo. The Mo, perhaps deceptively, is located at one corner of this latter cluster and seems well away from the site of likely action in the nitrogen fixation process (Fig. 16.20(b)). To what extent had these structures been anticipated by the spectroscopic studies on the system? Rather well—consider the more complicated Fe$_7$Mo cluster (Fig. 16.20(b)). It had proved possible to isolate this species, along with enough of its surrounding groups to preserve its integrity, although it has proved impossible to obtain the crystals needed for a structure determination. Has the extraction process destroyed the cluster? EPR spectra of the native protein and the extracted cluster were both similar and unusual; the cluster had survived. Analysis of the cluster species gave a Fe:Mo ratio of not less than 6:1 and not more than 8:1, in accord with the observed 7:1. Mössbauer spectra showed that most, if not all, of the Fe atoms were involved in the interactions which gave rise to the unusual EPR signal. Measurements related to the EPR showed that at least five different types of Fe atom are present; in the proposed structure each of the seven is unique, although this does not mean that they will necessarily be distinguishable. Finally, EXAFS not only showed that the nearest neighbours of the Mo are S and Fe but that it is also coordinated to two atoms which are either N or O (as mentioned above, EXAFS cannot distinguish). For the (average) Fe, EXAFS revealed, correctly, that each Fe is bonded to three S atoms. Actually, six of the seven iron atoms present are three-coordinate, a very unusual situation which must surely be connected with the coordination of N$_2$ as the first step of the fixation process. Not surprisingly, this is another case in which no model compounds have yet been prepared, although one may be confident that they will. This does not mean that the biological process will soon be replicated by some model system. In the biological system, only the protein containing the bridging ferredoxin [Fe$_4$S$_4$] cluster has been found capable of activating the protein containing the Fe$_7$Mo clusters. No substitute has been found and there is no detailed understanding of this selectivity.

16.7 Protonation equilibria in bioinorganic systems

The molecules involved in many bioinorganic systems can be very complicated, they can interact not only with metal ions but also other species, such as protons, in the system. It is all too easy for the system to be so complicated that it appears that progress and understanding are scarcely possible. Fortunately, the situation is not as bad as this. In this section just one example is considered but it is sufficient to demonstrate the fairly self-evident conclusion that the more data there are available and, equally important, the more accurate these data, the greater the progress that can be made.

In Section 5.3 there was described the classical approach to equilibria in systems involving complexes. Although the existence of chelate effects was recognized, the section was largely based on monodentate ligands coordinating to a single metal ion. However, in a typical amino acid there will be at least two potential donors—a nitrogen and an oxygen; in peptides a plethora of possibilities will exist. In real-life biological systems some of the potential donor sites are likely to be protonated and, indeed, the nature of the complexes formed is pH-dependent. This pH dependence may well be crucial to the biological role of the system and so the question arises as to how the metal–ligand interactions vary with pH and how these interactions change the pH-dependent properties of the ligand itself. Although the former question is the more relevant and more interesting it is inseparable from the latter and in this section only the later problem, the pH-dependent properties of the ligand itself will be considered. This is complicated enough and it illustrates the general approach.

No problem arises if the ligand protonation is step-wise. One site is essentially fully protonated before protonation of another starts. More interesting is the case in which several sites are being protonated simultaneously, for it is reasonable to expect the presence of metal ions to significantly modify such equilibria. Consider the case of three competing sites, A, B and C, arranged along a chain. Using the notation of Section 5.2 the (macro) stepwise protonation constants will be (omitting in-coming protons for simplicity)

$$[A, B, C] \overset{K_1}{\rightleftharpoons} [A, B, C]H \overset{K_2}{\rightleftharpoons} [A, B, C]H_2 \overset{K_3}{\rightleftharpoons} [A, B, C]H_3$$

However, these K values, whilst in principle obtainable by following pH changes as a function of added acid (or alkali), make no recognition of the various protonation sites. Let us assume that A, B and C behave as independent sites, that the protonation of the group A is independent of the protonation of the rest of the molecule, for example. Denote the protonation constant k^A

$$
\begin{bmatrix} A \\ -B \\ C \end{bmatrix} \underset{k^A}{\rightleftharpoons} \begin{bmatrix} AH \\ -B \\ C \end{bmatrix}
$$

The above equilibria may therefore be rewritten as

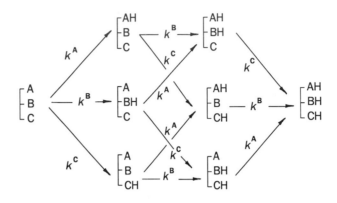

It follows that

$$K_1 = k^A + k^B + k^C$$

$$K_2 = 2(k^A + k^B + k^C)$$

$$K_3 = k^A + k^B + k^C$$

so that K_1, K_2 and K_3 are linearly related in this model. Clearly, determine these K values as accurately as we may, we will never be able to obtain the individual k values. Additional data on individual species—perhaps from NMR or UV spectroscopy—are needed. However, the errors in such additional data are very important. By definition of our problem, $k^A \approx k^B \approx k^C$ so the errors in the additional data must be small enough to enable the small difference between k^A and k^B, for example, to be determined.

Not surprisingly, the relationship $K_1 = 1/2K_2 = K_3$ is never accurately followed. The groups A, B and C are not totally independent. The protonation equilibrium at one site, A, say, is sensitive to the protonation at the others. Indicate the already protonated sites by suffixes so that instead of a simple k^A we now have k_-^A, k_B^A, k_C^A and k_{BC}^A (the dash indicates no protonated sites). We have been making the approximation that the group constant k^A is equal to all of the so-called *microconstants*, k_-^A, k_B^A, k_C^A and k_{BC}^A. Although the move to microconstants has increased the number of unknowns by a factor of four, the situation is not as bad as it may seem. Because of the change, K_1, K_2 and K_3 contribute three pieces of information, not just one. Further data are obtained from the detailed variation of the pH as acid is added in a titration. Similarly, the requirement that we reproduce not just one NMR spectrum but all its variants with pH, all provide more data,

$$4\text{-}(OH)C_6H_4CH_2CH(NH_2)\,CO_2H = [4\text{-}(OH)C_6H_4CH_2CH](NH_2)[CO]OH$$
$$= R(NH_2)OH$$

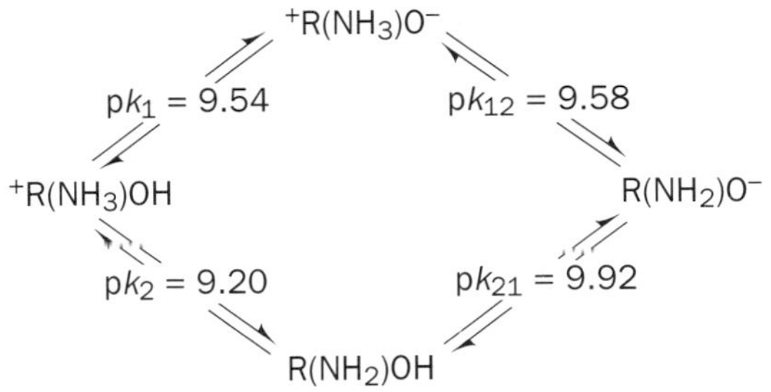

Fig. 16.21 Protonation of tyrosine.

$^+R(NH_3)O^-$

$pk_1 = 9.54$ $pk_{12} = 9.58$

$^+R(NH_3)OH$ $R(NH_2)O^-$

$pk_2 = 9.20$ $pk_{21} = 9.92$

$R(NH_2)OH$

although sometimes at the cost of yet more unknown—extinction coefficients in UV spectra for instance (although, for these, first approximations can be obtained from data on simpler molecules). Although to date they have been little exploited, infrared and Raman spectra, for instance, could provide yet more data. However, the differences between k_A^A, k_B^A and k_C^A, for example, are likely to be smaller than those between k^A and k^B. We need even more data at even greater accuracy to obtain yet more protonation constants. Lest the faint-hearted be tempted to give up the task, it should be pointed out that the group constants are of little use to us—the assumption that they exist is a negation of the hypothesis at the beginning of this section—that what happens at one particular donor site will influence the behaviour of the others.

In practice, of course, what happens is that the various data sets are brought together in a computer program, weighted according to their relative accuracies and some sort of least-squares fit made to these data. As more than one set of acceptable solutions may well be found, it is necessary to invoke some goodness-of-fit parameter to distinguish between the possibilities. It is here that errors become important—a unique solution is only possible if the noise (error) in the data fails to blur the distinction between it and alternative solutions. An example of such an analysis is for the protonation of tyrosine, $4\text{-}(OH)C_6H_4CH_2CH(NH_2)CO_2H$ (Fig. 16.21)—although the analysis above was for a more complicated example, most of the data available refer to, effectively, bidentate ligands.

The important lesson to be learnt from this example is the fact that the microconstants are, indeed, rather similar. It is not unreasonable to hope that when there is no alternative to assuming that such microconstants are identical that the error introduced is not too great.

Further reading

As has been made clear in the text, this chapter does not attempt a balanced overview of bioinorganic chemistry. Almost any current inorganic chemistry text will contain a section which details some of the areas excluded. A good, broad-based and forward-looking series of articles which is still worth reading is to be found in the November (pages 916–1001) issue of *J. Chem. Educ.* (1985) *62*. An overview of the subject is contained in *The Biological Chemistry of the Elements* by J. J. R. Fraústo da Silva and R. J. P. Williams, Clarendon Press, Oxford, 1991, although parts of the book are not really for the newcomer to the field. More information on the material in the present chapter will be found in the VCH series on *Physical Bioinorganic Chemistry*, A. B. P. Lever and H. B. Gray, Eds., particularly Volume 4, 1989. Other relevant data will be found in *Metal Complexes with Tetrapyrrole Ligands II* J. W. Buchler (Ed.), Springer-Verlag Berlin, 1991.

Other useful sources are

- 'Magnetic circular dichroism of hemoproteins' by M. R. Cheesman, C. Greenwood and A. J. Thomson, *Adv. Inorg. Chem.* (1991) *36*, 201.

- 'Natural and Magnetic Circular Dichroism, Spectroscopy on the Nanosecond Timescale' by R. A. Goldbeck and D. S. Klinger, *Spectroscopy* (1992) *7*, 17.

- *Biochemical Applications of Raman and Resonance Raman Spectroscopies* by P. R. Carey, Academic Press, New York, 1982.

- *EXAFS Spectroscopy, Techniques and Applications* B. K. Teo and D. C. Joy (Eds.), Plenum, New York, 1981.

- More recent is a review article 'EXAFS' by H. Bertagnolli and T. S. Ertel. *Angew. Chem., Int. Ed. Engl.* (1994) *33*, 45.

- 'Electronic Structures of Active Sites in Copper Proteins: Contributions to Reactivity' by E. I. Solomon, M. J. Baldwin and M. D. Lowery, *Chem. Rev.* (1992) *92*, 521.

- 'Active-site properties of the blue copper proteins' by A.G. Sykes, *Adv. Inorg. Chem.* (1991) *36*, 377.

- 'Long-range Electron-transfer in Blue Copper Proteins' by H. B. Gray, *Chem. Soc. Rev.* (1986) *15*, 17.

- 'The Synthetic Approach to the Structure and Function of Copper Proteins' by N. Kitajima, *Adv. Inorg. Chem.* (1992) *39*, 1.

- 'Calculating Equilibrium Concentrations for Stepwise Binding of Ligands and Polyprotic Acid–Base Systems' by E. Weltin, *J. Chem. Educ.* (1993) *70*, 568.

- 'X-ray Structure Analysis of FeMo Nitrogenase—Is the Problem of N_2 Fixation Solved?' D. Sellmann, *Angew. Chem., Int. Ed. Engl.* (1993) *32*, 64.

- 'The Iron–Molybdenum Cofactor of Nitrogenase' B. K. Burgess, *Chem. Rev.* (1990) *90*, 1377.

- *Biocoordination Chemistry: Coordination Equilibria in Biologically Active Systems* K. Burger (Ed.), Ellis Horwood, Chichester (1990).

Questions

16.1 Mushrooms of the *Amanita* family can contain surprisingly high concentrations of vanadium (in fly agaric up to $325 \, \text{mg kg}^{-1}$). It was originally believed that it was present as the compound shown in Fig. 16.22(a) but, currently, the compound shown in Fig. 16.22(b) is favoured. Given adequate supplies of the pure material, what measurements would enable a distinction between the two alternatives?

16.2 Two spectroscopic methods that have been used in bioinorganic chemistry but not discussed in the text are time-resolved infrared spectroscopy and time-resolved resonance Raman spectroscopy. Speculate on the types of investigation that might be made with these methods.

16.3 Starting with the A and C term spectra in Fig. 16.12 suggest how each would change as (a) the temperature of measurement is reduced, (b) the temperature of measurement is increased and (c) the magnetic field used in the experiment is increased.

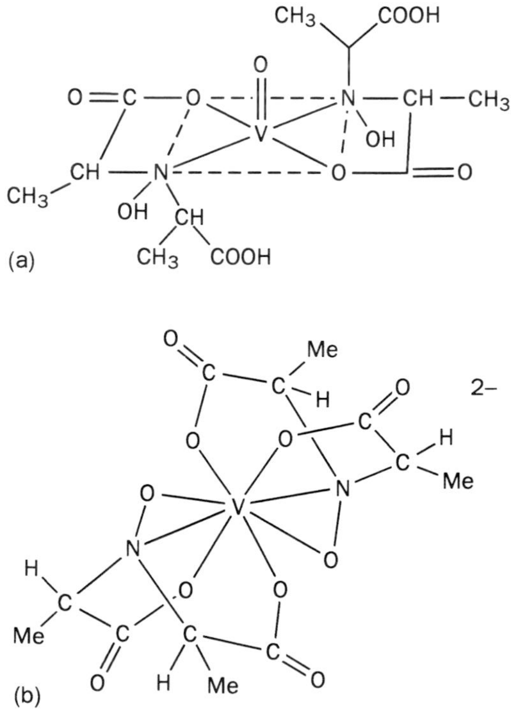

Fig. 16.22 Question 16.1.

17

Introduction to the theory of the solid state

17.1 Introduction

There have been many points in this book at which it has been apparent that the properties of aggregates of transition metal ions can differ from those of isolated molecules. This, in some measure, was a theme in the previous chapter and very much that of Chapter 15; it was met when discussing magnetic properties (Section 9.11), visible spectra (Section 8.11) and vibrational spectra (Sections 12.2.1 and 12.2.3). What, then, of the logical limit, the solid state? There has been an increasing interest in inorganic chemical aspects of the solid state in recent years, although it is a topic which is sometimes called material science. The subject covers the whole of the periodic table, not just transition metal species. Fortunately, the basic understanding and theory are common to all. It is the purpose of this chapter to give an introduction to this theory. Although it is a topic widely taught and studied, it seems the theory of the solid state is not always well understood by chemists. There are several reasons for this. First, crystalline materials have a great deal of symmetry, that of the translation operations that interrelate all the basic building blocks in the crystal (the term building blocks is preferred to molecules because not all crystals are molecular solids). The impact of group theory on chemistry is now so great that, given symmetry, chemists will look for the corresponding character table. Where, then, are the character tables appropriate to crystalline lattices? They are rather difficult to find. Not that they do not exist—a fragment of one is shown in Table 17.1—but the fact that, for an (idealized!) infinite crystal there is an infinity of translation operations, coupled with the fact that all character tables with which the chemist is familiar are square (as many rows as columns), means that we might well expect the full character table to be of dimensions infinity by infinity! Secondly, usually, chemists have learnt about crystals from crystallographers. This means that they have adopted

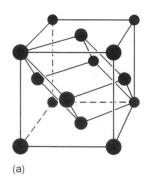

(a)

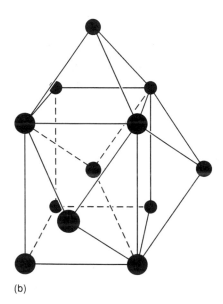

(b)

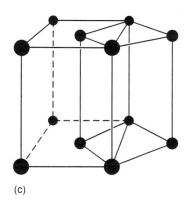

(c)

Fig. 17.1 Three centred lattices and associated primitive unit cells. The primitive is one-quarter of the volume of the face-centred cubic (a). For the body-centred tetragonal (b) and end-centred orthorhombic (c) it is one-half the volume.

Table 17.1 A fragment of the character table of a one-dimensional translation group T (not to be confused with the point group T). That for a three-dimensional group would be superficially similar. For the latter case the **k** and **t** would be vectors (a direction has to be specified in the three-dimensional case) and so they have been given as vectors in the table. The complex exponents indicate travelling waves (see the text); because the power of the exponentials must be (complex) numbers and **t** is a vector in real space, **k** must be in reciprocal space in order that the implied dot (scalar) products in the exponents (**k·t**) be dimensionless. The $t, t^2, t^3, \ldots$ at the head of each column are operators and correspond to translations **t**, 2**t**, 3**t**, ... The square, cube etc. form is used for the operators so that they operate properly on a function ψ, say. For the **k** irreducible representation above

$$t^2\psi = t(t\psi) = t(e^{i\mathbf{k}\mathbf{t}}\psi) = e^{i\mathbf{k}\cdot\mathbf{t}}\cdot e^{i\mathbf{k}\mathbf{t}}\psi = e^{2i\mathbf{k}\mathbf{t}}\psi$$

which is correct, whereas the use of the translation itself gives

$$2t\psi = 2\cdot e^{i\mathbf{k}\mathbf{t}}\omega$$

which is wrong (the answer has to be the character given for t^2 in the table)

T	E	t	t^2	t^3	$\ldots$
o	1	1	1	1	$\ldots$
k	1	$e^{i\mathbf{k}\mathbf{t}}$	$e^{2i\mathbf{k}\mathbf{t}}$	$e^{3i\mathbf{k}\mathbf{t}}$	$\ldots$
$-\mathbf{k}$	1	$e^{-i\mathbf{k}\mathbf{t}}$	$e^{-2i\mathbf{k}\mathbf{t}}$	$e^{-3i\mathbf{k}\mathbf{t}}$	$\ldots$
.	1	.	.	.	$\ldots$
.	1	.	.	.	$\ldots$

the language and approach of crystallographers. So, every self-respecting chemist knows the difference between a primitive cubic, a body centred cubic and a face centred cubic lattice. It perhaps comes as a surprise to learn that from the point of view of the theory of the solid state *all* lattices are primitive! The use of centred lattices is very convenient for the crystallographer but not for the theoretician or spectroscopist! As an illustrative example in Fig. 17.1 there are shown three representative centred lattices and the corresponding primitive unit cells. The present chapter has been written in such a way that this particular problem will be minimized. It is mentioned here because it is an important distinction, one that will certainly be encountered by the reader who wishes to pursue the subject in more detail. There remains the problem of the character tables of space groups and it is to these that we now turn. In doing so a rather unusual approach to the theory of the solid state will be followed but it is adopted in the hope that a different viewpoint will help circumvent the problems referred to above.

17.2 Nodes, nodes and more nodes

When discussing the bonding between planar, cyclic C_nR_n systems and transition metals, in outline in Section 10.1 and in more detail in Appendix 13, it was recognized that the π electron orbitals of the C_nR_n molecule are associated with different nodal patterns. Appendix 13 shows, but does not emphasize, the fact that different nodal patterns are associated with different irreducible representations of the relevant point group (different labels of symmetry species in Appendix 13). In fact, this is general. Symmetry labels—and character tables—describe different nodal patterns (generally, three-dimensional) in a very concise and convenient way. Given the problem of the very large size of the character tables of space groups (the infinite size problem disappears because real crystals are not infinite—but a full character table would still be enormous), can progress be made by talking in terms of nodal patterns, rather than character tables? Fortunately, the answer is yes.

In Section 10.1 (and Appendix 13) it was seen that as the number of CR units in a planar cyclic hydrocarbon C_nR_n increases, so too does the possible number of nodes in the p_π molecular orbitals. The maximum distinguishable number is that obtained when there is one node bisecting each C–C bond. Suppose that one C–C bond is broken and the C_nR_n molecule straightened out so as to become linear (Fig. 17.2). When the bond is broken, the corresponding nodal plane must be deleted and the maximum number becomes $(n - 1)$. As we shall see, there is a continuity between cyclic and linear C_nR_n systems in that just as the different nodal patterns in the cyclic system are associated with different symmetry species, so too for the linear—provided that the ends of the molecule can be ignored. Since, in the

Fig. 17.2 On the left are shown the nodal patterns of the six molecular orbitals of benzene (they are shown in more detail in Fig. 17.6). These patterns carry over into the diradical generated when a C–C bond is broken (centre) and thus into the linear structure (right). The actual numbers of nodes in the initial and final patterns should be compared.

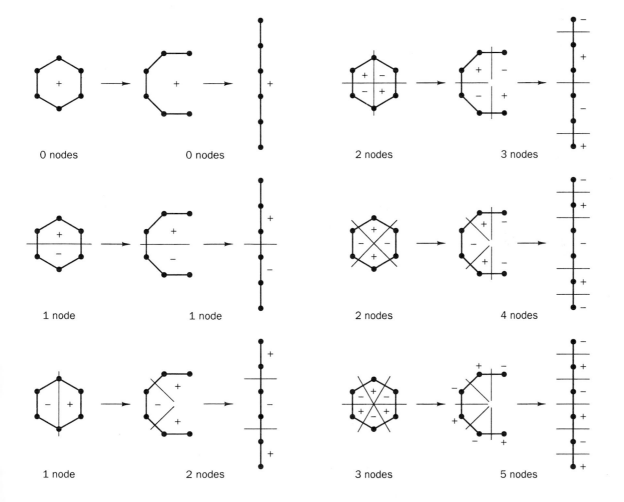

0 nodes   0 nodes

1 node   1 node

1 node   2 nodes

2 nodes   3 nodes

2 nodes   4 nodes

3 nodes   5 nodes

present case, in which our concern is with the solid state, the 'molecule' is an almost infinite lattice the ends can safely be ignored—as was done in Fig. 17.2. This figure contains the nodal patterns for a linear polyene and these will now be used to obtain the nodal patterns appropriate to crystals. The theory we shall develop is applicable to more than orbitals and electronic structure but, for simplicity, we will confine our discussion to this case and relate the crystal orbitals to the molecular orbitals of polyenes. Although the molecular orbitals of polyenes consist of p_π atomic orbitals there is no reason to expect that the crystal orbitals will also be π, but for the moment it is the nodal plane patterns that are of interest, not the precise nature of the orbitals involved. In the simple picture of a linear molecule presented above, all the nodal planes were parallel. So, too, in a crystal. Crystal molecular orbital nodes lie in parallel sets. Consider the two-dimensional crystal shown in Fig. 17.3(a). Despite its clearly finite size, it will be taken to be infinite; its rectangular shape is meant to indicate that the primitive translation vectors (the vectors that relate adjacent translationally related points) along the two different axis directions are not of the same size.

If a crystal molecular orbital had just one nodal plane it could be represented as shown in Fig. 17.3(b). Extending this pattern, two nodal planes could occur in the way shown in Fig. 17.3(c). Equally, however, a

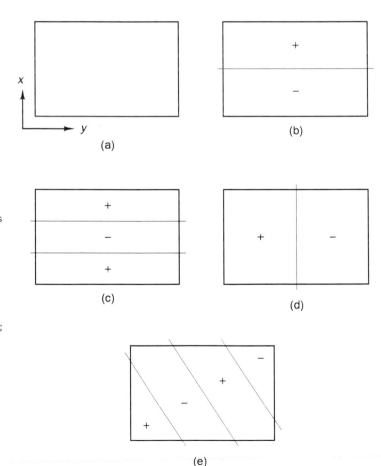

Fig. 17.3 Some of the possible nodal patterns in a two-dimensional crystal. The crystal is indicated in (a); the translation vector in the y direction has a magnitude which is one and a half times that in the x direction. If a phase (+ or −) were added to (a) it would show the no-node combination. One-node combinations are shown in (b) and (d), a two-node in (c). However, nodal patterns are not limited to those with nodal planes perpendicular to x or y; a three-node inclined pattern is shown in (e).

single nodal plane could be drawn as shown in Fig. 17.3(d)—or oriented in quite a different direction, as in Fig. 17.3(e) where, to be different, three nodal planes are shown. Are all of these possible alternatives equally acceptable? If so, how can this complexity be handled? It seems that we must not only somehow systematically classify a large number of different nodal plane patterns but also the large number of different possible directions in which they may lie. Following from the generalizations made at the beginning of this section, it seems likely that each different nodal pattern/direction combination corresponds to a different symmetry species. This speculation proves to be correct, so that we are moving in the right direction, but this recognition is no help in resolving the problem of classifying the multitude of different nodal patterns that can arise in a three-dimensional crystal. It is necessary to use some cunning!

Our concern is with the electronic structure of crystals and it is our hope that it will prove possible to describe the symmetry properties of the crystal orbitals of a crystal, for example, by something akin to the irreducible representations used to describe the molecular orbitals of molecules. By analogy with the molecular case it would be expected that the symmetry species of the crystal orbital which contains no nodes would be important (in the molecular case all spectroscopic selection rules are related to it, for instance). This will be the crystal counterpart of a point group totally symmetric irreducible representation (which contains no symmetry-required nodes). For the case of the crystal, represent this zero-node symmetry pattern by a dot on a piece of paper. Now do the same for corresponding parallel one-node, two-node, three-node and so on patterns (a direction will have to be chosen to which the nodes are all perpendicular but this apparent arbitrariness will pose no problem). Clearly, the dots should not be drawn randomly scattered around but, rather, related to one another in a sensible way. In Fig. 17.4 is shown how it can be done for the linear polyene of Fig. 17.2. The task is now to do the same for the crystal of Fig. 17.3. A difference is that each dot will now carry two indices, indicating the number of nodes in each of the two directions in the two-dimensional crystal of Fig. 17.3. This is easy when the nodes are perpendicular to either the x or y axes.

The general pattern is evident from Fig. 17.5(a), which should be carefully studied. It is easy to complete the pattern by extrapolation, as shown in Fig. 17.5(b), although extrapolation may not be enough. This is because although it is easy to say, for example, 'two nodes cutting the y axis and one cutting the x', it is much more difficult to picture this—how does one node disappear somewhere between the y and x axes? This is an important problem which will be explored shortly but it is helpful first to make some comments on Fig. 17.5(a). Notice that the greater the distance between nodes, the shorter the distance of the corresponding dot from the dot representing the origin (which we take to be 0, 0). Indeed, Fig. 17.5(b) could well have been drawn with an exactly reciprocal relationship. Had it so been drawn, then Fig. 17.5(b) cannot be quite correct, for it has already been seen (Fig. 17.3(a)) that in our two-dimensional crystal the translation vectors along y are longer than those along x. Evidently, to be accurately reciprocal, in our pattern of dots the separations along y must be appropriately shorter than those along x. Figure 17.5(c) is the corrected form of Fig. 17.5(b). Here, each point

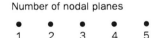

Number of nodal planes

0   1   2   3   4   5

Fig. 17.4 A dot map which can be used to symbolically represent the entire set of nodal plane patterns at the right-hand side of Fig. 17.2.

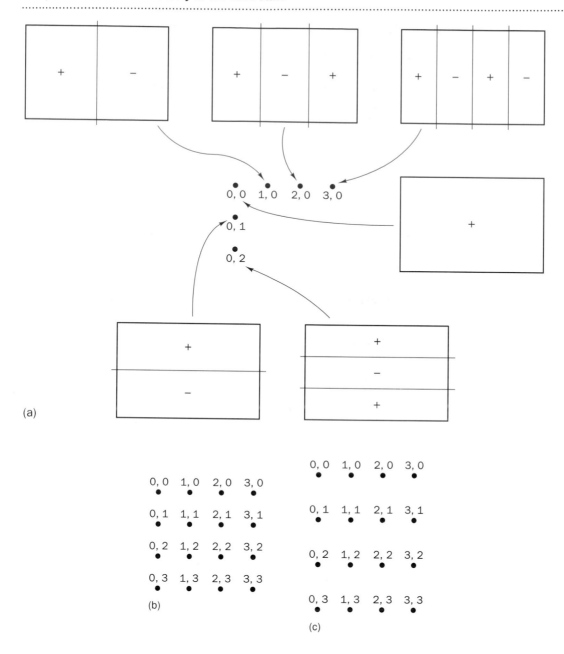

(a)

(b)

(c)

Fig. 17.5 (a) Some of the nodal patterns of the two-dimensional crystal of Fig. 17.3 with all nodes perpendicular to either x or y, together with a dot-map representation of them. (b) The dot map of Fig. 17.5(a) extended to include nodal patterns inclined to x and y. (c) Figure 17.5(b) corrected for the fact that the magnitude of the translation along y is 1.5 times that along x in Fig. 17.3. Here, the separations are in the inverse ratio, 0.67:1.

represents a unique nodal pattern and this, of course, is just what irreducible representations do also. It is not surprising to learn that there is a one to one correspondence between the dots and the irreducible representations of the translational group of the crystal. We may not have full details of these irreducible representations but, at least, we have a way of representing them; each irreducible representation corresponds to a dot in a figure such as Fig. 15.5(c). It is easy to see, in a general way, how to extend the pattern of Fig. 17.5(c) into three dimensions. We shall return to this task shortly, when it will lead us to the important concept of the Brillouin zone.

17.3 Travelling waves and the Brillouin zone

There is a major defect in our argument so far and removing it will also remove the problem of representing the 'two nodes along one axis, one along the other' nodal combination in dot patterns such as that in Fig. 17.5(c). This defect becomes evident when diagrams such as those in Fig. A13.1(b) are carefully studied. One of these is reproduced in Fig. 17.6; it shows the π molecular orbitals (MOs) in benzene. In Fig. 17.6 there are two singly degenerate MOs. In them, each carbon p_π orbital makes an equal contribution (albeit, with differences of phase). In the individual members of the other, degenerate pairs, of MOs, equal carbon p_π contributions do not occur. However, if the carbon p_π contributions to the degenerate pair, taken together, are calculated, then the carbon p_π orbitals *are* found to make equal contributions to the pair. To actually check this requires explicit expressions for the MOs and so it can only be qualitatively checked for Fig. 17.6.[1] An alternative, quantitative, check is provided by the knowledge that the one-node functions vary round the benzene ring as $\cos \theta$ and $\sin \theta$, respectively, so that the contribution of any carbon p_π orbital, the sum of the squares of the coefficients at that carbon atom, varies as

$$\cos^2 \theta + \sin^2 \theta = 1$$

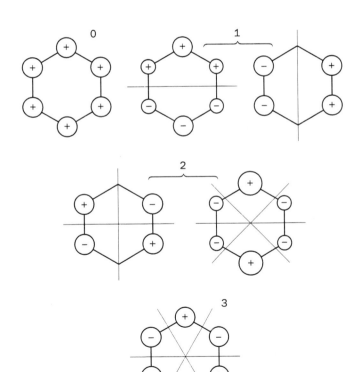

Fig. 17.6 The p_π molecular orbitals of benzene. The size of the circles approximately show the magnitude of the coefficient of the molecular orbital at each atom.

[1] For a linear polyene such as that shown in Fig. 17.2, some symmetry species appear more than once. It is then necessary to sum over all functions transforming as the same irreducible representation before all carbon p_π orbitals are found to make equal contributions.

which of course is a constant independent of θ (the two-node functions vary as $\cos 2\theta$ and $\sin 2\theta$ and so lead to the same conclusion)—and any concerns about how θ is defined vanish. This pattern is a very general one, applying whether it is a one-, two- or three-dimensional case which is under consideration. As an example of a three-dimensional case (and real crystals are three-dimensional), explicit expressions for the ligand group orbitals in an octahedral complex are given in Tables 6.1 and 6.2. If the reader wishes, the pattern can be checked for these (again, sum the squares of coefficients; there can be one, two or three orbitals contributing to a sum).

How does all this apply to the nodal patterns associated with the two-dimensional crystal under study? In a crystal molecular orbital, the relevant orbital on an atom that is situated close to a node will appear with a small amplitude, whereas those on atoms well away from nodes will have large amplitudes—amplitudes are what nodes are all about, be it here or in a vibrating violin string. How then, if they have different amplitudes, can all corresponding atomic functions make equal contributions, for their contribution is just the square of their amplitude? Within our model, they cannot, so something is wrong with the model. The way out is simple. The waves, and nodes, are travelling, not stationary (and so the wave-functions complex, involving i, $\sqrt{(-1)}$). So, over a period of time, as the waves pass through the crystal, all the atomic orbitals *do* make equal contributions.

There are two immediate consequences of this refinement. First, for every wave going in one direction there must be another, equivalent, going in the opposite direction—it leads to the same conclusion in terms of orbital involvement and so the two cannot be distinguished, one is as good as the other. Now, an important point. Although the waves are closely related, they do not transform as a pair (if they did, they would have a pattern that in a point group would carry an E irreducible representation label). Remember, we are working with translations and only translations. There is no combination of pure translations which either interchanges or mixes the two wave motions. Functions which transform as doubly degenerate irreducible representations of a point group are always mixed or interchanged by at least one operation of a group—this is why they are degenerate.[2] It follows that the two travelling waves, although related, transform as different irreducible representations of the translation group.

Secondly, the problem posed by a difference in the number of nodes along two perpendicular axes disappears. This is demonstrated in Fig. 17.7 which shows some time-sequenced pictures of a travelling $4 + 2$ nodal pattern, most readily seen for a finite crystal, although the general pattern can clearly be extended in principle to an infinitely large crystal. The meaning of the additional points included in Figs. 17.5(b) and (c) compared with Fig. 17.5(a) becomes clear; they correspond to travelling waves moving at an angle to both x and y axes. An important lesson, therefore, is that in a crystalline solid one should think in terms of travelling waves, not stationary ones, because it is possible to attach symmetry labels to travelling waves but not, in general, to stationary waves.

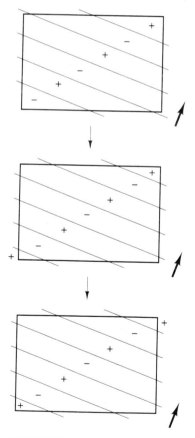

Fig. 17.7 Time-sequenced pictures of a travelling wave in the crystal of Fig. 17.3. Such travelling waves can be along axial directions as well as inclined to them.

[2] The case of separable degeneracy is not covered by this statement but it is true for all of the point groups that normally concern the chemist.

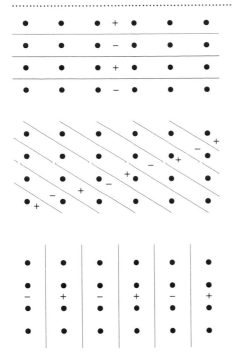

Fig. 17.8 Some standing-wave patterns for a two-dimensional crystal. The nodal planes lie at exactly half the distance between atomic layers.

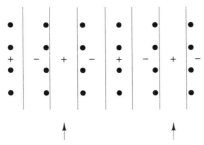

Fig. 17.9 A wave pattern in which the separation between nodes is less than the separation, in the same direction, between atoms. The result is that in two regions (arrowed) there are phases but nothing to which these phases can refer.

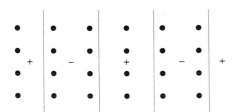

Fig. 17.10 A wave pattern which gives the same amplitudes as those in Fig. 17.9 but which has an internode separation which is greater than the interatomic separation in the same direction.

Can there ever be stationary waves in a crystalline solid? The answer is yes, under very special circumstances. So, if, along an axis, the nodal planes exactly bisect interatomic planes, there is no need to invoke travelling waves to obtain equal atomic orbital contributions. Each atom is at a point of greatest amplitude (which may be positive or negative) and so all orbitals appear with the same coefficient although this is either positive or negative, matching the sign of the amplitude. Pictures for three such waves in the two-dimensional case are shown in Fig. 17.8. All three satisfy the requirements imposed on functions which transform as an irreducible representation of a space group. The example considered is that of a two-dimensional space group but the conclusion is general. Standing wave irreducible representations exist only when there is a matching between the nodal separation of the wave and the magnitude of the corresponding translation vector.

An important point follows. Within the context of the present discussion, it is meaningless to have wavelengths shorter than those shown in Fig. 17.8 in any of the three nodal plane directions shown there. A shorter wavelength would mean that somewhere in space we would have a pattern akin to that shown in Fig. 17.9. However, in Fig. 17.9, for two of the four regions with positive amplitude there are no atoms within the region and so no orbitals that could have this amplitude! So, those amplitudes are meaningless, and with them the nodal pattern that gave rise to them. Another way of looking at this is to compare Fig. 17.9 with Fig. 17.10. Qualitatively, at least, the atomic orbital amplitudes are the same in Figs. 17.9 and 17.10, and a detailed study shows that it is not difficult to ensure that they are also quantitatively the same. If the pattern in Fig. 17.9 is not meaningless, it certainly is redundant. We see, then, that nodal patterns such as those shown in Fig. 17.8 are the limit that we need to consider, be it two-dimensional or three-dimensional space about which we are talking. Such nodal patterns can be incorporated in dot maps of the type given in Fig. 17.5(c)—enormously extended—and represent the limit of the dot patterns. Further dots can be added outside this limit but, within the present context are either meaningless or redundant.[3]

Wave patterns in very similar directions will terminate at very slightly displaced, adjacent, points. A family of such points will define a surface which turns out always to be planar, except where two or more such planes intersect. Most important is the three-dimensional case, which we now consider and which follows, without complications, by extension of the two dimensional. The three-dimensional limiting surfaces define some rather beautiful shapes, the shape depending on the particular lattice under consideration. Examples of some of them are given in Fig. 17.11. The shapes shown in Fig. 17.11 are pictures of the *Brillouin zones* of the solids. A Brillouin zone can be thought of as containing a multitude of closely packed points, each with a unique set of labels and each point corresponding to a unique

---

[3] The present context is one in which there is a single set of points, presumably representing atoms, interrelated by translation vectors. Had there been units made up of several points, grouped together and interrelated between themselves by, say, point group operations, then the 'empty spaces' in figures, such as Fig. 17.9 would not be empty. In such a case there would be a distinction between Figs. 17.9 and 17.10; the additional node patterns would be neither meaningless nor redundant. They are associated with the second, third, and so on, Brillouin zones.

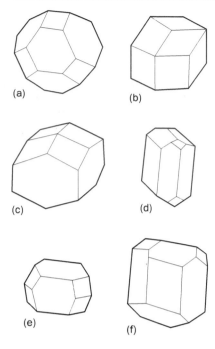

Fig. 17.11 Drawings of the shape of some Brillouin zones: (a) face centred cubic, (b) trigonal, (c) tetragonal, (d) orthorhombic, (e) monoclinic and (f) triclinic. In each case an axis of highest symmetry is arranged approximately vertically in the plane of the paper. In principle, there is one general shape of Brillouin zone for each of the 14 Bravais lattices but, because of the freedom allowed by changes in relative axis size for the lower symmetry groups, there is a total of 24 qualitatively different shapes.

(travelling) wave pattern. Each point corresponds to the translation group equivalent of an irreducible representation in a point group character table, so the Brillouin zone, containing all points, is, in a sense, the equivalent of a translation group character table. As we have seen, the further that one of the closely packed points is from the origin (the origin is at the centre of the Brillouin zone and is the no-node combination), the shorter the distance between nodes in the corresponding travelling wave pattern in real space. Because of this reciprocal relationship one commonly talks of *reciprocal space*; all Brillouin zones are in reciprocal space. Conventionally, the vector drawn from the origin to a particular point is called the '**k** vector' for that point. As we have also seen, for every vector **k** there exists a vector $-\mathbf{k}$. The origin is often referred to as $\mathbf{k} = 0$ in phrases such as 'At $\mathbf{k} = 0, \ldots$'. It is important because $\mathbf{k} = 0$ is the equivalent of the totally symmetric irreducible representation of a point group. As a result, just as for point groups, the spectroscopic selection rules for crystals correspond to non-zero integrals and this means integrals that they transform as $\mathbf{k} = 0$, which is why phrases such as at $\mathbf{k} = 0, \ldots$ will be encountered. The label **k**, with a suitable plethora of subscripts to indicate the point in the Brillouin zone to which it refers, is the equivalent of a point group irreducible representation label such as $A_1, B_{2g}, \ldots$ with which the reader will be familiar. Not surprisingly, instead of reciprocal space the term '**k** space' will be encountered. Just to add a bit of physics, and remembering that we are talking about travelling waves, the name momentum space will also be found in books on solid state physics.

It is to be emphasized that in the above discussion we have been talking about translations, and only translations. A particular pattern of translations may well lead to the automatic generation of rotational axes. So, it takes nothing but translation operations to generate a cubic lattice. However, when we look at the lattice so generated we find that it also has threefold, fourfold and twofold rotation axes. These are something additional. For instance, the way that a cubic lattice is brought into coincidence with itself by a C_4 rotation, the way that the points map onto one another, cannot be replicated by *any* combination of translation operations applied to the lattice. Inclusion of operations other than pure translation operations is important for those interested in the details of the theory of space groups, but is not essential for our discussion. The *lattices* are *always* of the highest symmetry compatible with any crystal type,[4] be it cubic, tetragonal, orthorhombic, hexagonal, trigonal, monoclinic or triclinic; addition of point group operations can never increase the symmetry although the absence of such operations can lower it. Clearly, it is convenient to restrict discussion to the high-symmetry cases. These cases have given us the Brillouin zone and a myriad of **k** vectors—and that's enough! The point group operations serve to interrelate **k** vectors that are otherwise distinct. So, in any cubic lattice, invariably of O_h symmetry, a **k** vector in a general position in **k** space has 47 symmetry-related equivalents (there are 48 operations in the O_h point group). Diagrams showing energy levels across a Brillouin zone (which will be met shortly) would only show one of the 48.

[4] For instance, *all* lattices are centrosymmetric, contain centres of symmetry, although not all crystals are centrosymmetric.

17.4 Band structure

The content of the three previous sections is sufficient to enable us to consider a specific example which will form the basis for application of the basic theory. Let us consider a crystal composed of a diatomic molecule AB, with just one molecule in a unit cell[5] (if there were several, they would be interrelated by point group-type operations and these we have just excluded). All of the A are interrelated by pure translations, as too are the B. We first consider just the A; most of the development we need will follow—the inclusion of B will not add greatly to it. Each and every atomic orbital of A spans the entire Brillouin zone of the space group. That is, there is a separate combination of each and every orbital of A corresponding to each and every possible nodal pattern, every $\mathbf{k}$ vector. As the nodal patterns vary, as $\mathbf{k}$ varies, so too the corresponding (crystal) molecular orbital energies must be expected to vary. Each and every single energy level in the isolated A atom gives rise to an energy band in a crystalline solid. Let us look at this in more detail; we do this by considering just one orbital of A, and select one that leads to the revelation of all of the important phenomena; we select a p_x orbital. To simplify the pictures that will arise, but for no other reason, the discussion will be confined to the xy plane, these two axes, of course, being mutually perpendicular. There is only one A atom in the (two-dimensional) unit cell and so just one p_x orbital in each unit cell.

Figure 17.12 shows pictures of the combinations of these p_x orbitals appropriate to three points in a two-dimensional $\mathbf{k}$ space. To avoid having to work with travelling waves the points have been chosen to be at the centre ($\mathbf{k} = 0$) and two surface points of the Brillouin zone, but the intermediate points follow by interpolation. At the centre of the zone, where, as shown in the left-hand diagram, the pattern is the same in all unit cells, along the y axis there is an in-phase π overlap; that is, a π bonding interaction. However, along the x axis the interaction is σ antibonding. The edge of the Brillouin zone along the x axis has a pattern in which there is an alternation of phase between unit cells along the x axis. This is pictured at the top right in Fig. 17.12. Examination shows that the interactions are now π bonding along the y axis (the same as before, nothing has changed along the y axis) but now σ bonding along x. Because a σ antibonding interaction has become bonding, the x axis zone edge pattern is lower in energy than that at the zone centre. Conversely, at the y axis zone edge (lower right in Fig. 17.12), the σ and π interactions are both antibonding and the pattern has an energy which is higher than that at the zone centre.

These energy patterns are brought together in Fig. 17.13 which shows plots along two directions in the Brillouin zone. These are the directions corresponding to the x and y axes of Fig. 17.12 but, because the Brillouin zone is in reciprocal space, they are called x^* and y^*, respectively, to

---

[5] This is the first time that the name unit cell has been used, a name that seems innocuous enough. In fact, some care is needed in its use because there is no unique definition of unit cell for any crystal structure. For *any* crystal structure there is an infinity of acceptable choices of unit cell—the only requirement on it is that it contains one primitive structural unit and that it generates the entire crystal by translation operations alone. A unit cell need not have six faces and its faces need not be planar (just as tiles with curved edges can cover a surface). It is often convenient to choose as unit cell a volume defined by the primitive translation vectors themselves but, equally, for other purposes this can be an inconvenient choice.

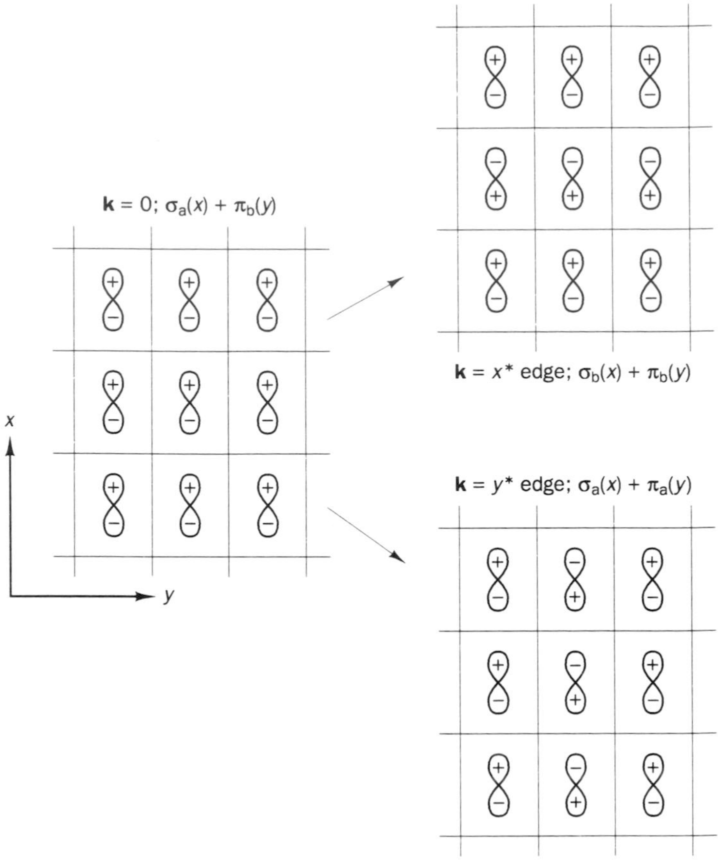

Fig. 17.12 Standing-wave patterns for the example described in the text. To the left is the zone-centre pattern and two zone-edge patterns to the right. The type of bonding (σ or π) is indicated together with whether the interaction is bonding ($_b$) or antibonding ($_a$) and the axis along which these interactions are directed.

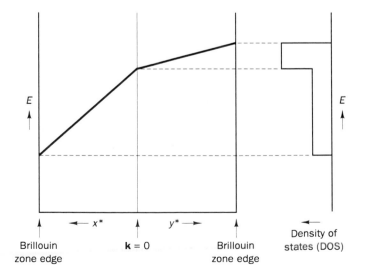

Fig. 17.13 An energy-level diagram for the two directions in the Brillouin zone shown in Fig. 17.12. For simplicity, a linear interpolation has been used between the points pictured in Fig. 17.12 and this gives rise to the simple block form of the density of states on the right. The reader is cautioned that there are good reasons why this linear interpolation is not valid; it is used here solely on grounds of simplicity.

distinguish them from x and y.[6] Because the change from bonding to antibonding is expected to have larger energetic consequences for σ than π interactions, the slope of the plot to the x^* zone edge is greater than that to the y^*. For simplicity, the interpolations from zone centre to edge have been made linear, although in reality they would be curved. To the right of Fig. 17.13 is shown the resulting density of states plots (DOS; these were first met in Fig. 15.25). It is clear that what was an individual level in the isolated atom has become a band in the solid—a somewhat unrealistic rectangular-sectioned band, but a band nonetheless. Further, the energy varies across the band (in jargon, the '*dispersion* of the band'); the energy varies with position in **k** space. The solid-state physics literature is full of plots of dispersion curves, usually along directions in reciprocal space that correspond to high-symmetry directions in the crystal; we shall meet examples shortly.

It is evident that the greater the interaction between adjacent translation-related orbitals, the broader will be the resulting band. Inner orbitals will usually give narrow bands, outer orbitals will tend to give broad bands. Increase in pressure on a crystal will tend to increase the interorbital interactions and thus to broaden the bands, this not only affecting the outermost orbitals but, with sufficient pressure, the inner ones too. So, it is calculated that the 1s orbitals of atomic hydrogen give rise to a band such that hydrogen becomes a metal if it is subjected to ca. 10^6 atmospheres (10^{11} N m^{-2}) pressure. Returning to our example, if the p_x orbital of A was empty, so too will be all levels of the band that have been generated from it. If the p_x orbital contained a single electron the corresponding band would be half-filled (if the bottom half of the band was filled then the electrons would be paired up). If the A p_x orbital were filled, so too will be the band. Unless, of course, something happens to change the situation.

There are two important things that can happen, the first of which will have to be discussed at some length. This arises from the recognition that bands adjacent in energy can overlap with each other. So, if a band corresponding to a doubly filled A orbital overlaps with that corresponding to an empty A orbital, what results is a single band (really, composed of two overlapping bands) which is half-filled. Although such overlapping is likely to be of common occurrence, as we shall see, one has to be careful—overlapping can be avoided. In the example considered above it was seen that interactions between p_x orbitals changes from antibonding to bonding with increase in **k** along the x^* axis. As shown in Fig. 17.14, the corresponding change in s orbitals is the opposite, from bonding to antibonding. Now, it is generally true that s atomic orbitals lie below the corresponding p, so the different behaviour of s and p_x along x^* means that they are potentially set on a collision course! In fact, they interact; this interaction can lead to a mutual repulsion, a so-called avoided crossing, as shown in Fig. 17.14 (the energy levels do not cross although the wavefunctions behave as if they did). In such cases, increase in pressure increases the separation between two bands. Although avoided crossings are not uncommon, more general is band overlap. Such overlap is not limited to pairs of orbitals—there is no limit on the number of energy bands that can overlap with each other. Just as in

---

[6] In the general case, where x and y are inclined at an angle of other than 90°, care has to be taken in defining the corresponding axes in **k** space.

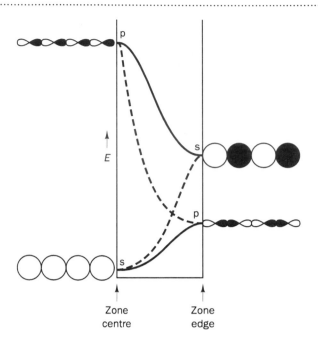

Fig. 17.14 An avoided crossing in a simple one-dimensional case. At **k** = 0 the interactions between adjacent s orbitals is bonding and between the p_σ it is antibonding. At the zone boundary the s orbital interaction is antibonding and the p_σ is bonding. Instead of the energy levels crossing, the s and p_σ orbitals mix so that in the middle of the diagram (at the mixture indicated by the crossing of the dotted lines) there are two sp hybrids involved. By the zone boundary they have become unmixed again.

a molecule there is a highest occupied molecular orbital (a HUMO) which is of importance, so there is a crystal equivalent, of no less importance.

The crystal equivalent is called the *Fermi Surface*. This may be pictured for the p_x example discussed above and which formed the subject of Figs. 17.12 and 17.13; such a picture is given in Fig. 17.15 for the case of an incompletely filled band. Notice particularly the way that the Fermi surface depends on direction in **k** space. In some directions the Fermi surface shown in Fig. 17.15 is quite close to **k** = 0; in others it is close to the zone surface. When the Fermi surface lies within a band a conductor results (travelling waves help to explain how electron migration occurs under the influence of an electrical field). When the Fermi surface occurs at the top of a band, the band is full and an insulator results. When there is a small gap between the top of a full band and the bottom of an empty one, a semiconductor results; semiconductors are usually classified in terms of their band gap. Two important semiconductors, Si and GaAs, have band gaps of 1.1 and 1.4 eV respectively; for comparison that of diamond is 6.0 eV.

The Fermi surface defined above was defined at a very low temperature, formally 0 K. As the temperature is raised the sharp horizontal line in Fig. 17.15 which represents the Fermi surface becomes increasingly blurred as electrons are thermally excited into levels which at 0 K were empty. In semiconductors, with a small band gap by definition, electrons are similarly thermally excited from the top of the highest filled band (the *valence band*) into the bottom of the lowest empty band (the *conduction band*). Their electrical conductivity increases with temperature. Add the possibility of incorporating into such a semiconductor impurity atoms which have energy levels which fall within the band gap, together with the possibility of bringing materials with different band gaps into physical contact, and one has the basis of modern semiconductor technology. Metals, of course, have conductivities which decrease with temperature. This may most readily be

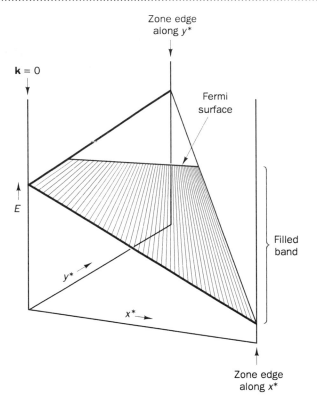

Fig. 17.15 A possible Fermi surface obtained by partially filling the band structure indicated in Fig. 17.13. Because the example is two-dimensional the Fermi surface is here a line. For simplicity the zone edge which runs from the x^* to the y^* axis has been shown as a straight line. Really, it should be more faceted, as a cross-section of any of the Brillouin zones shown in Fig. 17.11 demonstrates.

understood somewhat colloquially. Electrical conductivity occurs because the application of a voltage difference changes the energy level pattern at the Fermi surface, travelling waves that serve to transfer electron density towards the positive pole being preferentially populated. They have defined **k** values. One may think of one such wave setting out from the negative pole. Because of thermal vibrations of the atoms the lattice docs not quite have the regularity required for the accurate definition of its **k** value. Mismatches occur and the wave is attenuated on its way to the positive pole. Clearly, the higher the temperature the greater the thermal vibrations and the greater the attenuation. The same phenomenon occurs for semiconductors, of course, but the increase in thermal population of the conduction band with temperature dominates. One final word: the Fermi surface of Fig. 17.15 has been constructed in such a way as, hopefully, to make understanding easier. It is a highly distorted, unreal Fermi surface. Reality arrives in the next section!

So far we have restricted our discussion to the A atoms and considered those things that can modify the pattern originally derived, one in which the occupancy of a band is determined solely by the occupancy of the orbital from which it is derived. Of course, everything that has been said about the band structure derived from the orbitals of A hold for the orbitals of B also. Next it has to be recognized that there can be interaction between the bands derived from the orbitals of A and B—this is the equivalent of bonding in an isolated AB molecule. In molecules, interaction only occurs between orbitals of the same symmetry. So, too, in solids. Interaction between crystal orbitals can only occur when the crystal molecular orbitals (which are commonly referred to as *Bloch functions*, although the use of this term is not confined to electronic wavefunctions—it applies to any function which transforms as an irreducible representation of the translation group) have the same symmetry. That is, they have the same **k** vectors. It follows that such interactions are not required to be constant across the Brillouin zone. For example, the energy difference between two Bloch functions at **k** = 0, the zone centre, may be very different from the corresponding energy difference at a point on the surface of the Brillouin zone. It is worthwhile at this point to emphasize that a Brillouin zone is not, of itself, an energy surface, although such an interpretation is sometimes encountered. Equally, it is not a hollow shell, although sometimes drawn in this way. As we have seen, it is perhaps best regarded as a compact—and perhaps solid—mass of points of the sort shown in Fig. 17.5(c) in which each individual point represents a unique irreducible representation.

17.5 Fermi surface

In the above section the concept of the Fermi surface, as the solid state counterpart of a HOMO, was introduced. Just as HOMOs are important in molecules, so Fermi surfaces are of vital importance in crystalline solids. The object of the present section is to attempt to answer questions such as 'what does a real Fermi surface look like'? (the one introduced in the previous section was highly unreal) and 'how are they determined—if they can be'? It will avoid complications of the type encountered towards the end of the previous section if we consider only solids with a single type of atom. Because they show the way the bands are filled, it is more illuminating also to confine the discussion to metals. It is clear that the more electrons that there are and the more valence shell electrons, the more complicated will be the behaviour of energy levels and their occupancy over the Brillouin zone. We therefore start with a simple case, that of metallic sodium. Sodium has one relevant valence shell orbital, a 3s, and just one valence shell electron. The structure of the metal is body centred cubic (bcc), for which a picture of the Brillouin zone is shown in Fig. 17.16(a). We would expect the band to be half-filled and the electrons paired; this is just what happens. The Fermi surface is shown in Fig. 17.16(b); it is, with but minor deviations, spherical and occupies half of the volume of the Brillouin zone. As indicated in Fig. 17.14, s orbital interactions become increasingly antibonding with **k** and so the Brillouin zone is filled from the centre outwards; for other orbitals with other interaction patterns it may be the boundaries of the Brillouin zone which are filled first. Note from this last sentence how easy it is to equate Brillouin zone with energy, something which, as has been said, is to be avoided. There is one other lesson to be learnt from Fig. 17.16(b). This is that it is customary in solid-state theory to label various parts of the Brillouin zone. For all Brillouin zones the centre is labelled Γ; the other labels, like the labels for the irreducible representations of point groups, are perhaps best thought of as having a unique meaning for each Brillouin zone. For the bcc[7] Brillouin zone, the point P is on the zone boundary, along a threefold axis, and the point H is similarly on the boundary, along a fourfold axis.

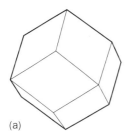

(a)

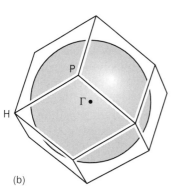

(b)

Fig. 17.16 (a) The Brillouin zone for a body centred cubic (bcc) lattice and (b) the Fermi surface of sodium, which has a bcc structure. This surface is essentially spherical and nowhere touches the zone surface. Notice how, to facilitate discussion of Brillouin zones and Fermi surfaces, labels are given to some points in the Brillouin zone (those shown are an incomplete list of those used for the bcc Brillouin zone).

---

[7] bcc = body centred cubic; even though one has to work with the corresponding primitive cells, the language of the crystallographer is used!

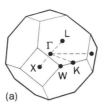

(a)

Fig. 17.17 (a) The face centred cubic (fcc) Brillouin zone together with some of the labels used for special points. Although special points may appear several times in the figure they are only labelled once. (b) A calculated band structure for metallic copper, which has a fcc structure. The Fermi level is shown dotted. For further comment on this figure see the text. (c) A picture of the Fermi surface for copper. It is to be remembered that such a Fermi surface is concerned with a large number of electrons. The electron density is not uniform throughout the Brillouin zone—for example, below the zone surface there are the electrons occupying the d orbital originating levels shown in Fig. 17.17(b). (d) Two cross-sections of the zone surface shown in Fig. 17.17(c). Figures (a) and (c), taken together, should enable this figure to be understood.

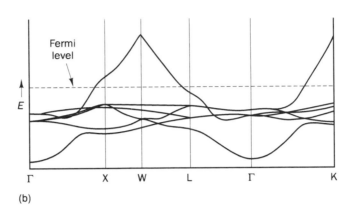

(b)

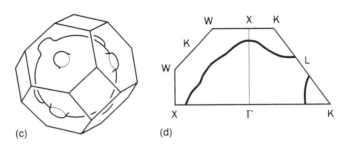

(c)　　　　(d)

The next example is more complicated; it is copper, each atom having six valence shell orbitals (five d and one s—the configuration of the isolated Cu atom is $3d^{10}4s^1$) to be included together with, of course, 11 electrons. In the language of Chapter 7, there is only one hole to consider. Copper has the face centred cubic (fcc) lattice, the Brillouin zone of which was shown in Fig. 17.11(a). This is repeated in Fig. 17.17(a) with the labels of special positions indicated. In Fig. 17.17(b) are shown the results of calculations on the energy bands in copper and their occupation at 0 K; the Fermi level is indicated. Figure 17.17(b) merits close inspection. Most evident are the peaks at W and K; they correspond to the antibonding combination of the 4s orbitals at the zone surface. The band of levels lying together originate in the 3d orbitals, although the way that they get mixed with the 4s is evident on close inspection. Thus, the line representing the peak at W is not continuous with that representing the peak at K; an avoided crossing has occurred. Even this is not simple—the single low-lying level at Γ involves the bonding combination of 4s orbitals. Of particular interest is the way that at L, all nearby points in the Brillouin zone lie below the Fermi surface, the only labelled point in Fig. 17.17(b) apart from Γ so to do. The consequence of this is evident in Fig. 17.17(c), which shows the Fermi surface for copper.

A double cross-section of this surface is given in Fig. 17.17(d), a figure which needs reference to Fig. 17.17(a) for its interpretation. In the L direction copper is an insulator! That is, if in some (thought) experiment, we took a perfect single crystal of copper and constrained electrons to attempt to move through the crystal in straight lines, then in the L directions (there are four such directions, corresponding to the four threefold axes of a cube) the material would be an insulator. Support for this comes from a comparison between the electrical conductivities of sodium and copper at very low temperatures. Sodium is the better conductor.

The diagram shown in Fig. 17.17(b) was referred to as calculated. The question immediately arises as to whether there exists any experimental method by which such calculations can be checked. The answer is that there is a host of methods which give information on the details of the Fermi surfaces of metals, at least. They are all methods which the chemist traditionally regards as in the province of physics, but it is only there that the answer to the question is to be found. Almost all relate to a remarkable and unexpected phenomenon. That single crystals of metals at low temperature and in strong magnetic fields display oscillatory behaviour. Oscillatory behaviour in almost any measurement that is made on them, the oscillations occurring as the strength of the magnetic field is varied and the value of the measurement plotted against the magnetic field strength (or, better, the reciprocal of the magnetic field strength). The most important and famous of these measurements is of the de Haas–van Alphen effect. This is the result of measurement of the magnetization induced within a single crystal of a metal as the magnetic field is varied (the phenomenon also varies with the sample orientation). The plot obtained for a copper single crystal when the magnetic field makes equal angles with each of the coordinate axes is shown in Fig. 17.18(a).

The simplest explanation of the phenomenon requires that the reader makes a brief reference back to Section 9.3 and the discussion there about the orbital magnets induced by a magnetic field, which has to be extrapolated to the case of solids.[8] In a solid, as the magnetic field is varied, the splittings into clockwise and counter-clockwise rotating components for each level also varies and so the precise nature of the level at the Fermi surface can change also. As the clockwise component is replaced by counter-clockwise, so the local field alternates between reinforcing the applied field and reducing it. So, oscillations are observed. The relevant crystal orientation— and the path of the induced currents—for the copper case shown in Fig. 17.18(a) is given in Fig. 17.18(b). As the crystal is rotated, the pattern of oscillations changes also, thus providing information on the Fermi surface.

17.6   Solid state and coordination compounds

Where do coordination compounds come in? At several points in this text there have been discussed either molecular aggregates—Chapter 15 is devoted to them—or crystalline species. In this section some general points

---

[8] One point of extrapolation is clear. The possibility that there be circular electron orbits in a solid has not been covered by the discussion in the text. However, since the Fermi surface, by definition, is a surface of equal energy, it is evident that, as required by the discussion in Section 9.3, there is no energetic barrier to such circular orbits.

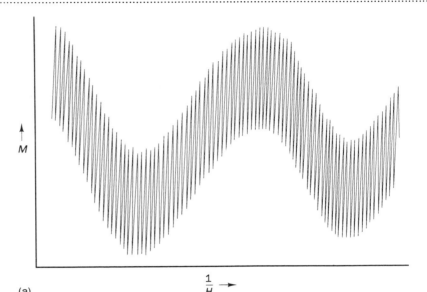

(a)

$\frac{1}{H}$ →

Fig. 17.18 (a) The de Haas–van Alphen oscillations in metallic copper when the magnetic field is orientated as indicated as shown in (b). The two periods of oscillation may be correlated with the concurrent existence of the two different circulatory pathways shown. A different orientation of the crystal with respect to the magnetic field would lead to a different oscillation pattern and thus provide information on the topology of the Fermi surface (adapted from Ashcroft and Mermin).

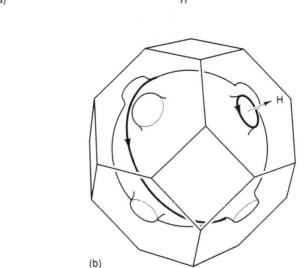

(b)

will be made and then the above discussion will be applied to some topics covered in earlier chapters. Clearly, when we turn to complexes, we have to start with the AB pattern, covered in the previous section, and develop it. The relevant situation is one in which either or both of A and B are coordination compounds. Were we to follow the pattern developed so far in the chapter, we could consider each atomic orbital of every atom of A and do the same for B. This, however, would not be sensible. A major theme of many earlier chapters has been the exploitation of the molecular orbitals appropriate to A and/or B when these are coordination compounds. The question which immediately faces us is the extent to which these discussions have to be modified by the introduction of interactions between translationally related sets of A and also of B. It is not difficult to see the general outcome. When an orbital of a coordination compound, A, is largely located on a central metal atom, then the intervening ligands serve to insulate these orbitals from each other. The orbitals do not interact and the discussion of this chapter becomes largely irrelevant. The crystalline material behaves like a set of superimposed atoms, each unaware of the others' existence. Such a behaviour is much less likely to occur for molecular orbitals delocalized over the whole molecule. Here, the contribution from orbitals on the periphery, on the 'outside' of the complex, is important. Such orbitals enable an interaction between adjacent molecules, the interaction giving rise to dispersion patterns of the type that have been discussed. Evidently, some—external—ligand orbitals will be the most affected.

We can at once understand two things. First, why ligand field theory works so well (in a typical coordination compound the d orbitals of the transition metal ions are physically well separated). Secondly, why our understanding of ligand orbitals is much less detailed and less general than that of metal ions. The energy levels that ligand orbitals subtend are likely to be environment-sensitive. The (complex) species B will only partially insulate the ligands of one (complex) A molecule from those of another. Change B and these residual A–A interactions will also change. Further, interactions (strictly, over $\mathbf{k}$-space) between ligand orbitals of A and external orbitals of B may occur.

With this discussion in mind, it is helpful to reconsider the distinction between charge-transfer bands introduced by Robins and Day, the class 1, class 2 and class 3 distinction that was discussed in some detail in Section 8.11. An example of class 2 behaviour (clear interaction but chemical identity retained) is provided by the coordination compound $KFe^{II}[Fe^{III}(CN)_6]$. It can at once be understood why the Fe^{II} and Fe^{III} ions retain their individuality. They are insulated from each other by the CN^- ligands. However, a little consideration suggests that it is a delicate matter. They have t_{2g}^6 and t_{2g}^5 configurations, respectively, and, following the discussion of Section 6.6, these t_{2g} electrons are partially delocalized into the CN^- π antibonding orbitals. It follows that they have a non-zero probability of being at the periphery of the complex. A mechanism for metal–metal interaction does exist; evidently, its effect is small. Of course, although this explains in outline why this compound is class 2 and not class 1 (no interaction), there is ample evidence for small metal–metal interactions of the type that have just been considered. Their effects are most important in magnetochemistry and were discussed in Section 9.11. A typical class 3 compound is Ag_2F (Table 8.4). The contrast with $KFe[Fe(CN)_6]$ is clear. In the crystal, the orbitals involving the Ag and Ag^+ ions must be delocalized over the entire 'molecule' (Ag_2F has an extended, not a molecular, structure). Interactions between the silver ions and atoms will surely occur, leading to dispersion, the observed broad band absorption and a black compound results. Any characteristics of Ag and Ag^+ are completely lost.

There are aspects of the discussion in the last paragraph which can be extended well beyond the compounds covered by the Robins and Day classification. Consider the following problem. NiO, when pure, is a pale solid. TiO and VO both have the same structure as NiO but are black and are almost metallic conductors of electricity, whereas NiO, when pure, is an insulator. That there is some electronic explanation for this pattern is made evident by the observation that when suitably doped with an impurity (Li_2O is the one usually cited) NiO also becomes black and highly electrically

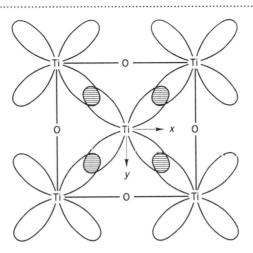

Fig. 17.19 Titanium(II), with a t_{2g}^2 configuration, in a simple structure such as that of NaCl (that adopted by TiO, shown here), has the possibility of direct metal–metal interaction through the overlap of t_{2g} orbitals as shown. Such interaction has the effect of giving a broad, incompletely filled, band and so the compound is a conductor of electricity and is black.

conducting. What is the explanation for this pattern? To avoid the problem of the electronic levels introduced by an impurity we will consider just the origin of the differences between TiO, VO and pure NiO. All three compounds crystallize in the NaCl, rock salt, structure—which means that each metal ion is octahedrally surrounded by oxide anions, so that the ideas of simple crystal field theory should be applicable. The electronic configurations of the three ions are d^2, d^3 and d^8, respectively. In an octahedral crystal field these will become t_{2g}^2, t_{2g}^3 and $t_{2g}^6 e_g^2$. Only the first two have incompletely filled t_{2g} shells. As Fig. 17.19 shows, in the TiO structure overlap can occur between such t_{2g} orbitals, leading to an incompletely filled band structure and an explanation for the electrical conductivity. For NiO, a comparable band structure will be less available. First, the partially occupied d orbitals are e_g, which do not overlap each other in the same way—they point towards an oxygen, not a nickel; the electrons are more likely to be localized on the metal atoms. Secondly, the band structure in NiO will involve participation of the O orbitals. Finally, it turns out that the Ni d orbitals are a bit more contracted than are the Ti or V, reinforcing the idea that they will tend to be relatively insulated from one another. We can confidently predict that NiO will have a lower electrical conductivity than the other two oxides. As we will see, the difference in colour also follows. One important point: because it was important to show the interaction between several metal atoms, Fig. 17.19 does not show just a single primitive unit cell. When multiples of the primitive unit cell are shown in this way it is a good idea to check whether the interaction between adjacent units is in-phase (as here). If so, then moving to the zone boundary the interactions will become out-of-phase and so the interaction shown in Fig. 17.19, at $\mathbf{k} = 0$, at Γ, is at the bottom of the band (assuming that overlap with other bands can be neglected).

Some generalizations have been made about those features which enhance the probability of band formation of the type just discussed in transition metal compounds:

1. the cation occurs early in the transition metal series;

2. the cation is in the second or third transition metal series;

3. the ligand is not too electronegative;

4. the actual charge on the cation is not great;

5. the ligand is small.

Of these features, 1, 2 and 5 are either evident from the discussion above or are simple extensions of that discussion. Points 3 and 4 derive from the fact that highly ionic materials (hard cations and anions) have large band gaps. That in NaCl, for instance, is ca. 8.5 eV; that in KF is approximately 11 eV. In contrast, a typical semiconductor such as GaAs, composed of soft species, has a band gap of 1.4 eV. In the present context, these generalizations find some reflection in the observation that unlike NiO, the compounds NiS, NiSe and NiTe are quite good electrical conductors and black in colour.

Finally, a comment on the content of Chapter 15. In Fig. 15.25, density of state diagrams for Ag and V were given and compared to M_6 clusters of the same elements. The general conclusion was that there are distinct similarities between M_6 and the bulk metal. Perhaps this was not altogether surprising. After all, the isolated-atom orbital energies are the same and similar metal–metal interactions are involved. The extrapolation enabled by our present discussion is not too surprising either. The more metal–metal interactions in a cluster, the more it is likely to resemble the solid metal. The data in Fig. 15.25 were for isolated octahedra. We may safely predict that in a solid composed of these octahedra the corresponding density of states would even more closely resemble those of the pure metal.

17.7 Spectra of crystalline materials

In the previous section the spectra of crystalline materials were briefly alluded to. In this section the key question to be addressed is what are the spectroscopic selection rules for a solid; how do they differ from those for isolated species?

When dealing with molecules, and therefore point groups, the general selection rule is simple: for a transition to occur the relevant *integral* must be totally symmetric—that is, transform as the irreducible representation of the relevant point group. The totally symmetric irreducible representation is the one which has a character of $+1$ for all operations of the point group (all other irreducible representations, effectively, have equal amounts of $+1$ and -1 values so that integrals having any of these other symmetries have integrals over all space of zero). The general selection rule in the solid state is similar and for the same reason. The relevant integral must have the symmetry properties of $\mathbf{k} = 0$. If it did not, in some parts of space the integral would be positive and in others negative. The orthogonality properties of the different irreducible representations ensures that in such an integral over all space the positive and negative contributions combine to give an answer of precisely zero. But this is not all. The vast majority of spectroscopies of interest to the chemist have one feature in common. This is that they involve radiation with wavelengths which are enormous compared with the magnitude of a typical translational vector of a crystal. Such a vector might be of the order of 5–10 Å, and that is the wavelength of a rather soft X-ray. The nearest that the chemist normally gets is in ultraviolet spectroscopy, with wavelengths of perhaps 1000 Å or more. So, whatever the light wave does to the contents of a unit cell (however the unit cell is defined), it will effectively be the same as that which it does in adjacent unit cells. It takes a great many unit cells before a difference becomes apparent. And only if our reference unit cell is sensitive to what happens thousands of Angstroms away will it make any difference. In general, no such sensitivity exists[9] so that, effectively, in all spectroscopies of interest to the chemist, all unit cells respond in the same way. The phenomena observed are *translationally invariant*, and transform as $\mathbf{k} = 0$.[10] However, it has just been seen that the entire integral must transform as $\mathbf{k} = 0$. Although for ease of explanation it was convenient to talk of 'the light wave' earlier in this paragraph, as far as an integral is concerned it is the corresponding operator which is of importance. It, too, must transform as $\mathbf{k} = 0$. We are left with the ground and excited state wavefunctions. The picture just developed clearly indicates that the most important transitions will be those in which they too transform as $\mathbf{k} = 0$ (since all unit cells respond in the same way).

One exception, which is more than a mathematical curiosity, is to be noted. If the ground state wavefunction transforms as $\mathbf{k}_i$ and the excited state wavefunction transforms as $-\mathbf{k}_i$, then the integral transforms as $\mathbf{k} = 0$. A travelling wave moving in one direction ($\mathbf{k}_i$) is cancelled out by a travelling wave moving in the opposite direction ($-\mathbf{k}_i$). It can at once be understood why the spectral bands of solids are so often broad (and the compounds are black if they are also electrically conducting—there is a low-lying excited band). For the ground state the $\mathbf{k}_i$ can span the entire Brillouin zone, together with all the attendant dispersion. Similarly, the excited state $-\mathbf{k}_i$ can span the entire Brillouin zone, together with *its* attendant dispersion. In Fig. 17.20 is shown for a one-dimensional example how this can even give rise to an apparent band splitting if ground and excited states have matching, but different, high densities of states in the appropriate regions of $\mathbf{k}$ space.

Although complications of the sort that have just been described are well known, they are least likely to occur when the ions or molecules making up the crystal are well separated from each other and/or the structural interactions between the structural units rather weak. This is the case with most coordination compounds. For 'concentrated' species such as anhydrous metal oxides and halides complications are expected and, indeed, occur—as was seen in the last section. What of the case where no complications are found? Here, the effective building block of the crystal is the (primitive) unit cell. Whatever happens in one unit cell happens in all others. This is the ultimate justification for the use of unit cells in diagrams such as Figs. 9.8 and 12.4. In considering the spectra of solids just the contents of one (primitive) unit cell need to be considered. When interactions between

---

[9] A problem arises for some members of a series of related crystal structures for some compounds— ZnS and CSi (carborundum) are well-known examples—in which translational vectors of up to thousands of Angstroms can occur (so-called polytypism). However, the long-range communication between unit cells which must be involved may be rather differently mediated although the mechanism is not known (for a speculation on one possibility see S. F. A. Kettle, *J. Cryst. Spect. Res.* (1990) *20*, 59).

[10] It is important to remember that the theory developed in this section concerns only pure translation operations. It contains no rotation or reflection operations; inclusion of these into the theory is required to explain the anisotropies associated with the absorption and emission of radiation by real crystals.

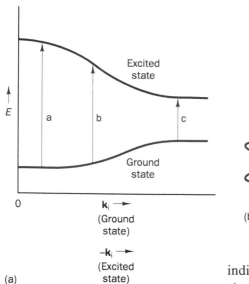

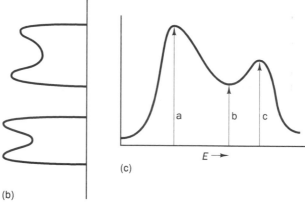

Fig. 17.20 (a) The rather different dispersions of the ground and excited states involved in an electronic transition, together with (b) the corresponding density of states. The resulting spectrum is shown as (c); the observed maxima and minima can be correlated with features in the dispersion curve. In principle, similar patterns apply to all forms of spectroscopy carried out on solids, not just electronic spectroscopy.

individual structural units becomes important (as in TiO and VO) the band structure of the solid means that quite a different approach has to be adopted, one that takes account of the dispersion of the energy bands across the Brillouin zone. Clearly, success in this latter endeavour is more likely to come for simple species—and these are just the ones most likely to show the phenomena which require it. Not surprisingly, the work published by those concerned with simple species tends to be rather different in content and interpretation to that published by those working with large coordination compounds.

Throughout this chapter the assumption of an infinite, perfect and pure crystal has been made. In general, these assumptions are sufficiently valid for the general results obtained to apply to real-life crystals. Two important exceptions are to be noted, however. The first has already been mentioned but not developed (it would require a chapter of its own). This is the presence of impurities. If these, usually by intention but sometimes by accident, have energy levels which interleave the band gaps in the host material then the properties of the host material are dramatically changed. The model developed in this chapter is an adequate starting point for a discussion of these changes but, of itself, is not adequate to deal with them. The assumption of an infinite and perfect crystal is usually valid. When the size either of the crystal or of the perfect blocks within a crystal become too small, however, dramatic changes occur. For instance, cadmium phosphide, Cd_3P_2, is a semiconductor with a band gap of only 0.5 eV. Not surprisingly, it is black (the dark colours of many main group element sulfides similarly reflect small band gaps). If small particles of the compound are prepared as colloids and then separated the colour of the product depends on the particle size. So, at 30 Å they are brown; as the particle size diminishes further, in rapid succession they become red, then orange, then yellow and finally white at about 15 Å. The band gap has increased to 4 eV. These colour changes have nothing to do with the lighter colours commonly displayed by coloured materials when they are finely ground (this is a purely optical effect which may be cancelled by changing the refractive index of the material surrounding the finely ground crystals). Such particle size effects, and the breakdown in the simple model presented in this chapter, are likely to become of increasing importance as the search continues for molecular electronic devices.

Further reading

The solid state is covered in most current inorganic texts but usually more at a factual level than a theoretical. Some parts of the present chapter would be explored better in texts on solid state physics. For rather obvious reasons, the two following references provide good stepping stones between the content of the present chapter and selected wider literature references (given in the reference sections of these papers):

- 'The Brillouin Zone, an interface between Spectroscopy and Crystallography' by S. F. A. Kettle and L. J. Norrby, *J. Chem. Educ.* (1990) *67*, 1022.
- 'Really, Mr. Bravais, your lattices are all primitive' by S. F. A. Kettle and L. J. Norrby, *J. Chem. Educ.* (1993) *70*, 959.

Also relevant (even though the title name has not been used in this chapter) is 'The Wigner–Seitz Unit Cell' by S. F. A. Kettle and L. J. Norrby, *J. Chem. Educ.* (1994) *71*, 1003.

An excellent and readable account which follows a development quite similar to that of the present chapter is given in *Solids and Surfaces, a Chemist's view of bonding in extended structures* by R. Hoffmann, VCH, New York, 1988.

A short, easy-to-read article which has many examples of band structures is 'Electronic Structure of Elemental Calcium and Zinc' by T. D. Brennan and J. K. Burdett, *Inorg. Chem.* (1993) *32*, 746. To fully understand this paper, it would be helpful to have available a picture of the first Brillouin zone of a fcc metal, such as that given in Fig. 17.11.

A good general reference is *Solid State Chemistry and its Applications* by A. R. West, Wiley, Chichester, 1984.

Fermi surfaces (and some beautiful diagrams of them) will be found in *The Fermi Surfaces of Metals* by A. P. Cracknell, Taylor and Francis, London, 1971.

Methods of determining Fermi surfaces are particularly well described in Chapter 14 of *Solid State Physics* by N. W. Ashcroft and N. D. Mermin, Saunders College, Philadelphia, 1976.

Band Theory of Solids. An Introduction from the Point of View of Symmetry by S.L. Altman, Oxford Science Publications, Oxford, 1991, develops the approach begun in this chapter.

The interface between metal clusters (Chapter 15) and bulk metals is explored in 'Bonding in Molecular Clusters and Their Relationship to Bulk Metals' by D. M. P. Mingos, *Chem. Soc. Rev.* (1986) *15*, 31 (this reference uses some of the language of Chapter 15), and 'Colloidal Semiconductor Q-Particles: Chemistry in the Transition Region Between Solid State and Molecules' by H. Weller, *Angew. Chem., Int. Ed. Engl.* (1993) *32*, 41.

Questions

17.1 In Fig. 17.21 is shown a centred two-dimensional lattice with two structural dots to each unit cell. Show that it is possible to redefine the array in terms of a lattice of which the (primitive) unit cell contains only a single dot.

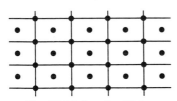

Fig. 17.21 Question 17.1.

17.2. When studied in the crystalline material the Raman band associated with the totally symmetric (breathing) v(Cr–CO) mode of $Cr(CO)_6$ was little different from the same band when the compound was studied in solution. What conclusions follow from these observations?

17.3. In the text two-dimensional crystals have often been discussed in preference to real-life three-dimensional crystals, simply because it is so much easier to draw and understand pictures of two-dimensional crystals. Fortunately, the extension to three-dimensional crystals is almost trivial—no new principles arise. You are given the task of rewriting Section 17.2 in a manner such that it deals exclusively with the three-dimensional case. Make a list of the *key* points at which you would need to make changes to Section 17.2 in such a rewriting.

17.4. The Brillouin zone for the simple (primitive) cubic lattice is a (simple!) cube. The Brillouin zone for the face-centred cubic lattice is pictured in Fig. 17.11; that for the body-centred cubic is shown in Fig. 17.16. Show that all three have O_h symmetry.

Appendix 1

Conformation of chelate rings

Consider the ligand ethylenediamine, en, which is typical of many. When it chelates to a metal atom a five-membered ring is formed, as shown in Fig. A1.1. Is this ring planar? The answer is no, it is puckered. It is easy to see why this should be so. In a regular pentagon each internal angle is 108°, not far from the regular tetrahedron value of 109.5°. However, if the chelate ring is part of an octahedral complex then the N–M–N angle will be close to 90°. This will open up the bond angles at the four atoms of the en ligand in the ring to an average of 112.5°; add to this the fact that the M–N bonds will be longer than the C–C and C–N and it is clear that some steric strain is likely to exist. This strain can be relieved at little or no cost if the ring puckers. Having recognized the possibility of puckering, the next thing is to recognize that two different puckered forms exist; much of this appendix is concerned with them.

Any three points in space define a plane, unless the three points lie in a straight line. Consider the plane defined by the metal atom and the two nitrogen atoms of an en ligand in a complex. Our concern is with the two carbon atoms in the ligand. Relative to the plane, these can be arranged in either of the two ways shown in the top part of Figure A1.2. The two configurations are denoted λ and δ, as indicated. An easy way of remembering which configuration is which is to view the chelate ring with the metal atom closest to the observer, as shown in Figure A1.2. From this viewpoint, in the δ configuration the bonds involving the two carbon atoms roughly trace out a δ and for the λ configuration they trace out a λ. This is shown at the bottom of Figure A1.2.

This is not the end of the conformational possibilities that have to be considered. In an octahedral complex there are two different ways of arranging three chelate rings, denoted by the upper case letters Λ and Δ. They are shown in the upper part of Figure A1.3. The simplest way of remembering the difference between Λ and Δ is to put arrows on the Δ in the natural, clockwise sense, as is done at the bottom of Figure A1.3. These arrows serve as a reminder that the Δ configuration resembles a normal, right-handed screw (the Λ configuration resembles a left-hand screw).

Finally, the possible configurations of the three individual ligands (λ, δ) are brought together with the two possible arrangements of a set of three of them in an octahedral complex (Λ, Δ). A set of eight possible

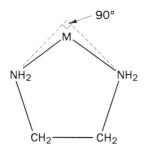

Fig. A1.1 Typical puckered five-membered ring formed by chelated en ligand.

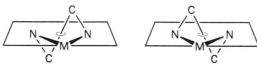

Fig. A1.2 The λ and δ ring configurations.

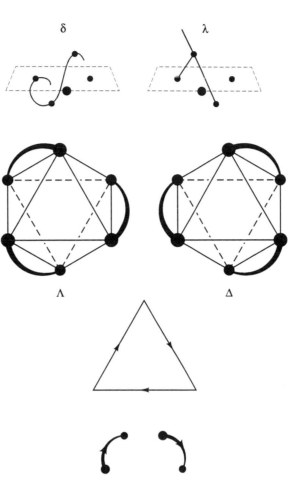

Fig. A1.3 The Λ and Δ configurations of three chelate rings in an octahedral complex.

configurations is obtained, all of which must be assumed to be present in a solution of a complex such as $[Co(en)_3]^{3+}$:

$\Lambda(\lambda\lambda\lambda)$	$\Delta(\delta\delta\delta)$
$\Lambda(\lambda\lambda\delta)$	$\Delta(\delta\delta\lambda)$
$\Lambda(\lambda\delta\delta)$	$\Delta(\delta\lambda\lambda)$
$\Lambda(\delta\delta\delta)$	$\Delta(\lambda\lambda\lambda)$

The first member of each pair (those on the same line) is the mirror image of the second (they are enantiomorphic pairs). Generally speaking, members of families of such compounds rapidly interconvert. However, they have been extensively studied, usually by NMR (often with somewhat more exotic ligands than en) because they provide an opportunity of investigating the effect of steric and other influences on the relative stabilities of octahedral complexes. In crystals of salts of the $[Co(en)_3]^{3+}$ cation, only one of the eight forms listed above will usually be present. But which, and why? These questions do not seem to have general answers, but they are fascinating nonetheless.

For simplicity, this appendix has considered only five-membered rings. Six-membered chelate rings are also quite common and follow a similar pattern; Figure A1.2 would be adapted to a six-membered ring by inclusion of an additional atom in the plane shown in each of the diagrams in this figure. The subsequent discussion would have been essentially unchanged.

Finally, the ligand that has been considered is a rather symmetrical one, en. A related but less symmetrical ligand has one C–H bond of en replaced by a C–CH$_3$. This ligand is 1,2-diaminopropane (usually denoted by pn); it contains an asymmetric carbon atom (that bonded to the methyl group). In principle, therefore, it is capable of separation into optically active isomers and our discussion could be elaborated to include this. This is not some academic exercise, a study of an increasingly remote problem. The naturally occurring ligands of bioinorganic chemistry are invariably optically active and an understanding of the properties of their complexes would not be possible if the possibilities for as ligand such as pn were not well understood.

The subject matter of this appendix is comprehensively treated by C. J. Hawkins in *Absolute Configuration of Metal Complexes*, Wiley, New York, 1971.

Appendix 2

Valence shell electron pair repulsion (VSEPR) model

This appendix contains an outline of the most successful of the simple approaches to molecular structure, one that is described in most current introductory and inorganic texts. In contrast to the treatment given in such texts, the present is more concerned with an analysis of it, in order to assess whether it correctly highlights the key factors involved in determining molecular geometry. Whether, for main group complexes at least, it may be said to provide an explanation (as opposed to a prediction) of their geometries. The model is based on the postulate that repulsion between valence electron pairs surrounding an atom lead to the observed atomic arrangement around this atom, an arrangement in which these electron pairs are as far apart as possible. That is, it is postulated that bonding forces are not the vital factor in determining molecular geometry. This view finds support in the observation that relatively little energy is required to excite those vibrations which correspond to bond angle changes. These, of course, change the geometry around the central atom. The actual energy is about one half of that required to excite the corresponding bond stretching vibrations.

The simplest way of looking at the VSEPR approach is to regard each valence electron pair as being represented by a point, the points being constrained to move over the surface of a sphere drawn around the central atom. As the points (i.e. valence shell electron pairs) repel each other, the most stable molecular arrangement will be that in which the points are as far apart as possible. If each valence shell electron pair is called P, then the following geometries of the Ps around the central atom M are at once predicted:

MP_2 linear
MP_3 equilateral triangular
MP_4 tetrahedral
MP_6 octahedral

Comparison with the geometries which are discussed in Chapter 3 will show that for main group elements these predictions are rather good. So, $[BF_3 \cdot NMe_3]$, in which there are four valence electron pairs around the boron, is approximately tetrahedral, the distortion being consistent with the fact that one of the valence electron pairs (that involved in the B–N bond) would not be expected to have exactly the same spatial distribution as the other three, so that (B–N)–(B–F) bond repulsions will be slightly different from (B–F)–(B–F) repulsions. Similarly, extending the discussion to a compound which would not normally be regarded as a complex, the oxygen atom in water is surrounded by four electron pairs (two from the O–H bonds and two from the lone pairs on the oxygen atom). Repulsion between these electron pairs leads to a tetrahedral distribution and, in accord with this, the H–O–H bond angle in water has roughly the tetrahedral value, the deviation from regularity being in accord with the suggestion that the centre of gravity of the electron density in the O–H bonds is further away from the oxygen atom than is the case for the lone pair electrons. Such arguments lead to the prediction that the relative magnitude of the electron repulsion is in the order:

lone pair–lone pair > lone pair–bonding pair > bonding pair–bonding pair

an order which almost invariably explains distortions from symmetric structures.

So far, coordination numbers five, seven, eight, nine and above have not been discussed. The case of nine-coordination is simple, for the predicted geometry is that discussed in Section 3.2.7, the three-face-centred trigonal pyramid, and is the one commonly observed. The others are not so clear-cut, because the valence electron pair arrangements which involve minimum electron repulsion are not intuitively obvious. This difficulty can be removed by assuming some form for the repulsive energy between the points constrained to move over the surface of a sphere (the points which represent the valence electron pairs). For instance, it could be assumed that the energy varies as $1/r^n$, where r is the interpoint distance, and the arrangement of lowest energy thus determined. Unfortunately, the results depend to some extent on the value given to n, (a value of between 6 and 12 is usually chosen) and in any case it is not certain that the real energy varies as $1/r^n$—an exponential might be more appropriate, for instance. In any case, it is far from certain that deformable and diffuse electron densities interact in a way that is well modelled by undeformable charges located at precise points. Even with the VSEPR model, it seems safest to conclude that, for coordination numbers five, seven, eight, ten and above, factors other than valence shell electron repulsions may play a part in determining molecular geometry. These are cases for which, as discussed in Chapter 3, there seems little energetic difference between several possible geometries.

Although the bones of the VSEPR model have been given above, it can be elaborated further—for instance by arguing that double bonds are bigger than single and that electronegative atoms suck electron density out of bonding pairs, making them smaller, so that they can be squeezed more tightly together. This last refinement is used to explain why the bond angle in NF_3 is smaller than that in NH_3, a problem to which we shall return.

On the whole, the VSEPR method predicts the geometries of main group compounds and complexes rather well. This is not the same thing as saying that it provides a correct explanation of molecular geometry. Indeed, in the opinion of some, its status and use is just that of prediction, that of an aid to getting the right answer. The VSEPR method only really works for main group elements (the d electrons of transition metals do not consistently have the steric effects envisaged by the model; the f electrons of the lanthanides and actinides have no generally recognized steric effects). Amongst those main group molecules for which it fails are some, such as Li_2O (gaseous), which is linear and not bent like water. The explanation which is offered is rather naive: these molecules are ionic and so the model does not apply. What such an explanation is really saying is that the geometry is determined by nuclear–nuclear repulsion (lithium–lithium in Li_2O; this repulsion is a minimum for the linear molecule) and this interaction is ignored in the VSEPR model. However, it is difficult to believe that this is the explanation for the observation that the anion BrF_6^-, with seven valence shell electron pairs around the bromine, is octahedral in both solution and solid. In contrast, the related species ClF_6^- is believed and IF_6^- known to be a distorted octahedron, as is XeF_6, all species with seven electron pairs around the central atom. For heavier elements, in particular, other effects come into play—correlation effects (which are mentioned frequently in this book) and relativistic effects (which are also mentioned, but less frequently). So, the fact that a species which may be regarded as derived from the NH_4^+ ion, $N(AuPH_3)_4^+$, is tetrahedral but its arsenic counterpart, $As(AuPH_3)_4^+$, is square pyramidal, is probably to be associated with the operation of these two effects in the latter compound (and not to Au–Au bonding, although this explanation has been suggested).

Although the VSEPR model is commonly applied to complexes formed by main group elements, data do not exist which would enable us to assess its accuracy for these. Instead, we are forced to use data on simpler molecules and we shall consider NH_3 and NF_3. Table A2.1 gives the results of some accurate calculations on these two molecules. Of particular interest is the origin of the energy barrier to inversion in them—at room temperature an isolated ammonia molecule turns inside out, like an umbrella in a high wind, many, many times a second. NF_3 does not. The problem is relevant to our discussion because at one point in the umbrella motion all of the atoms of ammonia are co-planar; at either extremity they have the observed pyramidal shape. Here, then, we have two geometries which can be, and have been, compared in detail. One is the most, and the other the least, stable of all of the entire set of geometries mapped out by the umbrella motion. The problem was investigated theoretically by carrying out accurate calculations on the planar molecules (corresponding to the top of the energy barrier) and also on the pyramidal species. In order to identify the origin of the energy

Table A2.1 Change in energies between planar and most stable pyramidal geometries of NH_3 and NF_3. Positive quantities mean that the pyramidal form is the more stable. All energies are in atomic units (1 au = 2626 kJ mol^{-1})

Molecule	Change in bonding energy	Change in repulsive energy	Change in electron–electron repulsion	Change in nuclear–nuclear repulsion
NH_3	−0.475	0.489	0.158	0.290
NF_3	7.626	−7.519	−3.789	−3.495

The 'Change in repulsive energy' entries are not the sum of the corresponding two following repulsion entries because of small changes in kinetic energy which have not been explicitly included. There are very small changes in bond lengths between the two geometry sets which have been ignored in the discussion in the text.
Data from A. Schmiedekamp, S. Skaarup, P. Pulay and J. Boggs, *J. Chem. Phys.* (1977) 66, 5769.

barrier—and hence the reason why the molecules are pyramidal rather than planar, for each of the three energy terms identified in this appendix—bonding, electron–electron repulsion and nuclear–nuclear repulsion—the change in going from a planar to a pyramidal molecule was calculated. These data, the numbers in Table A2.1, can then be used to understand why NH_3 and NF_3 are pyramidal rather than planar. In Table A2.1 a positive quantity means that this contribution stabilizes the pyramidal molecule relative to the planar.

The first thing to be noted about Table A.2.1 is that the entries in the 'change in bonding energy' column are non-zero. Contrary to the VSEPR model's assumption, bonding *is* a function of bond angle. Even more striking is the different pattern of the signs of the entries in the 'change in bonding energy' and 'change in repulsive energy' columns. The lesson is a salutary one. The sum of the two entries for NH_3 and also for NF_3 is positive—the pyramidal molecule is the stable one—but the two molecules achieve this result in quite opposite ways. Pyramidal NH_3 is stable because this geometry minimizes the repulsive forces—bonding is a maximum in the planar molecule. In contrast, NF_3 is pyramidal because this geometry maximizes the bonding; repulsive forces are a minimum in the planar molecule, in contradiction to the VSEPR model's predictions. The most charitable thing that can be said about the second set of entries in Table A2.1 is that they show that electron–electron and nuclear–nuclear repulsion energy changes are of comparable importance in determining molecular geometry (for NH_3 the nuclear–nuclear are the more important). It can be argued that the electron– electron terms in Table A2.1 include *all* the electrons in the molecules and not just those which are the focus of the VSEPR model. However, such an objection misses a key point. If other electron–electron repulsion interactions are important and yet ignored by the model then this, too, is a weakness. One should not argue too much from one example, yet that which we have given is a dramatic one—in both cases considered, electron–electron repulsion makes a smaller contribution to determining the molecular shape than do either bonding or nuclear–nuclear repulsion. One is forced to remember a difficult lesson; a model that leads to a correct prediction is not necessarily a correct model. A further indication of the existence of concealed assumptions within the VSEPR model is provided by some other molecules related to ammonia. The C–N–C bond angle in triisopropylamine, $N(CHMe_2)_3$, is 119.2°—the $N–C_3$ framework is essentially planar, presumably because of steric interactions between the isopropylamine groups. In trimethylamine, $N(CH_3)_3$, the C–N–C bond angle is 111° but in the corresponding fluorinated compound $N(CF_3)_3$ it is 118°. This change is surprising because fluorination has little effect on bond angles in alkanes. In $N(C_2F_5)_3$ the C–N–C bond angle is 119.3°. Clearly, there are factors of importance in determining these angles that are not included in the simple VSEPR model. Nonetheless, the value of this model as a usually correct predictor of the molecular geometries of compounds of main group elements is unquestioned. Further, the sort of analysis which has been given above for NH_3 and NF_3 is not the only one possible. It has been much more common to ask a different question: as to whether there is theoretical support for the electron repulsion inequality sequence:

lone pair–lone pair > lone pair–bonding pair > bonding pair–bonding pair

Most workers who have considered this question have concluded that there is. Crucial is the way that a single entity, the electron distribution in a molecule, is divided up to give distinct lone and bonding pairs. What has usually been done is to adopt some criterion which seems to lead to a division as close as possible to that made, more qualitatively, by the VSEPR model itself.

We have reached a point at which opinions differ; the reader is entitled to his or her own. Those who would like the benefit of a view which is intermediate between the two extremes presented in this appendix might find it helpful to read an article by A. Rodger and B. F. G. Johnson in *Inorg. Chim. Acta* (1988) *146*, 37. The most recent exposition of the VSEPR model is to be found in *The VSEPR Model of Molecular Geometry* by R. J. Gillespie and I. Hargittai, Prentice-Hall (Allyn and Bacon), New York, 1991.

Appendix 3

Introduction to group theory

Figure A3.1 shows a square planar complex and something of the symmetry which it possesses. Most evident are the pictorial representations of mirror planes—reflection in any of these infinitely thin, double-sided mirror planes turns the square planar complex into itself. The solid black shapes in the figure indicate rotational axes—◆ shaped for twofold and ■ for fourfold. Although Fig. A3.1 shows symmetry elements—mirror planes and rotational axes—in group theory it is the corresponding *operations* which are of interest; the consequences of the *act* of reflecting or rotating. If the operations indicated in Fig. A3.1 are counted one finds a total of 16. In Table A3.1 these 16 operations are listed across the top. Some of the operations which are very closely related are grouped together—rotation by $\pm 90°$ about the C_4 (fourfold rotation) axis; rotation by $180°$ about equivalent C_2 (twofold rotation) axes; reflection in equivalent mirror planes (denoted σ_d). Symmetry operations which are grouped together are said to be in the same class. Each number in the table is known as a *character* and the whole table is called a *character table*. This *group* of 16 symmetry operations is called the D_{4h} group, and the character table is the D_{4h} character table. Down the left-hand side of the character table are listed some of the symmetry symbols used in the text (in Table 7.2, for instance). All such symbols originate in character tables. Each symmetry symbol is associated with a unique row of characters—no

Table A3.1 The D_{4h} character table

D_{4h}	E	$2C_4$	C_2	$2C_2'$	$2C_2''$	i	$2S_4$	σ_h	$2\sigma_d$	$2\sigma_d'$
A_{1g}	1	1	1	1	1	1	1	1	1	1
A_{2g}	1	1	1	−1	−1	1	1	1	−1	−1
B_{1g}	1	−1	1	1	−1	1	−1	1	1	−1
B_{2g}	1	−1	1	−1	1	1	−1	1	−1	1
E_g	2	0	−2	0	0	2	0	−2	0	0
A_{1u}	1	1	1	1	1	−1	−1	−1	−1	−1
A_{2u}	1	1	1	−1	−1	−1	−1	−1	1	1
B_{1u}	1	−1	1	1	−1	−1	1	−1	−1	1
B_{2u}	1	−1	1	−1	1	−1	1	−1	1	−1
E_u	2	0	−2	0	0	−2	0	2	0	0

This, the axis of highest symmetry, is rather complicated because it is the axis of three distinct symmetry operations. The first is the operation of rotation by 90°, either clockwise or anticlockwise, each denoted by C_4. The second is the operation of rotation by 180°, denoted C_2. The third is a composite operation: rotate by 90° and then reflect in the σ_h mirror plane (see below). Denoted S_4, this operation takes the top of one ligand into the bottom of the next (cf. C_4, which sends top into top).

A pair of equivalent mirror planes, σ_d

A pair of equivalent mirror planes, σ'_d

Horizontal mirror plane, σ_h

The other C'_2

The other C''_2

A pair of equivalent two-fold axes, C''_2. The other is perpendicular to this one.

A pair of equivalent two-fold axes, C'_2. The second is on the other side of the diagram.

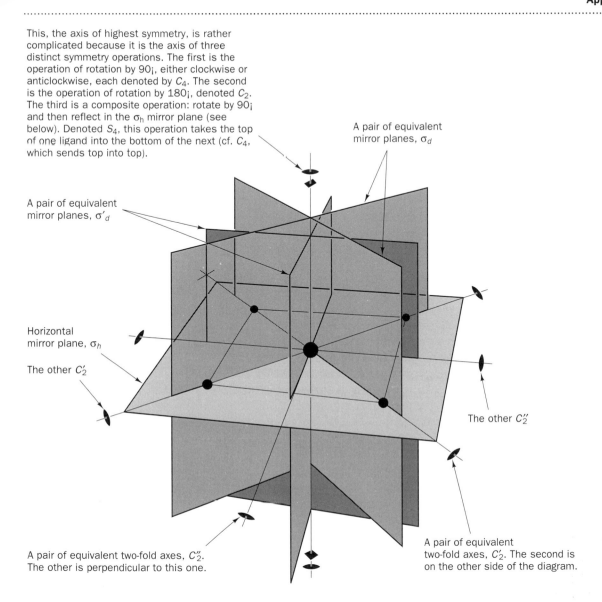

Fig. A3.1 Symmetry elements and operations associated with a square planar complex. There are two other symmetry operations, not shown on the diagram. The first, apparently trivial operation, is 'leave everything alone' and is denoted E. The other is the operation of inversion in the centre of symmetry (which is at the central atom) and is denoted i.

two rows are identical, although inspection reveals that the characters block into four 5×5 squares, the characters within one square being very simply related to the corresponding characters within another. This fundamental building unit of the D_{4h} character table is, in fact, the character table of the D_4 group (which has only the E, $2C_4$, C_2, $2C'_2$ and $2C''_2$ operations of the D_{4h} group). In Table A3.1 any one character corresponds to a symmetry symbol, (its row) and to one or more symmetry operations (its column). The character has the property of telling us how something which carries its particular symmetry symbol behaves under its particular symmetry operation. So, something of A_{2u} symmetry is multiplied by 1 (i.e. turned into itself) both by a C_4 rotation (be it clockwise or anticlockwise) and also by the C_2 rotation. However, it is multiplied by -1 (i.e. turned into itself with all signs reversed) by the C'_2 and C''_2 rotations (any of either pair of operations). The

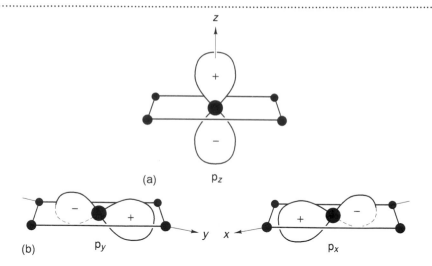

Fig. A3.2 (a) The p_z orbital has A_{2u} symmetry while (b) the p_x and p_y orbitals together have E_u symmetry. Because p_x and p_y are interconverted by some of the D_{4h} group (e.g. C_4), they must be considered as a pair.

(a) p_z

(b) p_y p_x

p_z orbital shown in Fig. A3.2(a) has A_{2u} symmetry and the reader should check that it behaves under these operations in the way that has just been described. Those symmetry symbols which, for the leave-alone operation (E), have characters of 2, describe the behaviour of *two* independent things simultaneously. The number 0, for these species, means either that the two things are interchanged in some way or that one is changed into itself ($+1$) and the other into the negative of itself (-1) so that, together, a character of zero is obtained ($+1 - 1 = 0$). The number -2 means that under the symmetry operation each of the two objects is turned into itself with all signs reversed. So, p_x and p_y (Fig. A3.2(b)) together have E_u symmetry. The reader should check that the D_{4h} character table gives an accurate description of their behaviour under the symmetry operations.

The *direct product* of two symmetry species, say $A_{2g} \times B_{1g}$, is formed by multiplying together in turn their characters under each symmetry operation, thus

D_{4h}	E	$2C_4$	C_2	$2C_2'$	$2C_2''$	i	$2S_4$	σ_h	$2\sigma_d$	$2\sigma_d'$
A_{2g}	1	1	1	-1	-1	1	1	1	-1	-1
B_{1g}	1	-1	1	1	-1	1	-1	1	1	-1
$A_{2g} \times B_{1g}$	1	-1	1	-1	1	1	-1	1	-1	1

Comparison with Table A3.1 then shows that the set of characters produced is, in fact, that labelled B_{2g}. We say that 'the direct product of A_{2g} and B_{1g} is B_{2g}'. For E-type symmetry species direct products are a bit more difficult. Consider $E_g \times E_u$:

D_{4h}	E	$2C_4$	C_2	$2C_2'$	$2C_2''$	i	$2S_4$	σ_h	$2\sigma_d$	$2\sigma_d'$
E_g	2	0	-2	0	0	2	0	-2	0	0
E_u	2	0	-2	0	0	-2	0	2	0	0
$E_g \times E_u$	4	0	4	0	0	-4	0	-4	0	0

This direct product does not correspond to a symmetry species in Table A3.1. It does, however, correspond to a sum of them. Consider the sum $A_{1u} + A_{2u} + B_{1u} + B_{2u}$:

D_{4h}	E	$2C_4$	C_2	$2C_2'$	$2C_2''$	i	$2S_4$	σ_h	$2\sigma_d$	$2\sigma_d'$
A_{1u}	1	1	1	1	1	−1	−1	−1	−1	−1
A_{2g}	1	1	1	−1	−1	−1	−1	−1	1	1
B_{1u}	1	−1	1	1	−1	−1	1	−1	−1	1
B_{2u}	1	1	1	−1	1	−1	1	−1	1	−1
$A_{1u} + A_{2u} + B_{1u} + B_{2u}$	4	0	4	0	0	−4	0	−4	0	0

This is the same as that generated by $E_g \times E_u$, so we can write:

$$E_g \times E_u = A_{1u} + A_{2u} + B_{1u} + B_{2u}$$

'the direct product of E_g and E_u is A_{1u} plus A_{2u} plus B_{1u} plus B_{2u}' (note that the product of suffixes $g \times u$ gives u). What have been called 'symmetry species' are often also called irreducible representations (of the group), so, 'the A_{2u} irreducible representation of the D_{4h} group'. The two expressions are interchangeable and often mixed with each other to avoid excessive repetitions. Similarly, one may well meet statements such as 'the p_z orbital of the central metal atom transforms as the A_{2u} irreducible representation'. The direct product $E_g \times E_u$ is an example of a reducible representation, which, as has just been demonstrated, may be expressed as a sum of irreducible components (which themselves cannot be reduced further, which is why they are called irreducible).

In Table A3.2 is given a complete table of direct products for the D_4 group. That for the D_{4h} group would be four times as large; it can be obtained from that for D_4 by adding suffixes according to the rules that

$$g \times g = u \times u = g; \qquad g \times u = u$$

Notice that in Table A3.2 A_1 only appears on the diagonal and that it appears on every diagonal element; had we been working in the group D_{4h} then it would have been the A_{1g} irreducible representation which would have behaved in this way. This is important; it can be seen from Table A3.1 that A_{1g} is the only symmetry species which never goes into its negative under a symmetry operation. If we were interested in an integration over all space and the mathematical function being integrated had A_{1g} symmetry then it

Table A3.2 Direct products for the D_4 group

D_4	A_1	A_2	B_1	B_2	E
A_1	A_1	A_2	B_1	B_2	E
A_2	A_2	A_1	B_2	B_1	E
B_1	B_1	B_2	A_1	A_2	E
B_2	B_2	B_1	A_2	A_1	E
E	E	E	E	E	$(A_1 + A_2 + B_1 + B_2)$

Table A3.3 The O character table

O	E	$6C_4$	$3C_2$	$6C_2'$	$8C_3$
A_1	1	1	1	1	1
A_2	1	-1	1	-1	1
E	2	0	2	0	-1
T_1	3	1	-1	-1	0
T_2	3	-1	-1	1	0

would not automatically equal zero. If it had any symmetry other than A_{1g} then the negative and positive contributions to the integral would exactly cancel (this is easy to see for all except the Es in Table A3.1; for all the other As and Bs there are as many -1 entries as there are $+1$; the statement is also true for the Es). This conclusion is very relevant to quantum mechanics for there one is frequently interested in integrals over all space. So, if we are interested in an integral such as

$$\int \psi_1 \psi_2 \delta\tau$$

which is an overlap integral, we can discuss its symmetry properties (in particular, whether it is zero or not) by considering the direct product of the symmetry species of the two orbitals concerned. In this particular case it leads to the important result that only orbitals of the same symmetry species have non-zero overlap integrals. So, in those places in the text that direct products have been used, it is because, really, we are interested in products of orbitals; this is true even in places such as Section 8.7 where, ostensibly, the subject is the intensities of spectral bands.

Finally, in Table A3.3, is given the character table for the octahedral group O. It can be used to check that tables such as Table 6.3 are correct. The full octahedral group, which has symmetry species with g and with u suffixes, is denoted O_h and is four times as large. The two groups are related in that the character table for the group O_h is built up from that of the group O in the same way that the character table of the group D_{4h} is built up from that for the group D_4.

The reader will find simple applications of group theory in Chapters 10 and 15. A non-mathematical introduction to the subject is given in *Symmetry and Structure* by S. F. A. Kettle, Wiley, Chichester and New York, 1995. Also relevant, useful and/or relatively non-mathematical are: *Symmetry in Coordination Chemistry* by J. P. Fackler Jr., Academic Press, New York, 1965; *Symmetry in Chemistry* by H. H. Jaffé and M. Orchin, Wiley, New York, 1965; *Molecular Symmetry and Group Theory* by A. Vincent, Wiley, London, 1977; *Symmetry in Chemical Bonding and Structure* by W. E. Hatfield and W. E. Parker, Merrill, Columbus, Ohio, 1974; and *Group Theory for Chemists* by G. Davidson, Macmillan, London, 1991. More mathematical is *Chemical Applications of Group Theory* by F. A. Cotton, Wiley, New York, 1990.

Appendix 4

Equivalence of d_{z^2} and $d_{x^2-y^2}$ in an octahedral ligand field

In Section 6.3 it was noted that the action of a C_3 rotation operation converted $d_{x^2-y^2}$ into $d_{y^2-z^2}$ and d_{z^2} into d_{x^2} (or, for a C_3 rotation in the opposite sense, into $d_{z^2-x^2}$ and d_{y^2}). In order to see how the new orbitals are related to $d_{x^2-y^2}$ and d_{z^2}, we must first recognize that the d_{z^2} is more correctly written as $d_{\sqrt{(1/3)}(2z^2-x^2-y^2)}$ (this gives it the central annulus that d_{z^2} does not). Working with the labels alone (this is a perfectly valid simplification) and after a lot of trial and error, knowing the answer in advance or by use of some systematic method, we find that

$$y^2 - z^2 = -\tfrac{1}{2}(x^2 - y^2) - \frac{\sqrt{3}}{2} \times \frac{1}{\sqrt{3}}(2z^2 - x^2 - y^2)$$

(hint: show that simplification of the right-hand side expression gives the left-hand side) and

$$\frac{1}{\sqrt{3}}(2x^2 - y^2 - z^2) = \frac{\sqrt{3}}{2}(x^2 - y^2) - \frac{1}{2} \times \frac{1}{\sqrt{3}}(2z^2 - x^2 - y^2)$$

That is,

$$d_{y^2-z^2} = -\tfrac{1}{2}d_{x^2-y^2} - \frac{\sqrt{3}}{2}d_{z^2}$$

and

$$d_{x^2} = \frac{\sqrt{3}}{2}d_{x^2-y^2} - \tfrac{1}{2}d_{z^2}$$

In other words the new orbitals are simply linear combinations of the old. The readers should check that this is also true of $d_{z^2-x^2}$ and d_{y^2} (they only have to change two of the signs in the right-hand side expressions given above).

A final word, for the reader of Appendix 6, who has returned to this appendix in search of a character of -1. As shown above, when $d_{x^2-y^2}$ is rotated into $d_{y^2-z^2}$ its contribution to the latter is $-\tfrac{1}{2}$. Similarly, d_{z^2} contributes $-\tfrac{1}{2}$ to d_{x^2}. As the reader may well have discovered, the same result is obtained if rotation into $d_{z^2-x^2}$ and d_{y^2} is considered. In both cases the aggregate character is -1.

Appendix 5

Russell–Saunders coupling scheme

Consider a free atom or molecule in which several electrons occupy but do not fill a set of orbitals, which need not all be degenerate. There will usually be many ways in which the electrons can be distributed, some of which will be of lower energy than others. The differences in energy will be determined by electron repulsion and the more stable arrangements will be those with the least electron repulsion destabilization. So, electron–electron repulsion will be relatively small if the electrons occupy orbitals which are spatially well separated; it will be reduced yet more if the electrons have a high spin multiplicity, i.e. have parallel spins, because two electrons with parallel spins can never be in the same orbital.

If we are to discuss the energetics of free atoms or ions we have to separate the possible electron arrangements into sets, the members of each set all being degenerate. There are two schemes commonly adopted for this. The first, which is the subject of the present appendix, is the Russell–Saunders coupling scheme. The second, the j–j coupling scheme, is used when an interaction which we shall neglect for the moment, spin–orbit coupling, is large. It is of much less general applicability than the Russell–Saunders scheme, although relevant to the content of Chapter 11.

In the Russell–Saunders coupling scheme, the orbital motion of the electrons are coupled together, as are their spins. Within the simplified scheme discussed here—neglect of spin–orbit coupling—there is no coupling between these two sets. For any particular arrangement of the electrons within the orbitals the coupling between their orbital motions gives rise to a resultant; similarly, their spins couple to give a resultant, the resultants simply being the vector sums of the individual orbital or spin motions. It turns out that it is possible to measure not only the magnitude of the resultants but also something about their orientation in space. For each atom or ion there is at any particular instant an axis—let us call it the z axis—along which the z components of the orbital and spin resultants have well-defined values (unless something happens, absorption or emission of radiation, for example, which changes them). Not any z component can exist, however. If the largest component is M_z, then the others are $(M_z - 1)$, $(M_z - 2), \ldots, (M_z - 2M_z)$. That is, the components run from M_z to $-M_z$, a total of $(2M_z + 1)$ values in all. Each resultant will correspond to a particular arrangement of electrons in the orbitals and, for resultants related in the way we have just described, the arrangements will also be related. Note that if we wish to hold the z axis fixed in space we have to apply some fixed axial perturbation, an electric field, for example.

So far, the nature of the vectors has not been specified. They represent angular momenta (there is more on this topic in Chapter 9) and one talks about the orbital and spin angular momenta, their magnitudes and their z components. The quantization of orbital angular momentum, introduced arbitrarily in the Bohr theory of atomic structure, appears in present-day quantum mechanics in the way just described. The various z components of orbital and spin angular momentum are, within our approximation, energetically unimportant. The energy of a particular arrangement is determined solely by the absolute magnitude of the orbital and spin angular momenta, and is independent of their z components. The reason that the z components have been introduced is the following. For each of the $(2L + 1)$ z components of an orbital angular momentum vector and for each of the $(2S + 1)$ z components of a spin angular momentum vector there is an individual orbital or spin wavefunction. Since a complete wavefunction contains both orbital and spin parts and any of the orbital functions may be combined with any of the spin functions, $(2L + 1)(2S + 1)$ distinct combinations exist. All of these wavefunctions are degenerate. It is the removal of this degeneracy by a crystal field which is discussed in Chapter 7, by a magnetic field (amongst other things) in Chapter 9 and by spin–orbit coupling in Chapter 11. An individual energy level of a free atom or ion could be characterized by specifying $(2L + 1)$ and $(2S + 1)$—they are both integers —or, alternatively and more simply by specifying L and S. L is always an integer but S may not be, so a combination of the two is used. L and $(2S + 1)$ are given. Rather than quoting L as a number it is replaced by a letter: $L = 0$ is indicated by S; $L = 1$ by P; $L = 2$ by D; $L = 3$ by F; $L = 4$ by G; $L = 5$ by H; $L = 6$ by I and so on. A parallel with orbitals should be evident; for example, there are five d orbitals; similarly, the label D indicates a $(2 \times 2 + 1) =$ five-fold degeneracy. The value of $(2S + 1)$ is given as a number, written as a pre-superscript to the L symbol. Thus, ^1S, ^3P, ^1D, ^4F, etc. When one is talking one says singlet, doublet, triplet, quartet, quintet . . . . So, ^2D is 'doublet dee'. One talks of a ^2D term. When there are, say, two electrons distributed within a set of three (degenerate) p orbitals, this is spoken of as a p^2 (pee two) configuration, the fact that all three p orbitals are involved being implicit. Such a configuration will usually give rise to several terms. The p^2 configuration gives rise to $^1D + {}^3P + {}^1S$ terms.

There is a simple check that can be applied to a list of terms such as this, which it is claimed, arise from a given configuration. Consider the p^2 configuration. The first of the two electrons can be fed into the set of three orbitals in any one of six ways (allowing for spin) and the second in any one of five (it can only go into the same orbital as the first electron with the opposite spin). Recognizing that we have counted each arrangement twice because ↑↓ is the same as ↓↑, the number of distinct arrangements, and therefore wavefunctions, is $(6 \times 5)/2 = 15$. Now, the ^1D, ^3P and ^1S terms correspond to 5, 9 and 1 wavefunctions respectively, again a total of 15. The number of wavefunctions is the same whichever way it is calculated, as it should be. In Table A5.1 the process is worked out in some detail for the d^2 configuration, because this is of importance in Chapter 7. The z component

Table A5.1 The terms arising from the d^2 configuration $^1G + {}^3F + {}^1D + {}^3P + {}^1S$

Left panel:

Function	^1G	^3F ↑↑	^3F ↑↓	^3F ↓↓	^1D	^3P ↑↑	^3P ↑↓	^3P ↓↓	^1S
2 2̄	•								
2 1		•							
2 1̄	•								
2 0		•							
2 0̄	•								
2 −1		•							
2 −1̄	•								
2 −2		•							
2 −2̄	•								
2̄ 1			•						
2̄ 1̄				•					
2̄ 0			•						
2̄ 0̄				•					
2̄ −1			•						
2̄ −1̄				•					
2̄ −2			•						
2̄ −2̄				•					
1 1̄					•				
1 0						•			
1 0̄					•				
1 −1						•			
1 −1̄					•				
1 −2		•							
1 −2̄			•						

Right panel:

Function	^1G	^3F ↑↑	^3F ↑↓	^3F ↓↓	^1D	^3P ↑↑	^3P ↑↓	^3P ↓↓	^1S
1̄ 0							•		
1̄ 0̄								•	
1̄ −1							•		
1̄ −1̄								•	
1̄ −2					•				
1̄ −2̄				•					
0 0̄									•
0 −1						•			
0 −1̄							•		
0 −2		•							
0 −2̄			•						
0̄ −1	•								
0̄ −1̄								•	
0̄ −2	•								
0̄ −2̄				•					
−1 −1̄					•				
−1 −2		•							
−1 −2̄	•								
−1̄ −2			•						
−1̄ −2̄				•					
−2 −2̄	•								

It is to be emphasized that this table is a book-keeping exercise only; it cannot be assumed that the actual functions are as indicated in it (the actual ^1S function, for instance, is a linear combination of all functions listed with resultant spin and orbital angular momenta of 0).

of the orbital angular momentum of each electron in each distinguishable arrangement is indicated by a number. Numbers without a bar indicate that the electron has α spin; numbers with a bar, such as $\bar{2}$, indicate a β electron spin. In the d^2 case, we have $(10 \times 9)/2 = 45$ distinguishable arrangements and so the left-hand column contains 45 entries In the body of the table each of these is allocated to a component of one of the terms arising from the d^2 configuration, $^1G + {}^3F + {}^1D + {}^3P + {}^1S$. A further discussion of this general topic is contained in Section 11.3. The job of working out the actual wavefunctions of the terms is more difficult than the job of determining just what these terms are—and it is the latter that we have done. The job of working out the relative energies of the terms is rather hard work.

A final word about spin–orbit coupling. Its effect is to cause small splittings between functions which we have so far regarded as degenerate. These splittings may be observed spectroscopically and values of spin–orbit coupling constants thus obtained (Section 9.3).

Further reading An excellent treatment which goes well beyond that given above but which is both easy and, often, fun to read is *Orbitals, Terms and States* by M. Gerloch, Wiley, Chichester and New York, 1986.

Appendix 6

Ligand σ group orbitals of an octahedral complex

In this appendix are presented three different methods of obtaining the linear combinations of Table 6.2, repeated as Table A6.1, the ligand σ orbital symmetry-adapted linear combinations for an octahedral complex. There is much to be learnt from a careful comparison of all three methods.

Method 1

The first method is scarcely a derivation. Knowing the symmetries spanned by the ligand σ combinations $(A_{1g} + E_g + T_{1u})$, one finds an object of each of the symmetries in turn and asks what ligand combination matches the nodal pattern of the object? The most obvious reference objects to choose are the appropriate orbitals of the central metal atom. Labelling the ligand σ orbitals as in Fig. 6.7, repeated as Fig. A6.1, and referring back to Figs. 6.8 (s orbital), 6.9 (p orbitals) and 6.10(a) (d orbitals) it is evident that the matching combinations are of the form

$$
\begin{aligned}
\text{s:} \quad & (\sigma_1 + \sigma_2 + \sigma_3 + \sigma_4 + \sigma_5 + \sigma_6) \\
\text{p}_x\text{:} \quad & (\sigma_1 - \sigma_6) \\
\text{p}_y\text{:} \quad & (\sigma_2 - \sigma_4) \\
\text{p}_z\text{:} \quad & (\sigma_3 - \sigma_5) \\
\text{d}_{x^2-y^2}\text{:} \quad & (-\sigma_2 + \sigma_3 - \sigma_4 + \sigma_5) \\
\text{d}_{z^2}\text{:} \quad & (2\sigma_1 - \sigma_2 - \sigma_3 - \sigma_4 - \sigma_5 + 2\sigma_6)
\end{aligned}
$$

remembering that the full label of d_{z^2} is $\text{d}_{2z^2-x^2-y^2}$. These apart from normalization, and in one case a -1 factor, are the expressions found in Table A6.1.

Method 2

When we have two equivalent atomic orbitals, ψ_1 and ψ_2 say, the former on atom 1 and the latter on a symmetry-equivalent atom, 2, symmetry demands that the electron density at a given point in ψ_1 is equal to the electron density at the corresponding point in ψ_2. That is, $\psi_1^2 = \psi_2^2$ at these

Table A6.1 Ligand group orbitals of six octahedrally orientated σ ligands

Symmetry	Ligand group orbital
a_{1g}	$\dfrac{1}{\sqrt{6}}(\sigma_1 + \sigma_2 + \sigma_3 + \sigma_4 + \sigma_5 + \sigma_6)$

t_{1u}

$$\begin{cases} \dfrac{1}{\sqrt{2}}(\sigma_1 - \sigma_6) \\[2ex] \dfrac{1}{\sqrt{2}}(\sigma_2 - \sigma_4) \\[2ex] \dfrac{1}{\sqrt{2}}(\sigma_3 - \sigma_5) \end{cases}$$

e_g

$$\begin{cases} \tfrac{1}{2}(\sigma_2 - \sigma_3 + \sigma_4 - \sigma_5) \\[2ex] \dfrac{1}{\sqrt{12}}(2\sigma_1 - \sigma_2 - \sigma_3 - \sigma_4 - \sigma_5 + 2\sigma_6) \end{cases}$$

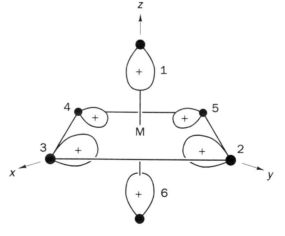

Fig. A6.1 Numbering system adopted for ligand σ orbitals in an octahedral complex.

points. It follows that $\psi_1 = \pm\psi_2$. In other words, ψ_1 and ψ_2 may be either in-phase or out-of-phase with each other. If ψ_1 and ψ_2 combine to form a molecular orbital, ϕ say, then an analogous argument shows that ϕ may either have the form $(\psi_1 + \psi_2)$ or $(\psi_1 - \psi_2)$, all other combinations leading to unequal electron densities on atoms 1 and 2. Normalizing, that is, multiplying by a factor so that the sum of the squares of the coefficients multiplying ψ_1 and ψ_2 equals unity, but neglecting overlap between ψ_1 and ψ_2, gives the orbitals

$$\phi_1 = \frac{1}{\sqrt{2}}(\psi_1 + \psi_2)$$

and

$$\phi_2 = \frac{1}{\sqrt{2}}(\psi_1 - \psi_2)$$

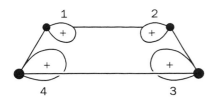

Fig. A6.2 Numbering system adopted for ligand σ orbitals of a square planar complex.

which, if atoms 1 and 2 are part of a larger, polyatomic, system are called group orbitals.

In what follows it will, for simplicity, be assumed

1. That all orbitals labelled ψ are of σ type with respect to a suitably placed metal ion, although the discussion has to be little modified to cover π and δ type interactions.

2. That all such orbitals have the same (positive) phase. That is, that

$$\phi_1 = \frac{1}{\sqrt{2}} (\psi_1 + \psi_2)$$

is a combination of two orbitals of identical phase; in

$$\phi_2 = \frac{1}{\sqrt{2}} (\psi - \psi_2)$$

ψ_2 has the opposite phase to ψ_1 because of the negative sign, not because it is inherently of opposite phase.

What if we have four identical σ orbitals ψ_1, ψ_2, ψ_3 and ψ_4 on atoms arranged at the corners of a square? The σ orbitals on the ligands of a square planar complex are an example of this situation. The simplest way of discussing this case is as follows. Label the orbitals cyclically as in Fig. A6.2 and consider the pairs ψ_1 and ψ_3, and ψ_2 and ψ_4. We can treat each *trans* pair as if the other were not present and use the discussion above to obtain group orbitals for each pair. These are

$$\phi_1 = \frac{1}{\sqrt{2}} (\psi_1 + \psi_3)$$

$$\phi_2 = \frac{1}{\sqrt{2}} (\psi_1 - \psi_3)$$

$$\phi_3 = \frac{1}{\sqrt{2}} (\psi_2 + \psi_4)$$

$$\phi_4 = \frac{1}{\sqrt{2}} (\psi_2 - \psi_4)$$

All that we have to do now is to form suitable combinations of ϕ_1, ϕ_2, ϕ_3 and ϕ_4 to obtain the ligand group orbitals that we are seeking. But how do we know which combinations to take? The answer is that the transformations of the final group orbitals must be described by the irreducible representations of the D_{4h} point group. If the language of the last sentence seems obscure, reread Appendix A3. This requirement boils down to the fact that any nodal planes contained in ϕ_1, ϕ_2, ϕ_3 and ϕ_4, must be compatible with each other for any interaction to occur. For example, neither ϕ_1 nor ϕ_3 contain any nodal planes; because they are equivalent to each other we simply, again, take their sum and difference

$$\theta_1 = \frac{1}{\sqrt{2}} (\phi_1 + \phi_3) = \tfrac{1}{2}(\psi_1 + \psi_2 + \psi_3 + \psi_4)$$

$$\theta_2 = \frac{1}{\sqrt{2}} (\phi_1 - \phi_3) = \tfrac{1}{2}(\psi_1 - \psi_2 + \psi_3 - \psi_4)$$

However, ϕ_2 contains a nodal plane which passes through ψ_2 and ψ_4 and so ϕ_2 cannot combine with the group orbitals involving these atoms, ϕ_3 and ϕ_4. Similarly, ϕ_4 has a nodal plane which means that it cannot combine with either ϕ_1 or ϕ_2. We therefore conclude that the ligand group orbitals of a square planar complex are

$$A_{1g}: \quad \theta_1 = \frac{1}{2}(\psi_1 + \psi_2 + \psi_3 + \psi_4)$$

$$B_{1g}: \quad \theta_2 = \frac{1}{2}(\psi_1 - \psi_2 + \psi_3 - \psi_4)$$

$$E_u(1): \quad \phi_2 = \frac{1}{\sqrt{2}}(\psi_1 - \psi_3)$$

$$E_u(2): \quad \phi_4 = \frac{1}{\sqrt{2}}(\psi_2 - \psi_4)$$

The reader is urged to check that these orbitals have the symmetries indicated by using the D_{4h} character table given in Appendix 3.

Finally, the problem of the ligand group orbitals appropriate to an octahedral complex ion. It is convenient to regard the octahedron as a synthesis of a square planar complex, with the combinations given above, and two additional ligands with orbitals ψ_5 and ψ_6, which give rise to group orbitals

$$\phi_5 = \frac{1}{\sqrt{2}}(\psi_5 + \psi_6)$$

$$\phi_6 = \frac{1}{\sqrt{2}}(\psi_5 - \psi_6)$$

Applying the nodal plane compatibility requirement it is found that only θ_1 and ϕ_5, neither of which has a nodal plane, combine together. What are the correct combinations? It is a general rule that, for the type of ligand group orbital which we have considered, there is always a combination

$$\frac{1}{\sqrt{n}}(\psi_1 + \psi_2 + \cdots + \psi_n)$$

which has the symmetry given by the first row of the appropriate character table (the totally symmetric combination, a row of 1s). In the present case this means that one combination is

$$\frac{1}{\sqrt{6}}(\psi_1 + \psi_2 + \psi_3 + \psi_4 + \psi_5 + \psi_6)$$

It is not difficult to show that this combination is simply

$$\sqrt{\left(\frac{2}{3}\right)} \times \theta_1 + \sqrt{\left(\frac{1}{3}\right)} \times \phi_5$$

There must be an orthogonal combination of θ_1 and ϕ_5 and this is

$$\sqrt{\left(\frac{1}{3}\right)}\theta_1 - \sqrt{\left(\frac{2}{3}\right)}\phi_5$$

(a combination of wavefunctions $(c_1\rho_1 + c_2\rho_2)$ is always orthogonal to $(c_2\rho_1 - c_1\rho_2)$ provided that ρ_1 and ρ_2 are, separately, orthogonal and normalized). Written in detail, this orthogonal combination is

$$\frac{1}{2\sqrt{3}}(\psi_1 + \psi_2 + \psi_3 + \psi_4) - \frac{1}{\sqrt{3}}(\psi_5 + \psi_6)$$

In summary, the ligand σ group orbitals of an octahedral complex ion are

A_{1g}: $\dfrac{1}{\sqrt{6}}(\psi_1 + \psi_2 + \psi_3 + \psi_4 + \psi_5 + \psi_6)$

$T_{1u}(1)$: $\dfrac{1}{\sqrt{2}}(\psi_1 - \psi_3)$

$T_{1u}(2)$: $\dfrac{1}{\sqrt{2}}(\psi_2 - \psi_4)$

$T_{1u}(3)$: $\dfrac{1}{\sqrt{2}}(\psi_5 - \psi_6)$

$E_g(1)$: $\frac{1}{2}(\psi_1 - \psi_2 + \psi_3 - \psi_4)$

$E_g(2)$: $\dfrac{1}{\sqrt{12}}(2\psi_5 - \psi_1 - \psi_2 - \psi_3 - \psi_4 + 2\psi_6)$

Again, the transformations of these combinations should be checked using the O character table in Appendix 3 (the g and u suffixes will have to be dropped from the expressions above). The only point of any difficulty in this check is the transformation of the two E_g orbitals under the C_3 rotation operations, where the character is -1. In fact, this is the same problem as that discussed in Appendix 4, to which reference should be made.

There is one final problem. Because of the square planar → octahedral sequence that was used above, the ligands are not labelled in the same way as in Chapter 3 (see Fig. 3.4) and in Chapter 6 (see Table 6.1 and Fig. 6.7). These latter were repeated as Table A6.1 and Fig. A6.1. The correlation between the ψs used above and the σs of Table A6.1 is

ψ_1 ψ_2 ψ_3 ψ_4 ψ_5 ψ_6
σ_5 σ_2 σ_3 σ_4 σ_1 σ_6

(just the labels 1 and 5 have to be interchanged between the two sets).

Method 3

The final method is group-theoretical, a full derivation has been given elsewhere.[1] That which follows is the shortest known to the author; we use Fig. A6.1. From the full octahedral group choose a set of operations consisting of complete classes, one of which is the identity, such that the total number of operations in the full group (48) is an integer times the number in the selected set. Further, choose a set such that the number of operations in the set is an integer times the number of objects to be considered (here, we have $\sigma_1 \to \sigma_6$ so this latter number is 6). This selection

[1] See S. F. A. Kettle, *J. Chem. Educ.* (1966) *43*, 21 or S. F. A. Kettle, *Symmetry and Structure*, Wiley, Chichester, 1995.

exercise is designed to make the final derivation involve the minimum of work. In the present case, a suitable set of operations is E, $8C_3$, $3C_2$, a total of 12. This set also falls within the group O so, again for simplicity, we will work in this smaller group. Now apply the formal projection operator method by considering σ_1 and applying the twelve operations to it in turn (which turns it into a σ which more often than not is one of the other σs) and then summing the results. This gives

E: (σ_1)

$8C_3$: $(2\sigma_2 + 2\sigma_3 + 2\sigma_4 + 2\sigma_5)$

$3C_2$: $(\sigma_1 + 2\sigma_6)$

Multiply the quantities in brackets by the appropriate characters from the O character table (Appendix 3) and add:

1. for A_1 the characters are E: 1, $8C_3$: 1, $3C_2$: 1, so we get

$$\sigma_1 + 2\sigma_2 + 2\sigma_3 + 2\sigma_4 + 2\sigma_5 + \sigma_1 + 2\sigma_6$$

and the sum is

$$2(\sigma_1 + \sigma_2 + \sigma_3 + \sigma_4 + \sigma_5 + \sigma_6)$$

2. for E the characters are E: 2, $8C_3$: -1, $3C_2$: 2, so we get

$$2\sigma_1 - 2\sigma_2 - 2\sigma_3 - 2\sigma_4 - 2\sigma_5 + 2\sigma_1 + 4\sigma_6$$

and the sum is

$$2(2\sigma_1 - \sigma_2 - \sigma_3 - \sigma_4 - \sigma_5 + 2\sigma_6)$$

3. for T_1 the characters are E: 3, $8C_3$: 0, $3C_2$: -1, so we get

$$3\sigma_1 - \sigma_1 - 2\sigma_6$$

and the sum is

$$2(\sigma_1 - \sigma_6)$$

Apart from normalizing factors, these are all entries in Table A6.1; the additional T_1 functions can be obtained by starting with σ_2 and σ_3 in place of σ_1. Either of these will also serve to generate the second E function but give ligand group orbitals corresponding to either d_{x^2} or d_{y^2}. The function corresponding to $d_{x^2-y^2}$ is obtained either by taking the difference between d_{x^2} and d_{y^2} or by the equivalent, but more formal, procedure described in Appendix 4. In O_h the A_1 function above becomes A_{1g}, the E becomes E_g and the T_1 becomes T_{1u}.

Appendix 7

Tanabe–Sugano diagrams and some illustrative spectra

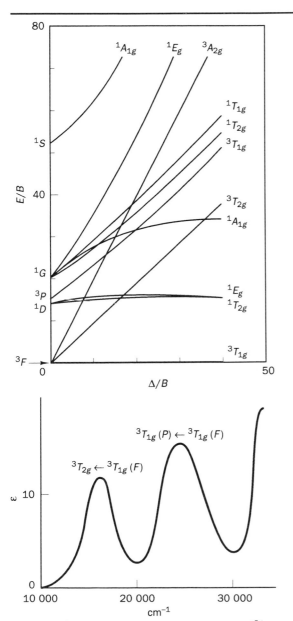

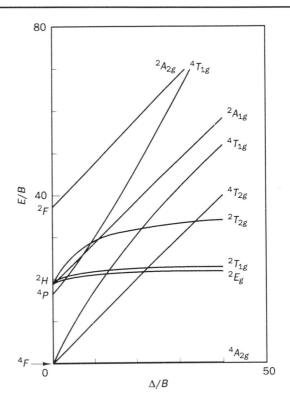

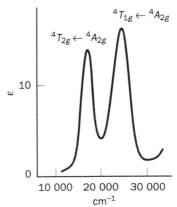

Fig. A7.1 d^2; $\gamma = 4.42$ and the spectrum of $[V(urea)_6]^{3+}$.

Fig. A7.2 d^3; $\gamma = 4.50$ and the spectrum of $[Cr(H_2O)_6]^{3+}$.

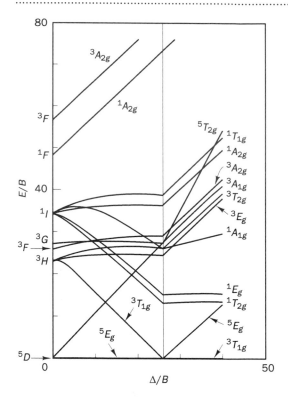

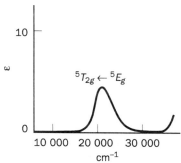

Fig. A7.3 d^4; $\gamma = 4.61$ and the spectrum of $[Mn(H_2O)_6]^{3+}$.

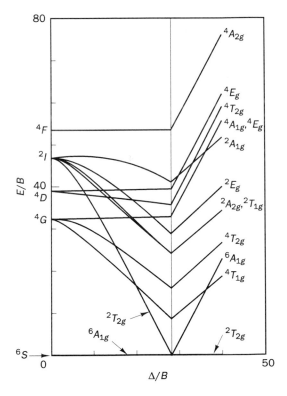

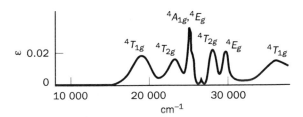

Fig. A7.4 d^5; $\gamma = 4.48$ and the spectrum of $[Mn(H_2O)_6]^{2+}$ (the ground state for all the transitions in the spectrum is $^6A_{1g}$).

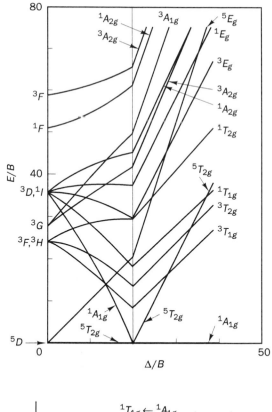

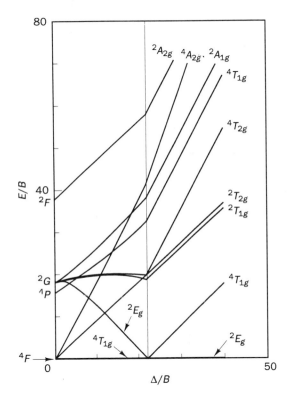

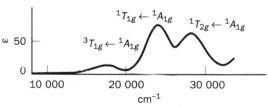

Fig. A7.5 d⁶; $\gamma = 4.81$ and the spectrum of $[IrCl_6]^{3-}$.

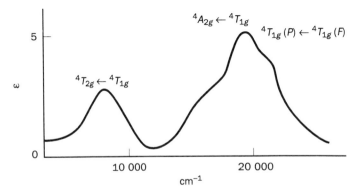

Fig. A7.6 d⁷; $\gamma = 4.63$ and the spectrum of $[Co(H_2O)_6]^{2+}$.

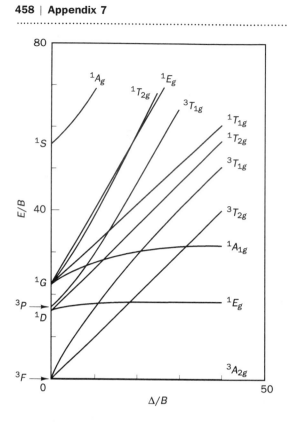

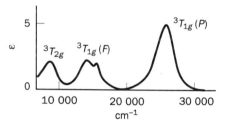

Fig. A7.7 d⁸; $\gamma = 4.71$ and the spectrum of $[Ni(H_2O)_6]^{2+}$.

Appendix 8

Group theoretical aspects of band intensities in octahedral complexes

The most general selection rule for electric-dipole transitions, the type of transition that is relevant to almost all of the bands observed in the electronic spectra of transition metal complexes, is a group theoretical statement of the excitation process described in qualitative terms in Section 8.1. It is the following:

> An electronic transition is electric-dipole allowed if the direct product of the symmetries of the ground and excited electronic terms of a molecule contains the symmetry species of one or more coordinate axes, both ground and excited terms having the same spin multiplicity.

This selection rule sounds somewhat formidable but in practice it is not difficult either to understand or apply. The coordinate axes come in because a $+ - -$ dipole has the same symmetry as a coordinate axis. Direct products arise because we need to consider, simultaneously, the ground and excited terms together with the mechanism of excitation (electric dipole). A triple direct product is therefore needed. The real selection rule is that this triple direct product must contain the totally symmetric irreducible representation—that for which all of the characters are $+1$ in the character table. For the case of octahedral complexes this is A_{1g}. However, the totally symmetric irreducible representation only arises when (any) irreducible representation is multiplied by itself. This means that we can split the triple direct product into two parts and compare them to see if the two parts contain the same irreducible representation. The most convenient division is to form the direct product of the ground and excited terms and to compare the result with the symmetry of the coordinate axes. Let us consider a specific example, the $^3T_{2g} \leftarrow \, ^3T_{1g}(F)$ transition in the d^2 case. Here the ground term orbital symmetry is T_{1g} and the excited term is T_{2g}. The direct product $T_{1g} \times T_{2g}$ is $A_{2g} + E_g + T_{1g} + T_{2g}$ (the relevant direct products are given in Table A8.1; g suffixes have to be added). Now, the Cartesian coordinate axes x, y, and z have the same symmetry properties as the corresponding p orbitals, which in our case is T_{1u}. As T_{1u} is not contained in the $T_{1g} \times T_{2g}$ direct product (this is $A_{2g} + E_g + T_{1g} + T_{2g}$) the $^3T_{2g} \leftarrow \, ^3T_{1g}(F)$ transition is orbitally forbidden. We could have seen this result without all of this work. The triple

Table A8.1 The O_h direct product table

O_h	A_1	A_2	E	T_1	T_2
A_1	A_1	A_2	E	T_1	T_2
A_2	A_2	A_1	E	T_2	T_1
E	E	E	$A_1 + A_2 + E$	$T_1 + T_2$	$T_1 + T_2$
T_1	T_1	T_2	$T_1 + T_2$	$A_1 + E + T_1 + T_2$	$A_2 + E + T_1 + T_2$
T_2	T_2	T_1	$T_1 + T_2$	$A_2 + E + T_1 + T_2$	$A_2 + E + T_1 + T_2$

This table is actually the product table for the O point group; for the point group O_h the usual suffix rule applies:
$g \times g = u \times u = g$; $g \times u = u$.

direct product must contain A_{1g} for the transition to be allowed. We focus on the g in this symbol. Both ground and excited terms are g and the coordinate axes u—but any triple direct product containing $g \times g \times u$ can never give g and so can never give A_{1g} and can never give an allowed transition. As shown in Section 8.7, a vibration of T_{1u} symmetry can remove this forbiddenness by mixing some p orbital component into the d.

To investigate this in more detail we have to answer the question 'what orbital can a vibration of T_{1u} symmetry mix with a d orbital?'. At a deeper level we are concerned with the question 'what excited terms can a vibration to T_{1u} symmetry mix with a ground term of T_{1g} symmetry?' again confining our discussion to the d^2 case. The answer to the first question will help with the second. A vibration will mix one orbital with a second if the direct product of the symmetry species of the vibration and the first orbital contains the symmetry species of the second orbital. Because the orbitals with which we are primarily concerned are d they are g and so a T_{1u} vibration can only mix a u orbital with them. The only ones available are p orbitals, with T_{1u} symmetry. As Table A8.1 shows, the direct product (vibration × orbital) is

$$T_{1u} \times T_{1u} = A_{1g} + E_g + T_{1g} + T_{2g} \text{ (adding suffixes)}$$

so we conclude that a T_{1u} vibration will mix some p orbital into a d, be the latter e_g or t_{2g} (since both of these feature in the direct product). That is, the vibration can introduce into a configuration such as t_{2g}^2 a tiny component of a configuration such as $t_{2g}^1 t_{1u}^1$. Into an excited state configuration such as $t_{2g}^1 e_g^1$ we can also introduce a small $t_{2g}^1 t_{1u}^1$ component and also an $e_g^1 t_{1u}^1$ component. Clearly, we could now extend our discussion to answer the same question about excited states by working out the terms arising from these configurations, much as is done for strong field complexes in Chapter 7, but we shall use a simpler approach. Again consider the $^3T_{2g} \leftarrow {}^3T_{1g}(F)$ transition in the d^2 case as an example. We can include the effect of T_{1u} vibrations by simply including T_{1u} in the direct product. That is, we now take the triple direct product

$$T_{1g} \times T_{2g} \times T_{1u} = (A_{2g} + E_g + T_{1g} + T_{2g}) \times T_{1u}$$

$$= (A_{2g} \times T_{1u}) + (E_g \times T_{1u}) + (T_{1g} \times T_{1u}) + (T_{2g} \times T_{1u})$$

$$= A_{1u} + A_{2u} + 2E_u + 3T_{1u} + 4T_{2u}$$

This final direct product contains T_{1u} three times so, as T_{1u} is the symmetry species of the coordinate axes to which the selection rule directs our attention, the intervention of the T_{1u} vibration not only makes the $^3T_{2g} \leftarrow {}^3T_{1g}(F)$ transition allowed but also splits it into three separate transitions. These will differ slightly in energy and so contribute to the breadth of the observed peaks.

The picture is further complicated when it is recognized that there are two distinct sets of T_{1u} metal–ligand vibrations in an octahedral complex and, further, there is a set of T_{2u} vibrations which may also give intensity to the d–d transitions by way of four subpeaks. We therefore conclude that we must expect the $^3T_{2g} \leftarrow {}^3T_{1g}(F)$ transition to be composed of no less than ten $(3 + 3 + 4)$ subpeaks, which may be separated by energies of the order of magnitude of the energy of metal–ligand vibrations (ca. $200\,\text{cm}^{-1}$ or more). Each of these peaks is broadened by the vibrational modulation of Δ discussed in Section 8.7 and also, probably, by spin–orbit coupling and by the Jahn–Teller effect. The large half widths observed in d–d spectra are understandable!

Appendix 9

Determination of magnetic susceptibilities

Recent years have seen a considerable increase in the number of techniques available for the determination of the magnetic susceptibility of a complex. Traditional methods involve a measurement of the change in weight of a sample when it is placed in a strong magnetic field, a field which can contain a uniform region (the Gouy method, to be described later) or have a constant, and large, field gradient (the Faraday method). A method based on the regular reversal of such a gradient by the use of electromagnets driven by an alternating current is the basis of the *alternating force magnetometer*. Induction methods are based on the measurement of the change in the magnetic field itself, which may be either static or alternating, when a sample is inserted. The *vibrating sample magnetometer* also measures a field change, this time when a sample is vibrated—for instance, by attaching it to the cone of a loudspeaker driven by an ac signal. More recently, very sensitive—and very expensive—measurements have become available by use of a super-conductive quantum interference device (SQUID). This depends on effects related to the magnetic fields generated within a superconducting coil and their variation when a sample is placed within them. Finally, for solutions, the effects on a reference NMR signal produced by the presence of dissolved paramagnetic species may be used to measure paramagnetism. The reference at the end of this appendix gives more details of most of these methods.

The method for determining magnetic susceptibilities most likely to be met by the reader is the Gouy method. The description which follows is for the conventional arrangement, although an inverted arrangement—in which the sample is fixed and the change in weight of the magnet measured—is becoming increasingly popular. In the Gouy method, a cylindrical sample of the complex under study is suspended between the poles of a magnet. The bottom of the cylinder is halfway between the two poles and on the interpole axis (see Fig. A9.1). It is thus at the position of maximum field intensity. Ideally, this maximum field is constant within a small volume, so that small changes in the position of the bottom of the cylinder can be ignored. The top of the cylinder is, essentially, out of the magnetic field. In this way the cylinder experiences the full field gradient, the observable quantity being an apparent change in weight of the sample when it is placed in the magnetic field, either by switching on an electromagnet or bringing up a permanent magnet. Paramagnetic materials are usually attracted into a magnetic field—

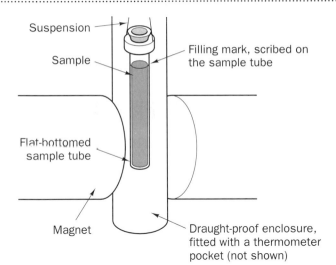

Fig. A9.1 Apparatus to determine magnetic susceptibility by the Gouy method.

Suspension

Sample

Filling mark, scribed on the sample tube

Flat-bottomed sample tube

Magnet

Draught-proof enclosure, fitted with a thermometer pocket (not shown)

they show an apparent increase in weight—the only exception being when the inherent diamagnetism of the sample swamps its paramagnetism.

Experimentally, the sample is either a solution or a finely ground solid within a silica or Perspex (Lucite) tube (ordinary glass contains paramagnetic impurities). The sample length is usually from 5 to 10 cm and its diameter 2 to 10 mm. It is suspended from the beam of a balance in a suitable draught-proof enclosure. More complicated arrangements are used to study magnetic behaviour over a temperature range. The sample tube is filled to the same height for all measurements and, for solids, every effort must be made to ensure that it is uniformly filled (it has been shown that loose packing can lead to the crystals twisting around in the magnetic field so that the assumption of random orientation is not valid; this can have a significant effect on the results). A compound with several unpaired electrons per molecule would be expected to show a greater paramagnetic effect than a compound with only one, assuming that the same number of molecules of each is studied. That is, the quantity

$$\frac{\text{weight increase in the field}}{\text{weight taken} \div \text{molecular weight}} = \frac{\Delta w}{w \div M}$$

would be expected to be larger for the former. The quantity on the right-hand side of this equation is proportional to the molar magnetic susceptibility of the compound, χ_M:

$$\chi_M = \frac{\gamma \Delta w \times M}{w}$$

The constant of proportionality varies from apparatus to apparatus—the more powerful the magnet used, the greater Δw. Hence, γ is called the tube calibration constant. One small addition has to be made to the equation for χ_M—the sample tube is diamagnetic and will lose weight in the field. The actual magnitude of this correction is determined by a trial run with the tube empty, when its weight drops by δ. The value of δ must be added to the experimental values of Δw so as to make Δw more positive (or less negative).

Table A9.1 Pascal's constants; diamagnetic corrections, all values × 10^{-6}/g atom

Li^+	0.6	F^-	11	H	3
Na^+	5	Cl^-	26	C	6
K^+	13	CN^-	18	N	2
NH_4^+	11.5	CNS^-	35	O	5
Fe^{2+}	13	SO_4^{2-}	40	S	15
Co^{2+}	12	CO_3^{2-}	29.5	P	26
Ni^{2+}	12	OH^-	12.0		

The corrected expression for χ_M is,

$$\chi_M = \frac{\gamma(\Delta w + \delta)M}{w}$$

It is convenient to measure Δw and δ in milligrams and w in grams. This, and multiplying γ by 10^3—so that it becomes of the order of unity—leads to a modified relationship which is the one usually quoted

$$10^6 \chi_M = \frac{\gamma(\Delta w + \delta)M}{w}$$

Note that this equation refers to the whole molecule. To find out the number of unpaired electrons on the transition metal atom in a complex an allowance must be made for all of the paired electrons present in the molecule. These effects are approximately additive and are listed for some of the more common species found in complexes in Table A9.1. A more complete list is given in a reference at the end of this appendix. To make the correction one sums the appropriate values. So, for $[Ni(H_2O)_6]^{2+}SO_4^{2-} \cdot H_2O$, the corrections are: Ni^{2+} 12; SO_4^{2-} 40; $7H_2O$ 7($2 \times 3 + 5$) = 77—a total of 129 ($\times 10^{-6}$/g atom). The correction calculated in this way, D, say, is added to χ_M and the sum is χ'_M, the susceptibility corrected for diamagnetism. That is,

$$\chi'_M = \chi_M + D$$

The diamagnetic correction can be more elaborate than indicated above, but in practice—and particularly when a complicated ligand is involved—it is best to measure the diamagnetic correction, either directly on the ligand (using Pascals constants for the remainder) or by measuring the susceptibility of an analogous but diamagnetic complex. The reader is reminded that in some cases (Co^{II} provides the most important example) it is better not to make a diamagnetic correction unless a similar allowance is being made for temperature independent paramagnetism (TIP), see the end of Section 9.3.

With a value for χ'_M, μ_{eff} is obtained using a relationship derived in Chapter 9

$$\mu_{eff} = \sqrt{(3R/N_A^2\beta^2)}\sqrt{\chi'_M \times T} = 2.83\sqrt{\chi'_M \times T}$$

μ_{eff} can then be treated using whichever approach is deemed appropriate. For example, if the spin only model is used, we have that

$$\mu_{eff} = \sqrt{n(n+2)}$$

Bringing all this together, gives

$$n(n+2) = 8.07T\frac{[\gamma(\Delta w + \delta) \times M \times 10^{-6} + D]}{w}$$

Since $n = 0, 1, 2, 3, \ldots$, the left-hand side of this equation equals 0, 3, 8, 15, ... All that remains is to determine γ; this can be done from measurement of the applied magnetic field and the length of the sample experiencing the field gradient, but simplest is to calibrate the apparatus using a standard material.

References

A good review of reasonably recent developments is C. J. O'Connor, *Prog. Inorg. Chem.* (1982) *29*, 203. There have been several useful articles dealing with the measurement of magnetic susceptibilities in *J. Chem. Educ.* (1972) *49*, pages 69, 117 and 505; (1983) *60*, pages 600 and 681.

Appendix 10

Magnetic susceptibility of a tetragonally distorted t_{2g}^1 ion

The object of this appendix is to expand the discussion in Section 9.9 on the magnetic susceptibility of a t_{2g}^1 ion such as Ti^{III}. It has been written in the expectation that the reader will be more interested in the general arguments than in the detail. As far as is possible at a simple level, the individual mathematical steps will be justified, although the reader should allow neither the mathematics nor the sometimes off-putting equations to inhibit a general scan of the appendix. The treatment is algebraic, although today a computer would be used to treat the problem and would do so more accurately than does the presentation in this appendix. However, the technical language used in discussions of magnetism is based on the algebraic approach and so an understanding of this approach is essential.

The example chosen is almost the simplest possible—a single d electron confined to the t_{2g} set of orbitals in a tetragonally distorted octahedral complex. The effect of the octahedral ligand field is to give three degenerate orbitals, each of which may be occupied by an electron with spin up or spin down. There are six wavefunctions, all degenerate. A tetragonal distortion is now applied and, as Table 6.5 indicates, this separates the d orbitals into the degenerate pair (d_{zx}, d_{yz}) and d_{xy}, taking the axis of distortion as the z axis. If the tetragonal splitting is t then retention of the centre of gravity requires that the energy of the degenerate pair be $-t/3$ and that of d_{xy} be $2t/3$, assuming that the distortion is such that the latter orbital is destabilized. The octahedral ligand field stabilization is the same for all three, $2/5\Delta$, and is omitted from this point on. Including spin, we have four wavefunctions of energy $-t/3$ and two of energy $2t/3$. Each successive step at each point in the development of the energy-level argument involves including the next largest interaction and at this point this is the spin–orbit coupling. The form of the result of including it follows from the rule given in Section 9.5. Spin–orbit coupling causes the degenerate orbitals d_{zx} and d_{yz} to interact and thus to lose their degeneracy. Further, each of these orbitals interacts with d_{xy}. Detailed calculations give the energies below—check that when ζ (zeta, the spin–orbit coupling constant) $= 0$ they give the energies above. Once spin–orbit coupling has been included it is no longer possible to refer to orbitals or to use labels such as d_{zx}.

$$E_3 = \tfrac{1}{2}[\tfrac{1}{2}\zeta + \tfrac{1}{3}t + (t^2 - t\zeta + \tfrac{9}{4}\zeta^2)^{1/2}] \quad \text{(doubly degenerate)}$$

$$E_2 = -\tfrac{1}{3}t - \tfrac{1}{2}\zeta \quad \text{(doubly degenerate)}$$

$$E_1 = \tfrac{1}{2}[\tfrac{1}{2}\zeta + \tfrac{1}{3}t - (t^2 - t\zeta + \tfrac{9}{4}\zeta^2)^{1/2}] \quad \text{(doubly degenerate)}$$

Each of the above levels is doubly degenerate, and each pair is a Kramers doublet, with a degeneracy that can only be removed by a magnetic field. The effect of such a magnetic field on this set of energy levels is conveniently divided into two parts. The first part is the effect on each individual degenerate pair; the second part is the magnetic-field-induced interactions between pairs, interactions which mix them. They are called the first- and second-order Zeeman effects, respectively. The first-order Zeeman effect leads to the following energy levels, where, for simplicity, it has been assumed that the field is applied along the z axis. The quantity k is the orbital reduction factor.

$$E_3\,(b) = \tfrac{1}{2}[\tfrac{1}{2}\zeta + \tfrac{1}{3}t + (t^2 - t\zeta + \tfrac{9}{4}\zeta^2)^{1/2}] + \frac{1 - w^2(1 + k)}{1 + w^2} \times \beta H$$

$$E_3\,(a) = \tfrac{1}{2}[\tfrac{1}{2}\zeta + \tfrac{1}{3}t + (t^2 - t\zeta + \tfrac{9}{4}\zeta^2)^{1/2}] - \frac{1 - w^2(1 + k)}{1 + w^2} \times \beta H$$

$$E_2\,(b) = -\tfrac{1}{3}t - \tfrac{1}{2}\zeta + (1 - k)\,\beta H$$

$$E_2\,(a) = -\tfrac{1}{3}t - \tfrac{1}{2}\zeta - (1 - k)\,\beta H$$

$$E_1\,(b) = \tfrac{1}{2}[\tfrac{1}{2}\zeta + \tfrac{1}{3}t - (t^2 - t\zeta + \tfrac{9}{4}\zeta^2)^{1/2}] + \frac{w^2 - (k + 1)}{1 + w^2} \times \beta H$$

$$E_1\,(a) = \tfrac{1}{2}[\tfrac{1}{2}\zeta + \tfrac{1}{3}t - (t^2 - t\zeta + \tfrac{9}{4}\zeta^2)^{1/2}] - \frac{w^2 - (k + 1)}{1 + w^2} \times \beta H$$

In these expressions, w is a numerical coefficient which is a complicated function of both ζ and t. H is the field strength—that is, the first-order Zeeman splitting is proportional to the field strength—and β is the Bohr magneton. The first-order Zeeman effect has removed all degeneracies.

Inclusion of the second-order Zeeman effect gives

$$E_3\,(b) = \tfrac{1}{2}[\tfrac{1}{2}\zeta + \tfrac{1}{3}t + (t^2 - t\zeta + \tfrac{9}{4}\zeta^2)^{1/2}] + \frac{1 - w^2(1 + k)}{1 + w^2} \times \beta H$$

$$+ \frac{1}{(t^2 - t\zeta + \tfrac{9}{4}\zeta^2)^{1/2}} \times \left[\frac{w(k + 2)}{1 + w^2}\right]^2 \times \beta^2 H^2$$

$$E_3\,(a) = \tfrac{1}{2}[\tfrac{1}{2}\zeta + \tfrac{1}{3}t + (t^2 - t\zeta + \tfrac{9}{4}\zeta^2)^{1/2}] - \frac{1 - w^2(1 + k)}{1 + w^2} \times \beta H$$

$$+ \frac{1}{(t^2 - t\zeta + \tfrac{9}{4}\zeta^2)^{1/2}} \times \left[\frac{w(k + 2)}{1 + w^2}\right]^2 \times \beta^2 H^2$$

$$E_2\,(b) = -\tfrac{1}{3}t - \tfrac{1}{2}\zeta + (1 - k)\,\beta H$$

$$E_2\,(a) = -\tfrac{1}{3}t - \tfrac{1}{2}\zeta - (1 - k)\,\beta H$$

$$E_1(b) = \tfrac{1}{2}[\tfrac{1}{2}\zeta + \tfrac{1}{3}t - (t^2 - t\zeta + \tfrac{9}{4}\zeta^2)^{1/2}] + \frac{w^2 - (k+1)}{1+w_2} \times \beta H$$

$$- \frac{1}{(t^2 - t\zeta + \tfrac{9}{4}\zeta^2)^{1/2}} \times \left[\frac{w(k+2)}{1+w^2}\right]^2 \times \beta^2 H^2$$

$$E_1(a) = \tfrac{1}{2}[\tfrac{1}{2}\zeta + \tfrac{1}{3}t - (t^2 - t\zeta + \tfrac{9}{4}\zeta^2)^{1/2}] - \frac{w^2 - (k+1)}{1+w_2} \times \beta H$$

$$- \frac{1}{(t^2 - t\zeta + \tfrac{9}{4}\zeta^2)^{1/2}} \times \left[\frac{w(k+2)}{1+w^2}\right]^2 \times \beta^2 H^2$$

These equations are quite horrific in appearance and we have considered a particularly simple case! The important thing to remember is that the second-order Zeeman effect is characterized by terms which depend on the square of the field strength. The splitting of the six t_{2g} functions we have just discussed are shown schematically in Fig. A10.1, which has previously appeared as Fig. 9.2.

The final step is to consider the thermal population of these energy levels. Clearly, given the fearsome equations we have obtained, it makes sense to simplify the problem—we have seen what the real-life algebraic problem looks like, and that is enough. First, we revert to an octahedral ligand field,

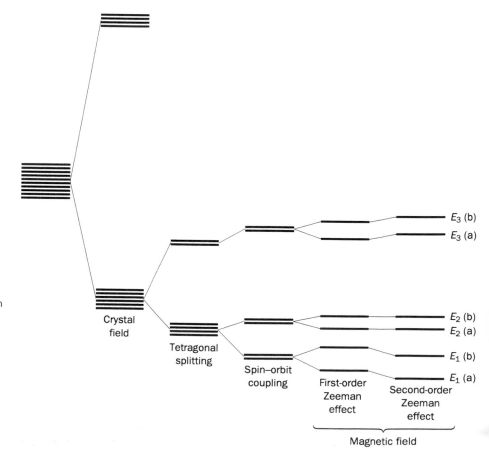

Fig. A10.1 Schematic energy level diagram illustrating how the various interactions discussed in this appendix determine the ground state derived from a t_{2g}^1 configuration in a magnetic field.

by setting $t = 0$; this simplification has the bonus that w becomes equal to 2. Secondly, we revert to a pure crystal field model by setting the orbital reduction factor equal to 1. With these simplifications, the energy levels become

$$E_3(b) = \zeta + \beta H + \frac{4\beta^2 H^2}{3\zeta}$$

$$E_3(a) = \zeta - \beta H + \frac{4\beta^2 H^2}{3\zeta}$$

$$E_2(b) = -\tfrac{1}{2}\zeta$$

$$E_2(a) = -\tfrac{1}{2}\zeta$$

$$E_1(b) = -\tfrac{1}{2}\zeta - \frac{4\beta^2 H^2}{3\zeta}$$

$$E_1(a) = -\tfrac{1}{2}\zeta - \frac{4\beta^2 H^2}{3\zeta}$$

These expressions were first derived by Kotani (by a less circuitous route!) and the simple theory which neglects distortion and the orbital reduction factor is called Kotani theory. Note that levels which are split apart in the more detailed analysis are degenerate in the Kotani treatment.

We now have to derive an expression for the temperature variation of the magnetic susceptibility in terms of the thermal population of these levels. To do this, the energy levels are conveniently rewritten in the form

$$E_n = E_n(0) + E_n(1)H + E_n(2)H^2 + \cdots$$

Table A10.1 gives the above set of energy levels in this format. We will return to this table when we have the mathematical expressions that require the data in this form.

An important property of the magnetic moment of a molecule, μ_n, is that it is proportional to the decrease in energy of the molecule with increase in the applied magnetic field. In atomic units the proportionality constant is unity:

$$\mu_n = -\frac{dE_n}{dH}$$

Combining the last two mathematical expressions gives

$$\mu_n = -E_n(1) - 2HE_n(2)\ldots$$

an expression that will be used shortly.

In Chapter 9 we saw that

$$\chi_M = \frac{N(\text{average magnetic moment per molecule})}{H}$$

The average magnetic moment per molecule will be given by an expression of the form

$$\frac{\sum(\text{magnetic moment of a molecule}) \times (\text{number with this moment})}{(\text{total number of molecules})}$$

Table A10.1 Set of energy levels corresponding to Kotani theory

E_n	$E_n(0)$	$E_n(1)$	$E_n(2)$
E_3 (b)	ζ	β	$4\beta^2/3\zeta$
E_3 (a)	ζ	$-\beta$	$4\beta^2/3\zeta$
E_2 (b)	$-\zeta/2$	0	0
E_2 (a)	$-\zeta/2$	0	0
E_1 (b)	$-\zeta/2$	0	$-4\beta^2/3\zeta$
E_1 (a)	$-\zeta/2$	0	$-4\beta^2/3\zeta$

where the sum is over all molecules. Because the population of the set of energy levels will be governed by a Boltzmann distribution law, it follows that

$$\chi_M = \frac{N \sum_n \mu_n \exp\left(\frac{-E_n}{kT}\right)}{H \sum_n \mu_n \exp\left(\frac{-E_n}{kT}\right)}$$

We now expand the exponentials as a power series in H, a quantity on which we know that the energy depends:

$$\exp\left(\frac{-E_n}{kT}\right) = \exp\left(\frac{-1}{kT}[E_n(0) + E_n(1)H + \cdots]\right)$$

$$= \exp\left(\frac{-E_n(0)}{kT}\right) \times \exp\left(\frac{-E_n(1)H}{kT}\right) \cdots$$

where further terms, omitted, will—except at very low temperatures or high fields—be approximately

$$\exp\left(\frac{-0}{kT}\right) = \exp(0) = 1$$

and so may be neglected. Using the expansion

$$\exp(x) = 1 + x + \frac{x^2}{2!} + \frac{x^3}{3!} + \cdots$$

and putting this, together with the expression for μ_n given earlier, into the equation for χ_M gives, approximately,

$$\chi_M = \frac{N \sum_n (-E_n(1) - 2HE_n(2))\left(1 - \frac{E_n(1)H}{kT}\right) \times \exp\left(\frac{-E_n(0)}{kT}\right)}{H \sum_n \exp\left(\frac{-E_n(0)}{kT}\right)}$$

We now expand the numerator but first note that when the magnetic field is zero the average magnetic moment per molecule must be zero and so the numerator must be zero also. It follows that

$$\sum_n -E_n(1) \times \exp\left(\frac{-E_n(0)}{kT}\right) = 0$$

We have, then,

$$\chi_M = \frac{N \sum_n \left[\frac{(E_n(1))^2 H}{kT} - 2HE_n(2) + \frac{2H^2 E_n(1)}{kT}\right] \times \exp\left(\frac{-E_n(0)}{kT}\right)}{H \sum_n \exp\left(\frac{-E_n(0)}{kT}\right)}$$

Cancelling the Hs and neglecting the term $2HE_n(1)/kT$, which will be very small except at low temperatures or high fields we have, finally,

$$\psi_M = \frac{N \sum_n \left[\frac{E_n(1))^2}{kT} - 2E_n(2) \right] \times \exp\left(\frac{-E_n(0)}{kT} \right)}{\sum_n \exp\left(\frac{-E_n(0)}{kT} \right)}$$

Using the relationship derived in Section 9.2:

$$\mu_{eff}^2 = \frac{3RT\chi_M}{N^2\beta^2} = \frac{3kT\chi_M}{N\beta^2}$$

we have, for μ_{eff}^2:

$$\mu_{eff}^2 = \frac{3kT \sum_n \left[\frac{(E_n(1))^2}{kT} - 2E_n(2) \right] \times \exp\left(\frac{-E_n(0)}{kT} \right)}{\beta^2 \sum_n \exp\left(\frac{-E_n(0)}{kT} \right)}$$

These expressions for χ_M and μ_{eff}^2 are known as the van Vleck relationships.

It is a simple matter to apply these equations to our problem; we have only to sum over the six different E_n states using the values for $E_n(0)$, $E_n(1)$ and $E_n(2)$ given in Table A10.1. This gives

$$\mu_{eff}^2 = \frac{3kT \left[2\left(\frac{\beta^2}{kT} - \theta\frac{\beta^2}{3\zeta} \right) \exp\left(-\frac{\zeta}{kT} \right) + 2(0 - 0) \exp\left(\frac{\zeta}{2kT} \right) + 2\left(0 + 8\frac{\beta^2}{3}\zeta \right) \exp\left(\frac{\zeta}{2kT} \right) \right]}{\beta^2 \left[2 \exp\left(-\frac{\zeta}{kT} \right) + 2 \exp\left(\frac{\zeta}{kT} \right) + 2 \exp\left(\frac{\zeta}{2kT} \right) \right]}$$

where the factors of 2 arise because $E_1(a)$ and $E_1(b)$, for example, make the same contribution to the summation. This expression simplifies to

$$\mu_{eff}^2 = \frac{\left(\frac{3\zeta}{kT} - 8 \right) \exp\left(\frac{-3\zeta}{2kT} \right) + 8}{\frac{\zeta}{kT} \left[\exp\left(\frac{-3\zeta}{2kT} \right) + 2 \right]}$$

or, by putting $\zeta/KT = x$,

$$\mu_{eff}^2 = \frac{(3x - 8) \exp\left(-3\frac{x}{2} \right) + 8}{x \left[\exp\left(-3\frac{x}{2} \right) + 2 \right]}$$

another equation first derived by Kotani and discussed in Section 9.9. This is the equation used to give the theoretical plot in Fig. 9.4(a); for Fig. 9.4(b), the corresponding but complete expression, including k and t, was used.

Appendix 11

High temperature superconductors

At the present time, several different, but frequently related, classes of high temperature (high T_c) superconductors are known. The first class that will be considered is based on La_2CuO_4. This compound has a structure which may be thought of as consisting of planes of octahedra of oxygen atoms, the octahedra sharing corners, with a copper atom at the centre of each octahedron, as shown in Fig. A11.1. The lanthanum ions are sandwiched between the oxygen/copper octahedra layers. The system becomes superconducting when Sr^{2+} ions are substituted for La^{3+}. To maintain overall charge neutrality[1] one Cu^{2+} has to become Cu^{3+} for every La^{3+} substituted by Sr^{2+}. Superconductivity occurs, it seems, because of the presence of Cu^{3+} ions in the copper–oxygen planes. It also seems that the superconductivity is localized in these planes.

The second class of superconductors, the so-called 1,2,3 compounds, are of approximate formulae $YBa_2Cu_3O_7$, although the oxygen content is again variable. Neither yttrium (Y^{3+}) nor barium (Ba^{2+}) have variable valency so charge neutrality requires that, for the above formula, of the three copper ions, two are Cu^{2+} and one Cu^{3+}. This class of compound is structurally more complicated than the first although they seem to have in common that the superconductivity is associated with the coexistence of Cu^{2+} and Cu^{3+} in an oxide lattice. In the 1,2,3 compounds the lattice is three-dimensional. Further, there are two distinct copper sites. One, in which the copper ions are square planar, seems an obvious place for the d^8 ion Cu^{3+} (the d^8 ions

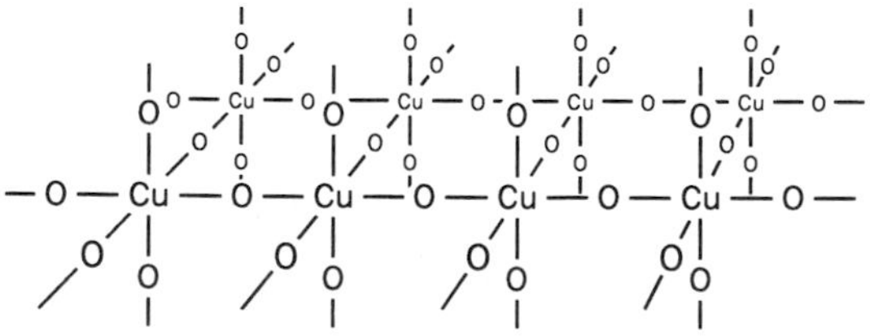

Fig. A11.1 Structure of La_2CuO_4 showing the octahedral copper environment.

[1] The present discussion neglects the fact that the substituted compound may lose oxygen, although the evidence is that this loss may be important in the phenomenon of high temperature superconductivity. At the present time, the oxygen loss is regarded as a fine-tuning mechanism; for every oxygen atom lost two Cu^{3+} must revert to Cu^{2+}, to compensate for the fact that $O^{2-} \rightarrow O$.

Pt^{II}, Pd^{II} and, to a lesser extent, Ni^{II}, characteristically form square planar complexes). In the second site the copper ions are five-coordinate, being surrounded by oxygens at the corner of a square pyramid. Such an environment is not uncommon in copper(II) compounds and so this seems to be an obvious site for the Cu^{2+} ions. The ratio of occurrence of the square to square pyramidal sites is 1:2. This is also the ratio of Cu^{3+} to Cu^{2+}, supporting the model presented for the location of the two different copper ions.

The relative complexity of the 1,2,3 and other, yet higher temperature, superconductor structures has meant that the majority of theoretical work has been concerned with superconductors based on La_2CuO_4, the $La_{2-x}Sr_xCuO_4$ superconductors, and only these will be considered in this appendix (the highest temperature superconductor reported at the time of writing this appendix contains Sr, Ca and Bi as well as Cu and O—it is reported to superconduct at $-23\,°C$). However, to highlight the problem of explaining the superconductivity of the high T_c materials we shall give an outline of model which for the past 30 years has been the accepted model of superconductivity. This is the *Bardeen, Cooper and Schrieffer* (BCS) model.

The simplest explanation of the BCS model is an anthropomorphic one. Suppose that the reader (an electron) enters at one corner of a large, but extremely crowded, room in which an informal party is in progress. The guests (atoms) are making small, to some extent random, movements (atomic vibrations). The reader (electron) is presented with the task of crossing to the opposite corner of the room (this simulates an electron's contribution to an electrical current). In crossing the room the reader will find themselves frequently deflected—this simulates electrical resistance. Now, suppose that an ultra important person (UIP)—much more important than a VIP— together with their partner enters the room, in place of the reader. As the pair cross the room they will experience no resistance, the guests part to allow them free passage. Their movement models the movement of super-conducting electrons through a lattice—these electrons also move as pairs, so-called Cooper pairs. The essential of the BCS model is a coupling between the movement of the superconducting electrons and the vibrational motion of the lattice through which they move. Normally, BCS theory is applied to materials which are superconducting at very low temperatures—at these temperatures the vibrational motion of the atoms will be very small anyhow. At higher temperatures, such as those relevant to the present discussion, the apparently[2] random motion of the atoms will be of much greater amplitude. Does the BCS model still apply? Apparently not, and therein lies the problem.

It is easy to demonstrate the problem of explaining the superconductivity of the high-temperature systems. In Fig. A11.2 is shown a copper–oxygen plane in La_2CuO_4; there is an additional oxygen above and below each copper which, for simplicity, are omitted. Each copper(II) ion in Fig. A11.2 has a single unpaired electron and these are antiferromagnetically coupled, just as is described in Section 9.11, giving the arrangement shown

---

[2] In fact, the atomic excursions may be expressed as a linear sum of thermally populated lattice modes combined with appropriate phase differences. In addition, there will be the zero point vibrational amplitude of each mode to be considered, but this contribution becomes less important as the temperature rises.

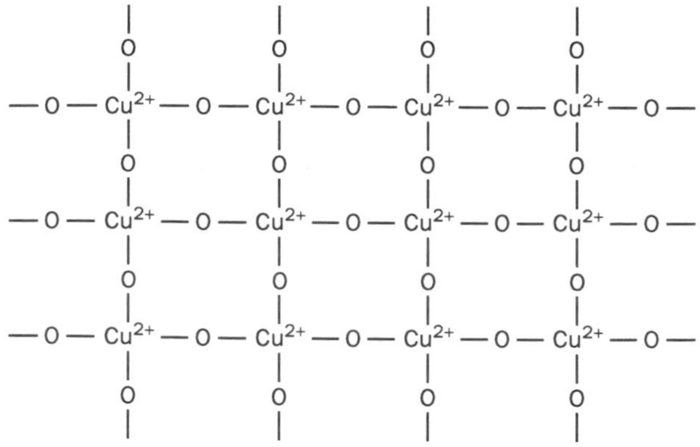

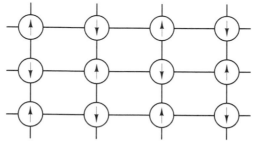

Fig. A11.2 The copper–oxygen plane in La$_2$CuO$_4$.

Fig. A11.3 Antiferromagnetic coupling between unpaired electrons on copper(II) rows in La$_2$CuO$_4$.

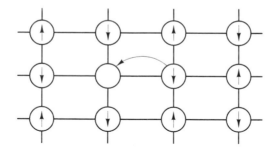

Fig. A11.4 Effect of Cu^{2+} becoming Cu^{3+}, leading to the possibility of an electron hop.

schematically in Fig. A11.3. When Sr^{2+} substitutes for La^{3+} this substitution occurs between the copper–oxygen layers—and this is rather remote from where the superconducting action takes place, or, rather, is believed to take place—remote from the copper-oxygen planes of Fig. A11.2. When Cu^{2+} becomes Cu^{3+} an electron is lost; consider just one such substitution, a substitution which leaves the structure shown schematically in Fig. A11.4. In all forms of electrical conductivity, superconductivity not excluded, electrons have to move from site to site; in the present case the obvious electron hop, such as that indicated by a curved arrow in Fig. A11.4 leads to the arrangement shown in Fig. A11.5. The key point is that as far as spin arrangements are concerned, Figs. A11.4 and A11.5 are not identical; parallel spins occupy adjacent sites in Fig. A11.5 but not in Fig. A11.4. The discussion of Section 9.11 makes it quite clear that the arrangement of Fig. A11.5 is energetically unfavourable. If it is correct, then superconductivity requires energy to make it occur. Of course, this is just the opposite of reality and so something is wrong with the ideas that led up to Figs. A11.4 and A11.5.

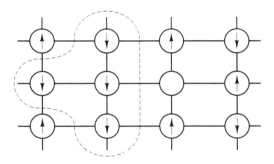

Fig. A11.5 The result of the electron hop shown in Fit. A11.4 is that parallel spins occupy adjacent sites.

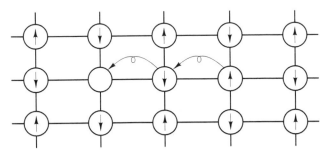

Fig. A11.6 Two-electron synchronous hop with spin change (a) before and (b) after.

(a)

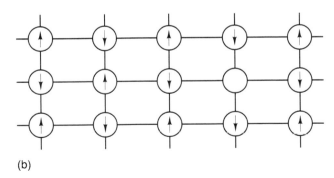

(b)

The model is wrong; the problem is how to correct it. At the present time the answer is not known, although it is possible to make some relevant comments.

First, the problem posed by Figs. A11.4 and A11.5 could have been avoided had the conduction electron hopped across the diagonal of a square of copper ions in Fig. A11.4 rather than along an edge. This alternative can be excluded because the theory is unambiguous—there has to be some overlap between the orbital of the before and after sites. The distance involved in a diagonal hop is too great, particularly when it is remembered that the orbitals of a Cu^{3+} ion will be contracted compared with those of a Cu^{2+}. Secondly, we have adopted a one-electron model. At several points in this book it has been recognized that a one-electron model may not provide an adequate explanation; is this another example? Almost certainly, the answer is yes, but there is no agreement on the model to be adopted. Perhaps the simplest two-electron model is illustrated in Fig. A11.6, which parallels Figs. A11.4 and A11.5. Two electrons hop synchronously, along the sides of a square, exchanging spins in the process (this spin change is indicated by a loop in Fig. A11.6). A single-spin flip, on its own, would require energy.

Finally, the discussion has implicitly considered Cu^{2+} in an octahedral environment of oxygen atoms. If the Cu^{3+} were in the same environment it would have two unpaired electrons (it would be $t_{2g}^6 e_g^2$). The electron lost in moving $Cu^{2+} \rightarrow Cu^{3+}$ would not be the one shown as lost in figure A11.4—because this would leave no unpaired electrons on the Cu^{3+}—but, rather, one of the other Cu^{2+} e_g electrons which have not been included in the figures. Alternatively, and much more likely, the Cu^{3+} would undergo a distortion (*not* a Jahn–Teller distortion because the Cu^{3+} would have a $^3A_{2g}$ ground state—see the detailed discussion of the d^8 case in Section 8.2) which splits the e_g orbitals. Presumably, such a distortion would be tetragonal, involving the oxygens above and below each copper ion in our figures, the oxygens that were omitted for simplicity. Such a distortion, which would be associated with a lattice vibration, brings us comfortably closer to the BCS model with which the appendix started but, on its own, it leaves unresolved the problem posed by Figs. A11.4 and A11.5.

Further reading

A rather different approach from that given above will be found in 'Three theories of superconductivity' by F. A. Matsen, *J. Chem. Educ.* (1987) *64*, 842. The structural patterns associated with superconductivity are well illustrated in 'The Crystal Chemistry of High-Temperature Oxide Superconductors and Materials with Related Structures' by H. Müller-Buschbaum, *Angew. Chem., Int. Ed. Engl.* (1989) *28*, 1472. A mixture of structure and some theory is to be found in a very readable, non-mathematical, article 'Some Structural-Electronic Aspects of High Temperature Superconductors', by J. K. Burdett, in *Adv. Chem. Phys.* (1993) *83*, 207.

A very readable review of recent development in both experimental and theoretical aspects of chemical superconductors is to be found in the September 1994 (page 722 on) issue of *Chemistry in Britain*.

Appendix 12

Combining spin and orbital angular momenta

In Section 11.3 the problem of combining spin and orbital angular momenta to give the correct resultant was encountered. Put another way, how do terms such as ^3H, as occurs for an f^2 ion such as Pr^{III} and ^4I, appropriate to an f^3 ion such as Nd^{III}, split when the spin and orbital motions couple together? The solution of the general problem will be indicated by considering these two examples in detail. This is done in Tables A12.1 and A12.2. In each table down the left hand side are given all of the components of the orbital angular momentum; at the top are listed all of the components of the spin angular momentum. The entries in the body of the tables are total angular momenta, obtained by adding the appropriate spin and orbital momenta (those at the top and side of the table). In Table A12.1 it is not difficult to see that all of the components of the state with total angular momentum of 6 ($^3H_6$) are covered by the left-hand column and final row entries.

If these entries are deleted from the table, the components of the state with total angular momentum 5 ($^3H_5$) are similarly obtained from the table remaining. Repeating this process gives the components of the total angular momentum 4 ($^3H_4$) state. That this pattern is not accidental is evident from Table A12.2, where it is repeated. Note particularly that each set of total angular momenta (the entries within the shaded panels in Tables A12.1 and A12.2) contain an entry from the figures in the first row of the table. Further, that these entries, together, are a complete listing of the magnitudes of the permissible angular momenta. It follows that, as described in Section 11.4, the allowed values of the total angular momentum are obtained by taking the maximum value of the orbital angular momentum and adding each component of the spin angular momentum to it.

One final word of caution, one that also appears in Appendix A14 and which explains the appearance in the tables of the word 'schematic'. What has been described in this appendix is really a method of counting; it is not a method of obtaining wavefunctions. So, it is not true that the component of the total angular momentum = 6 state which has an angular momentum of 0 along the z axis is the function with orbital angular momentum = -1 combined with spin = 1, although this is the association made in the table. The correct wavefunction is a linear combination of all three functions with

Table A12.1 The (schematic) origin of the $^3H_4$, $^3H_5$ and $^3H_6$ levels of the ^3H term arising from the f^2 configuration

^3H	Spin angular momentum		
	1	**0**	**−1**
5	6	5	4
4	5	4	3
3	4	3	2
2	3	2	1
Orbital 1	2	1	0 ⟵ $^3H_4$
angular 0	1	0	−1
momentum −1	0	−1	−2
−2	−1	−2	−3
−3	−1	−3	−4
−4	−3	−4	−5 ⟵ $^3H_5$
−5	−4	−5	−6 ⟵ $^3H_6$

Table A12.2 The (schematic) origin of the $^4I_{9/2}$, $^4I_{11/2}$, $^4I_{13/2}$ and $^4I_{15/2}$ levels of the ^4I term arising from the f^2 configuration

^4I	Spin angular momentum			
	3/2	**1/2**	**−1/2**	**−3/2**
6	15/2	13/2	11/2	9/2
5	13/2	11/2	9/2	7/2
4	11/2	9/2	7/2	5/2
3	9/2	7/2	5/2	3/2
2	7/2	5/2	3/2	1/2
1	5/2	3/2	1/2	−1/2 ⟵ $^4I_{9/2}$
Orbital 0	3/2	1/2	−1/2	−3/2
angular −1	1/2	−1/2	−3/2	−5/2
momentum −2	−1/2	−3/2	−5/2	−7/2
−3	−3/2	−5/2	−7/2	−9/2
−4	−5/1	−7/2	−9/2	−11/2 ⟵ $^4I_{11/2}$
−5	−7/2	−9/2	−11/2	−13/2 ⟵ $^4I_{13/2}$
−6	−9/2	−11/2	−13/2	−15/2 ⟵ $^4I_{15/2}$

Problem A12.1

Construct a table similar to Tables A12.1 and A12.2 for the ^5I (f^4) case and show that the levels $^5I_4$, $^5I_5$, $^5I_6$, $^5I_7$ and $^5I_8$ are given by the procedure described in this appendix (see also Section 11.4).

Appendix 13

Bonding between a transition metal atom and a C_nR_n ring, $n = 4$, 5 and 6

In this appendix are merged two closely related approaches to the bonding between a cyclic C_nR_n ($n = 4$, 5 and 6) unsaturated hydrocarbon and a transition metal atom. The first approach is that developed in Chapter 10. The axis between the centre of the C_nR_n ring and the metal atom is taken as z and the orbitals of the metal atom classified according to the number of planar nodes which they have and which *contain* the z axis (*cutting* the z axis does not count). The π orbitals of the C_nR_n ring are classified using the same criterion. Interaction—be it bonding or antibonding—only occurs between orbital pairs with the same nodal count. All interactions between orbitals with different nodal counts are zero. This approach is summarized in Figure A13.1. In this figure the metal orbitals are drawn in a direction perpendicular to the z axis (thus enabling their identity to be established, s, p_z and d_{z^2} look much the same when viewed down z). However, the hydrocarbon π molecular orbitals *are* viewed down z, because this enables their nodality to be clearly seen. As is the common pattern—from orbitals through to acoustics—the more nodes, the higher the energy. Of course, for this simple statement to be true the vibrations under discussion must be immediately related to each other. So, the statement is applicable to the π orbitals of hydrocarbons, because different combinations of carbon p_π orbitals are involved. It does not apply to the metal orbitals because these, s, p and d, have no such simple common basis. For unsaturated hydrocarbons it means that the orbitals of highest nodality—of highest energy—are empty in the isolated molecule. These higher orbitals, then, are regarded as π electron acceptors in a transition metal complex and the (occupied) orbitals of lower nodality are π electron donors. Schematic molecular orbital diagrams based on these arguments are given in Figure A13.2. We do not indicate the final occupancy of the molecular orbitals because this will depend on which particular metal atom is involved.

The second, equivalent, approach is group theoretical (see Appendix 4). The local symmetry of the transition metal—C_nR_n unit in a complex is C_{nv}. The transformation under the operation of the C_{nv} character table of the n carbon p_π orbitals in the ring gives the irreducible representations subtended by these p_π orbitals. The metal orbitals are similarly classified; inter-actions occur between metal and π molecular orbitals of the same symmetry species.

In Table A13.1 this general pattern is detailed for each of C_4R_4–M, C_5R_5–M and C_6R_6–M rings symmetrically bonded to a transition metal. Alongside each of the three character tables, for the C_{4v}, C_{5v} and C_{6v} point groups, are given the species which describe the symmetry of transition metal orbitals—so, s, p_z and d_{z^2} orbitals transform as A_1 in all three groups. The tables show that there is always a C_nR_n π molecular orbital of A_1 symmetry, so interaction between metal and ligand orbitals of A_1 symmetry can always occur. Similar arguments apply to the other orbitals; it can therefore be seen that the metal d_{xy} orbital is strictly non-bonding in C_4R_4–M complexes (there is no ligand orbital of B_2 symmetry). Conversely, the ligand orbital of B_2 symmetry in C_6R_6 complexes is non-bonding (there is no valence shell metal orbital of B_2 symmetry).

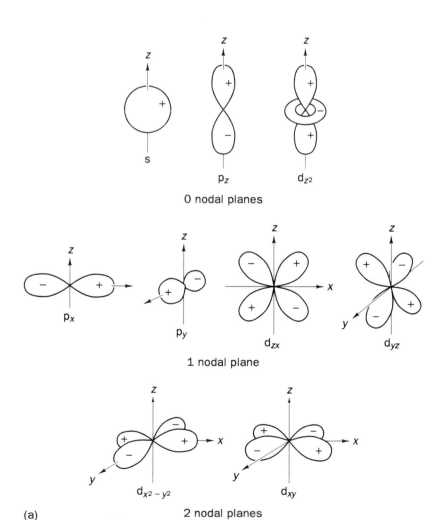

Fig. A13.1 Classification of (a) metal and (b) hydrocarbon π orbitals according to the number of nodal planes they have which contain the z axis (the axis of highest symmetry) (*continued*).

(a)

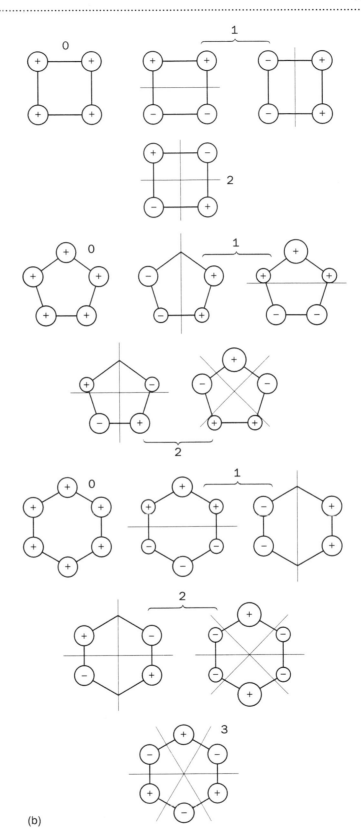

Fig. A13.1 (b) (continued).

(b)

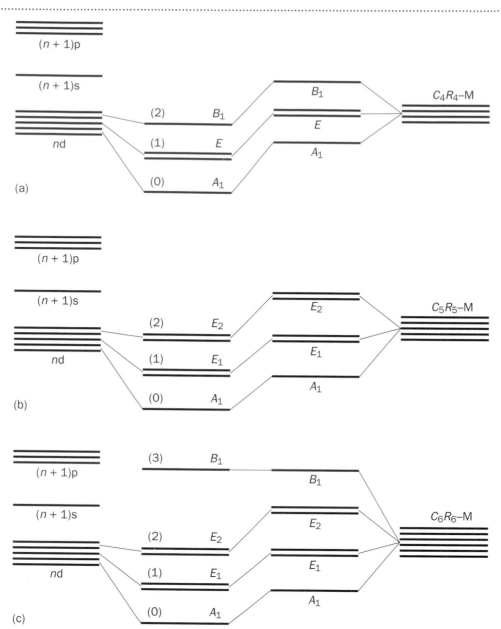

Fig. A13.2 Schematic molecular orbital energy level diagrams for transition metal complexes of (a) C_4R_4, (b) C_5R_5 and (c) C_6R_6. The ligand and final molecular orbitals are classified by their symmetry species under the relevant point group. The final molecular orbitals are also classified by the number of nodal planes they have which contain the z axis. In all cases, for simplicity, only interactions between d orbitals and the ligand orbitals are shown. However, as Table A13.1 shows, p orbitals must be expected to be involved in the (1)-node interactions and both p and s orbitals in the (0)-node interactions.

The connection between the two approaches described above is also shown in Table A13.1. At the right-hand side of each character table is listed the number of nodal planes containing the z axis associated with each type of interaction. There is something to be said in favour of each model in comparison with the other. Thus, the 'number of nodal planes' criterion does not discriminate between B_1 and B_2 (in the C_{4v} and C_{6v} cases), yet only one is involved in the bonding. Conversely, although in each case the number of nodal planes quoted is the lowest, for each case higher numbers exist also (we give these for the A_1 case). In such cases, the number of nodal planes criterion provides a distinction that the symmetry criterion does not. Only orbitals with the same number of nodal planes interact.

Table A13.1

(a) C_4R_4: p_π orbitals of C_4R_4 transform as $A_1 + E + B_1$ (taking the C_4 atoms to lie in the $2\sigma_v$ mirror planes)

C_{4v}	E	$2C_4$	C_2	$2\sigma_v$	$2\sigma_v'$	Metal orbitals	C_4R_4 π?	Nodal planes
A_1	1	1	1	1	1	s, p_z, d_{z^2}	yes	0 or 4
A_2	1	1	1	−1	−1	−	no	8
B_1	1	−1	1	1	−1	$d_{x^2-y^2}$	yes	2
B_2	1	−1	1	−1	1	d_{xy}	no	2
E	2	0	−2	0	0	p_x, p_y; d_{zx}, d_{yz}	yes	1

(b) C_5R_5: p_π orbitals of C_5R_5 transform as $A_1 + E_1 + E_2$

C_{5v}	E	$2C_5$	$2C_5^2$	$5\sigma_v$	Metal orbitals	C_5R_5 π?	Nodal planes
A_1	1	1	1	1	s, p_z, d_{z^2}	yes	0 or 5
A_2	1	1	1	−1	−	no	10
E_1	2	2 cos 72	2 cos 144	0	p_x, p_y; d_{zx}, d_{yz}	yes	1
E_2	2	2 cos 144	2 cos 72	0	$d_{x^2-y^2}$, d_{xy}	yes	2

(c) C_6R_6: p_π orbitals of C_6R_6 transform as $A_1 + E_1 + E_2 + B_1$ (taking the C_6 atoms to lie in the $3\sigma_v$ mirror planes)

C_{6v}	E	$2C_6$	$2C_3$	C_2	$3\sigma_v$	$3\sigma_v'$	Metal orbitals	C_6R_6 π?	Nodal planes
A_1	1	1	1	1	1	1	s, p_z, d_{z^2}	yes	0 or 6
A_2	1	1	1	1	−1	−1	−	no	12
B_1	1	−1	1	−1	1	−1	−	yes	3
B_2	1	−1	1	−1	−1	1	−	no	3
E_1	2	1	−1	−2	0	0	p_x, p_y; d_{zx}, d_{yz}	yes	1
E_2	2	−1	−1	2	0	0	$d_{x^2-y^2}$, d_{xy}	yes	2

Appendix 14

Hole–electron relationship in spin–orbit coupling

The argument used in this appendix is that of providing a simple illustrative example. The general procedure used follows that of Appendix 12, which, if necessary, should be read before proceeding further.

Consider the ^7F term arising from the f^8 configuration. We denote electrons by arrows and holes by circles. The septet spin state may be considered as arising from the following arrangement, either of eight electrons or of six holes:

Working with holes and following the procedures of Appendix 12 we have

^7F		Spin (holes) angular momentum							
		3	**2**	**1**	**0**	**−1**	**−2**	**−3**	
	3	6	5	4	3	2	1	0	← $^3F_0$
	2	5	4	3	2	1	0	−1	← $^3F_1$
Orbital	1	4	3	2	1	0	−1	−2	← $^3F_2$
angular	0	3	2	1	0	−1	−2	−3	← $^3F_3$
momentum	−1	2	1	0	−1	−2	−3	−4	← $^3F_4$
	−2	1	0	−1	−2	−3	−4	−5	← $^3F_5$
	−3	0	−1	−2	−3	−4	−5	−6	← $^3F_6$

The entries in this table can be classified according to the levels that result. So, the $^7F_6$ level has components with resultant angular momentum of 6, 5, 4, 3, 2, 1, 0, -1, -2, -3, -4, -5 and -6. Deleting entries with these values from the table leaves 5 as the highest value. This must come from a $^7F_5$ level, with components 5, 4, 3, 2, 1, 0, -1, -2, -3, -4 and -5. Proceeding in this manner it is concluded that the $^7F_6$, $^7F_5$, $^7F_4$, $^7F_3$, $^7F_2$, $^7F_1$ and $^7F_0$ levels of the ^7F term are obtained. Had the ^7F arising from the f^6 configuration been under discussion, we would have been working with electrons and the $^7F_0$ level would have been the most stable—the orbital and spin magnets would have been arranged in opposed fashion,

N–S
S–N

but because we are considering the f^8 configuration, and so working with holes, it is the $^7F_6$ level which is the ground state because it is this which has the most stable arrangement, that with orbital and (hole) spin magnets parallel.

Index